Historical
Encyclopedia of
Atomic Energy

Historical Encyclopedia of Atomic Energy

Stephen E. Atkins

Greenwood Press
Westport, Connecticut • London

Library of Congress Cataloging-in-Publication Data

Atkins, Stephen E.
  Historical encyclopedia of atomic energy / Stephen E. Atkins.
    p.   cm.
  Includes bibliographical references and index.
  ISBN 0–313–30400–9 (alk. paper)
  1. Nuclear energy—History Encyclopedias.   I. Title.
QC772.A87   2000
333.792'4'03—dc21   99–29618

British Library Cataloguing in Publication Data is available.

Copyright © 2000 by Stephen E. Atkins

Library of Congress Catalog Card Number: 99–29618
ISBN: 0–313–30400–9

First published in 2000

Greenwood Press, 88 Post Road West, Westport, CT 06881
An imprint of Greenwood Publishing Group, Inc.
www.greenwood.com

Printed in the United States of America

The paper used in this book complies with the
Permanent Paper Standard issued by the National
Information Standards Organization (Z39.48–1984).

10 9 8 7 6 5 4 3 2 1

To my father,
Francis (Frank) Eugene Atkins

His work at the uranium processing plant
at Oak Ridge, Tennessee, during World War II
probably saved his life.

Contents

Introduction   ix

The Encyclopedia   1

Chronology of Atomic Energy   411

Selected Bibliography   427

Index   445

Introduction

The story of atomic energy is both a tale of incredible discovery by a community of dedicated scientists and a lesson in how to adapt to a powerful new source of energy. In the first 40 years of this century theoretical and experimental physicists probed the atom to understand its secrets. These scientists operated within an international community where bits of knowledge and insight were immediately shared with colleagues. Slowly the outline of the structure of the atom was understood only to discover in an experiment by Otto Hahn in Germany in 1939 that the atom was a potential source of energy. Scientists from all over the world immediately understood the explosive potential of the atom, and many feared going further. This knowledge and the subsequent mobilization of scientists for work on an atomic weapon ended the first phase of the history of atomic energy.

Once governmental authorities in France, Germany, Great Britain, the Soviet Union, and the United States comprehended the potential of atomic energy as a military weapon, the nature of the debate over atomic energy changed dramatically. Suddenly, secrecy was the keynote rather than dissemination of information. In the beginning, this secrecy was self-imposed by scientists themselves, because they feared that German scientists might use the knowledge to give Adolf Hitler a bomb. Soon, however, the governmental agencies imposed secrecy in the interests of national security. Five countries seriously investigated the possibility of building an atomic bomb in the mid-1940s—Germany, Japan, the Soviet Union, the United Kingdom, and the United States. Only the United States, with considerable help from British and German emigrant scientists, was successful in building an atomic bomb in time for use in World War II. The detonation of two atomic bombs on the Japanese cities of Hiroshima and Nagasaki ended the second phase of the history of atomic energy.

The third phase of atomic energy lasted from 1945 to 1960. It was in this period that the United States lost its monopoly on atomic weapons and other countries joined the nuclear world. Despite an optimistic assessment by American politicians that it would take the Soviet Union a decade or more to become a nuclear power, the Soviet Union detonated a nuclear device in 1949. This knowledge spurred the U.S. government to proceed to build a hydrogen bomb, and the nuclear race was on. Soon the possession of extensive stockpiles of nuclear weapons made the Soviet Union and the United States superpowers. Great Britain and France soon joined the ranks of nuclear powers. Suddenly, the fear among the nuclear powers was over the proliferation of nuclear weapons. Efforts were made by the nuclear powers to control the spread of nuclear weapons, and nonnuclear countries resented these initiatives. China, and then India, tested nuclear devices almost to spite the established nuclear powers.

The fourth stage of atomic energy was the boom in nuclear reactors for use in domestic energy in the period from 1960 to 1980. Governments started out subsidizing nuclear power plants as a way to produce cheap energy and rid themselves of dependence on energy controlled by outside states. Optimistic projections of the cost of electricity from these plants were calculated. Commercial companies began to sell plants to utility companies. Promises were made of a golden future made possible by atomic energy. Toward the end of this period, these promises began to wear thin as utility companies experienced no such cost savings. Finally, the Three Mile Island accident in 1979 marked the beginning of the end of the boom phase.

The final phase has been a serious examination of atomic energy as a public good, and this period has lasted from 1980 to the present. During this phase, antinuclear and environmental activists have critiqued atomic energy and found it lacking. Both the Three Mile Island accident in 1979 and the Chernobyl nuclear accident in 1986 mobilized public opinion worldwide against atomic energy. National governments have been reluctant to accede to this type of pressure, and some, such as France, have been able to resist it. Atomic energy is still with us, but with much more modest pretensions. Radioactive waste has become a major issue both in the United States and elsewhere.

Part of this final phase has been in the form of new initiatives in fusion energy research. Fusion energy is less radioactive and produces less radioactive waste for disposal, but it takes high-temperature technology for it to work. The development of magnetic fusion reactors (tokamaks) and laser fusion have been major advances. The technology has been expensive, and governments have been reluctant to provide the resources for a full-scale effort. Moreover, the memory of the cold fusion fiasco has also been a retardant. Fusion energy is the logical successor to fission energy, but progress has been slow because of technological and political issues.

Atomic energy has been a success story, but much as in fairy tales, it almost lost its soul in the process. Energy can be generated safely, but the failure of its adherents to recognize the dangers to the public of nuclear accidents and radioactive fallout from testing almost cost atomic energy its good name. Cavalier treatment of scientific critics and the public in general by the Atomic Energy Commission (AEC) and the nuclear industry created an atmosphere that generated distrust directed against government agencies. Court cases showed governmental misconduct and produced disillusionment.

It is only fair to state where I stand on atomic energy. My approach is from the vantage point of a historian. I have tried to make each citation as objective as possible. My father did, however, work in the atomic energy field as a plumber/steam fitter. He was a member of the work crew at Oak Ridge, Tennessee, in 1944 and early 1945. He told me about the problem that he had with his draft board in Columbia, Missouri, because they couldn't understand what a plumber was doing in Oak Ridge, Tennessee, to earn a draft exemption. Although the draft board at Oak Ridge gave him six-month deferments, his local draft board made him return to Columbia, Missouri, each month and justify his exemption. My father worked at Y Plant, and he remembers seeing a courier carrying a lead box—with an escort of about a dozen men—leave the plant for an unknown destination. It was probably processed uranium for the first atomic bomb detonated over Hiroshima. When his job was finished, the draft board was notified, and my father was drafted into the U.S. Navy. After the war, he worked for about a year on the Savannah River Project building a plutonium processing plant. Again, my father never knew what he was working on while he was there. Late in life, he found himself at several nuclear power plants working for Bechtel at Russelville, Arkansas, and Hanley at Byron, Illinois. This is the extent of my personal association with atomic energy.

My interest in atomic energy has been more a product of my fascination with the intellectual process. My major at the University of Missouri at Columbia was European history, and my doctorate was in French history at the University of Iowa. In my classes in Western civilization at both schools, I tried to impart to my students the intellectual challenge posed by Albert Einstein, Niels Bohr, Enrico Fermi, and others in the hunt for scientific knowledge. Once, for relaxation, I read the autobiography of Max Born and found him a fascinating character. Both the successes and the failures of the search for the structure of the atom intrigue me. For all of these reasons, when the opportunity to do this book came along, I decided to do it.

A brief note on the use of terms is necessary. The term *atomic energy* was in vogue for much of the period until the early 1950s, when the prevalent term became *nuclear energy*. For all intents and purposes, both terms describe the same thing. Therefore, I have used both terms interchangeably throughout the book. My intention has been to use the terminology current at the time of the citation, but there may be a few lapses.

Another important point is that in the course of this work I make mention of the atomic bomb and the hydrogen bomb, or superbomb. These are the popular terms for these weapons, but as a physicist has warned these terms are not scientifically accurate. In truth, the atomic bomb should be called the "fisson bomb" or the "atomic bomb based upon nuclear fission." The "hydrogen bomb" should be called the "fusion bomb" or the "atomic bomb based upon fusion." While this distinction is important and should be mentioned, popular usage makes it necessary to retain for this work the names atomic bomb and hydrogen bomb.

This book is intended to provide both a ready reference and a guide for the further study of atomic energy. While no encyclopedia on such a controversial subject can claim to be definitive, the material has been selected to illustrate the depth and breath of each topic. This book is directed more toward the general reader and the high school or college student interested in the history of atomic energy than toward the specialist in the field. Furthermore, an effort has been made to be as objective as possible regarding the subject content of the entry. Each entry includes at least one suggested reading for those readers who want to inquire more deeply about the subject. Atomic energy is still a hot topic in the news and the background covered in this book will help readers understand current and future issues.

A

Abelson, Philip Hauge

Philip Abelson (1913–   ) was an American physicist active in the separation of isotopes of uranium for the Manhattan Project. He was born on April 27, 1913, in Tacoma, Washington. His undergraduate degree in chemistry was from Washington State University, Spokane, Washington. In 1939, Abelson received his Ph.D. in physics from the University of California at Berkeley. Although Abelson was working on experiments at the University of California that were leading him toward the discovery of nuclear fission, the Germans, under Otto Hahn, beat him by a few weeks. His entire professional career, apart from the war years, was spent at the Carnegie Institution in Washington, DC.

During World War II, Abelson worked for the Naval Research Laboratory in Washington, DC. On the eve of the war, he assisted Edwin McMillan in experiments at the Berkeley cyclotron that created neptunium. Next Abelson worked on the separation of isotopes of uranium for the U.S. Navy, beginning in 1941. Using the thermal diffusion process at the Philadelphia Navy Yard, he was able to isolate small amounts of uranium-235. The navy was interested in this process to generate atomic energy for future ship propulsion, especially submarines. This unexpected source of uranium-235 prompted General Leslie Groves, head of the Manhattan Project, on the recommendation of J. Robert Oppenheimer, to add Abelson's process to the mix of methods in order to acquire enough material for an atomic bomb. Because Abelson's research made it possible to consider isolating enough U-235 to make a bomb, a huge plant at Oak Ridge, Tennessee, was constructed to carry out this process.

After the war, Abelson returned to the Carnegie Institution. His research turned away from nuclear topics to biophysics projects. Abelson became director of the Geophysics Laboratory in 1953, and in 1971 he became president of the Carnegie Institution. Abelson became the editor of *Science* and found himself in the middle of the controversy over safe, permissible radiation levels. He also served on the Plowshare Committee, which promoted peaceful uses of nuclear energy.

Suggested readings: John Daintith, Sarah Mitchell, Elizabeth Tootill, and Derek Gjertsen, *A Biographical Encyclopedia of Scientists*, 2nd ed., vol. 1 (Bristol, UK: Institute of Physics, 1994); Richard Rhodes, *The Making of the Atomic Bomb* (New York: Simon and Schuster, 1986).

Acheson Board

In 1946 the Acheson Board was given the task by Jimmy Byrnes, the secretary of state in the Truman administration, to devise a plan to control atomic energy both domestically and internationally. Its members included Dean Acheson, chair, General Leslie R. Groves, Vannevar Bush, James Bryant Conant, and John J. McCloy. Acheson immediately appointed the Lilienthal Group to serve as a committee of expert consultants. This committee presented a report calling for the internationalization of atomic energy with UN control. Because members of the Acheson Board disagreed with several aspects of the Lilienthal Report, they tightened parts of it before it was turned over to Bernard Baruch, who also made major changes in the plan. Baruch was the U.S. representative to the United Nations, and it was his job to present this plan before the international body. The Baruch Plan was taken before the United Nations with restrictive provisions, and the Soviet Union refused to participate. Several members of the Acheson Board were unhappy about this turn of events, but distrust of Stalin and the Soviets was beginning to influence U.S. foreign policy.

Suggested reading: Steven M. Neuse, *David E. Lilienthal: The Journey of an American Liberal* (Knoxville: University of Tennessee Press, 1996).

Acheson-Lilienthal Report. *See* Acheson Board; Lilienthal Group

Advisory Committee on Atomic Energy

The Advisory Committee on Atomic Energy was formed in August 1945 to make recommendations on atomic energy general policy to the British government. Gen 75 approved the formation of this committee and appointed Sir John Anderson as its chair. Anderson had quasi-ministerial status and reported directly to the prime minister. During World War II, Anderson had held a leadership role in the British efforts to assist the Americans in building an atomic bomb. Membership on the original committee consisted of four scientists, two senior government officials, and two armed forces representatives. Later, three scientists, four senior government officials, and one armed forces representative also became members. For two years—from August 1945 to the middle of 1947—this committee was extremely active as a policymaking body helping to formulate British government policy on atomic energy. After this period, this committee no longer functioned as a policymaking body. After a few more months of meetings, this committee simply disappeared.

Suggested reading: Margaret Gowing, *Independence and Deterrence: Britain and Atomic Energy, 1945–1952*, vol. 2 (New York: St. Martin's Press, 1974).

Advisory Committee on Reactor Safeguards

The Advisory Committee on Reactor Safeguards (ACRS) was an independent body of scientists and engineers established to serve as a watchdog over the Atomic Energy Commission (AEC). It was formed in 1953 out of the Reactor Safeguard Committee, and its primary mission was to oversee reactor safety. It was a response to a widely held belief in the scientific community that the AEC was indifferent to nuclear safety concerns. Rogers McCullough was its first chairperson. The 15 members of the committee were selected from industry, national laboratories, and universities to provide a cross section of experience in nuclear energy. Meetings were held in secret, and their findings were published. It was the one important independent check on the AEC's regulatory program. The first instance of a substantive disagreement between the ACRS and the AEC was over the breeder reactor proposed by Detroit Edison. This reactor, also called the Fermi reactor, used highly concentrated plutonium as a fuel, and the ACRS was fearful of accidental detonation. Even after intensive lobbying, the members of the Advisory Committee recommended against the construction of the plant. The AEC ignored this recommendation and approved the building of the plant. A court case ensued, brought by the United Auto Workers (UAW), but the AEC won the four-year court battle. In the mid-1960s the ACRS became concerned about the building of nuclear power plants close to heavily populated areas, but again the AEC ignored its recommendations. The chair of the AEC, Glenn Seaborg, rejected these warnings because of the costs and economics of safety. Next the ACRS concerned itself with the size of reactors and the danger of a meltdown, the China Syndrome. Research was then directed toward emergency cooling systems to prevent such meltdowns. *See also* Reactor Safeguard Committee.

Suggested readings: Rodney P. Carlisle, *Supplying the Nuclear Arsenal: American Production Reactors, 1942–1992* (Baltimore, MD: Johns Hopkins University Press, 1996); Daniel F. Ford, *The Cult of the Atom: The Secret Papers of the Atomic Energy Commission* (New York: Simon and Schuster, 1982); Spencer R. Weart, *Nuclear Fear: A History of Images* (Cambridge: Harvard University Press, 1988).

Advisory Committee on Uranium. *See* Uranium Committee

African Nuclear Weapons Free Zone Treaty. *See* Treaty of Pelindaba

Aircraft Nuclear Propulsion

In the euphoria of the success of the atomic bomb program, the U.S. Air Force embarked on a program to build a nuclear aircraft. The Aircraft Nuclear Propulsion (ANP) program was originally called the Nuclear Energy Propulsion, Aircraft (NEPA) Program. Initial impetus for this program was a rumor that the Soviet Union was working on a similar project and nearing success. This program also became a pet of the powerful Joint Committee on Atomic Energy. An engineer, Donald Keirn, was placed in charge of the project in 1946. Much of the initial research was done at Oak Ridge, Tennessee, at a vacated plant. Design called for a nuclear-propelled bomber that could fly at least 12,000 miles at 450 miles per hour without refueling. The air force wanted an aircraft powerful enough to fly to the Soviet Union to deliver a payload of atomic weapons and return without refueling. First the Fairchild Corporation and later General Electric held the contract for the aircraft design. Interservice rivalry perpetuated continued research. After it became apparent that problems made it unlikely that a nuclear-driven aircraft was possible, the NEPA Project was canceled in 1951. Two main problems were never resolved: (1) material dense enough to shield the pilots but light enough for use on an aircraft, and (2) the issue of the radioactive dangers of an aircraft crash. The air force continued to explore the possibility, however, and contracts were given to aircraft companies to design engines. Finally, the air force determined that this aircraft might not have a military mission, but it could serve as a future prototype for space travel. In the years between 1946 and 1961, nearly $1 billion was spent by the U.S. government on a program to develop a nuclear propulsion system for manned aircraft. Insurmountable problems of weight, shielding, and design led to the final closing of this program in 1961. In total, the ANP program cost the American taxpayer around $4.6 billion with few concrete results.

Suggested readings: Kenneth Gantz, *Nuclear Flight: The United States Air Force Programs for Atomic Jets, Missiles, and Rockets* (New York: Duell, Sloan, and Pearce, 1960); Peter Pringle and James Spigelman, *The Nuclear Barons* (London: Michael Joseph, 1981).

Akers, Wallace Allen

Sir Wallace Akers (1888–1954) was a British industrial chemist who became active in the British nuclear program. He was born on September 9,

1888, in London, England. Akers's father was an accountant. After attending Aldenham School, he went to Oxford University. After completing his degree in chemistry, he joined the chemical firm Brunner Mond, where he worked as a chemist from 1911 to 1924. Four years' work with an oil company in Borneo was followed by a return to Brunner Mond in 1928. By this time, Brunner Mond had merged with Imperial Chemical Industries (ICI). In 1931, Akers became head of the Billingham Research Laboratory, and in 1941, he was promoted to director of all ICI research.

Akers's importance in the nuclear field was heading a program to gather and process uranium, or in British usage of the day, Tube Alloys. During the war, he worked closely with Sir John Anderson, the government head of the British atomic bomb project. He was placed in charge of a secret directorate, the Tube Alloys organization, in October 1941 to carry out this mission. Akers's role was to take charge of all aspects of atomic energy but especially uranium production. He led the mission of British scientists in 1943 to America to negotiate a collaboration pact with the Americans on the Manhattan Project. The Americans rejected him as the potential head of the British participation team because of his abrasive personality and his close ties to ICI. His connections with British business made it difficult for Americans to justify using U.S. taxpayer funds. This rejection led to James Chadwick replacing him as liaison head. Akers returned to the United Kingdom to continue work on the nuclear bomb issue.

After the war, Akers returned to ICI and directed its research activities. He was knighted in 1946. Among his postwar activities was to build a central research laboratory of ICI, which was later named the Akers Research Laboratory. He died on November 1, 1954, in Alton, England.

Suggested readings: Ronald William Clark, *The Birth of the Bomb* (New York: Horizon Press, 1961); John Daintith, Sarah Mitchell, Elizabeth Tootill, and Derek Gjertsen, *A Biographical Encyclopedia of Scientists*, 2nd ed., Vol. 1 (Bristol, UK: Institute of Physics, 1994); Bertrand Goldschmidt, *Atomic Rivals* (New Brunswick, NJ: Rutgers University Press, 1990).

Alamogordo. *See* Trinity

Aldermaston

Aldermaston in Berkshire is the site of the British atomic weapons production center. It was a former Royal Air Force base that had been built in 1941 but abandoned in 1948. Much of the initial atomic work was conducted at Ft. Halstead in Kent, but beginning on April 1, 1950, the construction of research facilities at Aldermaston was begun. Construction peaked in 1953, and Aldermaston became the new home of atomic weapons production. In a reorganization in 1950 the Atomic Weapons Research Establishment (AWRE) was separated from High Energy Research (HER). William Penney retained control of HER at Aldermaston to continue work on the British atomic bomb. The goal was to produce a plutonium bomb resembling the bomb that had been dropped by the Americans on Nagasaki, a goal accomplished with the detonation of a British atomic bomb on Monte Bello in October 1952.

Aldermaston has remained the center of the British nuclear weapons establishment. In the mid-1990s, it has two research reactors (VIPER and HERALD); a high-powered laser (HELEN); a Cray computer; and a high-intensity X-ray beam. As the British government reevaluates its nuclear program, the role of Aldermaston is under study, and the scientists there are uncertain about their future.

Suggested readings: Eric Arnett, ed., *Nuclear Weapons After the Comprehensive Test Ban: Implications for Modernization and Proliferation* (Oxford: Oxford University Press, 1996); Brian Cathcart, *Test of Greatness: Britain's Struggle for the Atom Bomb* (London: Murray, 1994); Margaret Gowing, *Independence and Deterrence: Britain and Atomic Energy, 1945–1952*, vol. 2 (New York: St. Martin's Press, 1974).

Allen v. United States of America

The court case of *Allen v. United States of America* was brought forth by 24 plaintiffs from southwestern Utah on August 30, 1979, charging injury and wrongful death from exposure to radioactive fallout from nuclear testing at the Nevada Test Site. They charged that agents of the federal government were negligent in conducting nuclear tests from 1951 to 1962. By the failure of the U.S. government to use "reasonable care," the plaintiffs both deceased and suffering from cancer had sustained severe damages. The discovery process took nearly two years before the case finally went to trial. Trial began on September 20, 1982, and lasted until December 17, 1982. After listening to the testimony and reading the transcripts, the federal judge, Bruce Jenkins, ruled against the U.S. government and held the government's operatives at the Nevada Test Site had acted negligently, causing death and injury to 10 of the 24 plaintiffs. Refusing to settle the case with the plaintiffs, federal attorneys appealed the judgment and the monetary damages. An appeals court overturned this decision in April 1987, maintaining that the government had sovereign immunity. In January 1988, the appellate court ruling was upheld by the U.S. Supreme Court. Regardless of the final outcome of the case, this trial was important because for the first time the U.S. government had to admit that the Atomic Energy Commission and its agents had been negligent in their conduct of nuclear testing.

Suggested reading: Howard Ball, *Justice Downwind: America's Atomic Testing Program in the 1950s* (New York: Oxford University Press, 1986).

Allier Affair

Jacques Allier had a small but important role in the history of both the Allied and German atomic bomb programs. French nuclear chemist Frederic Joliot-Curie recognized the importance of heavy water as a moderator for an atomic reactor. In the winter of 1939, Joliot-Curie sent a message to Raoul Dautry, the French minister of armaments, requesting him to buy up all the heavy water that the Norwegian company Norsk-Hydro could produce. Allier was a lieutenant in the French Secret Service, the Deuxieme Bureau, and a former banker with the Bank of Paris and the Low Countries; he learned from Norwegian contacts that the Germans wanted to buy the Norsk-Hydro stock of heavy water and sign a contract for future production. In March 1940 Allier traveled to Norway with the backing of the French government and under his mother's maiden name Freiss to purchase Norsk-Hydro's heavy water. He convinced Norsk-Hydro general manager Axel Aubert that it would be better to sell the 185 kilograms (407 pounds) of heavy water to the Allies rather than to the Germans. His cause was aided by the Bank of Paris's and the Low Countries' majority interest in the Norse-Hydro Company. A gentleman's agreement was signed on March 9, 1940, giving Norsk-Hydro 15 percent of any profits from technical developments from the loan of the heavy water to France. German fighters forced Allier's plane to land in Hamburg, but he had switched the 26 cans of heavy water into two loads and sent them to Montrose, Scotland. He then shipped the cans to Edinburgh and then to Paris by train and by the Channel ferry. The heavy water was transported to a bombproof vault near Joliot-Curie's laboratory where secret experiments were to be undertaken on a chain-reacting pile. When the Germans seized the Norwegian plant two months later, the supply of heavy water was almost nonexistent. The Paris supply was smuggled to the British in May 1940, depriving the Germans of access to it for their experiments. This secret intrigue deprived the Germans of a large supply of heavy water at a key stage of their planning for an atomic reactor. Soon after Allier smuggled the heavy water out of Norway, he attempted to warn the British government about German interest in building an atomic bomb. He had an interview with Sir Henry Tizard, the chairman of the Committee for the Scientific Survey of Air, who appeared unimpressed by his arguments. He gave Tizard a list of German scientists working on atomic energy projects, which Tizard passed on to the War Cabinet for further study. This meeting ended Allier's brief but important role in atomic energy, and he returned to intelligence work.

Suggested readings: Bertrand Goldschmidt, *Atomic Rivals* (New Brunswick, NJ: Rutgers University Press, 1990); Maurice Goldsmith, *Frederic Joliot-Curie: A Biography* (London: Lawrence and Wishart, 1976); Dan Kurzman, *Blood and Water: Sabotaging Hitler's Bomb* (New York: Henry Holt, 1997); Thomas Powers, *Heisenberg's War: The Secret History of the German Bomb* (New York: Knopf, 1993).

Alsos Mission

The Alsos Mission was a mixed civilian-military operation to learn how much the Germans knew about an atomic bomb and how far along their scientists were in constructing such a weapon. Colonel John Lansdale proposed the idea for such a mission to General Leslie Groves, head of the Manhattan Project in early 1943, and Groves accepted his plan. Colonel Boris T. Pash was its military chief, but General Groves directed its activities. In a sort of joke, *Alsos* is the Greek translation for "grove."

Colonel Boris Pash was the field commander of Alsos. A former high school teacher, Pash had been trained in security by the Federal Bureau of Investigation (FBI). Before assuming command of this mission, he had been chief of the Counter-Intelligence branch of the U.S. Army in San Francisco. One of his early missions had been to investigate J. Robert Oppenheimer's background for his security clearance to be the head of the Los Alamos Scientific Laboratory. Pash had fought in the Russian Civil War as a White Russian, and his father was a leader of the Russian Orthodox Church in exile. His hatred of the Bolsheviks extended to anyone who had ever displayed any sympathy with them. For this reason, he opposed Oppenheimer's security clearance. When the mission to investigate the progress of German scientists was accepted, Pash was given command of it. His aggressiveness on security matters had won him a considerable number of high-ranking officers as enemies, so the new assignment was timely.

Alsos's first assignment was to go to Italy in 1943 to collect scientific information about German atomic research at Italian universities. This mission produced little useful information, but it reinforced the need for scientific expertise. Samuel Goudsmit, a Dutch-born physicist, was recruited to be the scientific adviser on the mission. While only a limited number of people knew about the existence of the Allied atomic bomb program, the scientific members soon learned more about it. Soon after the invasion of Europe, the mission established its headquarters in Paris. Colonel Pash had a letter of personal authority from the U.S. secretary of war, Henry Stimson, ordering everyone to assist Pash in his mission. Agents, however, worked on their own and traveled quickly to liberated areas to find personnel and documents. At one time it was considered to either capture or kill Werner Heisenberg, but this action was rejected because it might warn the Germans what the Allies were up to at Los Alamos. Another part of the mission was specially adapted A-26 aircraft flying over Germany using a radiological air-sampling system designed by physicist Luis Alvarez. These flights were made to detect if the Germans had developed a radioactive bomb, but no such evidence was found. Documents captured at the University of Strasbourg soon convinced Goudsmit and his fellow scientists by December 1944 that the Germans were nowhere near building an atomic bomb. After this was determined, Groves gave Alsos the mission of rounding up German nuclear physicists to keep them out of Russian hands. This policy worked well, and the Americans picked up most of the German scientists with knowledge of atomic research. General Groves also ensured that the Soviets would not obtain uranium supplies accumulated by the Germans by bombing a metal refining plant in Oranienburg in 1945. Another effort recovered much of Belgium's prewar uranium supply from a Stassfurt factory in April 1945. Some of this ore was used for Little Boy.

Suggested readings: Jeremy Bernstein, *Hitler's Uranium Club: The Secret Recordings at Farm Hall* (Woodbury, NY: American Institute of Physics, 1996); Samuel A. Goudsmit, *Alsos* (New York: Henry Schuman, 1947); David Irving, *The German Atomic Bomb: The History of Nuclear Research in Nazi Germany* (New York: Simon and Schuster, 1967); Thomas Powers, *Heisenberg's War: The Secret History of the German Bomb* (New York: Knopf, 1993); Charles A. Ziegler and David Jacobson, *Spying Without Spies: Origins of America's Secret Nuclear Surveillance System* (Westport, CT: Praeger, 1995).

Alvarez, Luis Walter

Luis Alvarez (1911–1988) was an American physicist whose work on the atomic bomb and research on subatomic particles made him one of the top nuclear physicists in the postwar United States. He was born on June 13, 1911, in San Francisco, California. His father was a physician who later worked at Mayo Clinic. Two marriages, Geraldine Smithwick (1937) and Janet L. Landin (1958), produced two sons and two daughters. He attended the Polytechnic High School in San Francisco, California, and Rochester High School in Rochester, Minnesota. He entered the University of Chicago in 1928 with the desire to major in chemistry. In his junior year, Alvarez changed his major to physics. Alvarez received his bachelor of science degree in 1932, his master's in 1934, and his doctorate in 1936 from the University of Chicago. His adviser was Arthur Compton, and Alvarez's research was with a modified Geiger counter measuring cosmic rays. His research found him in the middle of the controversy between Compton and Robert Milikan at the California Institute of Technology. Alvarez traveled back to California to the University of California at Berkeley for a teaching job in 1936. By 1945, he was a full professor.

Alvarez was always a productive experimental researcher. In 1939, he worked with Felix Bloch to measure the magnetic movement of the neutron. During the war, he divided his time between the radiation laboratory of the Massachusetts Institute of Technology (MIT) (1940–1943) and the Los Alamos Scientific Laboratory, New Mexico (1944–1945). At MIT, Alvarez worked on radar and other applied military science projects including radar and a radiological air-sampling method, but at Los Alamos his attention was on building an atomic bomb. Perhaps his greatest contribution was the exploding-wire electric detonators to help solve the implosion problem.

Among his other contributions was a way to measure the bomb yield, so he rode the backup B-29, the *Great Artiste*, to witness the Hiroshima bomb.

After the war, Alvarez continued his research on nuclear problems. His first task was to build the first linear accelerator for protons and then to design a bubble-chamber technique for observing electrically charged sub-atomic particles. For this type of high-level research, Alvarez received the 1968 Nobel Prize for Physics. Alvarez was also a strong proponent of build-ing the hydrogen bomb. For this reason, he, along with Edward Teller and Ernest O. Lawrence, testified against J. Robert Oppenheimer in his security clearance reinstatement trial. He died on September 1, 1988, in Berkeley, California.

Suggested readings: Luis W. Alvarez, *Alvarez: Adventures of a Physicist* (New York: Basic Books, 1987); Peter W. Trower, ed., *Discovering Alvarez: Selected Works of Luis W. Alvarez* (Chicago: University of Chicago Press, 1988).

American Nuclear Program

The American nuclear program was contained in the Manhattan Project until the end of World War II. Once this project had accomplished its goal of building an atomic bomb the next stage was uncertain. It was not until the debate over civilian versus military control ended with the passage of the Atomic Energy Act in August 1946 that was it possible to move ahead. The shift from military to civilian control also created dislocation. Most of the top scientists left Los Alamos to return to their peacetime academic posts. Finally, a national policy regarding atomic research and testing was lacking. The new Atomic Energy Commission (AEC) had to inform Presi-dent Harry Truman that its arsenal of atomic weapons was almost depleted. It took nearly two years from the end of the war before the necessary guid-ance was forthcoming and new bombs were tested at Eniwetok in the spring of 1948.

A major problem in the late 1940s and 1950s was the supply of uranium. Due to military needs and the burgeoning reactor program, the United States hoarded all the uranium it could find. This policy continued until 1959 when the AEC had all the uranium it needed for the foreseeable fu-ture. After shifting the testing site from the Pacific islands to the Nevada Test Site, the U.S. government instituted a series of tests lasting almost a decade. After atmosphere testing was banned, the testing went under-ground. The civilian program of atomic energy was much slower to develop than weapons research.

Weapons research dominated the American nuclear program for the first two postwar decades. A series of research centers were founded, including the Brookhaven National Laboratory and the Lawrence Livermore National Laboratory, and added to the existing research establishments. Soon after the atomic weapons program was reestablished, attention turned to the possibility of a superbomb, or a hydrogen bomb. Edward Teller and Ernest Lawrence were the strongest proponents, and they were able to persuade President Truman to authorize the building of a hydrogen bomb despite opposition from many in the scientific community. This crash program produced a hydrogen bomb before the Soviet Union, but the race ended up closer than expected. Even the first venture into civilian nuclear power generation at Shippingport, Pennsylvania, was an outgrowth of the navy's ship propulsion program.

The boom period for civilian nuclear power ran between 1960 and 1975. Optimistic economic predictions for nuclear-powered generation of power fueled this boom. Stimulating the market by use of the turnkey approach, American companies specializing in the building of nuclear reactors sold public utilities on larger and more cost-efficient plants. Soon every large power company showed interest in investing in nuclear power. At first it was relatively easy for the plants to gain licensing and build, and the AEC made it easy to do so. Problems began to develop, however, as the antinuclear lobby started generating negative publicity for nuclear plant shortcomings. The *Calvert Cliffs* decision in 1971 making the AEC conform to environmental laws changed the complexion of the licensing debate. Suddenly the cost of pushing through licensing and possible legal entanglements made power companies more hesitant. This might have been more acceptable if the bottom-line cost of nuclear power had been better. Except at the most efficient plants, it cost more to produce nuclear-generated power than coal- or hydro-generated power. The final straw, however, was the Three Mile Island incident. Negative publicity combined with a lukewarm economic picture to kill the growth of civilian nuclear power. American utility companies canceled plans for nuclear plants as soon as they could. Consequently, no nuclear facilities have been built in the United States in over a decade, and none are in the planning stage. Many of the existing plants have either been decommissioned or are in the planning stages for decommissioning. As of December 31, 1966, all 14 government plutonium-tritium reactors have been mothballed, and 16 commercial reactors have been shut down. Yet in 1996 the United States had 110 nuclear reactors in operation that produced nearly 22 percent of the nation's electricity.

The slowdown in civilian nuclear power has been matched by a reduction in the U.S. government's nuclear weapons industry. Although the United States still spends in the neighborhood of $25 billion a year on nuclear weapons, the government allocates $8 billion on dismantlement and environmental cleanup. Much of the rest of the budget ends up in administrative costs. Cutbacks at the national laboratories have produced low morale and uncertainty about the fate of the nuclear weapons program. Furthermore, weapon control agreements have resulted in a sharp drop in nuclear weapons stockpiles. In the period from 1990 to 1995 the American nuclear stockpile

Figure 1
Nuclear Power Plant Simplified Schematic

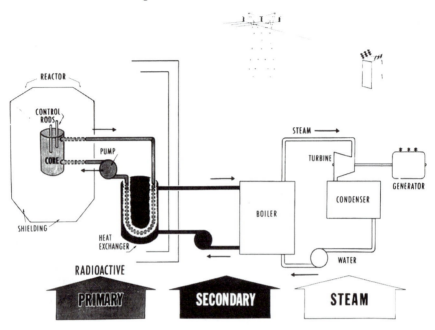

This illustration prepared by the Atomic Energy Commission shows how a nuclear power plant operates in theory.

was reduced from 21,000 to 14,000. Further reduction in agreements and budget cuts will continue to make inroads on the American nuclear arsenal. Efforts by government and military officials in 1997 and 1998 to renew underground testing of nuclear weapons were part of a campaign to modernize nuclear weapons in this shrinking environment.

Suggested readings: Corbin Allardice and Edward R. Trapnell, *The Atomic Energy Commission* (New York: Praeger, 1974); Frank G. Dawson, *Nuclear Power: Development and Management of a Technology* (Seattle: University of Washington Press, 1976); *Nuclear Power Reactors in the World* (Vienna, Austria: International Atomic Energy Agency, 1997).

American Nuclear Society

The American Nuclear Society (ANS), founded in 1954, is a professional society of scientists and engineers interested in the advancement of atomic energy. Its headquarters is in La Grange Park, Illinois. ANS works to advance the cause of science and engineering in the nuclear industry. With a membership of around 16,000 and a staff of 63, most of its activities concern specialized publications on nuclear energy subjects. Membership is divided into 59 local groups for more specialized meetings. These local groups conduct seminars in their localities on nuclear energy issues. It holds a semiannual conference for its membership. Considerable attention is paid to publishing books on the history of atomic energy.

Suggested reading: Earl R. Kruschke and Byron M. Jackson, *Nuclear Energy Policy: A Reference Handbook* (Santa Barbara, CA: ABC-CLIO, 1990).

American Nuclear Testing Program

The American Nuclear Testing Program began in 1945 and lasted until a testing moratorium was announced by President George Bush in the early 1990s. Pressure built up in early 1945 among the administrators and scientists involved in the Manhattan Project to test the plutonium bomb, the Fat Man, because there was a fear that the device might be a dud. This test, code name Trinity, at the Alamogordo Test Site, New Mexico, proved beyond a doubt that the design for the plutonium bomb worked. A duplicate of this bomb was used on August 9, 1945, on the Japanese city of Nagasaki. While it was acknowledged by scientists that further testing was needed to develop more sophisticated atomic bombs, it was not until after the war in 1947 that the Crossroads series of atomic weapons tests at the Bikini Atoll in the Marshall Islands of the Pacific were conducted. These tests took place in the middle of a power struggle between the air force and the navy over the role of atomic weapons versus conventional naval forces. Atomic bombs used in these tests were only slightly improved models over those developed for use against Hiroshima and Nagasaki.

The intensification of tensions between the Soviet Union in the Cold War and the Korean War caused a renewed interest in the testing of atomic weapons. Spurred by the expense and distance of Pacific testing, Atomic Energy Commission (AEC) administrators, military leaders, politicians, and scientists agreed on the necessity for a North American testing site. After a survey of possible sites, a south central Nevada bombing grounds was selected to become the Nevada Test Site. With a continental testing site for testing smaller-size nuclear weapons and the Pacific sites of Bikini and Eniwetok for larger-yield weapons, the United States conducted demonstrations of both tactical atomic weapons and hydrogen bombs. Both types of testing continued until international and domestic public pressure combined to limit atmospheric testing. After atmospheric testing ended in the aftermath of the Limited Test Ban Treaty in 1963, the United States conducted an extensive series of underground nuclear tests.

Altogether the United States has held 942 tests, of which 725 have been underground, 203 in the atmosphere or on the surface of land or water, 9 in space, and 5 underwater. Although most of the domestic tests were conducted at the Nevada Test Site, tests were also held at 10 other sites—

Alamorgodo, New Mexico; Carlsbad, New Mexico; Hattiesburg, Missississippi; Farmington, New Mexico; Grand Valley, Colorado; Fallon, Nevada; Rifle, Colorado; and two other remote sites in central Nevada. Pacific test sites were at Bikini Atoll, Christmas Island, Eniwetok Atoll, and Johnston Atoll.

The United States began a series of subcritical nuclear tests in 1997. These tests were part of a program to revitalize the nuclear bomb arsenal called the Science Based Stockpile Stewardship (SBSS) program. A total of 8 underground tests were conducted at the Nevada Test Site beginning in June 1997 and ending in September 1997. Peace groups charged that those subcritical nuclear tests were in violation of the Comprehensive Test Ban Treaty, but the tests were carried out nevertheless. As part of the same program, the U.S. government announced in 1998 that it was changing from actual testing to using higher-powered lasers and supercomputers to simulate testing. Three nuclear weapons laboratories, at Los Alamos, Sandia, and Livermore, will share the $45 million ten-year program.

Suggested readings: International Physicians for the Prevention of Nuclear War, *Radioactive Heaven and Earth: The Health and Environmental Effects of Nuclear Weapons Testing in, on, and Above the Earth* (New York: Apex Press, 1991); Robert L. Miller, *Under the Cloud: The Decades of Nuclear Testing* (New York: Free Press, 1986).

Anderson, Carl David

Carl Anderson (1905–1991) was an American experimental physicist whose discovery of the positron helped further knowledge about the atom. He was born on September 3, 1905, in New York City. His parents were immigrants from Sweden. Later his father was a small restaurant entrepreneur in Los Angeles. In 1946, he married Lorraine Elvira Bergman, and they had two sons. He was educated at the California Institute of Technology, where he received his bachelor of science degree in 1927 in physics and his Ph.D. in physics in 1930 under the direction of Robert Millikan. Soon after graduation, he began his academic career at California Institute of Technology (Caltech). He was appointed an assistant professor in 1933 and advanced up the ranks until he became professor in 1939. There he taught and conducted research until he retired in 1978.

Anderson's experimental research concerned the study of cosmic rays by photographing their tracks in a cloud chamber designed by Charles Wilson to study the path of atomic particles. Robert Millikan introduced him to the cosmic rays problem in the middle of the controversy between Millikan and Arthur Compton. On August 2, 1932, Anderson observed evidence of particles of positive charge, which he named positrons. After careful review, he announced his findings to a skeptical scientific community. Although Anderson was unaware of it, he had confirmed Paul Dirac's theory about such a particle. The next year, other scientists—Guiseppe Occhialini of the University of Rome and Patrick Blackett of Cambridge University—confirmed his discovery. Anderson's reward was the 1936 Nobel Prize in Physics. He continued to conduct research on particles, and at one time it was thought that he had found a heavier particle, the mu-meson. Research is unclear about the nature of the meson.

In Anderson's subsequent career, he continued to look at new lines of research. During World War II, he was active in the Office of Scientific Research and Development (OSRD), coordinating research programs at Caltech with military research projects. Anderson died in January 1991.

Suggested readings: Henry A Boorse, Lloyd Motz, and Jefferson Hane Weaver, *The Atomic Scientists: A Biographical History* (New York: Wiley Science Editions, 1989); Emilio Segre, *From X-Rays to Quarks: Modern Physicists and Their Discoveries* (San Francisco: W. H. Freeman, 1980); Anthony Serafini, *Legends in Their Own Time: A Century of American Physical Scientists* (New York: Plenum Press, 1993).

Anglo-American Declaration of Trust

The necessity to gain access to strategic ores for the war effort caused representatives from the United Kingdom and the United States to negotiate an agreement. On June 13, 1944, the governments of the United Kingdom and the United States signed a Declaration of Trust to ensure cooperation between them on obtaining uranium and thorium ores. At the time, only Canada and the Belgian Congo had mines with sufficient uranium ore for the Manhattan Project. This agreement was a further refinement of the Quebec Agreement of August 19, 1943. This issue of access to strategic ores became a larger issue after the end of World War II, and the agreement was abrogated by the United States. This decision upset Great Britain because simultaneously the British discovered that the United States had contracted for all the uranium ore produced in Canada. This virtual monopoly of international uranium ore production by the United States in the period just after World War II made it difficult for Britain to launch its own atomic energy program.

Suggested reading: Phillip L. Caneton, Richard G. Hewlett, and Robert C. William, *The American Atom: A Documentary History of Nuclear Policies from the Discovery of Fission to the Present*, 2nd ed. (Philadelphia PA: University of Pennsylvania Press, 1991).

Antarctic Treaty

Among international concerns in the 1950s was the possibility that the arms race might extend to the Antarctic Region. Its intent was to demili-

tarize and denuclearize the Antarctic and promote scientific research without
dealing with questions of territorial sovereignty. Twelve countries signed the
Antarctic Treaty on December 1, 1959, to make the Antarctic a neutral
zone for scientific research. The original twelve were Argentina, Australia,
Belgium, Chile, France, Japan, New Zealand, Norway, South Africa, Soviet
Union, United Kingdom, and the United States. Four other countries have
subsequently joined this treaty guaranteeing a nuclear free Antarctica—Brazil, India, Poland, and Germany. Article V has a provision to prohibit the
introduction of any nuclear device or radioactive waste into the Antarctic
Region. This treaty was the first agreement on banning nuclear development
negotiated on a regional level, and it was the first treaty establishing a nuclear free zone. On March 4, 1962, the United States circumvented this
treaty by installing a portable nuclear reactor at the U.S. Navy's McMurdo
Air Base, nickname Nukey Poo. Because this reactor had numerous problems, it was shut down in September 1972 and shipped back to the United
States. This treaty is still in force and no further attempts have been made
to introduce nuclear weapons or nuclear power plants in the Antarctic.

Suggested readings: *The Antarctic Treaty System* (Canberra: Australian Govern-
ment Publishing Service, 1983); Peter Beck, *The International Politics of Antarctica*
(New York: St. Martin's Press, 1986); Cornnie Brown and Robert Munroe, *Time
Bomb: Understanding the Threat of Nuclear Power* (New York: William Morrow,
1981); David Pitt and Gordon Thompson, eds., *Nuclear Free Zones* (London: Croom
Helm, 1987).

Antiballistic Missile Treaty

The Antiballistic Missile Treaty was negotiated in 1972 between the Soviet Union and the United States to control the delivery of nuclear weapons
because an agreement to ban nuclear weapons was impossible to conclude.
Both countries subscribed to the mutual assured destruction (MAD) strategy in which either side was capable of inflicting devastating destruction on
the other. Building ballistic missile defense systems upset this balance and
was recognized as a destabilizing factor. It would allow a first strike capability without the fear of retaliation. This treaty limited each superpower to
only two antiballistic missile (ABM) sites: its capital city and at a selected
intercontinental ballistic missile (ICBM) site. A 1974 protocol restricted the
ABM deployment to one site. The Soviet Union picked Moscow and the
United States its Safeguard site in North Dakota. This latter choice proved
to be a mistake on the part of the United States because the Safeguard
system was abandoned as too costly in 1976.

Suggested reading: William J. Durch, *The Future of the ABM Treaty* (London:
International Institute for Strategic Studies, 1987).

Antinuclear Movements. *See* Antinuclear Movements in the United States; Ban the Bomb Movement; SANE

Antinuclear Movements in the United States

Protest against nuclear energy was until the early 1970s concerned with its military uses and disarmament. Beginning with reservations by scientists working on nuclear problems in the late 1940s, the military uses debate peaked in the late 1950s and early 1960s over radioactive fallout from nuclear weapons tests. The disarmament movement picked up steam in the mid-1950s until it became apparent that the movement lacked the political clout to change the Cold War mentality of either the Western Allies or the Soviet Bloc. Slowly, however, aspects of the antinuclear movement's program on limiting nuclear weapons were negotiated into treaties. A different tact emerged in the 1980s with the emphasis now on a nuclear freeze and nuclear free zones. Again this movement ran into the entrenched power of the state.

The civilian nuclear energy protest movement emerged in the late 1960s and was directed against individual nuclear sites. By 1974, national antinuclear organizations were formed. These organizations attacked nuclear energy over health hazards, safety hazards, cost overruns, and waste disposal problems. But it took the Three Mile Island incident in 1979 and its aftermath to mobilize widespread opposition toward nuclear energy. A de facto moratorium on nuclear plant construction after 1979 ended this phase of the antinuclear campaign. Opposition then turned in the 1980s toward individual plants with the nuclear energy industry on the defensive. By the 1990s it had become apparent that the United States had abandoned dependence on nuclear technology as the basis for future electricity production. Many of the adherents of the antinuclear movement have moved to the environmental movement.

Suggested readings: Christian Joppke, *Mobilizing Against Nuclear Energy: A Comparison of Germany and the United States* (Berkeley: University of California Press, 1993); David S. Meyer, *A Winter of Discontent: The Nuclear Freeze and American Politics* (New York: Praeger, 1990); Jerome Price, *The Antinuclear Movement* (Boston, MA: Twayne, 1982); Wolfgang Rudig, *Anti-Nuclear Movements: A World Survey of Opposition to Nuclear Energy* (Harlow, UK: Longman Current Affairs, 1990).

Apex Committee

Fearful of losing its big power status, the British government undertook a research campaign to build and test an atomic device in the late 1940s and early 1950s. The Apex Committee was formed in March 1952 by the new Churchill government with the intention of overseeing the testing of the British atomic bomb at Monte Bello, Australia. Its members were Sir Anthony Eden, the foreign secretary, Lord Cherwell, pay-master general and scientific adviser to Winston Churchill, Duncan Sandys, minister of supply, and Earl Alexander, defense minister. Meetings of the Apex Committee were monthly, and William Penney, head of the bomb testing program, and his fellow scientists were often present to give progress reports. Once it became apparent that the British atomic program was going to be successful, the Apex Committee faded away. It was replaced by the Atomic Energy Authority, which still administers atomic energy in the United Kingdom.

Suggested reading: Brian Cathcart, *Test of Greatness: Britain's Struggle for the Atom Bomb* (London: Murray, 1994).

Appel des 400

The French government initiated a major construction of nuclear power plants in the 1970s. In the midst of the French debate over more nuclear reactors in the Messmer Plan, a petition appeared in February 1975, signed by 400 French scientists, criticizing the Messmer Plan. Most of the leadership of this opposition came from the scientists at the University of Grenoble. They were especially upset about the fast breeder project at Creys-Malville. The signatures on the petition eventually grew to number 4,000. Despite this vocal opposition, the French government built its energy infrastructure on nuclear power plants. An outgrowth of this petition campaign was the formation of the French scientific group Groupement d'information scientifique sur l'énergie nucléaire modeled on the American organization the Union of Concerned Scientists to present arguments against the civil use of nuclear power in France. This movement indicated strong opposition to the French government's nuclear energy program, but this movement has failed to influence the government. In 1999, France obtained 78 percent of its electricity from nuclear power plants.

Suggested reading: Helena Flam, ed., *States and Anti-Nuclear Movements* (Edinburgh, Scotland: Edinburgh University Press, 1994).

Arco Reactor Test Site. *See* National Reactor Testing Station

Ardenne, Manfred von

Manfred von Ardenne (1907–1997) was a German scientist–engineer who ran an independent atomic energy research project during the Nazi regime. He was a scientific entrepreneur who built a laboratory in the Lichterfeld suburb of Berlin in 1928. His income came from government and private business contracts and patents on inventions. Atomic energy came of interest to him soon after he learned of the discovery of fission. Ardenne built an electron microscope at his laboratory, and German scientists began to flock to his laboratory to use it. Lacking funds for a full-scale atomic weapons project, Ardenne was able to attract the patronage of a friend of the family, Wilhelm Ohnesorge, an official in the German Post Office. Armed with a large government contract from the Post Office, Ardenne hired German physicist Fritz Houtermans to conduct work on chain reaction theory. Houtermans soon realized the possibility of plutonium, but since he was a dedicated anti-Nazi, Houtermans was reluctant to proceed further. Nevertheless, Ardenne had his laboratory work on the electromagnetic separation of isotopes, and he believed it possible to separate a few kilograms of U-235. His conversations with Werner Heisenberg, Otto Hahn, and Ernst von Weizsacker discouraged him from continuing along these lines. Nevertheless, Ardenne had built a Van de Graaff electrostatic machine and started work on building a 60-ton cyclotron. Despite lukewarm support from other German physicists, Ardenne continued to conduct basic research on materials for an atomic reactor.

After the German defeat, Soviet authorities convinced Ardenne to continue his scientific work in the Soviet Union. His laboratory and most of his scientists were transferred to the Soviet Union, and he spent the next decade working on the magnetic separation of isotopes for the Soviets. Most of his work was in Sinop near Sukhumi on the southern Caucasian coast of the Black Sea. On the suggestion by Soviet authorities, Ardenne then shifted his emphasis to the study of plasma research and controlled fusion. In 1955 he returned to East Germany and established a private institute in Dresden. Ardenne stayed in East Germany. The institute now specializes in the treatment of cancer. He died in May 1997.

Suggested readings: David Irving, *The German Atomic Bomb: The History of Nuclear Research in Nazi Germany* (New York: Simon and Schuster, 1967); Thomas Powers, *Heisenberg's War: The Secret History of the German Bomb* (New York: Knopf, 1993).

Argentine Nuclear Program. *See* South American Nuclear Program

Argonne National Laboratory

The Argonne National Laboratory was founded on July 1, 1946, to conduct research on radiation and radioactive materials for both military and peaceful uses. Work on the Argonne site, however, had begun during World War II. It was the dream of Arthur Compton, dean of Physical Sciences at the University of Chicago. He selected the site at Argonne Forest about 25 miles southwest of Chicago. Only a labor dispute retarding construction in the late 1942 prevented the testing of the first atomic reactor in December 1942. A reactor was built there in early 1943, and crucial experiments were undertaken there in 1943 and 1944 that proved useful in the construction of the atomic bombs in the Manhattan Project. After the war, the Metallurgical Laboratory of the University of Chicago was merged into the new Argonne National Laboratory. The University of Chicago administered the activities of the Argonne National Laboratory under contract with the Atomic Energy Commission (AEC). Argonne's direct was Walter Zinn, a physicist with extensive experience in reactor design. It was the intention of the AEC to use the laboratory as a place to centralize nuclear reactor research under Zinn's expertise. This decision was controversial, and other laboratories protested vigorously, but the AEC stood firm. During the 1950s and 1960s, the Argonne Laboratory experimented with breeder reactors in an effort to achieve technical leadership in this type of nuclear technology. Its first success was the Experimental Breeder Reactor (EBR-1), which started operations in 1951. Subsequent research was also done with the next version, the EBR-2. Walter Zinn became concerned about the building of breeder reactors in a populated area so he proposed a testing facility in Idaho. The National Reactor Testing Station (NRTS) was started on April 4, 1949, near Arco, Idaho. Argonne labs used this facility to test most of its prototypes in the next 50 years. It was at the NRTS that Argonne personnel tested the reactor for the USS *Nautilus* in 1953. In the next two decades the Argonne National Laboratory devoted most of its efforts to perfecting the EBR-2. This reactor established the feasibility of a fast breeder reactor, but in 1994 the president and Congress ordered Argonne labs to cease experimenting with advanced reactor technology.

Suggested readings: Frank G. Dawson, *Nuclear Power: Development and Management of a Technology* (Seattle: University of Washington Press, 1976); Jack M. Holl, *Argonne National Laboratory, 1946–96* (Urbana: University of Illinois Press, 1997).

Armament Research Department

The Armament Research Department (ARD) was given the mission by the British government to work on the military applications of atomic energy. ARD's first home was at Fort Halstead near Sevenoaks in Kent, and it had played an important role in developing British armaments in World War II. When this establishment was turned over by the British government to atomic bomb research, the resident scientists were hesitant about staying. It was only after William Penney took over in 1946 that morale improved. Penney had been a British representative to the American Manhattan Project, and he was a specialist on explosives and the blast damage. His expertise on nuclear blast damage had been appreciated by the Americans. ARD also had a presence at the Woolwich Arsenal where precision engineering was a specialty. In 1948 the Armament Research Division was renamed the Armament Research Establishment (ARE). Later in the early 1950s, ARE and High Energy Research (HER) were divided, and HER moved to Aldermaston, where its personnel continued to work on building British atomic weapons.

Suggested readings: Brian Cathcart, *Test of Greatness: Britain's Struggle for the Atom Bomb* (London: Murray, 1994); Margaret Gowing, *Independence and Deterrence: Britain and Atomic Energy, 1945–1952,* vol. 2 (New York: St. Martin's Press, 1974).

Army Ordinance Bureau

The Army Ordinance Bureau was the military side of the German effort to build an atomic bomb in World War II. This section was originally headed by General Erich Schumann, a professor of military physics at the University of Berlin, director of the Second Physics Institute at the University of Berlin, and an army general, and his chief assistant, Kurt Diebner. Both were considered mediocre physicists, especially Schumann, by their scientific colleagues. General Schumann received a letter dated April 24, 1939, in which Paul Harteck and Wilhelm Groth informed him of the possibility of an atomic bomb. Schumann turned this information over to Kurt Diebner for further study. On September 16, 1939, in a secret conference in Berlin, Diebner and Erich Bagge brought together a select group of German scientists to discuss the feasibility of an atomic weapons project. In a second meeting also in Berlin on September 26 of an enlarged group, the scientists studied ways to make fission useful. Later in 1939, Diebner began uranium bomb research at the army ordinance proving grounds at Kummersdorf, near Berlin. At the same time, the Army Ordinance Bureau took over the Kaiser Wilhelm Institute for Physics in Berlin to serve as a research headquarters. The bureau's efforts in 1940–1941 were devoted to two lines of research: development of a chain-reacting reactor and the separation of U-235. By the end of 1942, however, Schumann became discouraged about the lack of progress and transferred the uranium project over to the Reich's Research Council.

Suggested readings: Thomas Powers, *Heisenberg's War: The Secret History of the German Bomb* (New York: Knopf, 1993); Richard Rhodes, *The Making of the Atomic Bomb* (New York: Simon and Schuster, 1986).

Aryan Physics Movement

The Aryan Physics movement was an attempt by disaffected elements in the German scientific community who distrusted new theories of physics to return Germany to traditional experimental physics. It was part of an ideological war that broke out in the early 1920s between the Aryan Physics adherents and those physicists who supported Einstein's theory of relativity and Planck's quantum theory. This type of warfare lasted until the end of the Nazi regime. The supporters of Aryan Physics came from three segments of German scientific life: conservative scientists opposed to modern physics; anti-Semitic scientists opposed to Albert Einstein and other Jewish scientists; and right-wing, nationalistic scientists opposed to the internationalism espoused by Einstein and others. Battles took place in the pages of German journals over atomic physics and over appointments to prestigious chairs at universities. At its peak of influence in 1939, however, the adherents of Aryan Physics could muster only 6 out of the 81 physics positions at the university level.

This atmosphere of conflict had a deleterious impact on German physics research. Even before the Nazis had consolidated power in 1933, many Jewish physicists were becoming uncomfortable with the anti-Semitic tone of the followers of Aryan Physics. Many of the top names began to sound out their contacts in Great Britain and in the United States for possible teaching positions. When the supporters of Aryan Physics and Nazi educational authorities concluded agreements to rid the German academy of scientists of Jewish descent, numerous top names found academic posts abroad almost immediately. Many of these same scientists made major contributions to the building of the Manhattan Project's atomic bomb. Those German physicists and chemists who remained under the Nazi state and were conducting advanced research into atomic problems still had to cope with their opponents holding key administrative posts. It took the influence of Heinrich Himmler, the head of the Schutzstaffel (SS), to break the power of the Aryan Physics movement; he backed Werner Heisenberg following a charge in an SS journal that Heisenberg was a Jewish sympathizer, or a "white Jew." This support by the political head of the political police meant that a distinction could be made between Einstein the Jewish physicist and the scientific merit of the theory of relativity. Once the new physics was accepted as beneficial to the Nazi state, then the political power of Aryan Physics diminished. Most of the damage had been done by this time, and German scientists never caught up with their colleagues on the Allied side in either nuclear theory or applied nuclear weapons development. In the postwar reconstruction of Germany, adherents of Aryan Physics were quickly removed from German universities.

Suggested readings: Alan D. Beyerchen, *Scientists Under Hitler: Politics and the Physics Community in the Third Reich* (New Haven, CT: Yale University Press, 1977); Mark Walker, *Nazi Science: Myth, Truth, and the German Atomic Bomb* (New York: Plenum Press, 1995).

Arzamas-16

Arzamas-16 is the Soviet equivalent of the American nuclear research center at Los Alamos. Soviet authorities selected the town of Sarov as its research site on April 2, 1946. At the time Sarov was a small town of several thousand inhabitants whose claim to fame was its large monastery. This monastery was appropriated for the use of the scientists. The site is 400 kilometers east of Moscow, and a small munitions plant had operated there during World War II. This secret site was selected by General Pavel M. Zernov and Yuli Khariton because it was both isolated and protected. Zernov was appointed the head of operations at Arzamas-16, and Khariton was chosen to be scientific director. Other important appointments were Veniamin Zukerman, the director of the Explosion Diagnostics Division, and Lev Altshuler, director of the Implosion Hydrodynamics Division. In March 1949, Stalin directed that Arzamas-16 become the major weapons production site. Facilities were added to produce 20 atomic weapons annually. The closed city, renamed Kremlev, has a modern population of over 80,000 inhabitants, of which around 25,000 work at the All-Russian Scientific Research Institute of Experimental Physics. Besides basic research on nuclear weapons research, this facility has had the mission to design and produce nuclear warheads. Until the early 1990s, about 60 percent of the activity at Arzamas-16 was with nuclear weapons–related work, but this type of work has decreased steadily since the mid-1990s. Khariton remained scientific director until 1992, and Viktor Mikhailov was his successor.

Suggested reading: Thomas B. Cochran, Robert S. Norris, and Oleg A. Bukharin, *Making the Russian Bomb: From Stalin to Yeltsin* (Boulder, CO: Westview Press, 1995).

Aston, Francis William

Francis Aston (1877–1945) was a British chemist and physicist whose discovery of the diversity of atoms was a major breakthrough in nuclear physics. He was born on September 1, 1877, in Harborne, a suburb of Birmingham, England. His father was a wealthy metal merchant. Aston remained a lifelong bachelor. He attended Mason College (later to become the University of Birmingham), Birmingham, where he studied chemistry, beginning in 1893. In 1896, Aston began using a loft in his home as a private laboratory and workshop where he conducted much of his research. From 1898 to 1900, Aston did research at Mason College under P. F. Frankland. His first job was with the British Brewery at Wolverhampton, where Aston remained for three years. His expertise on complex research problems brought him to the notice of J. H. Poynting, who invited him to become a researcher at the University of Birmingham. Aston stayed at Birmingham until 1910, when he moved to the Cavendish Laboratory at Cambridge University to work as a research assistant to Sir J. J. Thomson. Aston remained at Cambridge University for the rest of his professional career except for the period 1914–1918 when he worked as an aeronautical engineer testing dopes and fabrics for aircraft coverings at the Royal Aircraft Establishment at Farnborough.

Aston's fame came from his use of and refinements to the spectrograph. Sir J. J. Thomson had invented a primitive spectrograph, but Aston's refinements allowed his new spectrograph to calculate the mass of ions accurately. He conceived the idea during his military duty in World War I. His first spectrography was ready by 1919, and his first projections showed that isotopes are not restricted to radioactive elements but are common throughout the periodic table. In the first six years, 50 elements were analyzed. Further modification in 1927 permitted him to resurvey atomic weights and correct some minor discrepancies. He identified 212 of the 281 naturally occurring isotopes in his career. This research gained him the 1922 Nobel Prize for Physics.

Aston's distinguished scientific career made him aware of the promise and the dangers of atomic energy. He used this knowledge to advocate peaceful uses of it. Aston died on November 20, 1945, in Cambridge, England.

Suggested readings: Henry A. Boorse, Lloyd Motz, and Jefferson Hane Weaver, *The Atomic Scientists: A Biographical History* (New York: Wiley Science Editions, 1989); John Daintith, Sarah Mitchell, Elizabeth Tootill, and Derek Gjertsen, *A Biographical Encyclopedia of Scientists*, 2nd ed., vol. 1 (Bristol, UK: Institute of Physics, 1999); Richard Rhodes, *The Making of the Atomic Bomb* (New York: Simon and Schuster, 1986).

Atomic Bomb

The atomic bomb was the popular name given to the type of military atomic devices dropped on Japanese cities to end World War II. This bomb uses a trigger to unleash a chain reaction to produce a controlled nuclear explosion. Scientists had agreed upon the theory behind an atomic bomb as early as 1939 when it became known that fission produced excess energy, but the engineering required to build such a bomb took much longer and $2 billion. Work on the bomb soon produced a difference of opinion on what type of bomb to build. The two options were a uranium bomb depending on U-235 or a plutonium bomb necessitating a new type of detonator. After considerable debate and experimentation, the scientists in the Manhattan Project built both. Only the plutonium bomb was tested in the Trinity Test on July 16, 1945. One of each type was detonated over Hiroshima and Nagaski in August of 1945.

Research for several years after the war centered on building bigger and better atomic bombs. The first postwar replacement was the Mark-4, which was a plutonium Fat Man–type bomb with improved aerodynamic design and easier service accessibility. About 250 Mark-4s were produced, and these bombs stayed in the nuclear inventory until 1953. Its successor, the Mark-6, was introduced in 1951 and was the first mass-produced atomic bomb. It was designed as a strategic weapon, and its yield was twice that of its predecessor. The Mark-5 was developed about the same time as the Mark-6, but it was designed as a lightweight strategic bomb. Both the Mark-5 and Mark-6 became the standard atomic bombs until hydrogen bombs replaced them as primary strategic weapons. Another atomic bomb, the Mark-7, was the first tactical nuclear weapon to be made available to all three military services. Moreover, it was a uranium-235 gun-type model similar to the Little Boy design. This bomb and its improved versions remained in the American nuclear arsenal longer than any other type, lasting from 1952 to 1967. The last atomic bomb to enter the nuclear arsenal was the super-lightweight Mark-12, which could be delivered at supersonic speeds. Soon after the advent of the hydrogen bomb, the atomic bomb reverted to a more modest role. Most scientific effort was then directed toward the more powerful hydrogen bomb, and atomic bombs were left to be used as tactical nuclear weapons for the battlefield. *See also* Fat Man; Little Boy.

Suggested readings: Ronald William Clark, *The Birth of the Bomb* (New York: Horizon Press, 1961); James N. Gibson, *Nuclear Weapons of the United States: An Illustrated History* (Atglen, PA: Schiffer Publishing, 1996); Dennis Wainstock, *The Decision to Drop the Atomic Bomb* (Westport, CT: Praeger, 1996).

Atomic Bomb Casualty Commission

The Atomic Bomb Casualty Commission (ABCC) was established on November 16, 1947, by Presidential Directive to study the health and medical records of the survivors of the atomic bomb blasts in Japan. Since testing had been so limited, little knowledge of the effects of the bomb on people existed. It was known, however, from the Trinity Test in July 1945 that there were three waves of radiation: radiation gamma wave, thermal radiation, and finally, shock wave. The initial reaction of American scientists to the news of radiation injuries at Hiroshima and Nagasaki was to discount it as Japanese propaganda because they believed that the atomic bomb killed only by explosive blast. When horror stories continued, General Leslie Groves asked Los Alamos scientists to check on Japanese radiation victims' claims. He believed them to be a hoax. A radiation survey team under General Thomas Farrell and Dr. Stafford Warren was sent to Hiroshima on September 8, 1945, and Nagasaki on October 8 and reported on conditions when it returned to the United States on October 15. This survey noted the range of health problems associated with the atomic bomb, but it was only an after-action report. Soon it became apparent that a more long-term study of the health of the victims was necessary. In the winter of 1946, Austin Brues and Paul Henshaw, two American scientists, traveled to Japan, and they produced the Brues-Henshaw Report which called for a comprehensive study of the long-term health effects of the atomic bombs on Japanese civilians.

The resulting Atomic Bomb Casualty Commission was a civilian commission organized by the Atomic Energy Commission (AEC) but operated by the National Academy of Science and the National Research Council with the mission to study medical and genetic consequences of atomic warfare. Data and materials had been collected beginning in 1945 and consolidated in 1947 under the auspices of the Atomic Bomb Casualty Commission. An ABCC clinic was built on a hill overlooking Hiroshima, and around 75,000 young Japanese were studied for radiation damage. Researchers found an elevated incidence of leukemia during the first decades after the bombing and higher incidence of thyroid, breast, and lung cancers. Because this clinic only examined and did not treat victims, it became so unpopular by the mid-1950s that people stopped going to it. This caused a crisis in the ABCC organization, and a more patient, friendly attitude was adopted. In 1974, this commission was replaced by the Radiation Effects Research Foundation (RERF). Although the RERF is technically a joint American and Japanese organization, it is under Japanese control.

Suggested readings: Corinne Browne and Robert Munroe, *Time Bomb: Understanding the Threat of Nuclear Power* (New York: William Morrow, 1981); Congressional Quarterly, *The Nuclear Age: Power, Proliferation and the Arms Race* (Washington, DC: Author, 1984); Sue Rabbitt Roff, *Hotspots: The Legacy of Hiroshima and Nagasaki* (London: Cassell, 1996); William J. Schull, *Effects of Atomic Radiation: A Half-Century of Studies from Hiroshima and Nagasaki* (New York: Wiley-Liss, 1995).

Atomic Bomb Victims

Studies have been done on Japanese civilians to study the impact of the atomic bomb on survivors. The most comprehensive of these studies was completed by the Atomic Bomb Casualty Commission (ABCC) on the survivors of the Hiroshima and Nagasaki atomic bombs. Besides learning about the short-term and long-term health of these survivors, the commission has devoted much of its energy to studying the psychological impact of surviving an atomic bomb blast. Many of the victims had disfiguring injuries that necessitated plastic surgery and other medical procedures. The most famous case has been the story of the "Hiroshima Maidens," who were brought to the United States in the 1950s and 1960s for reconstructive surgery for disfigurement. While these and other survivors have been able to rebuild their lives, they have had difficulty in dealing with certain physical and psychological problems. They have lived with the fear that their health will suffer from delayed radiation effects. This fear has also extended to genetic damage to their children and grandchildren. Economic instability is another concern because delayed radiation effects may decrease their ability to work or care for themselves. Finally, discrimination against them by nonvictims has made their lives more difficult. The isolation caused by their experiences has made many of the atomic bomb victims feel pessimistic about their futures. This pessimism has translated into their becoming passive and withdrawn.

Suggested readings: Committee for the Compilation of Materials on Damage Caused by the Atomic Bombs in Hiroshima and Nagasaki, *Hiroshima and Nagasaki: The Physical, Medical, and Social Effects of the Atomic Bombings* (New York: Basic Books, 1981); William J. Schull, *Effects of Atomic Radiation: A Half-Century of Studies from Hiroshima and Nagasaki* (New York: Wiley-Liss, 1995).

Atomic Energy Act of 1946

On October 3, 1945, President Harry Truman sent a message to the U.S. Congress calling for legislation to created an Atomic Energy Commission (AEC) to set policy for the domestic control of atomic energy. After the failure of the May-Johnson Bill, a substitute bill, the McMahon bill, was substituted by the Senate Special Committee on Atomic Energy under the leadership of Senator Brien McMahon on December 20, 1945. Hearings were held between January 22 and April 8, 1946. President Truman approved the legislation and signed the Atomic Energy Act on August 1, 1946. It established the Atomic Energy Commission as a government agency. Key points of this legislation were that it formed a full-time civilian commission to oversee atomic energy issues, and it was the product of a far-ranging congressional debate. The law mandated a complete government monopoly over aspects of atomic energy including ownership of nuclear materials and facilities to make or use them. It also created a legislative body, the Joint Committee on Atomic Energy (JCAE), to oversee the legislative side of atomic energy. Furthermore, an amendment by Senator Arthur Vandenburg gave the Atomic Energy Commission's Board of Military Advisors a tacit veto over all AEC decisions. Corporate America found this law disappointing, because these corporations wanted unfettered opportunities to exploit nuclear energy. This act also contained a clause that prohibited the exchange of information with other nations for the use of atomic energy. This clause violated several secret agreements between the United States and the United Kingdom over the sharing of atomic energy information and caused diplomatic complications with Great Britain.

Suggested readings: Corbin Allardice and Edward R. Trapnell, *The Atomic Energy Commission* (New York: Praeger, 1974); Frank G. Dawson, *Nuclear Power: Devel-*

opment and Management of a Technology (Seattle: University of Washington Press, 1976).

Atomic Energy Act of 1954

The Atomic Energy Act of 1954 allowed private corporations to build and own nuclear plants. It ordered the Atomic Energy Commission (AEC) to cooperate with private industry and provide them information about nuclear technology. This act was the culmination of a long lobbying campaign by several large corporations to enter the nuclear business. They had always been disappointed in total government control of nuclear energy, and their political clout in a Republican-controlled Congress allowed them to intervene. Their lobbying had also made a convert of President Dwight Eisenhower, and on February 17, 1954, he made a proposal to Congress to change the Atomic Energy Act of 1946 to include wider participation of industry in nuclear power development. It had been the long-standing policy of the AEC to encourage private enterprise, but with severe restrictions on ownership and access to atomic materials. Representative Sterling Cole of New York, chair of the Joint Committee on Atomic Energy (JCAE), and Senator Bourke Hickenlooper of Iowa, vice chair of the Joint Committee on Atomic Energy, wrote the legislation modeling it on the wording of the Federal Communications Act of 1934. This law mandated that the AEC actively support these ventures. The law also boosted the powers of the Joint Committee to initiate and then serve as conference committee on nuclear legislation. Another provision of the law promoted international cooperation between the United States and foreign governments on atomic energy. This legislation significantly increased the powers of both the AEC and the JCAE. This legislation and the Price-Anderson Bill to provide government liability insurance for the nuclear energy industry were the key components of the government drive to privatize atomic energy.

Suggested readings: Frank G. Dawson, *Nuclear Power: Development and Management of a Technology* (Seattle: University of Washington Press, 1976); Daniel F. Ford, *The Cult of the Atom: The Secret Papers of the Atomic Energy Commission* (New York: Simon and Schuster, 1982); Mark Hetsgaard, *Nuclear Inc.: The Men and Money Behind Nuclear Energy* (New York: Pantheon Books, 1983).

Atomic Energy Authority

The Atomic Energy Authority (AEA) is the body in Great Britain responsible for all nuclear activity. It was formed in 1954 with the mandate to coordinate all nuclear energy initiatives in the British Isles. Since it monopolized expert knowledge in the nuclear field, government policy followed the lead of the AEA. It was the AEA that designed and promoted the advanced gas-cooled reactor (AGR), sometimes against the wishes of electricity companies. The AEA was also a strong advocate of the fast breeder reactor development, and a prototype was built on the northern edge of Scotland in the early 1970s. Both the gas-cooled reactor and the fast breeder reactor program proved to be failures. Efforts to build an export market for both types of reactors collapsed in the 1960s. The AEA still retains control of the nuclear industry in Great Britain, but it has become a target of antinuclear activists.

Suggested reading: Helena Flam, ed., *States and Anti-Nuclear Movements* (Edinburgh, Scotland: Edinburgh University Press, 1994).

Atomic Energy Bill of 1946

The Atomic Energy Bill of 1946 provided the legal authority for the British government to have a monopoly over atomic energy in the United Kingdom. This legislation was initiated in early 1946 and introduced to the House of Commons on May 1, 1946. It became a law on November 6, 1946. Both the relevant parliamentary committees and private industry had been consulted before this bill was presented to Parliament. Among its provisions were authority for the Ministry of Supply for promoting and controlling the development of atomic energy. Other provisions gave the Ministry of Supply the right to acquire by compulsion atomic materials and plants working with atomic energy. Two other major provisions were the requirement of a license from the minister of supply for any atomic energy activity and restrictions on the disclosure of information about atomic energy plants, with harsh penalties for disclosure. This last point was the most controversial part of the bill. A lack of controversy over the bill was partially due to the extreme secrecy by the British government concerning atomic energy. Both the Parliament and the press were given a minimum of information.

Suggested reading: Margaret Gowing, *Independence and Deterrence: Britain and Atomic Energy, 1945–1952,* vol. 2 (New York: St. Martin's Press, 1974).

Atomic Energy Commission

The Atomic Energy Commission (AEC) was a five-member full-time civilian commission established in August 1946 by the Atomic Energy Act to direct the nuclear program of the United States. Each commissioner was appointed by the president of the United States for a five-year term, and appointments were staggered so that a new appointment took place each year. They had to be confirmed by the Senate. One of the five was designed by the president as chair, and he reported to the president on atomic energy business. The original commission had a membership of David Lilienthal, a Democrat and former head of the Tennessee Valley Authority (TVA); Summer Pike, a Republican and former member of the Securities and Exchange Commission; William Waymack, a Republican and newspaper editor from Des Moines, Iowa; Robert Bacher, a Democrat and a physicist who had worked on the Manhattan Project; and Admiral Lewis Strauss, a Republican who was a reserve naval officer and former investment banker. The AEC had four operating committees—Advisory Committee on Reactor Safeguards, General Advisory Committee, Military Liaison Committee, and the Patent Compensation Board. The two most important committees were the General Advisory Committee and the Military Liaison Committee. Nine scientists and advisers were members of the General Advisory Committee, and their role was to advise the commission on scientific and technical matters involving nuclear energy. Because of the military implications of atomic weapons, the Military Liaison Committee became the most powerful of the committees, and its relationship with the Military Applications Division in approving atomic weapons development was the key factor.

Since the AEC shared authority with the Joint Committee on Atomic Energy (JCAE) on the oversight of atomic energy, the working relationship between these two bodies became crucial. Often the relationship improved or worsened by the personal interaction between the chair of the AEC and the chair of the JCAE. Sometimes partisan politics intervened, as was the case between David Lilienthal, a liberal Democrat, and Senator Bourke Hickenlooper, a conservative Republican from Iowa. Later, Lewis Strauss, a conservative Republican former banker, and Representative Chet Holifield, a liberal Democrat from California, experienced the same type of hostility.

At its first meeting on January 2, 1947, the AEC assumed control from the U.S. Army of its installations and personnel devoted to atomic energy. This organization assumed control of 37 installations in 19 states and Canada, 1,942 army personnel, 3,950 government workers, nearly 38,000 contractor employees, and an inventory of facilities and property of approximately $2.2 billion. The annual budget of the AEC started at $300 million in 1947, and by 1953, it had reached more than $4 billion. Despite this apparent prosperity, the AEC had two significant problems: the exodus of scientific talent at the end of the war back to peaceful pursuits and a depleted nuclear weapons stockpile. Both problems continued to haunt the AEC until late in the 1940s.

Because of their access to the president and their influence with Congress, the chairpersons of the AEC became political powers in their own rights. The first chairperson was David E. Lilienthal, and he lasted from November 1, 1946, to February 1950. His successor was Gordon Dean, who was chairperson from July 11, 1950, to June 30, 1953. Lewis L. Strauss succeeded him on July 2, 1953, and remained chair until June 30, 1958. John A. McCone served as chairperson from July 14, 1958, to January 20, 1961. Glenn T. Seaborg held this position from March 1, 1961, to August 16, 1971. James R. Schlesinger was the next chairperson, and his job lasted from

August 17, 1971, to January 26, 1973. The last chairperson was Dixy Lee Ray, who assumed the chair on February 6, 1973, and remained until the AEC's functions were transferred to other agencies.

Despite the success of its nuclear policies, the AEC came under attack in the late 1960s and 1970s. A critic of the AEC once described the commission as a secretive agency located in the executive branch and too heavily influenced by the military. The refusal of the Atomic Energy Commission to respond to the scientific studies and public outcry over radioactive fallout from nuclear tests eroded public confidence in it. Too often AEC officials reacted negatively toward scientific evidence about the effects of iodine-131 and strontium-90. The AEC's hands-off approach toward the regulation of safety at private nuclear facilities also alarmed critics. Scientists had come to believe that the AEC was an inept bureaucracy that demonstrated indifference to the questions of radiation safety. AEC procedures also discourage critical public scrutiny of nuclear power plants in the 1950s and 1960s by controlling all information. In the end, the AEC lost credibility and had to be replaced. Consequently, in 1974 the Atomic Energy Commission was terminated, and its functions were assumed first by the Energy Research Development Administration and later by the Department of Energy. Its regulatory authority over the nuclear industry was taken over by the Nuclear Regulatory Commission.

Suggested readings: Corbin Allardice and Edward R. Trapnell, *The Atomic Energy Commission* (New York: Praeger, 1974); Daniel Ford, *The Cult of the Atom: The Secret Papers of the Atomic Energy Commission* (New York: Simon and Schuster, 1982); Harold Orlans, *Contracting for Atoms: A Study of Public Policy Issues Posed by the Atomic Energy Commission's Contracting for Research, Development, and Managerial Services* (Washington, D.C.: Brookings Institution, 1967).

Atomic Energy Commission, United Nations

On January 24, 1946, the United Nations created the Atomic Energy Commission (UNAEC) with its members taken from the nations on the Security Council and Canada. Soon after the end of World War II, several initiatives were undertaken to bring atomic energy under international control. In the Washington Declaration of November 1945 the United States, the United Kingdom, and Canada called for the formation of a UN committee to study the peaceful uses of atomic energy. The Council of Foreign Ministers meeting in Moscow in December 1945 also called for a commission to investigate the international control of atomic energy. With this high level of support from the major powers, the United Nations set up a commission with representatives from 11 countries represented on the Security Council and Canada with the responsibilities to provide a forum for the exchange of basic scientific information, peaceful uses of atomic energy, and the elimination of atomic weapons from the inventories of member states.

Figure 2
Production of Fissionable Materials

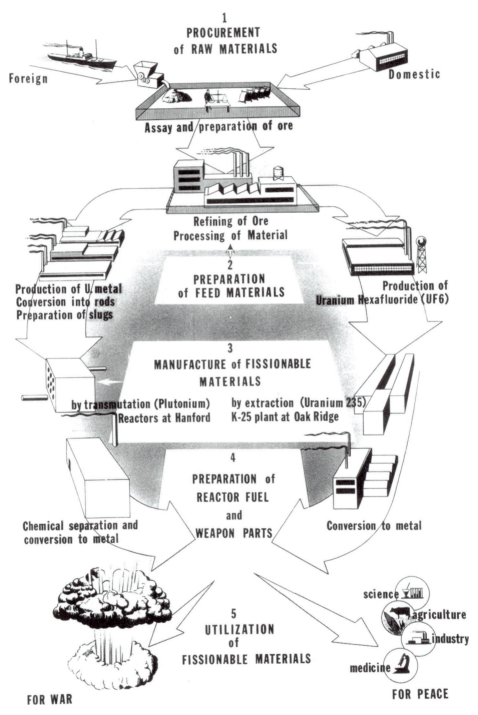

1
PROCUREMENT
of RAW MATERIALS

Foreign

Domestic

Assay and preparation of ore

Refining of Ore
Processing of Material

2
PREPARATION
of FEED MATERIALS

Production of U. metal
Conversion into rods
Preparation of slugs

Production of
Uranium Hexafluoride (UF6)

3
MANUFACTURE of FISSIONABLE
MATERIALS

by transmutation (Plutonium)
Reactors at Hanford

by extraction (Uranium 235)
K-25 plant at Oak Ridge

4
PREPARATION of
REACTOR FUEL
and
WEAPON PARTS

Chemical separation and
conversion to metal

Conversion to metal

science
agriculture
industry
medicine

5
UTILIZATION
of
FISSIONABLE MATERIALS

FOR WAR

FOR PEACE

This chart of the production of fissionable materials was produced by the Atomic Energy
Commission and shows the dual functions of the final product.

It was before the inaugural session of the UNAEC on June 12, 1946, that the Baruch Plan was presented by the United States and was quashed by the Soviet Union's mistrust of American motives. Initial hopes on the effectiveness of the UNAEC died slowly, but by early 1948, it was no longer considered a viable place for international cooperation on atomic energy. After two years' work and over 200 meetings, the UNAEC informed the Security Council in 1948 that it was no longer able to function and suggested disbanding. This suggestion was adopted by the Security Council.

Suggested readings: Bertrand Goldschmidt, *Atomic Rivals* (New Brunswick, NJ: Rutgers University Press, 1990); *International Atomic Energy Agency: Personal Reflections* (Vienna, Austria: International Atomic Energy Agency, 1997).

Atomic Energy Control Board

The Atomic Energy Control Board (AECB) was created in May 1946 by the Canadian Parliament under the Atomic Energy Control Act to oversee the development and control of atomic energy in Canada. This board had a full-time president and four part-time members. General A. G. L. Mc-Naughton was the first president, but in 1948 he was replaced by Chalmers Jack Mackenzie, who remained at this post until 1961. The AECB had the authority to sponsor research, to acquire radioactive materials, to make regulations, and to form or acquire companies in the nuclear field. Despite this far-reaching authority, the AECB delegated most of its functions to other government bodies or agencies. Nevertheless, the AECB served as the unifying body for all atomic energy activity in Canada for over five years. On April 1, 1952, the AECB was incorporated into a combined crown company, the Atomic Energy of Canada Limited, but the AECB retained much of its former structure.

Suggested reading: Robert Bothwell, *Nucleus: The History of Atomic Energy of Canada Limited* (Toronto, Canada: University of Toronto Press, 1988).

Atomic Energy of Canada Limited

The crown company Atomic Energy of Canada Limited (AECL) was formed by the Canadian government on April 1, 1952, to coordinate Canadian atomic energy activities. It assumed most of the responsibilities of the Atomic Energy Control Board (AECB) without abolishing that body. A board of directors included Chalmers Jack Mackenzie, president of the AECB; V. W. Scully, member of the AECB; Bill Bennett, member of AECB and director of Eldorado Mining Company; Geoffrey Gaherty, president of the Montreal Engineering Company; Andrew Gordon, head of the University of Toronto's chemistry department; René Dupuis, administrator at Quebec Hydro; Huet Massue, senior engineer with Shawinigan Water and

Power Company; and Richard Hearn, chief engineer for Ontario Hydro. Since the AECL had no sustaining budget, it had to obtain an annual vote of funding from the Canadian Parliament. This funding issue was never resolved satisfactorily and made the AECL an active participant in Canadian politics. In 1955, the AECL board proposed a civilian natural uranium, heavy water reactor of a 100-megawatt (electric) size to be built by a Canadian company, Canadian General Electric, for the electricity company, Ontario Hydro. This project was approved by the Canadian government in March 1955 and called the NPD (Nuclear Power Demonstration). After cost delays and other design problems, the original NPD program was scrapped in April 1957 and replaced by a larger reactor design (PD-2). This reactor was finally in operation on April 11, 1962. The AECL has continued to advocate civilian atomic energy projects in Canada until the present.

Suggested reading: Robert Bothwell, *Nucleus: The History of Atomic Energy of Canada Limited* (Toronto, Canada: University of Toronto Press, 1988).

Atomic Industrial Forum

The Atomic Industrial Forum (AIF) was a nonprofit trade association that developed a membership from across the spectrum of academic and private enterprise. The AIF was formed as an international association and at one time represented 625 organizations. Its mission was to encourage private initiative in the development of peaceful uses of atomic energy. A key figure in the organization in the 1950s was Walker Cisler, an engineer by training and the president of Detroit Edison Company and a champion of an atomic age under free enterprise. He used his position as president of the AIF to advance this viewpoint. Originally its headquarters was located in New York City, but in 1975 the main office moved to Washington, DC to improve its influence with Congress. Because the AIF's status as a nonprofit trade association limited direct lobbying, its leadership founded the American Nuclear Energy Council to lobby Congress. In the 1970s this organization took the lead in counterattacking the antinuclear movement and accusing these groups of creating unemployment and obstructing a clean and safe energy source. The AIF used mass media outlets to convince decision makers in both the federal and state government to support nuclear energy. One of its media efforts in the 1970s was the annual survey comparing the cost of coal and nuclear-generated electricity with atomic power, always more economical. Critics have accused the AIF of distorting the costs. It was the Three Mile Island accident, however, that put the AIF on the defensive, and their first forecast after the incident was to suggest that the accident had set the nuclear industry back from one to three years. In retrospect, this forecast was too optimistic. In the 1980s it became apparent that nuclear energy was in political trouble, and this organization lost much of its clout with government officials. In 1987, the AIF was absorbed by the Nuclear Energy Institute. This organization continues the pro-nuclear stance of the AIF and with a membership of around 400, a staff of 70, and a budget of nearly 20 million.

Suggested readings: Frank G. Dawson, *Nuclear Power: Development and Management of a Technology* (Seattle: University of Washington Press, 1976); Jerome Price, *The Antinuclear Movement* (Boston, MA: Twayne, 1982).

Atomic Reactors. *See* Chicago Pile One; Gas-Cooled Reactors; Light Water Reactors

Atomic Scientists' Association

The Atomic Scientists' Association (ASA) was a British association of individuals who had worked on the British, Canadian, or American bomb projects during World War II. It was founded in the spring of 1946 with Nevill Mott as its first president. Joseph Rotblat was one of the initiators of the establishment of the ASA after negotiations between the British Association of Scientific Workers (AScW) fell through. The scientists believed that the trade union and the left-wing orientation of the 15,000-member AScW would prove to be a detriment. By 1950 the ASA had a membership of 140 full members and 500 associate members. Most of the prominent atomic scientists in Great Britain belonged, crossing the political spectrum from pacifists to hawks. It served the same function in the United Kingdom as the Federation of American Scientists in the United States. Through its journal *The Atomic Scientists' News*, the ASA informed the British public on atomic energy news. Since most of its members believed in the internationalization of atomic energy, the ASA soon joined the movement against atomic weapons and for disarmament. By 1953 the ASA had begun to lose steam, with a steep membership decline. Only the activity to keep the journal alive sustained the organization. While the organization still exists, it remains only a shadow of its former self.

Suggested reading: Lawrence S. Wittner, *The Struggle Against the Bomb: One World or None: A History of the World Nuclear Disarmament Movement Through 1953*, vol. 1 (Stanford, CA: Stanford University Press, 1993).

Atomic Scientists of Chicago

The Atomic Scientists of Chicago was a group of scientists with strong reservations about the building and use of atomic weapons. Leo Szilard, a physicist, and James Franck, a physical chemist, were early leaders of a body of scientists at the Chicago Metallurgical Laboratory concerned about the use of the atomic bomb against the Japanese and the secrecy of the Man-

Scientists at the University of Chicago were instrumental in preparing the way for the building of the atomic bomb and several were active in the postwar world working on the hydrogen bomb. These scientists sitting from left to right were W. H. Zachariason, Harold C. Urey, Cyril Smith, Enrico Fermi, Samuel K. Allison, and those standing left to right were Edward Teller, T. Hogness, Walter Zinn, Clarence Zener, Joseph E. Mayer, Philip W. Schultz, R. H. Christ (Columbia University), and Carl Eckhart (University of California at Berkeley).

hattan Project. Szilard circulated a petition to President Harry Truman with 69 signatures from scientists objecting to the military use of the atomic bomb. Franck presented a report calling for much the same thing. Neither effort produced tangible results, and the scientists were surprised by the use of the atomic bomb on Hiroshima. Dissatisfaction with government policy led the Committee on Social and Political Implications at the Chicago Metallurgical Laboratory, which had produced the Franck Report, to form the Atomic Scientists of Chicago. The original executive committee had John Simpson, chair, Austin Brues, David Hill, J. J. Nickson, Eugene Rabinowitch, Glenn Seaborg, and Leo Szilard. It goals were reformation of public opinion toward atomic weapons, formulation of a domestic policy for atomic energy, and lobby for international control of atomic energy. At one time this organization claimed to have the support of over 90 percent of the scientists in the Manhattan Project. This organization proved to have a limited history because in November 1945 it merged with the Federation of Atomic Scientists. Soon afterward its journal, the *Bulletin of Atomic Scientists*, became the official journal of the new federation.

Suggested reading: Donald A. Strickland, *Scientists in Politics: The Atomic Scientists Movement, 1945–46* (West Lafayette, IN: Purdue University Studies, 1968).

Atomic Veterans

Atomic veterans is the name given to the civilians and the military men who were exposed to ionizing radiation without their knowledge or consent during the American nuclear testing. Approximately 250,000 military personnel and 150,000 civilians took part in the 626 U.S. nuclear detonations between 1945 and 1979. Although the Atomic Energy Commission (AEC) had authority over the testing of nuclear weapons, AEC officials abdicated responsibility to the military on safety conditions. The army formed the Human Resources and Research Office in 1951 to gather information on the training methods necessary for troops to function on the nuclear battlefield. In the drive to develop tactical nuclear weapons for use on the battlefield, safety precautions were overruled in the interest of national defense. A 1980 report by the Centers for Disease Control in Atlanta disclosed that the incidence of leukemia among the 2,595 atomic veterans in the study was nearly three times the level expected. When these veterans started experiencing signs of cancer, they applied to the Veterans Administration for service-connected disabilities. The Veterans Administration denied their claims based on reasonable doubt about the origin of their cancers. These veterans reacted by forming the National Association of Atomic Veterans in March 1979 to lobby the U.S. government for their cause. One of the founders of this group was Orville Kelly, who had served as an enlisted man in the U.S. Army during nuclear testing in the Pacific and who suffered from cancer. Kelly died in the summer of 1980.

Suggested readings: Howard L. Rosenberg, *Atomic Soldiers: American Victims of Nuclear Experiments* (Boston, MA: Beacon Press, 1980); Thomas H. Saffer and Orville E. Kelly, *Countdown Zero* (New York: Putnam's Sons, 1982).

Atoms for Peace

President Dwight Eisenhower in 1953 proposed the Atoms for Peace project to encourage the peaceful uses of nuclear energy. This announcement on December 8, 1953, at the UN General Assembly came unexpectedly, with few members of his administration knowing about it beforehand. President Eisenhower had become perturbed by the implications of the emphasis on the destructive nature of nuclear energy and looked for a way to reinforce its constructive uses. He decided to emphasize the positive side and proposed sharing basic information and technical assistance with the world. This idea was enthusiastically received worldwide, but little happened until the 1955 Geneva Conference on the Peaceful Uses of Atomic Energy convened in August. Besides numerous papers proposing new uses for nuclear energy, the prime importance of the meeting was the realization that Soviet scientists were as current on nuclear theories as their American counterparts. A side effect of this program was to allow the United States limited control over nuclear research in nonnuclear countries as a way to prevent nuclear weapons development. This was soon apparent, and many in the Third World resented this type of control. Nevertheless, the policy change to share atomic technology continued through the next four presidential administrations.

Suggested readings: *Atoms for Peace: An Analysis After 30 Years* (Boulder, CO: Westview Press, 1985); David Woodbury, *Atoms for Peace* (New York: Dodd, Mead, 1955).

Australian Atomic Weapons Test Safety Committee

The Australian Atomic Weapons Test Safety Committee (AWTSC) was founded in July 1955 to give the Australian government veto power over nuclear bomb testing in Australia. Prior to the formation of this committee, British atomic bomb testing in Australia had been at the discretion of the Australian prime minister. Both the atomic bomb tests at Monte Bello Islands and at Emu in southern Australia had taken place under this type of agreement. This committee, under the chairmanship of Sir Ernest Titterton, was intended to examine the safety procedures for both man and beast being taken by the British before each test. After hearing testimony of witnesses, the committee rejected reports of radioactive fallout damaging the health of animals and humans in Australia. Their findings were communicated to the prime minister along with recommendations for alternative or more extensive safety regulations. This committee, however, was handicapped because of British secrecy and because of the pro-British composition of the committee. Some critics have charged that the AWTSC's sole purpose was to reassure the Australian public all was well and not to protect the public from radioactive fallout. Later, when incidents of radiation-related injuries began to surface, this committee came under critical scrutiny by an Australian Royal Commission. This Royal Commission's findings contradicted the earlier conclusions of the AWTSC and found evidence of radioactive fallout in northern Australia and significant health risks to both humans and animals.

Suggested readings: Denys Blakeway and Sue Lloyd-Roberts, *Fields of Thunder: Testing Britain's Bomb* (London: George Allen and Unwin, 1985); International Physicians for the Prevention of Nuclear War, *Radioactive Heaven and Earth: The Health and Environmental Effects of Nuclear Weapons Testing in, on, and Above the Earth* (New York: Apex Press, 1991).

ß

Bainbridge, Kenneth

Kenneth T. Bainbridge (1904–1996) was an American physicist who directed the atomic tests at the Alamogordo Bombing Range in World War II. He was born on July 27, 1904, in Cooperstown, New York. His college education was at the Massachusetts Institute of Technology (MIT); Graduate school was at Princeton University, where he earned a M.A. in 1927 and a Ph.D in 1929. For five years after graduation, Bainbridge worked first at the Bartol Research Foundation and then at the Cavendish Laboratory at Cambridge University. Beginning in 1934, he was a Harvard physics professor. He remained attached to Harvard University until his retirement in 1975.

His primary scientific achievement was the development of high-resolution mass spectrometry. In 1936 he perfected a device that used successive electric and magnetic fields to separate ions according to their mass. This machine has become one of the most useful analytical tools in physics, chemistry, geology, meteorology, biology, and medicine.

Bainbridge participated in both major American scientific projects in World War II. His early war work was with I. I. Rabi at the MIT Radiation Laboratory on the radar project. In March 1941 Bainbridge traveled to England to study radar developments there and attended a meeting of the MAUD Committee, where he first learned about the details of a possible atomic weapon. On his return to the United States, he reported his findings to Vannevar Bush. His job at Los Alamos Scientific Laboratory was with George Kistiakowsky's division planning for the first atomic test. In March 1945, as the test assumed more importance, Bainbridge was placed in charge of a special division, Project TR, for a trial blast of several tons of TNT for calibration purposes. This successful test took place on May 7. Most of his time, however, was spent identifying test sites in the Southwest, and it was his recommendation to select the desert area of Journey of Death in New Mexico. He died in Lexington on July 14, 1996.

Suggested readings: David Abbott, ed., *The Biographical Dictionary of Scientists: Physicists* (London: Blond Educational, 1984); Ferenc Morton Szasz, *The Day the Sun Rose Twice: The Story of the Trinity Site Nuclear Explosion, July 16, 1945* (Albuquerque: University of New Mexico Press, 1984).

Ban the Bomb Movement

"Ban the Bomb" was the unifying cry of most of the antinuclear movements from the 1950s onward. It served as the heart of the Campaign for Nuclear Disarmament (CND) in Great Britain in the late 1950s and early 1960s. The slogan first appeared in banners on the first CND-sponsored Aldermaston March in 1958. In time the Ban the Bomb slogan developed a more broad appeal to antinuclear activists around the world. While this battle cry came out of the pacificist and disarmament movement, it lost some of its luster in the 1960s when Communist front organizations adopted the slogan for their peace activities against the North Atlantic Treaty Organization, (NATO). After it became apparent in the 1970s that banning nuclear weapons lacked any possibility of success, antinuclear groups resorted to other strategies including nuclear free zones. *See also* Campaign for Nuclear Disarmament.

Suggested readings: Milton S. Katz, *Ban the Bomb: A History of SANE, the Committee for a Sane Nuclear Policy, 1957–1985* (Westport, CT: Greenwood Press, 1986); Richard Taylor, *Against the Bomb: The British Peace Movement, 1958–1965* (Oxford, UK: Clarendon Press, 1988).

Baruch Plan

Bernard Baruch was an American economist who was appointed as the American delegate to the United Nations' Atomic Energy Commission by President Harry Truman. The Lilienthal Group had presented a plan for international control of atomic energy to the U.S. government. Baruch took this plan and changed it by modifying the controls and requiring a survey of uranium sources including those in the Soviet Union. Members of the Lilienthal Group were disappointed in the changes. Baruch's revised plan called for strict penalties to be levied against violators of a nuclear treaty. This plan was part of a U.S. diplomatic campaign to include the Soviet Union in an international alliance on peacetime applications of atomic energy. The Soviet distrusted the entire idea, but they were most unhappy about the punitive parts of the plan. After a meaningless counteroffer from the Soviet Union, which was rejected by the U.S. government, the Baruch

Plan became a failure and with it the first real chance of international co-operation on atomic energy issues. The collapse of the Baruch Plan was not unexpected since key senators opposed it, making it uncertain as to whether the U.S. Senate would have approved it. Moreover, the British government was also lukewarm about the plan.

Suggested readings: Corbin Allardice and Edward R. Trapnell, *The Atomic Energy Commission* (New York: Praeger, 1974); Steven M. Neuse, *David E. Lilienthal: The Journey of an American Liberal* (Knoxville: University of Tennessee Press, 1996).

Becquerel, Antoine Henri

Antoine Becquerel (1852–1908) was a French physicist whose discovery of radioactivity changed the course of nuclear research. He was born in Paris on December 15, 1852. His father came from a long line of physicists, and he specialized in research on the problems of fluorescence and phosphorescence. He married twice—Lucie-Zoe-Marie Jamin in 1874 and M. Lorieux in 1890—and had one son. Becquerel had his early education in science and engineering at the Ecole Polytechnique (1872) and the Ecole des Pontes-et Chaussees (1874). Because of his training he became an engineer in 1877. In 1876, he started teaching at the lowest level as a lecturer, or a démonstrateur, at the Ecole Polytechnique. Becquerel received his doctorate in 1888 from the Ecole des Pontes-et-Chaussees for a thesis on the absorption of light, and the next year he became a member of the French Academy of Sciences. In 1878, he was appointed professor at the Museum of Natural History, where he remained until 1895. Then he obtained a professorship at the Ecole Polytechnique and stayed there until 1908.

By the early 1890s, Becquerel was deep in research on fluorescence, following closely in the footsteps of his father's research. After hearing about the discovery of X-rays by Roentgen, Henri Poincare suggested that he investigate further. Becquerel tried experimenting with salts of uranium and in the process discovered radioactivity. He found that the uranium salt emitted invisible radiation capable of exposing photographic film, even if the film is wrapped in black paper that completely blocks out sunlight. His discovery in 1896 opened the door for further research on the nature of the atom. Becquerel received the 1903 Nobel Prize in Physics for his research on radioactivity. Intrigued by the process, Becquerel continued to study the properties of radiation until his sudden death of a heart attack on August 25, 1908, at Le Croisic, France.

Suggested readings: Henry A. Boorse, Lloyd Motz, and Jefferson Hane Weaver, *The Atomic Scientists: A Biographical History* (New York: Wiley Science Editions, 1989); John Daintith, Sarah Mitchell, Elizabeth Tootill, and Derek Gjersten, *A Biographical Encyclopedia of Scientists*, 2nd ed., vol. 1 (Bristol, UK: Institute of Physics,

1994); Emilio Segre, *From X-Rays to Quarks: Modern Physicists and Their Discoveries* (San Francisco: Freeman, 1980).

Beloyarsk Nuclear Reactor Fire

The Beloyarsk Nuclear Reactor fire incident in December 1978 was the most serious nuclear accident in the Soviet Union before Chernobyl. This complex was situated about 35 miles from Sverdlovsk, and it contained two RBMK reactors and a brand-new BM600 fastbreeder reactor. An explosion in the machine hall caused the roof to fall into the building, making cables short out. In case of fire, the operators were to shut down both RBMK reactors, but because the temperature outside was −60 degrees Centigrade the reactor cooling systems would freeze and allow the reactor cores to overheat. Fearing a double meltdown, the operators decided to shut down only one reactor. Fire crews had arrived, but it looked like the situation was out of control with the fire raging. As the fires reached the computer room, the firefighters gained control of the fires, and the meltdowns were avoided. While no radioactivity escaped from the reactors, this accident was a close call for a major disaster.

Suggested reading: John May, *The Greenpeace Book of the Nuclear Age: The Hidden History, the Human Cost* (New York: Pantheon Books, 1989).

Beria, Lavrentii

Lavrentii Beria (1899–1953) was the head of the Soviet Secret Police, and after World War II he headed of the Soviet Atomic Bomb Program. He was born on March 29, 1899, at Merkheuli, near Sukhumi, in Georgia, Russia. His job as head of the People's Commissar of Internal Affairs (NKVD) made him the second most powerful figure in the Soviet Union during Joseph Stalin's regime. Beria was one of the few administrators trusted by Joseph Stalin and early in the war gave him various tasks—evacuating wartime industries from war zones, overseeing industrial conversion, and delivering materials and troops to the fronts. In his capacity as head of the NKVD, he began to hear reports in 1940 about work on an atomic bomb from spies and other sources. At first Beria believed much of this news was disinformation, but gradually he realized that the Allies were making progress in developing an atomic bomb. In March 1942, Beria sent a report to Stalin suggesting the start of a Soviet atomic bomb project, but Stalin was still cautious about the size and expense of such a project. Although Vyacheslav Molotov, the Soviet foreign minister, was the wartime head of the newly launched Soviet Atomic Bomb Program, Beria followed foreign progress through the reports of his spies until Stalin, furious over the successful detonation of the American atomic bombs, made him the head of the Special

Committee on the Atomic Bomb on August 20, 1945. Beria distrusted scientists because he was unable to understand them, and soon he came into conflict over the direction of the atomic program with the famous physicist Pyotr Kapitsa. Kapitsa wanted to take information from the American project and build a better bomb, but Beria merely wanted to copy a successful version. Beria ran the Soviet atomic program through a series of competent assistants, and he tolerated no failures. Despite his success as head of the Soviet Nuclear Program, Beria depended upon Stalin for his authority. Stalin's death and the fear by other political leaders that Beria might seize power and depose his enemies made the Soviet leadership decide to purge him. He was arrested on June 26, 1953, and shot on December 23, 1953.

Suggested readings: David Holloway, *Stalin and the Bomb: The Soviet Union and Atomic Energy, 1939–1956* (New Haven, CT: Yale University Press, 1994); Amy Knight, *Beria: Stalin's First Lieutenant* (Princeton, NJ: Princeton University Press, 1993).

Berkeley Summer Colloquium of 1942

At the Berkeley Summer Colloquium in July–August 1942, a select group of scientists discussed the key issues of building an atomic bomb. During the discussion the development of a possible hydrogen, or superbomb, was brought up. Among the theoretical physicists at the Berkeley, California, gathering were J. Robert Oppenheimer, Hans Bethe, Edward Teller, Felix Bloch, Robert Serber, Emil Konopinski, and John H. Van Vleck. Meeting in LeConte Hall on the University of California, Berkeley, campus, they already understood that an atomic bomb based upon nuclear fission was possible from U-235 and from plutonium. The first few days there were discussions about technical details. Then Edward Teller introduced the topic of nuclear fusion and argued for its feasibility to produce a weapon. Nuclear fusion is a thermal reaction, and since it requires no critical mass, Teller argued that it had unlimited potential. Once ignited, its explosive potential would only be limited by the volume of deuterium supplied. Moreover, deuterium was much easier to produce and less expensive than extracting U-235 or plutonium. It did take an atomic explosive to produce a high enough temperature and density to trigger the fusion reaction. The only danger was the limits of the explosion, but calculations by Hans Bethe rejected the idea that such an explosion would ignite the atmosphere. Despite the appeal of the hydrogen bomb and agreement on general theory, the scientists sent forward a recommendation that it would take a major scientific and technical effort to build an atomic bomb but that such an effort should be undertaken.

Suggested readings: Lillian Hoddeson, Paul W. Henriksen, Roger A. Meade, and Catherine Westfall, *Critical Assembly: A Technical History of Los Alamos During the*

Oppenheimer Years, 1943–1945 (Cambridge: Cambridge University Press, 1993); Richard Rhodes, *The Making of the Atomic Bomb* (New York: Simon and Schuster, 1986); Robert W. Seidel, *Los Alamos and the Development of the Atomic Bomb* (Los Alamos, NM: Otowi Crossing Press, 1995).

Bethe, Hans Albrecht

Hans Bethe (1906–) is a German-American physicist whose knowledge of nuclear physics made him a leader in the theory of nuclear reactors. He was born on July 2, 1906, in Strasbourg, Germany. Bethe's father was a physiologist at the University of Strasbourg, then the University of Kiel, and finally, at the University of Frankfurt. His mother's father was a professor of medicine at the University of Strasbourg. His marriage to Rose Ewald in 1939 produced a daughter and a son. All of his academic training was in Germany at the universities of Frankfurt and Munich. He took a course from Walther Gerlach at Frankfurt and was impressed with him. Because of his clumsiness in the laboratory, he was advised to transfer to another school and study theoretical physics. He left in 1926 for the University of Munich and Arnold Sommerfeld. His Ph.D. in physics was from the University of Munich in 1928, where he was considered one of the finest students of Arnold Sommerfeld. He held teaching posts at the University of Frankfurt in 1928, the University of Stuttgart in 1929–1930, and the universities of Munich and Tübingen between 1930 and 1933. Part of 1930–1931 Bethe spent at Cambridge University and then in Rome with Enrico Fermi. Bethe was one of the German physicists who lost their teaching positions in 1933 because of the Nazi decree against Jews in government posts.

Bethe's reputation in the world of physics made it possible for him to find a teaching position abroad. His first post was a one-year replacement position at the University of Manchester in 1933, and he soon moved to the University of Bristol. In 1935, Bethe received an offer from Cornell University for an appointment as a theoretical physicist. Except for frequent visiting appointments, he stayed at Cornell until retirement in 1975. In March 1941, Bethe became an American citizen.

Bethe made a number of contributions both in nuclear physics and in the theory of energy production of stars. His first interest was in studying the forces within the nucleus of the atom. He worked with Rudolf Peierls and, by using the Schrödinger equation, solved some of the questions about proton and neutron interaction. Another of his contributions was in the theory of nuclear energy. In a series of three review articles published in the *Review of Modern Physics* in 1936 and 1937 he compiled a summary of the state of subatomic physics. These surveys of the field were called "Bethe's bible." Later he collaborated with George Gamow and Ralph Alpher in a paper on the origin of chemical elements. His paper "Energy Production in Stars" in 1939 first suggested a mechanism (carbon cycle) for the production of stellar energy. His and his colleagues' work became the justification for the Big Bang Theory of the creation of the cosmos. For these contributions, Bethe received the 1967 Nobel Prize for Physics. In the postwar world, Bethe continued to contribute important papers on solid state physics, meson physics, and astrophysics.

During World War II, Bethe became involved in war work. For the first couple of years, he stayed at Cornell and investigated theoretical physics problems on radar. In 1943, he was invited to participate in the Manhattan Project, where he served as director of the Theoretical Physics Division at Los Alamos. His stay at Radiation was from 1942 to 1943. Then he moved to the Los Alamos Scientific Laboratory, where he stayed from 1943 to 1946. Always affable and outgoing, his nickname at Los Alamos was the "Battleship" because of the methodical way he worked through problems. He also had a good reputation as a teacher. He stayed at Los Alamos until January 1946 when, despite pressures from Edward Teller to stay, he returned to Cornell University.

Besides his scientific contributions, Bethe played an active role in the postwar nuclear energy debate. Edward Teller tried to convince him to work on the hydrogen bomb, but Bethe was never convinced of its necessity. He used the forum of the American Physical Society to speak out against the H-bomb, but research on it continued anyway. Once Stanislaw Ulam found a way to make the bomb work, Bethe became convinced that there was no way to stop it, so in 1952 he returned to Los Alamos and worked on the bomb. Bethe's friendship with Edward Teller never recovered from Teller's testimony before the Gray Committee that revoked J. Robert Oppenheimer's security clearance. He became especially active in the 1960s and 1970s, advocating international control of nuclear energy and rejecting testing and development of nuclear weapons. Bethe served as a delegate to the first International Test Ban Conference at Geneva in 1968. In the 1980s, he became a vocal critic of the Reagan administration's Intercontinental Ballistic Missile (ICBM) Defense System.

Suggested readings: Jeremy Bernstein, *Hans Bethe, Prophet of Energy* (New York: Basic Books, 1980); Henry A. Boorse, Lloyd Motz, and Jefferson Hane Weaver, *The Atomic Scientists: A Biographical History* (New York: Wiley Science Editions, 1989); John Daintith, Sarah Mitchell, Elizabeth Tootill, and Derek Gjertsen, *A Biographical Encyclopedia of Scientists*, 2nd ed., vol. 1 (Bristol, UK: Institute of Physics, 1994); Anthony Serafini, *Legends in Their Own Time: A Century of American Physical Scientists* (New York: Plenum Press, 1993).

Bhabha, Homi Jehangir

Homi Bhabha (1909–1966) was the longtime head of India's atomic energy program and its most prominent advocate. He was born in Bombay, India, on October 30, 1909. Bhabha was from a privileged economic and social background with family connections to the famous steel-making Tata family. After attending schools in Bombay, he entered Cambridge University in 1927 to study mechanical engineering. Among his teachers in mathematics and theoretical physics was Paul Dirac. While at Cambridge, he began to conduct physics research at the Cavendish Laboratory. Bhabha received his doctorate in physics from Cambridge University in 1935. It was at Cambridge that he made extensive contacts with British and Canadian physicists that would prove so useful later in his career. On returning to India in 1939, he soon found a job at the Bangalore Institute of Science conducting research on cosmic rays. During the war, the British authorities were reluctant to use him on war-related projects. In 1945, Bhabha was appointed director of the Tata Institute of Fundamental Research in Bombay after using family contacts and friends in the Bombay government.

After Indian independence in 1947, Bhabha became a prominent scientific administrator. Shortly after independence Bhabha traveled to Canada, France, and the United Kingdom as an unofficial representative of the new Board of Atomic Energy Research and made extensive contacts with officials of these governments leading to the exchange of materials and scientists. Partly as the result of this successful mission, he was appointed by Prime Minister Jawaharlal Nehru in 1948 to be head of India's Atomic Energy Commission. This three-person commission was given total control of India's atomic energy program with a large budget and complete secrecy. Bhabha soon developed a strong personal relationship with Prime Minister Nehru, so government authorities listened to his advice on atomic energy. In 1954, he was appointed permanent secretary of the Department of Atomic Energy while retaining his position on the Atomic Energy Commission. It was his negotiations with the British and Canadian governments in the 1950s that produced first a research reactor and then a CANDU power reactor at Trombay, outside of Bombay. Plutonium for this later reactor produced the material for India's first atomic bomb tested in 1974. In 1955, he served as president of the United Nations Conference on the Peaceful Uses of Atomic Energy in Geneva, Switzerland. Bhabha was killed in an aircraft accident in Switzerland on January 24, 1966, when his commercial aircraft crashed into Mont Blanc.

Suggested readings: David Abbott, ed., *The Biographical Dictionary of Scientists: Physicists* (London: Blond Educational, 1984); Itty Abraham, *The Making of the Indian Atomic Bomb: Science, Secrecy and the Postcolonial State* (London: Zed Books, 1998); Shyam Bhatia, *India's Nuclear Bomb* (Ghaziabab, India: Vikas House, 1979).

Bikini Tests

The Bikini Atoll in the Marshall Islands was the site of a series of tests of American nuclear weapons in the immediate postwar period. In 1946 the

The underwater atomic blast in the Bikini Lagoon on July 25, 1946, produced this mushroom shaped cloud formation which proved to be highly radioactive. Around the cloud stem are the naval vessels used to test the power of the blast. (Photo reproduced from the Collections of the Library of Congress)

United States used the Bikini Test Site as part of Operation Crossroads to explode the first peacetime atomic bomb as part of a full-scale naval exercise. Despite some reluctance from the navy, naval ships were used to test the power of the bomb on a naval fleet. Twenty-two foreign representatives from the United Nations Atomic Energy Commission were invited as observers by the U.S. State Department. Two bombs—one airborne and one underwater—did considerable damage to the ships anchored in the Bikini Atoll. This demonstration was done more for international public relations than as a serious study of the effects of atomic bombs, but it was successful on both accounts. From 1946 to the end of testing in 1958, 66 nuclear devices were detonated on the Bikini Atoll. In 1964, the U.S. government announced that the natives of Bikini Atoll could start returning to the islands and live there. Topsoil was removed and new coconut trees planted for the islanders. About 100 natives of the Bikini Atoll returned over the next seven years, but when authorities tested the environment in 1975, they found that radiation levels exceeded federal guidelines. Finally, in 1978 tests on the inhabitants found that radiation levels were so high that the Bikinians were again evacuated for health reasons.

Suggested readings: Jane Dibblin, *Day of Two Suns: U.S. Nuclear Testing and the Pacific Islanders* (London: Virago Press, 1988); Peter Pringle and James Spigelman, *The Nuclear Barons* (London: Michael Joseph, 1981); Jonathan Weisgall, *Operation Crossroads: The Atomic Tests at Bikini Atoll* (Annapolis, MD: Naval Institute Press, 1994).

Blackett, Patrick Maynard Stuart

Patrick Blackett (1897–1974) was a British physicist who is most famous for developing a technique to test theories on the nature of the atom. He was born on November 18, 1897, in London, England. His father was a stockbroker. Blackett married Constanza Bayon in 1924, and they had a daughter and a son. His education was at the Royal Naval College at Dartmouth. After receiving his commission as an officer in the Royal Navy, he was on ships at both the Battle of the Falklands and the Battle of Jutland. Surviving the war, he resigned his naval commission and entered Cambridge University to work on science degrees. Blackett obtained his B.A. in 1921 and his M.A. in 1924. Beginning in 1923, Blackett started conducting research in the Cavendish Laboratory. He remained at Cambridge until 1933.

It was during his stay at the Cavendish Laboratory that Blackett made his most important contributions to nuclear science. He started using the Wilson Cloud Chamber to carry out experiments on alpha particles. Then, in 1933, an improved cloud chamber allowed Blackett to confirm Carl Anderson's discovery of the positron. His research accomplishments earned him the 1948 Nobel Prize for Physics.

Despite his fame as a researcher at the Cavendish Laboratory, Blackett made several professional moves. In 1933, he was appointed to a professorship at London University. Then he moved to the University of Manchester in 1937 and stayed there until 1953. During the war, Blackett served as the director of Operational Research (1942–1945). Among his accomplishments was traveling to India and encouraging scientific developments there. After the war, Blackett became active in the antinuclear weapons movement, serving as the president of the British Association of Scientific Workers (AScW), a trade union organization of 15,000 members. Finally, in 1953 Blackett made his final professional change to take a chair in physics at the Imperial College of Science and Technology, London. He retired in 1963 and died on July 13, 1974, in London.

Suggested reading: John Daintith, Sarah Mitchell, Elizabeth Tootill, and Derek Gjertsen, *A Biographical Encyclopedia of Scientists*, 2nd ed., vol. 1 (Bristol, UK: Institute of Physics, 1994).

Bock's Car

Bock's Car was the name of the B-29 bomber that detonated the second atomic bomb over the Japanese city of Nagasaki. Its name came from its captain, Frederick Bock. Major Charles Sweeney was the pilot, and the original target was Kokura. His bomber was much more heavily loaded than the *Enola Gay*, because the plutonium bomb, Fat Man, was considerably larger and heavier than the uranium bomb, Little Boy. Fat Man's arming was also much more complex. It was to be detonated by a radar proximity fuse that made everyone in the plane nervous. Almost as soon as the bomber was airborne, the bomb-signaling device started malfunctioning, but it took only a few minutes to find the problem—two wires crossed. It was also determined that a fuel transfer valve problem reduced the amount of fuel available for the mission. The Kokura target was obscured by smoke from a fire-bombing two days before, so the alternative target of Nagasaki was picked. A radar approach was made, and the bomb was detonated about a mile and a half off target. *Bock's Car* landed on an airfield in Okinawa with fuel gauges on empty.

Suggested reading: Kenneth K. Nichols, *The Road to Trinity* (New York: Morrow, 1987).

Bohr, Niels Henrik David

Danish physicist Niels Bohr (1885–1962) was one of the most important theoretical physicists in the twentieth century. He was born in Copenhagen, Denmark, on October 7, 1885. His father was a professor of physiology at the University of Copenhagen, and his brother, Harold, was a world-class mathematician. He married Margrethe Norlund in 1911, and they had five sons. All of his degrees were from the University of Copenhagen: baccalaureate in 1903; M.S. in 1909; and Ph.D. in 1911. He studied physics under C. Christiansen. His first position was as a researcher at the Cavendish Laboratory of the University of Cambridge, England, in 1911. J. J. Thomson was his supervisor, but Bohr found him unsympathetic. After only one year there, Bohr moved to the University of Manchester to be close to British experimental physicist Ernest Rutherford. His friendship with Rutherford began at this time and lasted throughout Rutherford's lifetime. Again after one year, he returned to the University of Copenhagen. In 1914, Bohr headed back to England, but this time to Victoria University. Finally, in 1916, Bohr made his last move back to the University of Copenhagen. This time he stayed there until the end of his career in 1962. Despite these numerous moves, Bohr's theoretical investigations into the structure of the atom resulted in his receiving the 1922 Nobel Prize for Physics.

Bohr is recognized as the second most influential physicist in the twentieth century, with only Albert Einstein ahead of him. He always seemed to be on the cutting edge of atomic theory. The first occasion was his 1913 theory of the "Bohr atom" in which an electron bound to a nucleus can only occupy a discrete set of orbits. The theory explained the discrete lines observed in atomic spectra in terms of electron transitions between Bohr orbits. His liquid-drop theory of the atomic nucleus proved invaluable in understanding nuclear fission, discovered by Otto Hahn. He also quickly figured out in 1940 the importance of U-235 in slow-neutron fission in uranium.

Niels Bohr, a Danish physicist, was a top theorist on atomic energy. He ran an institute in Copenhagen, Denmark that became the center of theoretical research on the structure of the atom.

Almost as significant as his theorizing about the nature of the atom was Bohr's missionary efforts to spread knowledge about progress in the nuclear field. Bohr was a great talker, and he was always happy posing questions for other physicists to ponder. Mathematics was not one of his strengths, but he was always good with ideas. After solving one of physics' big questions on the nature of the hydrogen atom, Bohr became the director of the Institute for Theoretical Physics in Copenhagen. He used his position as head of the Institute as a forum for conferences and workshops. Although Bohr wrote relatively little, partly because writing was always difficult for him, his influence as a leading light of intellectual inquiry into the nature of the atom as well as his being from a neutral country, made him the spokesperson for a generation of nuclear scientists.

The German occupation of Denmark in 1940 was dangerous for Bohr, because his mother was Jewish. Also, in 1938, Bohr had spoken out against Nazi racism at an International Congress of Anthropological and Ethnological Sciences in Elsinore, Denmark. Learning that he and his brother were scheduled to be arrested and deported to Germany in late September 1943, the Bohr brothers escaped to the Allies in an open-boat landing in Sweden. He was flown by a Mosquito bomber to England on October 6. Bohr was at the forefront of atomic theory, so he soon was serving as an adviser with the Manhattan Project. His fears about the atomic bomb made him a spokesperson for widespread dissemination of information to include the Soviet Union. He contacted various political figures, including President Roosevelt and Prime Minister Churchill. His reception from Roosevelt was cordial, but Churchill distrusted his motives from the beginning. A provision of a memorandum between Roosevelt and Churchill was that Bohr be isolated. Nevertheless, Bohr became a key consultant on the Manhattan Project.

After the war, Bohr returned to Copenhagen and continued his activities to broaden understanding of the dangers of nuclear war. In 1955, he organized the first Atoms for Peace conference in Geneva, Switzerland. He died on November 18, 1962.

Suggested readings: Henry A. Boorse, Lloyd Motz, and Jefferson Hane Weaver, *The Atomic Scientists: A Biographical History* (New York: Wiley Science Editions, 1989); Anthony French and P. J. Kennedy, ed., *Niels Bohr: A Centenary Volume* (Cambridge: Harvard University Press, 1985); Ruth E. Moore, *Niels Bohr: The Man, His Science, and the World They Changed* (New York: Knopf, 1996); Abraham Pais, *Niels Bohr's Times: In Physics, Philosophy, and Polity* (Oxford: Oxford University Press, 1991); Stefan Rozental, ed., *Niels Bohr: His Life and Work as Seen by His Friends and Colleagues* (Amsterdam, Holland: North-Holland, 1967); C. P. Snow, *The Physicists* (Boston, MA: Little, Brown, 1981).

Bohr Festival

Each June in the 1920s, Niels Bohr came to the University of Göttingen, Germany, to lecture on theoretical physics. At the time of these lectures Bohr was at the forefront of experimental and theoretical research on the nature of the atom. Although Bohr was not a dramatic lecturer, the content of his lectures was exciting. Consequently, these lectures drew an audience from around the world, but they became especially important to young German scientists. It was at one of these festivals that German physicist Werner Heisenberg first made contact with Bohr. Bohr ended up inviting him to attend his institute in Copenhagen for advanced study on the nature of the atom. The popularity of these types of scientific gathering and the development of scientific stars show the intensity and importance of the debate over the nature of the atom.

Suggested readings: Thomas Powers, *Heisenberg's War: The Secret History of the German Bomb* (New York: Knopf, 1993); Richard Rhodes, *The Making of the Atomic Bomb* (New York: Simon and Schuster, 1986).

Bohr-Heisenberg Conversation, September 1941

This highly controversial meeting of Niels Bohr and Werner Heisenberg in late September 1941 in Copenhagen, Denmark, has become a part of the folklore of the atomic bomb. Denmark was under German military occupation, and the Germans were on the outskirts of Moscow when Heisenberg requested a meeting with his old friend and mentor Niels Bohr. Since no minutes were kept and only Heisenberg's treatment has survived, no consensus has been established about the conversation except that Bohr's friendship with Heisenberg was seriously affected. Heisenberg evidently wanted to inform Bohr about the German atomic bomb program and somehow work out a deal with the Allies to forestall the future use of such a weapon. He made a vague reference to scientists of all countries avoiding bomb work for the duration of the war. If this was the crux of Heisenberg's arguments, his conversation was a failure, because Bohr distrusted Heisenberg because of his support for Germany. It is possible that Bohr interpreted Heisenberg's proposal as an attempt to gauge the progress of the Allied atomic bomb program. In any event, the meeting was a total failure, and Heisenberg's reputation in the postwar physics community suffered for it. Two features of the meeting did have long-range consequences: Heisenberg admitted German interest in an atomic bomb, and he had sketched out for Bohr a picture of a reactor. Both bits of information eventually came to the attention of American scientists after Bohr escaped from Denmark in 1943.

Suggested readings: Ruth E. Moore, *Niels Bohr: The Man, His Science, and the World They Changed* (New York: Knopf, 1966); Thomas Powers, *Heisenberg's War: The Secret History of the German Bomb* (New York: Knopf, 1993).

Born, Max

Max Born (1882–1970) was a German mathematician–physicist whose mathematical calculations helped advance atomic theory. He was born on December 11, 1882, in Breslau, Germany. His father was a distinguished professor of anatomy at the University of Breslau. In 1913, Born married Hedwig Ehrenberg, and they had two daughters and one son. His education was at the universities of Breslau, Heidelberg, and Zurich before ending up at the University of Göttingen, where he received a Ph.D. in physics in 1907. He traveled to Caius College, Cambridge University, for six months in 1907–1908 for advanced study at the Cavendish Laboratory. After a brief stay at the University of Breslau, Born was invited to return to the University of Göttingen for research and was soon offered a lectureship. Born stayed at Göttingen until he received an offer for a teaching chair at the University of Berlin in 1914. After the war broke out, Born was commissioned in the artillery in 1915 and assigned to a project to study sound ranging. In 1919,

Born exchanged teaching positions with Max von Laue and spent 2 years at the University of Frankfurt. Then Born returned to the University of Göttingen to take the post of director of the Physical Institute. He stayed at Göttingen for the next 12 years.

Born's early scientific work was on crystals, but later he turned to quantum theory. His study of the vibrations of atoms in crystal lattices resulted in his part of the Born-Haber cycle, which notes the reactions and changes by which it is possible to calculate the lattice energy of ionic crystals. His familiarity with mathematic theory helped him advise Werner Heisenberg on the matrix mathematics to explain quantum mechanics. On several other occasions, Born assisted other physicists to formulate the mathematical theory behind their experimental findings. He was notorious for his brilliance at the blackboard and ineptness in the laboratory.

Born's status as a Jewish professor made it imperative for him to find a job outside of Germany after the Nazi regime assumed control of the German state. Born was a natural target of the Nazis because of his strong support and friendship with Albert Einstein. In 1933, Born landed a position at Cambridge University as Stokes Lecturer. He stayed at Cambridge for three years before moving to the University of Edinburgh, Scotland, as the Tait Professor of Natural Philosophy. Born remained at Edinburgh until his retirement in 1953.

It was at Born's retirement that his longtime and close friendship with Albert Einstein was severely tested when Born informed him that he was returning to Germany for retirement. Soon after retirement, the Nobel Prize Committee presented Born with the 1954 Nobel Prize for Physics. He was cited for his work on statistical interpretation of the quantum theory. Always intellectually active and a great lover of music, Born was also asthmatic and frequently ill, but he rarely let his health interfere with his physical activities. Born died on January 5, 1970, at Göttingen, Germany, and on his gravestone is engraved his fundamental equation of matrix mechanics.

Suggested readings: Henry A. Boorse, Lloyd Motz, and Jefferson Hane Weaver, *The Atomic Scientists: A Biographical History* (New York: Wiley Science Editions, 1989); Max Born, *My Life: Recollections of a Nobel Laureate* (New York: Scribner, 1978); John Daintith, Sarah Mitchell, Elizabeth Tootill, and Derek Gjertsen, *A Biographical Encyclopedia of Scientists*, 2nd ed., vol. 1 (Bristol, UK: Institute of Physics, 1994); Max Jammer, *The Conceptual Development of Quantum Mechanics* (New York: McGraw-Hill, 1966).

Bothe, Walther Wilhelm Georg Franz

Walther Bothe (1891–1957) was a German physicist who is famous for his experimental research on electrons and for his work on the German atomic program in World War II. He was born on June 8, 1891, in Oranienburg, Germany. Bothe's father was a merchant. His marriage to Barbara Below in 1920 produced two daughters. Bothe studied physics under

Max Planck at the University of Berlin, receiving his Ph.D. in 1914. As soon as war broke out in 1914, Bothe joined the German army and served as a machine-gunner. He was taken captive by the Russians in 1915 and stayed in a Siberian prisoner-of-war camp until 1920. He returned to the University of Berlin, where he worked in Hans Geiger's radioactivity laboratory. Bothe remained in Berlin until 1931, when he was appointed a professor at Giessen University. Bothe's coolness toward the Nazis resulted in his losing a professorship at the University of Heidelberg in 1933.

His opposition to the Nazis produced a mental strain that soon caused his physical health to fail, and he spent time in 1933 in a Badenweiler sanatorium. In 1934, Bothe moved to the Max Planck Institute in Heidelberg, where he spent the rest of his career as a researcher and eventually as director of the Institute.

Bothe established his reputation as an experimental physicist at the University of Berlin. His use of the "coincidence method" of detecting the emission of electrons by X-rays using adjacent Geiger tubes showed that momentum and energy are conserved at the atomic level. A subsequent use of the same technique in 1929 for the study of cosmic rays indicated that they consisted of massive particles rather than photons. It was this research furthering Arthur Compton's scattering principle for which Bothe shared the 1954 Nobel Prize for Physics with Max Born.

Bothe was recruited early for the German atomic bomb program. One of his early unsuccessful experiments with graphite turned the German scientists away from using graphite as a moderator for an atomic reactor. This step proved fatal, because it delayed the building of reactors by making the availability of heavy water so important. Later in the war, he supervised the construction of a cyclotron. War delays caused materials to be delivered slowly; its magnet did not arrive until March 1943. The cyclotron was finished in 1944, almost too late to be used for the German atomic bomb experiments.

After his release by the Allies, Bothe returned to the University of Heidelberg as a professor of physics. He used the cyclotron for medical studies. Still active in research, he published several books and articles. Bothe retained his position at the university until his death in Heidelberg, Germany, on February 8, 1957.

Suggested readings: Alan D. Beyerchen, *Scientists Under Hitler: Politics and the Physics Community in the Third Reich* (New Haven, CT: Yale University Press, 1977); John Daintith, Sarah Mitchell, Elizabeth Tootill, and Derek Gjertsen, *A Biographical Encyclopedia of Scientists*, 2nd ed., vol. 1 (Bristol, UK: Institute of Physics, 1994).

Bradbury, Norris

Norris Bradbury (1909–1997), an American physicist, was the successor to Robert Oppenheimer as the director of the Los Alamos Laboratory and

Norris Bradbury was the second head of the Los Alamos National Laboratory, succeeding J. Robert Oppenheimer in 1946. His efforts to retain the primacy of the Los Alamos facility in the nuclear program were only partially successful.

he was instrumental in the building of the hydrogen bomb. He was born on May 30, 1909, in Santa Barbara, California. His academic training had been at Pomona College, California, and his graduate education at the University of California at Berkeley. After a two-year stint at the Massachusetts Institute of Technology (MIT) as a research fellow, Bradbury found a teaching position at Stanford University. His field of expertise was in conduction of electricity in gases, properties of ions, and atmospheric electricity. Once World War II started, Bradbury received a naval commission as lieutenant commander and worked on ordnance research at the Dahlgren Proving Ground of the U.S. Navy Ordnance Bureau with Captain William S. Parsons.

Bradbury was recruited for work on the Manhattan Project in June 1944. He served as associate leader of the Explosives Division under George B. Kistiakowsky. J. Robert Oppenheimer, the head of the Los Alamos Scientific Laboratory, selected Bradbury to succeed him at Los Alamos in 1945. Shortly after the war, key personnel began to leave Los Alamos, returning to prewar or new positions. Bradbury had replaced Oppenheimer at the time when it was uncertain whether the Los Alamos Laboratory would continue in any form. He recruited scientists to stay and work on improving the atomic bomb. Edward Teller was recruited to work on the hydrogen bomb, but he began feuding with Bradbury over its lack of priority and progress. Although Bradbury had strong support in the scientific community, Teller marshaled enough political support to start a new laboratory in California. Bradbury came into competition for funding and projects with Ernest Lawrence and Edward Teller at the Livermore Laboratory. Despite political setbacks, Bradbury remained director at Los Alamos until 1970. He died in August 1997.

Suggested reading: Richard Rhodes, *Dark Sun: The Making of the Hydrogen Bomb* (New York: Touchstone, 1995).

Bragg, William Henry

Sir William Henry Bragg (1862–1942) was a British physicist whose study of X-rays and use of a X-ray spectrometer made important discoveries on the nature of radiation. He was born on July 2, 1862, in Westwood, England. His father was a merchant seaman. In 1889, he married Gwendoline Todd and they had two sons and a daughter. His eldest son, William Lawrence Bragg, also became a famous physicist. After attendance at a variety of schools, Bragg graduated from Cambridge University in 1884 with a M.A. in physics. After a year's research at the Cavendish Laboratory under Sir J. J. Thomson, he accepted an offer of a chair of mathematics and physics at the University of Adelaide, Australia, beginning in 1886. Bragg stayed there until 1909 when he assumed the post of professor of physics at Leeds University. After six years at Leeds, Bragg moved to the University of London. His stay at the University of London lasted ten years before accepting a post with the Royal Institution, London, where he remained as a professor and administrator until 1942.

Bragg started his research on X-rays in 1904. At first most of his attention was on investigating alpha particles, but soon he turned to studying X-rays. Bragg concluded X-rays were waves so in 1915 he constructed a X-ray spectrometer to measure wavelengths of X-rays. It was for this research that Bragg shared the 1915 Nobel Prize for Physics with his son. Later at the Royal Institution, Bragg conducted experiments for the British Admiralty on hydro phones. He also worked hard to reestablish the research capacity of the Royal Institution.

Bragg's reputation for research and leadership made him a leader in the British scientific community. While he continued to do scientific research, most of his activities after his arrival at the Royal Institution were devoted to directing the research efforts of others. He died March 12, 1942, in London.

Suggested readings: Henry A. Boorse, Lloyd Motz, and Jefferson Hane Weaver, *The Atomic Scientists: A Biographical History* (New York: Wiley Science Editions, 1989); Gwendolyn M. Caroe, *William Henry Bragg, 1862–1942: Man and Scientist* (Cambridge: Cambridge University Press, 1978); John Daintith, Sarah Mitchell, Elizabeth Tootill, and Derek Gjertsen, *A Biographical Encyclopedia of Scientists*, 2nd ed., vol. 1 (Bristol, UK: Institute of Physics, 1994).

Bragg, William Lawrence

William Lawrence Bragg (1890–1971) was a British physicist who followed in his father's footsteps in studying the characteristics of X-rays. He was born on March 31, 1890, in Adelaide, Australia. His father was the famous physicist William Henry Bragg. The younger Bragg received his education first at the University of Adelaide and then at Cambridge University. After graduation, he taught at Cambridge as a lecturer. His research on X-ray diffraction by crystals and his discovery of ways to measure wavelengths of X-rays, which is now known as Bragg's Law, led to his sharing with his father the 1915 Nobel Prize for Physics. In 1919, he was appointed professor of physics at Manchester University. Bragg remained at Manchester until 1937, when he became the director of the National Physical Laboratory. After only one year there, however, he was appointed head of the Cavendish Laboratory at Cambridge University, replacing Ernest Rutherford. Bragg retained this position until 1953, when he became director of the Royal Institution in London. He remained there until his retirement in 1961. Bragg died on July 1, 1971, in London.

Suggested readings: Henry A. Boorse, Lloyd Motz, and Jefferson Hane Weaver, *The Atomic Scientists: A Biographical History* (New York: Wiley Science Editions, 1989); John Daintith, Sarah Mitchell, Elizabeth Tootill, and Derek Gjertsen, *A Biographical Encyclopedia of Scientists*, 2nd ed., vol. 1 (Bristol, UK: Institute of Physics, 1994); Frank N. Magill, ed., *The Nobel Prize Winners: Physics*, vol. 1 (Pasadena, CA: Salem Press, 1989).

Bravo Nuclear Test

Bravo was the code name for a hydrogen bomb test off Namu Island in the Bikini Atoll on February 28, 1954, that provoked an international incident over radioactive fallout. An earlier test of hydrogen principles had proved that a hydrogen-type bomb would work, but the Bravo test was the real demonstration of the Edward Teller–Stanislaw Ulam thesis. The scientists in charge, however, were from Los Alamos. An atomic bomb triggered the tritium and lithium-6-deuteride, allowing fusion to take place. Moreover, the Bravo bomb was a practical weapon because it could either be dropped by an aircraft or delivered by a missile. The device was placed inside a protective building, and the scientists were located on the control island of Enyu, which was 20 Miles away from the blast site. It was detonated early in the morning, and the yield was in the 15-megaton range; this yield exceeded expectations, which had been in the 6-megaton range. But what surprised the scientists was its high radioactivity level. Radiation levels began to climb, and the high readings made the scientists at Enyu realize the danger. The island of Enyu was 10 times hotter than the ground-zero area had been for Ranger's Shot Able. A hasty evacuation saved the scientists, but the fallout—in the form of white snowflakes—hit the Japanese fishing trawler the *Lucky Dragon* and the inhabitants of the island of Rongelap. Most of the international attention fell on the radiation sickness experienced by the crew of the *Lucky Dragon*, but long-term consequences were worse for those natives on Rongelap. Radioactive fallout in the form of pale powder reached one and a half inches deep on Rongelap. The Bikini testing range was so radioactive that further tests scheduled there had to be moved elsewhere. Further analysis confirmed that lethal fallout had covered a 7,000-square-mile area of the Pacific. This test had been a potential public relations disaster for the Atomic Energy Commission (AEC), but the most damaging information was covered up in a public relations blitz. The AEC, however, could not cover up the exposure of the Japanese crew of the fishing boat *Lucky Dragon*. Moreover, scientists at both Los Alamos and Livermore were alarmed by their miscalculations of the power of the nuclear device and became more cautious in subsequent testing.

Suggested readings: International Physicians for the Prevention of Nuclear War, *Radioactive Heaven and Earth: The Health and Environmental Effects of Nuclear Weapons Testing in, on, and above the Earth* (New York: Apex Press, 1991); Robert L. Miller, *Under the Cloud: The Decades of Nuclear Testing* (New York: Free Press, 1986).

Brazil Nuclear Program. *See* South American Nuclear Program

Breeder Reactors

The breeder reactor was a second-generation fission reactor that became popular with government and power utilities because of its economic potential. This type of reactor depends on fission-producing neutrons. These neutrons are captured by heavy nuclei in the fuel, producing more fissionable material, with energy as the result. The breeder reactor needed more uranium to operate, but the excess plutonium produced could be used to fuel light water reactors and additional breeder reactors. Because this type of reactor could operate indefinitely, nuclear engineers found it intriguing and worked hard in perfecting it. A shortage of uranium in the immediate postwar world also made this type of reactor attractive. The General Electric (GE) Company used much of its capital and energy building the Intermediate Power Breeder Reactor (IPBR) in the period after World War II. It was a liquid metal reactor and used no water. Since the uranium is converted into fissionable plutonium thus creating more fuel than it burns, the Atomic Energy Commission (AEC) invested nearly half a billion dollars in research and development by the early 1960s. It was this characteristic that attracted so much interest among possible investors. News of a possible worldwide uranium shortage caused a renewal of interest in breeder reactors in the early 1960s. An experimental commercial breeder reactor, Fermi I, was built on the shores of Lake Erie and run by Detroit Edison. On October 5, 1966, an accident took place at Fermi I when a piece of metal clogged the flow of the core coolant, producing overheating. This accident showed the danger of this type of reactor, but it did not lessen the enthusiasm for breeder reactors. Much of the research on breeder reactors in the United States was conducted at the Argonne National Laboratory and at the National Reactor Testing Station in Arco, Idaho. In 1977 France started building a sodium-cooled fast breeder reactor, the Superphenix, based on the design of smaller breeder reactors.

Breeder reactors have the reputation of being difficult to build and maintain. Any flaw in the cooling or control system can cause catastrophic results, because the reactions take place at a high rate of speed. Because of this, they have proved to be expensive to build, to have records of excessive downtime, and to raise safety concerns. Moreover, critics pointed out that the AEC's cost-benefit analysis was faulty, and that the costs of building and maintaining these reactors made them uneconomical. By the 1990s, Britain, France, Germany, and the United States had all become disenchanted with breeder reactors. Only Japan and, to a lesser extent, Russia, still retain interest in breeder reactors. As of December 31, 1996, only five breeder reactors were in operation (France—2, Japan—1, Kazakstan—1, and Russia—1), with two under construction in Russia. This lessening of interest in breeder reactors has important proliferation ramifications, because weapon-grade plutonium is a by-product of breeder reactors.

Suggested readings: Thomas B. Cochran, *The Liquid Metal Fast Breeder Reactor: An Environmental and Economic Critique* (Washington, DC: Resources for the Future, 1974); Richard Kokoski, *Technology and the Proliferation of Nuclear Weapons* (Oxford: Oxford University Press, 1995); Richard S. Lewis, *The Nuclear-Power Rebellion: Citizens vs. the Atomic Industrial Establishment* (New York: Viking Press, 1972); *Nuclear Power Reactors in the World* (Vienna, Austria: International Atomic Energy Agency, 1997).

British Nuclear Program

The British became the junior partners to the Americans in the development of the atomic bomb, but in the period from 1940 through 1941 the British effort was ahead of the Americans'. Among key policymakers it was at least considered to try and retain a monopoly on atomic processes, but this was decided impractical, considering the engineering problems of building a working atomic bomb. Once the decision was made to transfer their information and key British scientists to the United States, the British nuclear program fell behind. Several agreements had been negotiated during the war to soothe possible wounded feelings. Winston Churchill was adamant about becoming a partner, because he believed atomic energy was essential to Great Britain's postwar military defense. The transfer of the Halban team to Canada in 1942 to continue to work on a uranium heavy water system was important because the Americans were going in a different direction.

The British nuclear program found itself abandoned in 1945 when its ally, the Americans, cut British physicists off from all sources of information. Moreover, the new Labor government had no idea about an atomic program, because Churchill had informed only seven members of his government about the Anglo-American atomic bomb program. None of these seven had contacts with Labor politicians. An attempt by the Attlee government for further Anglo-American cooperation failed in 1946. Although it was British physicists who had provided the initial plans, British scientists were forced to leave all classified documents at Los Alamos when the war ended. Besides, the policy of compartmentalization kept British scientists isolated so that none of them had been exposed to key parts of the process of building either a uranium or a plutonium bomb. Fortunately, Klaus Fuchs was able to reconstruct some of the engineering plans from his photographic memory. Still, the British scientists were deficient in knowledge about the fabrication of electric detonators, plutonium metallurgy, exterior ballistics, arming, and fusing. Nevertheless, in early 1947 the Attlee government via a secret ministerial body, the Gen 173, decided to develop the capability to build nuclear weapons. William Penney was placed in charge of building the British atomic bomb. Unlike in the United States, the British atomic energy program was exclusively a government program with little help from private industry.

The first step in the British nuclear program was to build atomic reactors. Christopher Hinton was assigned to this task. Harwell was to remain a research establishment, so three additional complexes were necessary: a refinery to extract and refine uranium from uranium ore, the reactor itself, and a plutonium refinery. Risley in south Lancashire was founded to be near the south Lancashire engineering industry for engineering support, because its mission was industrial-scale engineering projects. The first refinery was built at Springfields, near Preston in Lancastershire. A reactor was constructed at Sellafield and later called the Windscale. Finally, the plutonium refinery was built at Windscale. Then in 1946, the government decided on the need for a gaseous diffusion plant to enrich uranium, and it was approved to be built at Risley Cheshire.

In 1946, the British government decided to separate civilian from military applications. Civilian applications of nuclear energy would remain under

John Cockcroft's supervision of the Atomic Energy Research Establishment (AERE) at Harwell. Development of an atomic bomb was given to the Armament Research Department under William Penney at Fort Halstead near Sevenoaks in Kent. It was soon decided that an atomic weapons production facility was needed, and a new facility at Aldermaston was built in the early 1950s. Lord Portal resigned his post as head of the atomic program in 1951 and was replaced by an obscure general, Sir Frederick Morgan. This change of leadership had little impact, but the victory of the Churchill government in 1951 did accelerate matters. The first British atomic bomb was detonated at Monte Bello Islands off Australia on October 3, 1952. Other tests took place in central Australia in 1953. Then in 1956 the British built a nuclear test site at Maralinga, Australia. Tests were conducted at Maralinga for six years, with seven atomic devices detonated.

Much as in other countries, the British civilian nuclear program alternated between euphoria about nuclear energy to reevaluation due to cost and benefit. In 1954 the Atomic Energy Authority was formed to coordinate British atomic energy activities. It immediately started looking for industrial as well as military projects. The Calder Hall reactor in the north of England came on-line in October 1956. While it was a dual-purpose reactor used for both civilian and military purposes, the British claimed it to be the first reactor providing electricity for civilian uses. Seven more of the same type were brought into service over the next three years. Although the British government planned for 16 more units to be devoted solely to the production of electricity, only 3 of them had been built by 1964. Plans were made for further expansion in the 1960s by the awarding of contracts to five industrial groups, but these groups lost interest. Because of the lack of economic benefit and increasing public discontent, the British nuclear program went into decline in the late 1960s.

The election of the Thatcher government in 1979 revitalized the British nuclear power industry. She was known as a strong supporter of nuclear power, and an ambitious nuclear energy plan was forthcoming. This policy stirred up opposition among British antinuclear groups, but the government was able to bypass the opposition. In 1988 the Thatcher government considered privatization of the British nuclear industry, but this scheme failed because of the huge costs and uncertainty about nuclear power. It became apparent that even the Conservative Party had reservations about nuclear power and abandoned its support.

Suggested readings: Brian Cathcart, *Test of Greatness: Britain's Struggle for the Atom Bomb* (London: Murray, 1994); Helena Flam, ed., *States and Anti-Nuclear Movements* (Edinburgh, Scotland: Edinburgh University Press, 1994); Bertrand Goldschmidt, *The Atomic Complex: A Worldwide Political History of Nuclear Energy* (La Grange Park, IL: American Nuclear Society, 1982); Margaret Gowing, *Independence and Deterrence: Britain and Atomic Energy, 1945–1952*, 2 vols. (New York: St. Martin's Press, 1974).

Brockett Report

The Brockett Report was a two-volume study of reactor safety research conducted at the National Reactor Testing Station (NRTS), Arco, Idaho, during the late 1960s. George Brockett, head of engineering at NRTS, started tests by building a small test reactor to study possible nuclear cooling accidents. These tests revealed that the fuel in the reactor could get much hotter, much faster, than safety systems had been designed to handle. Moreover, the tests found that the standard emergency cooling system would be unable to handle a meltdown emergency. These conclusions reached the Atomic Energy Commission (AEC) and caused considerable unhappiness. A task force headed by Stephen Hanauer was formed to review the emergency cooling problem indicated by the NRTS research results. All members of the task force were administrator–engineers. The task force met with representatives of the nuclear industry, who assured the task force that their experiments proved that their emergency cooling system worked. Brockett responded with his two-volume report in 1971, in which he took issue with the industry's interpretations. The most controversial part of the report was over how efficiently emergency cooling water would remove heat from hot uranium fuel rods. Reacting to this controversy, the AEC commissioned a series of tests called Full-Length Emergency Cooling Heat Transfer Tests (FLECHTs), but they were given to Westinghouse and General Electric to perform. The subsequent tests were found to be deficient by the NRTS staff. However the Brockett Report was ignored by the AEC and never published. Its recommendations about reactor safety were also set aside, and industry standards approved. During the public hearings on the emergency cooling system, in 1971–1973, the Brockett Report was leaked to the public, but again it was ignored.

Suggested reading: Daniel F. Ford, *The Cult of the Atom: The Secret Papers of the Atomic Energy Commission* (New York: Simon and Schuster, 1982).

Brookhaven National Laboratory

The Brookhaven National Laboratory was the third in a series of national research laboratories established by the U.S. government to conduct advanced research on atomic energy. Soon after the Argonne National Laboratory near Chicago had been set up, a movement started among universities on the East Coast for a research facility near them. Harvard University and the Massachusetts Institute of Technology lobbied for such a center to be located in the Boston area, but I. I. Rabi at Columbia University countered with support from institutions in New York, New Jersey, and Pennsylvania for the laboratory to be established on Long Island, New York. In 1946 a consortium of nine universities formed the Associated Universities Inc. (AUI) to administer the new laboratory. It was decided to locate the Brookhaven National Laboratory on Long Island on the site of 6,000 acres of a former army camp, Camp Upton. The laboratory opened in 1947 and soon had a staff of nearly 1,000 in departments of physics, biology, medicine, and engineering. In 1952 the Cosmotron, the first particle accelerator operating at over a billion volts, started operations. In 1958 the laboratory opened the Medical Research Center to conduct basic research on nuclear medicine. In 1991 the Radiation Therapy Facility opened. Besides conducting advanced research on nuclear problems, the scientists at Brookhaven were given special tasks by the Atomic Energy Commission (AEC) and its successor, the Department of Energy (DOE). One such task was the 1964 WASH-740 Update on nuclear safety. Brookhaven's scientists produced a report that the AEC refused to accept, because it concluded that as many as 45,000 people might be killed in a major reactor accident. In 1997 the Department of Energy terminated its contract with the Brookhaven National Laboratory. In 1998 the Brookhaven Science Association (BSA) assumed the administration of the laboratory.

Suggested reading: Daniel J. Kevles, *The Physicists: The History of a Scientific Community in Modern America* (Cambridge, MA: Harvard University Press, 1987).

Brown's Ferry Nuclear Power Station Incident

On March 22, 1975, an electrical fire at the Tennessee Valley Authority's Brown's Ferry Nuclear Power Station showed the inadequency of safety regulations at nuclear power plants. Both the Atomic Energy Commission (AEC) and its successor, the Nuclear Regulatory Commission (NRC), had adopted a policy of delegating safety to the operators of the nuclear facilities (industry self-regulation) with the assurances to the public that major accidents were not possible. The Brown's Ferry Nuclear Power Station had been under construction since 1966 near Decatur, Alabama. On the afternoon of March 22, electricians started an electrical fire that burned uncontrolled for seven and a half hours. Around 1,600 electrical cables, including most of the safety systems' cables, were destroyed. Plant operators were able to shut down one reactor, but the burned cables prevented the other reactor from being shut off. This second reactor went out of control, but a meltdown was avoided. This incident received major media attention, but after the news died out, the NRC relegated it as something that needed future attention. Critics of nuclear plant safety, however, asked, If this type of accident could happen, what would be the consequences of larger safety problems?

Suggested reading: Daniel F. Ford, *The Cult of the Atom: The Secret Papers of the Atomic Energy Commission* (New York: Simon and Schuster, 1982).

Bulletin of the Atomic Scientists

A group of scientists from the Chicago area opposed to the use of nuclear weapons formed the group the Atomic Scientists of Chicago in early 1945. Eugene Rabinowitch was the journal's first editor, and his philosophical position for the journal was "to preserve our civilization by scaring men into rationality." Rabinowitch had experience with irrational regimes, coming from the Soviet Union and then being exiled from Nazi Germany by the anti-Jewish laws. The *Bulletin of the Atomic Scientists* became the official publication of this group. When the Atomic Scientists of Chicago organization merged into the Federation of American Scientists (FAS), the journal followed it into this new body and soon served as its official publication. Over the next 50-plus years, the *Bulletin of the Atomic Scientists* became one of the most influential journals of its type in the world. Its official clock on the closeness of the world to nuclear war has long been its trademark.

Suggested reading: Milton S. Katz, *Ban the Bomb: A History of SANE, the Committee for a Sane Nuclear Policy* (Westport, CT: Greenwood Press, 1986).

Bush, Vannevar

Vannevar Bush (1890–1974) was a prominent scientist who used his prestige to convince the U.S. government to support scientific research on atomic energy. He was born on March 11, 1890, in Everett, Massachusetts. His father was a minister in the Universalist Church. After attending Tufts as an undergraduate, Bush received his Ph.D. in engineering from the Massachusetts Institute of Technology (MIT) and Harvard University in 1916. His first teaching job was at Tufts, but in 1916 he accepted a professorship in the Department of Electrical Engineering at MIT. During World War I, Bush worked at New London, Connecticut, developing a submarine detector. Bush spent most of his career associated with MIT. He had been the dean of the School of Engineering and the vice president of MIT in the late 1930s. Then in 1939, he became president of the Carnegie Institution in Washington, DC.

Bush's research was in the area of using advanced mathematics to analyze power circuits. In 1925, Bush and fellow scientists at MIT constructed a machine for solving differential equations. This analog computer proved to have industrial and scientific uses, and Bush became well known in engineering fields. In succeeding years, Bush continued to conduct research on computer theory.

As the war in Europe came closer, Bush decided to work toward military preparedness. His first position was as president of the National Advisory Committee for Aeronautics (NACA). Bush had discussions with Harry Hopkins, secretary of commerce, about mobilizing American scientific talent for

Two administrative leaders in the American effort to build the atomic bomb were Vannevar Bush (left), head of the Office of Scientific Research and Development, and James B. Conant, chairman of the National Defense Research Committee. Here they are testifying in a postwar hearing before the House Military Affairs Committee on Atomic Energy.

the war effort. Hopkins introduced Bush to President Franklin Roosevelt and proposed an agency for this purpose. Soon afterward, in 1940, Bush was selected by President Roosevelt to be head of the National Defense Research Committee (NDRC). It was his task to mobilize scientific talent for wartime research. Also in his field of responsibility was research on uranium. Despite the recommendations of several American physicists on the feasibility of the building of an atomic bomb, Bush was reluctant to launch an expensive program until the MAUD Report from British scientists confirmed the likelihood of success. On October 9, 1941, Bush met with President Roosevelt and Vice President Henry Wallace at the White House and recommended building an atomic bomb based on information from the MAUD Report. President Roosevelt approved this recommendation, and the Manhattan Project was launched. Bush, in the meantime, had become the head of the Office of Scientific Research and Development (OSRD), which had replaced the NDRC. Unrest from scientists at the University of Chicago's Metallurgical Laboratory and his belief in an efficient authoritarian organization made Bush place the building of the atomic bomb under the control of the U.S. Corps of Engineers. Bush selected the army over the navy, because of its more positive attitude toward scientists and their participation in decision making. He had no part, however, in the selection of General Leslie R. Groves as head of the project, and their initial encounter was hostile. After the Manhattan Project had been firmly established, Bush's role in atomic energy development diminished. He still had access to President Roosevelt, but Roosevelt's death in 1945 ended Bush's ready access to the president. Bush remained active in scientific politics after the war. He supported the May-Johnson Bill for military control of atomic energy. As his influence waned in government service, Bush decided to resign and devoted himself to public service in academia. He died in 1974.

Suggested readings: Daniel J. Kevles, *The Physicists: The History of a Scientific Community in Modern America* (Cambridge: Harvard University Press, 1987); Martin J. Sherwin, *A World Destroyed: The Atomic Bomb and the Grand Alliance* (New York: Knopf, 1975).

Buster-Jangle Nuclear Tests

Buster-Jangle was the combined code name for the second series of nuclear tests held at the Nevada Test Site in October–November 1951. For both political and safety reasons, the site of the tests was moved 20 miles north from Frenchman's Flat to Yucca Flat. This new series consisted of seven detonations—four airdrops, one tower shot, one surface shot, and one underground shot. The scientists were under pressure from the military to develop tactical atomic weapons, and these detonations were to test their feasibility. Military leaders wanted to commit troops to training near blast zones, but the scientists were concerned about their safety. The first test was Shot Able, and it was a tower shot. It was detonated on October 22, 1951, its blast disappointingly small. Scientists had been trying to shrink atomic weapons to tactical size by reducing the amount of plutonium in them, but they were too far with this test. Shot Baker was an airborne drop, and its blast was much larger than the first shot. Shot Charlie was another airdrop, and its blast was even larger than Shot Baker's. Shot Dog was the largest test yet, and it had the active participation of the U.S. Army. Army equipment was left in the blast area, and 2,796 personnel accessed damage and conducted maneuvers near the blast area. Shot Easy was the last of the Buster series.

The Jangle series followed closely upon the Buster series. Shot Sugar was the first test, and the device was at ground level because scientists wanted to examine ground damage. As in several of the tests, live animals, dogs and sheep, were placed at various distances from ground zero. This device was exploded on November 19, 1951, and it formed a huge radioactive crater. It was estimated by scientists that a person standing in the crater would receive a fatal dose of radiation in only four minutes. Shot Uncle was the last test of the Jangle series, and it was unique because it was an underground detonation. The device was placed 17 feet below the surface for the November 29, 1951, test. Again, a huge, highly radioactive crater was

C

Calvert Cliffs Decision

In 1971 in the *Calvert Cliffs Coordinating Committee v. Atomic Energy Commission*, a U.S. Appeals Court judge ruled that the Atomic Energy Commission (AEC) was subject to the National Environmental Policy Act. The case concerned the proposed building of a nuclear plant in Calvert Cliffs, Maryland. Environmental groups accused the AEC of not considering the potential environment consequences of thermal pollution. The Calvert Cliffs Coordinating Committee took the AEC to court, and on July 23, 1971, a three-judge panel of the U.S. Court of Appeals, District of Columbia Circuit, ruled that the AEC had to consider all environmental issues in its review process. This decision meant that the AEC needed to develop environmental impact analyses for each nuclear plant project. Until this decision, the statutory basis for issuing a construction permit for a nuclear plant was twofold—for the "common defense and security" and for the "health and safety of the public." The AEC had interpreted this to mean that only the safety issue could be raised as an objection to a construction or operating license hearing. This interpretation was put in peril by the passage of the National Environmental Policy Act on January 1, 1970. In an attempt to subvert this requirement and speed up the licensing process, the AEC tried to restrict public hearings. The judicial decision ended this strategy. Despite earlier opposition, the AEC did not appeal the decision. This court decision made it difficult for nuclear plants to pass environmental standards and caused lengthy delays before construction of plants could begin. A standstill on reactor licensing ensued for 18 months as the AEC restructured its licensing process. Critics of these restrictions lobbied Congress, and a bill to give the AEC the authority to issue temporary licenses for construction was passed and signed into law on June 2, 1972, by President Richard Nixon. A ruling by the conservative Burger Supreme Court in 1978 further curtailed some of these restrictions. In this ruling the Court indicated that government agencies did not have to comply strictly with the National Environmental Policy Act.

Suggested readings: Frank G. Dawson, *Nuclear Power: Development and Management of a Technology* (Seattle: University of Washington Press, 1976); Richard S. Lewis, *The Nuclear-Power Rebellion: Citizens vs. the Atomic Industrial Establishment* (New York: Viking Press, 1972).

Campaign for Nuclear Disarmament

The Campaign for Nuclear Disarmament (CND) was the leading British protest group campaigning against nuclear weapons in the late 1950s and early 1960s. Formed in 1956 by a handful of north London housewives to protest testing of nuclear weapons, the National Council for the Abolition of Nuclear Weapons Tests (NCANWT) was the forerunner of the CND. This group intended to raise the consciousness of the general public about atomic energy, and one of their early actions was to protest the Christmas Island hydrogen bomb test in May 1957. Most of the leaders of the NCANWT were unknowns, so in late 1957 a number of famous personalities—Bertrand Russell, J. B. Priestly, Jacquetta Hawkes, George Kennan, PMS Blackett, and Denis Healey—met at Kingsley Martin's flat to discuss forming a national campaign to fight for nuclear disarmament. The next day calls were made to the NCANWT leadership, inviting them to join the new movement. On January 16, 1958, the Campaign for Nuclear Disarmament was formed at an organizational meeting held in the home of Canon Collins. A new executive committee with members of the NCANWT and others was agreed upon, and Bertrand Russell became the first president. Canon Collins was subsequently elected the chairperson of CND. A public meeting at Central Hall, London, on February 17, 1958, attracted over 5,000 people, and funds were raised to support CND's activities. Its call for unilateral nuclear disarmament turned the CND into a mass movement.

The peak of CND's effectiveness was in the years between 1959 and 1961. An Aldermaston March on the nuclear weapons production center on Easter 1959 attracted between 20,000 and 25,000 participants, with full press coverage. Another Aldermaston March 1960 was a further triumph, with between 60,000 and 100,000 present at the end of the march in Trafalgar Square. Success as a mass party soon created organizational problems, with charges that the CND's leadership was elitist. In 1961 a national council, the Committee of 100, assumed control of CND. In a policy spat over the use of civil disobedience, Bertrand Russell resigned as president in October 1961. But it was the Cuban missile crisis of 1962, which showed the growing anti-Americanism of the CND, that eventually weakened the movement. Internal disputes became more serious and began to hurt local organizing. As more members of the CND converted to direct action, often using violent tactics, the CND started a leftward drift. This new direction and isolation from the Labor Party led by 1965–1966 to a substantial decline in the CND as a mass movement. After a hiatus of nearly 20 years, the CND revived itself in the early 1980s for another go at nuclear disarmament.

Suggested readings: John Mattausch, *A Commitment to Campaign: A Sociological Study of CND* (Manchester, UK: Manchester University Press, 1989); Richard Taylor, *Against the Bomb: The British Peace Movement, 1958–1965* (Oxford, UK: Clarendon Press, 1988).

Canadian National Research Council

The Canadian National Research Council was the organization under which the British-Canadian research on atomic energy program resided. It was founded in 1916, and its scientific mission was secondary to a primary mission to assist industry and promote Canadian trade. The British had inherited the team of Hans von Halban and Lew Kowarski from France whose interest was heavy water technology. Halban had continued his research on energy transference from fission and heavy water at Cambridge University. His team of six physicists—five of them either German or central European in origin—was making headway on an atomic reactor when the British decided to send Halban and his Cambridge team to Chicago to work on the Metallurgical Project. Partly as a result of Halban's protests and the Americans' greater interest in graphite rather than heavy water research, the Cambridge team was sent to Canada. It was placed under the organization umbrella of the Canadian National Research Council, with its headquarters in Montreal. Halban was appointed the director and given the go-ahead to concentrate on developing a heavy water reactor. The tightening of the exchange of information with the Americans on atomic technology caused serious problems for the scientists working in Canada, and the British government was concerned. When the Americans gained complete control of the Canadian uranium market by signing a contract for all the uranium ore from 1942 to 1946, the Canadian effort ground to a halt. Churchill constantly reminded Roosevelt about American secrecy hurting the war effort, and these complaints led to the Quebec Agreement in August 1943. In the spring of 1944, the decision was made to build a large heavy water reactor in Canada. British physicist John D. Cockcroft replaced Halban as the director of the Anglo-Canadian project. Lew Kowarski had built a small nuclear reactor, ZEEP (Zero Energy Experimental Pile), at Chalk River, on the banks of the Ottawa River, and the new large reactor was to be constructed on the same site.

Suggested readings: Robert Bothwell, *Nucleus: The History of Atomic Energy of Canada Limited* (Toronto, Canada: University of Toronto Press, 1988); Bertrand Goldschmidt, *The Atomic Complex: A Worldwide Political History of Nuclear Energy* (La Grange Park, IL: American Nuclear Society, 1982); Bertrand Goldschmidt, *Atomic Rivals* (New Brunswick, NJ: Rutgers University Press, 1990).

Canadian Nuclear Energy Program

The Canadian Nuclear Energy Program made Canada the third member of the nuclear club, behind the United States and the United Kingdom, but Canada had the advantage that the other members of the club—the Americans, the British, the French, the Soviets—never had, which was no pressure to develop a nuclear weapons program. Canada found itself in the middle of atomic weapons initiatives from both the American and British governments in World War II. In response, a team of British and French scientists were sent to Montreal in 1944 to work on experiments with heavy water. Soon Canada built a series of experimental natural uranium–heavy water reactors at a research center at Chalk River, Ontario. These reactors—Zero Energy Experimental pile (ZEEP), National Research X-perimental (NRX) Reactor, National Research Universal (NRU) Reactor—showed the potential for natural uranium–heavy water reactors for civilian power generation. An alliance was negotiated by the Atomic Energy of Canada Limited (AECL), which was a Canadian government corporation, for Canadian General Electric (CGE) to build and Ontario Hydro to operate facilities using Canadian nuclear reactors. Efforts to market Canadian nuclear technology abroad ran into stiff competition from reactors designed in the United States and the United Kingdom. Canada's reaction was to build a third-generation reactor—the Canadian deuterium-Uranium (CANDU). It was only with India that the Canadian government was able to negotiate a major deal, and it was for a CANDU reactor. Further sound relations ended with India, however, after the Indians exploded a nuclear device in May 1974 from plutonium gathered from the CANDU reactor. While minor deals were later worked out with Taiwan and Argentina, Canadian efforts to export their nuclear technology were largely unsuccessful. In the mid-1960s, Canada established another nuclear research center at Whiteshell, Manitoba. A series of power generating nuclear plants were built around Canada between 1960 and 1985—Rolphton, Douglas Point, Pickering, Bruce, and Gentilly I. Canada still has an active nuclear program, but it has been scaled down as nuclear energy has undergone troubled times.

Part of the reason for Canada's long participation in nuclear energy was its access to uranium ore. In the 1930s and 1940s, the Eldorado Mine near Great Bear Lake in the Northwest Territory was one of the three main sources for uranium ore in the world. While Canada contracted most of this ore to the United States for the Manhattan Project, it still made Canada a major player in the atomic energy business. Subsequent discoveries of uranium deposits at the Blind River area of Ontario in 1953 reinforced this status. This strike was estimated at the time to contain over a hundred million tons of ore. Although the fall in the price of uranium and the cancellation of contracts with the United States in the mid-1960s constituted a blow to the Canadian uranium industry, the accessibility of Canada to a ready supply of uranium ore has been a key factor in its continuing interest in the development of nuclear energy.

Suggested readings: Ronald Babin, *The Nuclear Power Game* (Montreal, Canada: Black Rose Books, 1985); Robert Bothwell, *Nucleus: The History of Atomic Energy of Canada Limited* (Toronto, Canada: University of Toronto Press, 1988); Wilfrid Eggleston, *Canada's Nuclear Story* (Toronto, Canada: Clarke, Irwin, 1965); Ron Finch, *Exporting Danger: A History of the Canadian Nuclear Energy Export Programme* (Montreal, Canada: Black Rose Books, 1986).

CANDU Reactor

The CANDU (Canadian Deuterium-Uranium) reactor was constructed at the Douglas Point nuclear power facility in the early 1960s. It was a pressure-tube, natural uranium fueled horizontal system of 200 megawatts (electric). This reactor used a uranium dioxide fuel with a zircaloy cladding. The site of this reactor was at Douglas Point on Lake Huron, about 10 miles north of the town of Kincardine, Ontario. Construction on the CANDU was begun in early 1960, but because of a series of construction delays, it started operations over six years later on November 15, 1966. The Douglas Point facility operated from 1966 until it was decommissioned in 1986. Ontario Hydro operated the reactor, but it refused to buy the facility from the Atomic Energy of Canada Limited (AECL). The CANDU reactor had extensive downtime during its lifetime because of design problems. Its problems were studied by nuclear engineers to find a way to avoid future design errors. They were successful, because it was a modified CANDU reactor that the Canadian government sold to the Indian government in the mid-1960s.

Suggested reading: Robert Bothwell, *Nucleus: The History of Atomic Energy of Canada Limited* (Toronto, Canada: University of Toronto Press, 1988).

Cavendish Laboratory

The Cavendish Laboratory at the University Cambridge has had a colorful past in research on the atom. Responding to the need for a chair of physics and an experimental laboratory, the Duke of Devonshire, William Cavendish, donated a large sum of money to Cambridge University in 1870 for both. James Clerk Maxwell was persuaded to assume the first Cavendish

Chair of Physics in 1871. The laboratory itself was under construction from 1871 to 1874 and cost considerably more than first envisaged, but the Duke of Devonshire covered the additional expense. The Cavendish name was decided upon to honor both William Cavendish and his relative Henry Cavendish, and eminent scientist. While only a limited number of researchers used the Cavendish Laboratory in its first decade, James Clerk Maxwell played an active role in supervising experiments. Maxwell's death in 1879 ended the first stage of the history of the Cavendish Laboratory.

The second phase was under the leadership of John William Strutt, Lord Rayleigh. Lord Rayleigh was a landed aristocrat with political connections to the top politicians of the day. His research had always been conducted in his private laboratory on his estate, but agricultural prices had collapsed, and he need additional income. His major contributions during his five-year tenure were to open the laboratory to women researchers and to develop a system of group classroom teaching to train experimenters. Lord Rayleigh resigned in 1884 to return to his estate and his private research projects.

Sir Joseph John Thomson became the director of the Cavendish Laboratory in 1885 and helped establish its worldwide scientific reputation. Thomson was a surprise choice as the Cavendish Chair, because of his youth and lack of experience as an administrator. In the first 10 years of Thomson's leadership, several promising lines of research were followed by his staff, but no major discoveries were forthcoming. The volume of work and number of research workers, however, had doubled. It was beginning in 1895 that the earlier research efforts started to produce results. For the next the years, major discoveries came routinely out of the laboratory, with Thomson receiving the Nobel Prize for his discoveries. At the same time, more research students started coming from abroad to study at the increasingly famous Cavendish Laboratory. Among those physicists not studying there in Thomson's day were Niels Bohr and Ernest Rutherford. The pace became so hectic that Thomson was unable to supervise researchers at his former high level, and several of his top researchers left to go elsewhere. By the last few years leading up to World War I, the leading edge of atomic research had shifted from Thomson at the Cavendish Laboratory to Rutherford at the University of Manchester.

The war almost shut down the Cavendish Laboratory. Because the scientists at the Cavendish had been engaged in pure, not applied, research, it did not occur to the British government to utilize the scientific talent there for the war effort. Individual scientists were sent to various scientific projects working on military weapons for the war effort, but others were simply wasted in the military. Only toward the middle of the war did government authorities begin to understand the need for close cooperation between scientific institutes and military projects. By the end of the war, the Cavendish Laboratory was at a low ebb, with few experiments going and outdated lab equipment. Moreover, in 1918 Thomson was appointed the master of Trinity College, and after some reluctance, he decided to give up the Cavendish professorship.

Thomson's successor in 1918 was Sir Ernest Rutherford, whose research and recruitment of key physicists made the Cavendish Laboratory one of the top research centers of its type in the interwar era. Rutherford's fame as an experimental researcher made him an obvious choice for the Cavendish Chair. The Cavendish Laboratory was showing its age and a lack of modern equipment when Rutherford took over in 1919. Rutherford attempted to persuade the university and the British government of the need for better financial support, but little was forthcoming from either source. Experiments at Cavendish during Rutherford's tenure continued to be simple; cheap equipment was designed by the researchers themselves. Despite its Spartan equipment, some of the great and near-great nuclear physicists conducted their experiments at the Cavendish. Some of the prominent Cavendish "boys" were Patrick Blackett, James Chadwick, John Douglas Cockcroft, Peter Kapitsa, Marcus Oliphant, and P. I. Dee. The Cavendish Laboratory continued to be an important center of research until World War II. During and after the war, the leading edge of atomic research transferred to the United States.

Suggested readings: James Gerald Crowther, *The Cavendish Laboratory, 1874–1974* (New York: Science History Publications, 1974); John Hendry, ed., *Cambridge Physics in the Thirties* (Bristol, UK): Adam Hilger, 1984); Alexander Wood, *The Cavendish Laboratory* (Cambridge UK: Cambridge University Press, 1946).

CERN

CERN (Conseil Européen pour la Recherche Nucléaire, or European Organization for Nuclear Research) is the European high-energy physics establishment housed in Geneva, Switzerland. It was formed on February 15, 1952, by a consortium of 11 European countries to coordinate advanced research in Europe on atomic energy. The American physicist, I. I. Rabi had proposed such an institute at the UNESCO (United Nations Educational, Scientific, and Cultural Organization) Assembly in Florence, Italy, in June 1950, and a resolution was passed by the assembly. Original members were Belgium, Denmark, France, Greece, Italy, the Netherlands, Sweden, Switzerland, West Germany, and Yugoslavia. Great Britain was initially reluctant to join, but it did so, and its scientists from Harwell helped in planning for CERN's high-particle linear accelerators. Soon the experiments at CERN allowed Europe to attain a status comparable to the United States in particle physics. It was further helped by the construction of large accelerators, culminating in a 500 GeV accelerator in 1977. This institute continues to be the leading research center in Europe for the study of subatomic physics.

Suggested reading: John S. Rigden, *Rabi: Scientist and Citizen* (New York: Basic Books, 1987).

Chadwick, James

Sir James Chadwick (1891–1974) was an eminent British physicist who worked on atomic research at the Cavendish Laboratory and later on the Manhattan Project. He was born on October 20, 1891, at Bollington in Cheshire, England. His father ran a laundry business in Manchester, and Chadwick's grandmother raised him. In 1925, he married Aileen Stewart-Brown, and they had two daughters. He won scholarship competitions to attend the University of Manchester. His inclination was to study mathematics, but he wandered into the wrong line and ended up in a physics course, and he stayed in physics. His baccalaureate was earned with first-class honors in 1911 from the University of Manchester; a master of science followed in 1912. After graduation, Chadwick served as a research assistant under Sir Ernest Rutherford. In 1913, Chadwick traveled to Berlin to work under Hans Geiger, but when the war broke out, he was interned at Ruhleben for the duration of the war. He and a fellow internee set up a small research laboratory at Ruhleben with the help of some German scientists. After the war, Chadwick returned to the University of Manchester before moving to Cambridge with Rutherford. Chadwick stayed at Cambridge as a researcher from 1919 to 1935.

It was in 1932 that Chadwick conducted the experiments that led to the discovery of the neutron. Rutherford had postulated the neutron's possible existence, but Chadwick proved its existence. On February 27, 1932, Chadwick published his discovery in the British journal *Nature*. For his research leading to the discovery of the neutron, Chadwick received the 1935 Nobel Prize for Physics. In 1935, the University of Liverpool offered him a professorship, and he remained there until 1948. It was at Liverpool soon after his arrival that Chadwick built Britain's first cyclotron. In the middle of his Liverpool stay, however, he served as an administrator with the British Mission to the Manhattan Project. He developed a strong friendship with General Leslie Graves, and the good relationship between American and British scientists working on the Manhattan Project was partly the result of his efforts. After the war, he became an influential adviser in the British effort to build an atomic bomb. In 1948, Chadwick returned to Cambridge University, where he stayed until his retirement in 1959. He also served as a part-time member of the United Kingdom's Atomic Energy Authority. On July 24, 1974, Chadwick died in Cambridge, England.

Suggested readings: Henry A. Boorse, Lloyd Motz, and Jefferson Hane Weaver, *The Atomic Scientists: A Biographical History* (New York: Wiley Science Editions, 1989); Margaret Growing, *Independence and Deterrence: Britain and Atomic Energy, 1945–1952*, vol. 2 (New York: St. Martin's Press, 1974); Emilio Segre, *From X-Rays to Quarks: Modern Physicists and Their Discoveries* (San Francisco: W. H. Freeman, 1980).

Chain Reaction

A nuclear chain reaction enables the energy stored in a fission fuel to be released in either a rapid or a controlled manner. In its natural environment, uranium releases its stored energy very slowly, over a period of millions of years, by spontaneous fission of the heavy uranium nuclei into two lighter nuclei, plus a few neutrons. Uranium occurs naturally in a mixture mainly of two isotopes, or forms: U-235 and U-238. U-235 has the special property of undergoing immediate fission when it comes in contact with a neutron. The leftover neutrons from the fission of one U-235 nucleus can stimulate the fission of a second one, and so on. On average, 2.5 neutrons are released in every U-235 fission. If one spontaneous fission occurs in a sufficiently pure sample of U-235, and it gives rise to 2 to 3 neutrons that stimulate fissions of their own, and each of these generate 2 to 3 more, in turn, before 100 such steps, enough fissions have taken place to consume all the U-235 present in the sample. This is a rapid chain reaction, such as takes place in an atomic bomb. A controlled chain reaction is achieved by reducing the size of the sample to allow some neutrons to escape and surrounding the sample with a material that absorbs neutrons, thus reducing the average number of neutrons from one fission available to cause reactions at the next step, to down near the critical value of 1. If the number of neutrons available to induce fissions at the next step is less than an average value of per decay, then the sample is subcritical. If the number is more than, a rapid chain reaction will result in an explosion. For a controlled chain reaction, that number must be maintained very close to the critical value of 1.

One of the key problems of building and exploding an atomic device was determining what could cause a reaction and how the chain reaction could be limited. The problem that bothered physicists in the late 1930s and early 1940s was the theory that once a chain reaction commenced, unless an inhibitor could be introduced, a chain reaction would go critical. This issue became even more worrisome in the theory of a fusion explosion. Until this problem was worked out in theory, some scientists were hesitant to continue research. The first self-sustaining chain reaction was Enrico Fermi's Pile No. 1 at the University of Chicago in 1942, and it proved that a chain reaction could be controlled. It takes around 30 pounds of uranium-235 and about 5 pounds of plutonium-239 to constitute a critical mass.

Suggested reading: Richard Wolfson, *Nuclear Choices: A Citizen's Guide to Nuclear Technology* (Cambridge: MIT Press, 1991).

Chalk River Facility

The Chalk River Facility at Petawawa, about 140 miles west of Ottawa near the small town of Deep River in Ontario, Canada, is the site of an experimental nuclear reactor run by the Canadian government. This facility had originally been a joint project of Canada and Great Britain under the directorship of British physicist John Cockcroft. He entrusted Lew Kowarski with directing a team to construct a small research reactor. After Cockcroft left in September 1946, the Canadian government assumed full control of the facility. In September 1945, a small experimental reactor, ZEEP, or Zero Energy Experimental Pile, had been built under the supervision of Lew Kowarski. Then in 1947 a larger test reactor, the National Research X-perimental (NRX) reactor, was constructed. The NRX was a reactor that used natural uranium and heavy water. This reactor contained about 10 tons of heavy water and about the same amount of uranium metal. Only a few grams of plutonium were available at the time in 1944, because American authorities refused to share the processes with the British or the Canadians. It was at this site that Admiral Rickover had his experimental nuclear design for a nuclear-powered submarine tested. During the experiments, the existence of a buildup of a black deposit on the surfaces of the fuel element was discovered. This deposit was named *crud* (Chalk River unidentified deposit). It took awhile, but the crud problem was eventually solved. In December 1952, a nuclear reactor accident involving the heavy water research reactor took place as a result of an unplanned power increase caused by operator error leading to a partial meltdown. The accident caused enough damage that the reactor had to be shut down for two years for repairs. Chalk River continued to be Canada's leading reactor research center for the next three decades. Among the reactor designs tested here were the NRU (National Research Universal Reactor) and the CANDU (Canadian Deuterium-Uranium Reactor).

Suggested readings: Bertrand Goldschmidt, *Atomic Rivals* (New Brunswick, NJ: Rutgers University Press, 1990); Theodore Rockwell, *The Rickover Effect: How One Man Made a Difference* (Annapolis, MD: Naval Institute Press, 1992).

Chazhma Bay Submarine Accident

A Soviet submarine, Echo II class K-431, had a nuclear accident during reactor refueling at the naval shipyard on Chazhma Bay near the Siberian town of Dunay on August 10, 1985. An uncontrollable spontaneous chain reaction caused an explosion that killed 10 people in the reactor room of the submarine, and around 290 people were exposed to radiation. Radiation caused 10 people to come down with acute radiation sickness and another 39 with a severe reaction. The submarine experienced major damage, with bulkheads destroyed and the hull breached in the aft section of the reactor compartment. Large areas of water around the submarine became contaminated with radioactivity, but most of the radioactive debris sank to the bottom or was dispersed in ocean currents. The submarine was so contaminated that the reactor compartment was sealed with concrete and the hole in the hull welded shut. It was decommissioned and is currently at Pavlovsk.

Suggested reading: Thomas B. Cochran, Robert S. Norris, and Oleg A. Bukharin, *Making the Russian Bomb: From Stalin to Yeltsin* (Boulder, CO: Westview Press, 1995).

Chelyabinsk-40

Chelyabinsk-40 is one of the most important plutonium production centers in the Soviet Union and the site of a nuclear waste explosion. It is located about 15 miles east of Kyshtym and near Kasli besides Lake Kyzyltash in Chelyabinsk province. A large facility and electrical power were already there in November 1945, because there had been a factory producing tanks during the war. Moreover, near Kyshtym was a major gulag station with 12 labor camps full of slave labor. The site was selected by Avraami Zaveniagin, Deputy People's Commissar of Internal Affairs (NKVD) under Beria. A closed city first named Beria and then changed to Ozersk was built to house the Chelyabinsk-40 workforce. As many as 70,000 prisoners worked at one time to build the facilities. This facility was under the administrative control of the Mayak Chemical Combine. Construction of converting the plant to a plutonium production and extraction facility began in earnest in late November 1945 under the command of a major-general in the NKVD, Iakov Rappoport. A lake was drained and a storage facility was built, and then the lake was refilled. Nuclear reactors were built in proximity to the underwater storage facility. Igor Kurchatov and Boris Vannikov moved to Chelyabinsk-40 to oversee the construction of the first reactor. The first plutonium-production reactor (A reactor) was operational in June 1948. Soon, however, the reactor had to be shut down until the swelling of uranium metal slugs in high-flux reactors was solved. This cost the Soviets almost a year before full production could start again. A radiochemistry laboratory was also constructed (B reactor), along with a metallurgical laboratory (V reactor).

Chelyabinsk-40 always had a problem with nuclear waste products. Radioactive wastes were placed in storage tanks or into a nearby pit. During 1957, the plutonium processing plant was kept at full capacity. On September 29, 1957, the nuclear waste reacted to moisture from a failed cooling water malfunction and went critical. A nuclear explosion of about 70–80 tons of highly radioactive waste resulted in which most of the staff was killed outright and others died of radiation sickness shortly thereafter. Huge amounts of strontium-90 and saaesum-137 were blown over a 750–1,500 mile radius. Hundreds of Soviet scientists and technicians died, and the facility had to be abandoned. An exclusion zone had to be established and an indeterminate number of habitants were relocated away from the contaminated zone. A report of this nuclear accident reached American intelligence circles not long after it happened. The report stated that sometime in the winter of 1957–1958 a nuclear explosion of an indeterminant nature had taken place at a nuclear storage installation at Kasli. A report of the Kasli disaster circulated in American intelligence circles but was never verified until an article in 1976 by Soviet scientist A. Zhores Medvedev mentioned it. The Kasli-Kyshtym area was one of the mission targets of Francis Gary Powers (American Central Intelligence Agency [CIA] pilot) when his U-2 plane was shot down over the Soviet Union on May 1, 1960.

Later this facility was rebuilt, and in 1990, it was renamed Chelyabinsk-65. Between 1948 and November 1990, Chelyabinsk-40 produced plutonium for nuclear weapons by five water-cooled, graphite-moderated production reactors. These reactors have all been decommissioned. Since 1990 the scientific emphasis at Chelyabinsk-65 has changed to fabrication of naval reactor fuel and civilian projects.

Suggested readings: Thomas B. Cochran, Robert S. Norris, and Oleg A. Bukharin, *Making the Russian Bomb: From Stalin to Yeltsin* (Boulder, CO: Westview Press, 1995); David Holloway, *Stalin and the Bomb: The Soviet Union and Atomic Energy, 1939–1956* (New Haven, CT: Yale University Press, 1994); Zhores Madvedev, *The Legacy of Chernobyl* (New York: Norton, 1990); Nikolaus Riehl and Frederick Seitz, *Stalin's Captive: Nikolaus Riehl and the Soviet Race for the Bomb* (Washington, DC: American Chemical Society, 1996).

Chelyabinsk-70

Chelyabinsk-70 is a nuclear weapons design laboratory that fulfilled the same function in the Soviet Union as the Lawrence Livermore National Laboratory in the United States. It is the home of the All-Russian Scientific Research Institute of Technical Physics. Chelyabinsk-70 was founded in 1955 near the town of Snezhinsk, which is located between Lakes Sinara and Silach, just east of the Urals. A gulag-administered scientific research facility was there with the designation of Site 21. The facilities were appropriated, and a large number of scientists from Sarov were relocated to Chelyabinsk-70. At one time the institute employed as many as 16,000 people, of which about a fourth were scientists. Unlike at Sarov, there had been no production facilities at Chelyabinsk-70. From 1955 to 1988, the facilities concentrated almost exclusively on nuclear weapons design, but since then the institute has begun a conversion to nonweapons work. Most recent research has been nonmilitary commercial projects in fiber optic communications, nuclear medicine, and industrial diamond manufacture. Kirill I. Shchelkin was the first scientific director at Chelyabinsk-70.

Suggested reading: Thomas B. Cochran, Robert S. Norris, and Oleg A. Bukharin, *Making the Russian Bomb: From Stalin to Yeltsin* (Boulder, CO: Westview Press, 1995).

Chernobyl Nuclear Power Station Accident

The nuclear accident at the Russian Chernobyl Nuclear Power Station on April 25–26, 1986, was the most serious accident in nuclear history. This plant is located on the banks of the Prypiat River in northern Ukraine and on the border with Byelorussia, about 90 miles north of Kiev. Initial construction of the facility had begun in the 1970s with units coming on-line at intervals: Unit 1 in 1977, Unit 2 in 1978, Unit 3 in 1981, and Unit 4 in 1983. By the mid 1980s, the plant had four graphite-moderated, water-cooled, and uranium-fueled (low-enriched uranium oxide) reactors, or RBMK-1000 type (reactor high-power boiling channel type), and two more under construction. In case of overheating in the reactor core, each Chernobyl reactor had an emergency cooling system to inject cold water to any part of the core and to shut the reactor down. There was also a computer system to monitor problems. Despite these safety features, Soviet authorities had been warned about design problems in the RBMK-1000 reactors. One problem with this design is its vulnerability to leads and faults in its thousands of pipes and welded joints. Another is its vulnerability to interruption of electricity. The biggest weakness, however, is the lack of a protective containment structure to control damage in case of an accident. Regardless of these deficiencies, at the time when the experiment on the reactor was scheduled, neither the plant director, Viktor Bryukhanov, nor the chief engineer, Nikolai Fomin, were present. The plant had been left in the hands of an electrical engineer without previous experience at a nuclear power plant. It has also been suggested but not proven that this test was part of the initial safety tests required before commencing operations in 1984 that had been postponed for political reasons.

A safety experiment with Unit 4 on Friday April 25, 1986, near midnight to correct an electricity gap precipitated the accident. The original testing schedules started the test around 1:00 P.M. in the afternoon. Consequently, the operators turned off the emergency water-cooling system so that it would not unnecessarily flood the core and shut down the reactor. An order from the Kiev regional grid controller curtailed the experiment because of the need for more electricity, but the emergency water-cooling system was not switched back on. The experiment started again at 11:10 P.M., but this time with a new crew of operators. This new crew made the mistake of not setting the automatic power despite knowledge that below a certain level the reactor became unstable. As the power level dropped, the operators tried to force it back up by withdrawing the control rods. They tried to use more

Both the size of the Chernobyl Nuclear Power Station and the extent of the damage to Unit 4 can be seen in this photograph. The nuclear meltdown at Chernobyl was the most serious accident in nuclear history. (Photo courtesy of AP/Wide World Photos)

steam pressure which also backfired. They then disengaged the emergency protection control attached to the steam drums. By 1:22 A.M., April 26, the computer system informed the operators that the reactor had to be shut down. The operators ignored this warning and initiated the original experiment, but at the same time they blocked the trip that would shut down the reactor when the Number 8 turbine generator was switched off for the experiment.

Almost as soon as the experiment commenced, a sudden power surge came from the reactor. In a panic now, the operators hit the emergency button to lower all the control rods into the core and shut down the reactor. Although the rods fell, they only went in partially. It was already too late, because the 1,661 fuel rods broke containment and the interaction between the hot fuel and water created a huge explosion. With the core now exposed to air from the explosion, a second explosion took place, and a ball of fire burst over the reactor core. It has been estimated that a quarter of the 2,000-ton core was blown outside the reactor vault. Since there is evidence from the fallout of traces of ruthenium (Ru-103 and Ru-106), which has a melting point of 2,225.0 degrees Celsius, the reactor core must have reached this temperature in causing a meltdown.

Now the biggest problem was an uncontrolled fire, which threatened the reactor in Unit 3. Firefighting teams were rushed in, but because of the high temperatures in the burning core the water had no impact. Gradually, however, they were able to contain the spread of the fire. It took 37 fire crews with 186 firemen and 81 machines to put out the spreading fires. Many of the firemen died of radiation poisoning in the next few weeks. To further control the burning core, the Russian Air Force bombed the exposed core with lead, sand, clay, dolomite, and boron carbide. By May 2, the air force had dropped more than 5,000 tons of materials. It also took other steps including draining water close to the reactor core. None of the steps seemed to help as the core continued to heat up. Unless the temperature could be contained, there was the danger of a further meltdown. Russian authorities managed to contain this threat, but only barely. Only after May 10 did the temperature begin to drop. This temperature drop puzzled the scientists, but they welcomed it nevertheless. Only after building a sealed containment building around Unit 4 in the next few weeks was the problem solved permanently.

The next problem facing Soviet authorities was the danger of radioactive fallout. It has been estimated that somewhere in the neighborhood of 396 MCI (Ci is a curie, which equals 1 billion nuclear disintegrations per second, and MCI is million curies) of radioactive materials escaped into the environment. Of this total, 276 MCI left the reactor in the form of radioactive gases and stayed in the atmosphere. Radioactive particles, which fell back to earth ranged between 25 and 75 MCI. The entire reactor had an activity of 5,750 MCI.

A large radioactive cloud formed over Chernobyl and began to drift slowly over the northern part of eastern Europe. Several eastern European countries started detecting the radioactive cloud, but for political reasons no announcement was forthcoming. Only when the cloud reached Sweden did an outcry result. Swedish scientists noted that radioactive plume at 7:00 A.M. on April 28 and immediately demanded an explanation from the Soviet government. From the composition of the clouds, these scientists were able to conclude that a nuclear reactor meltdown had occurred. Eventually the cloud extended over almost all of northern Europe, including most of Germany, Norway, Sweden, Denmark, Great Britain, and part of France. A late cloud covered most of southeastern Europe, including all of the Balkan countries and Greece.

Immediately after the accident Soviet authorities began a campaign to control information. After casualties at Chernobyl had reached 31 deaths, no further victim rates were announced. All such death rates became classified, even though rough estimates place the death toll as high as 2,000. These casualties were attributed to radiation exposure in the first few months after the accident. Estimates vary widely about the long-term health effects, but the slowness of Soviet authorities to evacuate civilians near Chernobyl will have an impact in the future. By 1990, the death total of those who had worked closely in the initial action against the accident had risen to over 5,700 and health problems were increasing for the survivors. Moreover, 130,000 people had to be resettled from the Chernobyl area and a large-scale reclamation of the land surrounding the site was undertaken. An exclusion zone of around 25 miles was established around Chernobyl on May 3, 1986.

An investigation to claim responsibility for the accident was launched by Soviet authorities soon after the immediate crisis was over. Several top administrators of the Soviet atomic energy program lost their jobs. In July 1987, six senior officials held directly responsible for the Chernobyl disaster were tried and convicted of criminal negligence. The former director and his two primary assistants were sentenced to 10 years of hard labor, and the other three to lesser sentences. All of the operators in the crew conducting the experiment had died with days of the accident. A further inquiry about the future of the RBMK reactor led to it being phased out of production.

Suggested readings: Viktor Haynes and Marko Bojcun, *The Chernobyl Disaster* (London: Hogarth Press, 1988); David R. Marples and Marilyn J. Young, eds., *Nuclear Energy and Security in the Former Soviet Union* (Boulder, CO: Westview Press, 1997); Grigori Medvedev, *No Breathing Room: The Aftermath of Chernobyl* (New York: Basic Books 1993); Grigori Medvedev, *The Truth About Chernobyl* (New York: Basic Books, 1991); Zhores Medvedev, *The Legacy of Chernobyl* (New York: Norton, 1990); Alla Yaroshinska, *Chernobyl: The Forbidden Truth* (Oxford, UK: Carpenter, 1994).

Chicago Pile One

Chicago Pile One (CP-1) was the first operational atomic reactor. It was built at the University of Chicago's squash courts under the west sidelines of the Stagg Football Stadium. Original plans had been to build the reactor at the new Argonne facility, but a construction strike ended that possibility. Arthur Compton, the head of the Metallurgical Laboratory, approved the transfer back to the University of Chicago campus without informing the president of the university, Robert Maynard Hutchins. Construction started on November 16, 1942, under the direct supervision of Enrico Fermi, with Walter Zinn in charge of building the pile. Two shifts were instituted: a day shift under Walter Zinn and a night shift under Herbert Anderson. It took 771,000 pounds of graphite, 80,590 pounds of uranium oxide, and 12,400 pounds of uranium metal to build the pile. Purified uranium was placed in the middle and uranium oxide on the outer positions. Two crews worked around the clock in the November cold, because the scientists refused to ask for heat for security reasons. The pile was ready on the morning of December 2, 1942. Fermi had the rod controlling the chain reaction slowly removed over a period of several hours. Each time the rod was moved he would make calculations with his slide rule before allowing the next move. By the afternoon 42 onlookers had assembled to watch the experiment. Finally, around 3:53 P.M., Fermi allowed the reactor to become self-sustaining. He let the pile run for 4.5 minutes before shutting it down and proclaiming the experiment a success. It had cost about $1 million to build and test the reactor, but it had produced a chain reaction. The participants celebrated with a bottle of Chianti. Compton informed James Conant shortly afterward that the experiment had been a success. This success caused the political authorities to decide to proceed with the Manhattan Project.

Suggested readings: Stephane Grouef, *Manhattan Project: The Untold Story of the Making of the Atomic Bomb* (London: Collins, 1967); Richard Rhodes, *The Making of the Atomic Bomb* (New York: Simon and Schuster, 1986).

China Syndrome

The China Syndrome is the popular idea that a nuclear reactor could go out of control and become so hot that it could eat into the earth's core. This concept had been studied seriously by the Advisory Committee on Reactor Safeguards (ACRS) in the mid-1960s. Several scenarios had been considered, such as bolts breaking and allowing the top head of the reactor to fly off, control rod mishaps, or an untested emergency core cooling system. Members of the ACRS wanted the nuclear energy industry to develop a better emergency system to prevent the possibility of a China Syndrome incident. It was this type of meltdown that was the source of a successful movie also entitled the *China Syndrome*. The movie was loosely based on several accidents that took place in the early 1970s at the Rancho Seco nuclear power plant in California. It was a hit and was still playing in theaters around the country when the Three Mile Island accident took place. Its coauthor, Mike Gray, had a nuclear engineering background. He reported on the Three Mile Island accident for *Rolling Stone* magazine. After the Three Mile Island accident, the movie received so much notoriety that it ended up grossing more than $100 million in the United States.

Suggested reading: Daniel F. Ford, *The Cult of the Atom: The Secret Papers of the Atomic Energy Commission* (New York: Simon and Schuster, 1982).

Chinese Nuclear Program

The Chinese nuclear program began in the mid-1950s with the help of the Soviet Union. Chinese scientists were trained at the Nuclear Research Laboratory at Dubna starting in 1954. As many as a thousand Chinese scientists and technicians passed through this laboratory. Initial impetus came from the transfer from the Soviets of a 6.5-megawatt nuclear reactor and plans for a gaseous diffusion uranium enrichment plant at Lanchow. Both plants were under construction in the period from 1955 to 1958. A falling out between the two allies in 1958 caused the Soviets to suspend technical assistance, but the Chinese were able to finish both facilities. It was materials from these plants that permitted the Chinese to test atomic weapons in

October 1964. China detonated an atomic bomb on October 15, 1964, at Lop Nur in the Sinkiang desert in northeastern China. Unlike the tests of other countries, this bomb contained uranium-235. This surprising development meant that the Chinese had discovered large quantities of uranium within China and that the Chinese were further along in nuclear technology than outsiders believed. Further tests in 1965 and 1966 reinforced both views. Then on June 17, 1967, the Chinese exploded an airborne hydrogen bomb. The leading scientists in these efforts were French-trained physicist Ch'ien San-chang and German-trained physicist Wang Kan-chang. Nieh Jung-chen became the military administrator of the Chinese Atomic Bomb Program in 1955. Nieh's military orientation characterized the Chinese program, because not until the 1970s did the Chinese devote any attention to civil power for electricity. China had three commercial reactors in operation in 1996 and two others under construction to be completed sometime in 2002. At present, reactors provide just a little over 1 percent of China's electricity. China's last atmospheric nuclear test was in October 1980. Throughout the 1990s, the Chinese have conducted underground nuclear tests at the Lop Nur Testing Site. Its last nuclear test took place on July 29, 1996, with the announcement that China would stop testing as soon as the Comprehensive Test Ban Treaty went into effect. This last test was the forty-fifth nuclear test in the series.

Suggested reading: Eric Arnett, ed., *Nuclear Weapons After the Comprehensive Test Ban: Implications for Modernization and Proliferation* (Oxford: Oxford University Press, 1996); Bertrand Goldschmidt, *The Atomic Complex: A Worldwide Political History of Nuclear Energy* (La Grange Park, IL: American Nuclear Society, 1982); *Nuclear Power Reactors in the World* (Vienna, Austria: International Atomic Energy Agency, 1997); Peter Pringle and James Spigelman, *The Nuclear Barons* (London: Michael Joseph, 1981).

Christmas Island Tests

Christmas Island in the Pacific was the site of British hydrogen bomb testing in 1957 and 1958. Earlier atomic bomb testing had been held in Australia, but part of the agreement between Great Britain and Australia was that hydrogen bomb testing be prohibited in Australia. Testing of the fission device to trigger the hydrogen bomb was tested in Australia at Maralinga and Monte Bello sites. Christmas Island is part of the Northern Line Island group, situated just south of the equator in the middle of the Pacific Ocean. Hawaii is 1,000 miles to the north, and Fiji is 1,500 miles to the southwest. The island is the largest coral atoll in the Pacific, and it had been discovered in 1777 by captain Cook. A base for hydrogen bomb testing was established here beginning in June 1956. Malden Island, which is 400 miles south of Christmas Island, was used as a subbase camp. Britain's first hydrogen bomb test

Great Britain conducted a series of hydrogen
bomb tests in the Christmas Islands region of the
Central Pacific in the mid-1950s. This test was
the third in the series and took place in June
1957. (Photo reproduced from the Collection of
the Library of Congress)

(Operation Grapple) was on Malden Island on May 15, 1957. It was a bomb
dropped from a Valiant bomber and detonated at 15,000 feet. Two other
hydrogen devices were detonated on Malden Island in June 1957. The next
series of hydrogen bomb tests moved to Christmas Island. On November
8, 1957, a hydrogen bomb was detonated over Christmas Island, and it was
much more powerful than previous devices. A total of seven tests of hydrogen bombs and two atomic bombs took place at Christmas Island beginning
on November 8, 1957, and ending on September 23, 1958. Smaller atomic
tests continued until 1963. A series of 24 American tests, Operation Dominic, were conducted between April and July 1962. All personnel left the
Christmas Islands in 1964, and British cleanup of radioactive materials lasted
until 1967. A number of British servicemen sued the British government
for health problems resulting from their participation in these tests, but the
courts were reluctant to award damages.

Suggested readings: Denys Blakeway and Sue Lloyd-Roberts, *Fields of Thunder:
Testing Britain's Bomb* (London: George Allen and Unwin, 1985); John May, *The*

Greenpeace Book of the Nuclear Age: The Hidden History, the Human Cost (New York: Pantheon Books, 1989).

Churchill, Winston

In his role as British prime minister and politician Winston Churchill (1874–1965) had little understanding of atomic theory, but he championed the building of an atomic weapon because of scientific advice and foreign policy considerations. In 1940, his scientific adviser and good friend Frederick Lindemann, Lord Cherwell, took the MAUD Report and reduced and simplified it for Churchill. Lindemann's only modification was to recommend that development take place in Great Britain rather than in the United States. Great Britain was in no economic or military shape to carry out such an ambitious program, and Churchill knew this. Churchill appointed Sir John Anderson to direct the British atomic energy program at the ministerial level. Wallace Akers, a senior Imperial Chemical Industries (ICI) administrator, headed the administrative side of this program. An effort by Niels Bohr to persuade Churchill to internationalize the atomic energy after the war and to inform the Soviets about the atomic bomb program were repudiated by Churchill. Churchill used his close ties with President Roosevelt to make the British a junior partner with agreements to share in the postwar benefits of atomic energy. Two factors intervened: Roosevelt's death and Churchill's loss of office in 1945. After it became apparent that the Americans were going to keep their nuclear secrets, Churchill lobbied the British government to launch its own nuclear program. It was not until he regained office that Churchill found out that his advice had been followed. Churchill supported the British nuclear weapons program because he considered it an indicator of great power status.

Suggested readings: Bertrand Goldschmidt, *Atomic Rivals* (New Brunswick, NJ: Rutgers University Press, 1990); Richard Rhodes, *The Making of the Atomic Bomb* (New York: Simon and Schuster, 1986).

Civil Defense

Civil Defense was a federal program to find ways for the American general public to survive a nuclear attack. After World War II, the U.S. government realized that atomic warfare was possible. Then the Soviet Union detonated an atomic bomb in 1949. Authorities in the United States soon thereafter set up the Federal Civil Defense Agency to plan for nuclear attack and other disasters. The Korean War only intensified disquiet over civil defense. Feeble efforts to implement ways for the American public to survive were attempted, but there was little funding for more than planning and shelter signs. In 1954 the Eisenhower administration conducted Operation Alert, which was an exercise drill in over 50 cities. Perhaps more important was a government pamphlet entitled *Survival Under Atomic Attack*. This pamphlet was used as a fourth-grade textbook in public schools around the nation. Then in the early 1960s the fallout shelters craze hit the United States, fueled by President John Kennedy's veiled reference to the need of such shelters in a Berlin Crisis speech. President Kennedy was opposed to a multibillion dollar government-sponsored program, but he could encourage Americans to take care of themselves. After this fad fell in popularity, little was done on civil defense in the next decade. In 1979 the Federal Emergency Management Agency (FEMA) replaced the Federal Civil Defense Agency, but little changed between the two organizations.

Suggested readings: John Dowling and Evans M. Harrell, eds., *Civil Defense: A Choice of Disasters* (New York: American Institute of Physics, 1987); Spencer R. Weart, *Nuclear Fear: A History of Images* (Cambridge: Harvard University Press, 1988).

Clamshell Alliance

The Clamshell Alliance was the most important antinuclear group in New England, and it served as a model for protest tactics against the Seabrook Nuclear Power Station. It was formed on July 11, 1976 at Rye, New Hampshire, under the auspices of Guy Chichester and was a coalition of opponents of the Seabrook plant. Somewhere between 50 and 60 persons attended the organizational meeting, but they were united in a desire to engage in protest activity. They named their group the Clamshell Alliance to symbolize local fears that pollution from the plant would destroy nearby clam beds and to emphasize an organization of independent entities. The Clamshell Alliance had no formal officers but did have a hired staff to coordinate communications between groups. Decisions were made by a coordinating committee with reference to regional bodies and sometimes took weeks or months. From the beginning the organization espoused nonviolent civil disobedience as the most effective tactic against the Seabrook plant. In a series of small demonstrations starting in August 1976 the members of the Clamshell Alliance learned nonviolent techniques. The climax came when the Clamshell Alliance organized a mass protest and occupation at the Seabrook Nuclear Power Plant in April 1977 that led to the arrest of 1,414 people for criminal trespass. This action led to national media exposure, and soon groups around the country began imitating the tactics of the Clamshell Alliance. Ultimately it was the success of the Clamshell Alliance tactics bringing in new supporters and the resulting internal division on continuing civil disobedience in the face of increased hostility from state authorities that caused the Alliance to fractionalize into competing groups. The Three Mile Island nuclear accident in March 1979 revitalized the Clamshell Alliance temporarily, but it also placed nuclear power on the national agenda and hurt the regional antinuclear groups like the Clamshell Alliance. By early 1980, enough members had left the Clamshell Alliance for other antinuclear organizations that it ceased to exist except in name only. Its legacy remains, however, as the most successful of the early antinuclear organizations in defining strategies and tactics.

Suggested readings: Ethan M. Cohen, *Ideology, Interest Group Formation, and the New Left: The Case of the Clamshell Alliance* (New York: Garland Publishing, 1988); Michael Morgan and Susan Leggett, eds., *Mainstream(s) and Margins: Cultural Politics in the 90s* (Westport, CT: Greenwood Press, 1996).

Clarendon Laboratory

The Clarendon Laboratory at Oxford University rivaled Cambridge's Cavendish Laboratory in experimental physics in the late 1930s and in the 1940s. Funds for the Clarendon Laboratory came from the sale of the Clarendon State Papers in 1751, but it took over 100 years after the sale before the laboratory opened in 1870. R. B. Clifton was appointed professor of experimental philosophy at Oxford University in 1865 and assumed the head of the laboratory at its opening. Clifton's emphasis on teaching over experimental research and his lengthy tenure of over 50 years made the Clarendon Laboratory almost a joke as a research facility. The change came when Frederick Lindemann assumed the professorship of experimental philosophy in 1919 and took charge of the Clarendon Laboratory.

Lindemann spent the next 20 years attempting to turn the Clarendon Laboratory into a world-class research facility. In the early years, progress was slow, but Lindemann had many contacts in government and business. At a time when the Cavendish Laboratory was world famous for its nuclear research, Lindemann was trying to find a scientific niche for the Clarendon. He had even suggested to Rutherford that the two laboratories specialize in different types of experimentations, but nothing came of these conversations. It was not until after the Nazis gained control of Germany in 1933 and the resulting flood of Jewish physicists that Lindemann was able to attract the top names for the Clarendon. To do this, however, Lindemann had gone to the Imperial Chemical Industries (ICI) and convinced the directors to set up a grant program. These funds allowed him to appoint these emigrant physicists. Beginning in the mid-1930s and lasting to the mid-1950s, the Clarendon became heavily involved in nuclear physics. Lindemann's absence during World War II as Churchill's science adviser hurt the functioning of the Clarendon Laboratory little.

Suggested readings: Roy Harrod, *The Prof: A Personal Memoir of Lord Cherwell* (London: Macmillan, 1959); C. P. Snow, *Science and Government* (Cambridge: Harvard University Press, 1961).

Clinton Laboratories. *See* Oak Ridge National Laboratory

Cockcroft, John Douglas

Sir John Cockcroft (1897–1967) was a British physicist whose work on the transmutation of atomic nuclei made him one of the top researchers on the structure of the atom in Great Britain. He was born in Todmorden, Yorkshire, England, on May 27, 1897, into a family of mill owners. His marriage to Eunice Elizabeth Crabtree in 1925 produced four daughters and one son. Shortly after entering the University of Manchester in 1914 to study mathematics, Cockcroft joined the British army, serving mostly as an enlisted man in the field artillery for the duration of the war. By the end of the war, he was a second lieutenant with several decorations. After the war, he worked as an apprentice with the engineering firm Metropolitan Vickers. This company sponsored him to study electrical engineering at the Manchester College of Technology. After obtaining his engineering degree, Cockcroft studied at Cambridge University and received a degree in mathematics. He joined the Cavendish Laboratory as a researcher working under Ernest Rutherford. It was at the Cavendish Laboratory that he conducted experiments with E. T. S. Walton on building a voltage multiplier to bombard the nuclei of lithium with protons. In 1932 the team was successful in bringing about the first nuclear transformation by artificial means. Both Cockcroft and Walton shared the 1951 Nobel Prize for Physics for this research. Cockcroft specialized in translation of scientific research into engineering projects.

During World War II, Cockcroft was busy with war research. He was instrumental in the development of radar for the British military forces. In 1940 Cockcroft was a member of a visiting mission to the United States to negotiate exchanges of military and scientific information. It was on this mission that Cockcroft began to develop good connections with American scientists that were to help Anglo-American cooperation on atomic energy during the war. In 1944 he became the director of the Anglo-Canadian Atomic Energy Commission in Montreal. Cockcroft was also instrumental in building the Chalk River Nuclear Facility in Canada.

After World War II, Cockcroft was named director of the Atomic Energy Research Establishment (AERE) at Harwell. He assumed the directorship in 1946 and remained in charge until 1959. He left this position to become master of Churchill College, Cambridge University. Cockcroft died in Cambridge on September 18, 1967.

Suggested readings: Henry Boorse, Lloyd Motz, and Jefferson Hane Weaver, *The Atomic Scientists: A Biographical History* (New York: Wiley Science Editions, 1989); Guy Hartcup and T. E. Allibone, *Cockcroft and the Atom* (Bristol, UK: Adam Hilger, 1984); Spencer R. Weart, *Scientists in Power* (Cambridge: Harvard University Press, 1979).

Code 390–590

Code 390 (Nuclear Power Division) was the US Navy's organization in charge of developing nuclear propulsion systems for submarines and surface vessels. It was formed by the navy's Bureau of Ships in the summer of 1948 and placed under the command of Captain Hyman Rickover. Rickover used this organization to form a team of naval officers and civilians to become experts in the design and operation of nuclear reactors for ship propulsion. He recruited men with technical expertise and engineering backgrounds and then subjected them to rigid training in nuclear technology. All of them attended the Oak Ridge School of Reactor Technology. Most of the naval officers trained in this fashion obtained important positions in the nuclear navy in their subsequent careers or had successful careers in the nuclear industry.

Rickover oversaw all aspects of Code 390, from training to reactor design. He conducted meetings during which technical expertise rather than naval rank determined decisions. Although this management system often produced friction among its participants, it also allowed a careful examination of technical issues. It soon became apparent that there were two candidates for a submarine reactor: Westinghouse's Mark I and General Electric's Mark A. The Mark I was a pressured water thermal reactor, and the Mark A was a sodium, beryllium intermediate reactor. Both designs had problems such as the Mark I's piping requirements and the Mark A's sodium instability with water. Rickover decided in 1950 to build and test both reactors, and he selected the Electric Boat Company of Groton, Connecticut, to build the submarines. Testing of both reactors took place at the National Reactor Testing Station in Idaho. Because the Mark I was further along in development and a simpler reactor in design, it was tested first and with success. A modified Mark I, the Mark II, was the reactor installed in the first nuclear submarine, the SSN *Nautilus*. An upgraded Mark A, the Mark B, was used in the second nuclear submarine, the SSN *Seawolf.* In early 1950, the Code 390 was changed to Code 490 by the Bureau of Ships, and then in July 1954 to Code 590. Rickover retained his position as its head and received a promotion to admiral in 1955.

Suggested reading: Richard G. Hewlett and Francis Duncan, *Nuclear Navy, 1946–1962* (Chicago: University of Chicago Press, 1974).

Cold Fusion

The major scientific scandal of the 1980s was the scientific claim of the discovery of cold fusion. Two professors, B. Stanley Pons, a professor of chemistry at the University of Utah, and Martin Fleishmann, a professor at the University of Southhampton (England), conducted an experiment with heavy water that they claimed produced fusion. By achieving fusion in a room-temperature jar of heavy water by passing a current between electrodes, their experiment promised a new age for the generation of energy. It meant that expensive plants and heavy financial investment would not be necessary to produce energy. The two professors announced their discovery in March 1989 at a press conference, instead of in a scientific paper. Scientists attempted to duplicate their cold fusion experiment with mixed results. Despite the lack of confirmation, research funds flowed into cold fusion research for several years before it became apparent that concrete results were not forthcoming. By 1992 it was apparent that cold fusion had come to a dead end, but the Japanese government refused to give up. They invested $20 million in further research before giving up hope of concrete gains in the summer of 1997.

Suggested readings: F. E. Close, *Too Hot to Handle: The Race for Cold Fusion* (Princeton, NJ: Princeton University Press, 1991); John R. Huizenga, *Cold Fusion: The Scientific Fiasco of the Century* (Rochester, NY: University of Rochester Press, 1992); Gary Taubes, *Bad Science: The Short Life and Weird Times of Cold Fusion* (New York: Random House, 1993).

Cold War

The Cold War describes the era between 1947 and 1991 when a state of undeclared war existed between the Soviet Union (and its allies) and the United States (and its allies). This rivalry stimulated the development of nuclear weapons and various doctrines for the use of these weapons of mass destruction. Since the United States had a monopoly over atomic weapons between 1945 and 1947, the US government attempted to use this advantage for diplomatic advantage. Joseph Stalin, the head of the Soviet government, refused to acknowledge this superiority, frustrating American plans for a postwar settlement. Instead, he pushed for a crash program to build a Soviet atomic bomb. Once the Soviet Union successfully tested an atomic bomb in 1947, a political and strategic stalemate existed between the two superpowers. Next, both powers engaged in a race to build and test the hydrogen bomb. An arms race ensued for the next two decades, with both sides building immense nuclear weapons stockpiles.

During this period, various strategies for use of nuclear weapons were considered by strategic planners on both sides. American military planners toyed briefly after 1945 with a strategy of first strike. First strike was a plan to use nuclear weapons in a preemptive attack to cripple the Soviet Union before the Soviets could acquire atomic weapons. For a variety of reasons, including diplomatic and humanitarian ones, this plan was never adopted. American planners did come up with the idea of containment of the Soviet expansion with nuclear weapons as a deterrent. This strategy lasted with refinements throughout the Cold War. The major refinement came in the strategy of mutual assured destruction (MAD). As the nuclear arsenals of both sides grew, the size of the yields of the nuclear weapons increased, and the delivery system became more sophisticated, the idea that a nuclear war would destroy both countries and possibly make the planet unable to sustain life gained currency. Planners on both sides used MAD as a deterrent to nuclear war. A balance of assured destruction lasted until the end of the Cold War. With the dismantling of the Soviet Union beginning in 1991 and the disintegration of the Soviet Bloc, the rivalry of the Cold War ended abruptly. Now both Russia and the United States are downsizing the nuclear arsenals they built during the Cold War.

Suggested readings: S. J. Ball, *The Cold War: An International History, 1947–1991* (London: Arnold, 1998); Timothy J. Botti, *Ace in the Hole: Why the United States Did Not Use Nuclear Weapons in the Cold War, 1945 to 1965* (Westport, CT: Greenwood Press, 1996); William Hyland, *The Cold War: Fifty Years of Conflict* (New York: Times Books, 1991); Richard Alan Schwartz, *The Cold War Reference Guide: A General History and Annotated Chronology with Selected Biographies* (Jefferson, NC: McFarland, 1997); Martin Walker, *The Cold War: A History* (New York: Holt, 1994).

Combined Policy Committee

The Combined Policy Committee (CPC) was formed during World War II by President Roosevelt and Prime Minister Churchill to coordinate the atomic energy plans of the United States, the United Kingdom, and Canada. It was negotiated as part of the Quebec Agreement on August 19, 1943. Membership consisted of three high-level American officials, two high-level British officials, and a high-level Canadian official. The Americans were Secretary of War, Henry Stimson, Vannevar Bush and James B. Conant. Two military men, Field Marshall Sir John Dill and Colonel J. J. Llewelin, represented the British. C. D. Howe was the Canadian official. Meetings were to be held in Washington, DC and in secrecy. Beside coordinating atomic energy plans, this committee was to settle questions or disputes between the parties. This committee proved effective during the war, but in the postwar world, it lost out to domestic political issues in the United States and loss of trust between the United States and the United Kingdom. This proved true especially after the exposure of Donald McLean as a spy, because he had been the British secretary to the CPC in 1947 and 1948.

Suggested readings: Thomas B. Cochran, Robert S. Norris, and Oleg A. Bukharin, *Making the Russian Bomb: From Stalin to Yeltsin* (Boulder, CO: Westview Press, 1995); Margaret Gowing, *Independence and Deterrence: Britain and Atomic Energy, 1945–1952*, vol. 2 (New York: St. Martin's Press, 1974).

Commercial Nuclear Ships

Soon after the success of building nuclear-powered naval ships, attention turned to the feasibility of a fleet of commercial nuclear-powered ships. The first such ship was the Soviet icebreaker *Lenin*, launched in 1960. During the winter of 1966–1967, an accident with one of its three reactors killed between 27 and 30 sailors and led to the *Lenin*'s being out of service for almost two years. Next was the nuclear-powered merchant ship, the *Savannah*, which was completed in 1962. It served as a commercial cargo vessel for the American Export Line and sailed nearly half a million miles without problems until it was pulled out of service for economic reasons. The Soviet Union launched two more icebreakers, the *Arctic* in 1975 and the *Siberia* in 1977. Other countries also experimented with nuclear-powered vessels— Germany with the *Otto Hahn*, which lasted from 1969 to 1979, and Japan with the *Mutsu*. The German ship proved uneconomical to operate, and the Japanese ship had such severe design problems that it was out of service for almost a decade. As it became apparent to shipping companies that these ships were not economical, interest lagged in them.

Suggested readings: Bertrand Goldschmidt, *The Atomic Complex: A Worldwide Political History of Nuclear Energy* (La Grange Park, IL: American Nuclear Society, 1982); John May, *The Greenpeace Book of the Nuclear Age: The Hidden History, the Human Cost* (New York: Pantheon Books, 1989).

Commissariat à l'Énergie Atomique

The Commissariat à l'Énergie Atomique (CEA) was established at the end of World War II, and its mission was to oversee the development of the French nuclear industry. On October 18, 1945, Charles de Gaulle, the president of the provisional government, formed the CEA and gave it responsibility for all aspects of atomic research, including military uses. It was placed under the prime minister to shelter the CEA from ministerial politics. Two officials administered the affairs of the CEA—the high commissioner and the administrator-general. The high commissioner was always a scientist who was in charge of scientific matters, and the administrator-general was a high-level administrator to handle administrative details. An executive body consisted of the high commissioner, the administrator-general, the prime minister, and three others, usually, but not restricted to, scientists. Raoul Dautry was the first administrator-general. Frederic Joliot-Curie was the first high commissioner, but he was dismissed in April 1950 for his close ties with the Communist Party. After this controversy, the office of high commissioner became subservient to the office of administrator-general as a purge of left-wing followers of Joliot-Curie took place. Their place was taken by graduates of the prestigious Corps des Mines. For the next 20 years, the CEA worked with varying degrees of success with the Electricité de France (EDF) on a common nuclear program until the CEA was reformed in 1970. The dual system of management was done away with, and its mission changed to general guidance rather than responsibility for operations. Moreover, the CEA no longer reported to the prime minister but to the minister of industry.

Suggested readings: Irvin C. Bupp and Jean-Claude Derian, *Light Water: How the Nuclear Dream Dissolved* (New York: Basic Books, 1978); Bertrand Goldschmidt, *Atomic Rivals* (New Brunswick, NJ: Rutgers University Press, 1990).

Committee for a Sane Nuclear Policy. *See* SANE

Committee for Nuclear Responsibility

The Committee for Nuclear Responsibility was a group of academics, politicians, and scientists active in the dissemination of antinuclear information to the public. It was founded in California by Lenore Marshall in the mid-1970s. Its board of directors had a number of distinguished names including four Nobel laureates: Linus Pauling (chemist), Harold Urey (physicist), James D. Watson (biochemist), and George Wald (medical researcher). Others included Robert Bellman (mathematics), Ramsey Clark (former U.S. attorney general), John Edsall (biochemist), Paul Ehrlich (biologist), John Gofman (medical physicist), Ian MacHarg (architect), and Richard Max McCarthy (politician). The goal of this organization was to publicize its antinuclear views to the public. Although the Committee for Nuclear Responsibility never espoused direct action, the prestige of its members made the committee a significant force in the antinuclear movement.

Suggested reading: Jerome Price, *The Antinuclear Movement* (Boston, MA: Twayne, 1982).

Committee on Biological Effects of Ionizing Radiation Report

The Committee on Biological Effects of Ionizing Radiation submitted a report (BEIR V) in 1990 on the health effects of low-level radiation. This report was published by the National Research Council of the U.S. National Academy of Sciences. Four earlier committee reports had supported the thesis that there was a threshold below which radiation was no longer a health risk. Using mostly data gathered from Japanese survivors of Hiroshima and Nagasaki, BEIR V concluded that radiation exposure regardless of the size of the dose caused physical damage. The extent of the damage was in proportion to the exposure. A formula was devised to calculate the death rate using the amount of the radiation dose relative to the size of the population exposed. Moreover, low doses of radiation caused higher incidences of tumors and leukemias than previous committees had estimated. Finally, the scientists found that the radiation doses of the survivors with cancer at Hiroshima and Nagasaki were less than previously thought.

Suggested reading: International Physicians for the Prevention of Nuclear War, *Radioactive Heaven and Earth: The Health and Environmental Effects of Nuclear Weapons Testing in, on, and Above the Earth* (New York: Apex Press, 1991).

Comprehensive Test Ban Treaty

The Comprehensive Test Ban Treaty (CTBT) was signed by 146 countries in New York City on September 24, 1996, and it culminated nearly 40 years of efforts to ban nuclear testing. This treaty prohibits all nuclear weapons testing and any other nuclear explosions without exception and in every environment. An agreement to proceed on such a treaty came out of an April 4, 1993, summit meeting in Vancouver, Canada, between William Clinton, president of the United States, and Boris Yeltsin, president of Russia. The United Nations General Assembly (UNGA) passed a resolution on December 16, 1993, supporting the multilateral negotiation of a CTBT. Negotiations for this treaty started in January 1994 with representatives from 38 states, but by the time of the final agreement, 61 states had participated in the talks. The United States had taken the lead in these negotiations because in 1992 President George Bush had stopped all nuclear testing by the United States and signed legislation to this effect. At the same time the Committee on Disarmament (CD) of the United Nations had also been active in working on this issue since the early 1980s. In 1996 the CD had a package of agreements, but India blocked consensus action, so the members of the CD took it before the 50th General Assembly for consideration. There it passed by a vote of 158 to 3 (Bhutan, India, and Libya) with 5 abstentions (Cuba, Lebanon, Mauritius, Syria, and Tanzania). On September 24, 1996, President Clinton became the first world leader to sign the CTBT.

Besides numerous stipulations banning nuclear tests, the treaty established a Comprehensive Nuclear Test Ban Treaty Organization located in Vienna, Austria, with the charge of carrying out the terms of the treaty. Various bodies were set up to carry out the terms of the treaty. A Conference of States Parties had the responsibility for general oversight of the implementation of the treaty. An Executive Council with 51 representatives from the member states has executive oversight. Finally, a Technical Secretariat supervises the monitoring and on-site verification requirements. This treaty is to go into force 180 days following the deposit of the last signature, or two years after the signature opening date (September 24, 1998). President Clinton submitted this treaty to the U.S. Senate on September 22, 1997, for its approval.

Suggested reading: Eric Arnett, ed., *Nuclear Weapons After the Comprehensive Test Ban: Implications for Modernization and Proliferation* (Oxford: Oxford University Press, 1996).

Compton, Arthur Holly

Arthur Holly Compton (1892–1962) was an American physicist who combined experimental research on the wavelengths of X-rays with leadership in the American scientific community. He was born on September 10, 1892, in Wooster, Ohio. His father was a dean and professor of philosophy at the College of Wooster. His brother Karl was also a prominent physicist. He married Betty Charity McCloskey, and they had two sons. Compton attended the College of Wooster, where he received his B.S. in 1913. His M.A. in 1914 and Ph.D. in 1916 were from Princeton University, where he studied under Owen W. Richardson. After an academic year as a professor at the University of Minnesota, Compton worked as a researcher at the Westinghouse laboratories for two years, from 1917 to 1919. In 1919, he traveled to England to study experimental methods under Ernest Rutherford at the Cavendish Laboratory, Cambridge University. After returning from England in 1920, Compton found a professorship in physics at the University of Washington, St. Louis, where he stayed until 1923. He then moved to a physics professorship at the University of Chicago at the invitation of A. A. Michelson and remained there until 1945. It was at the University of Chicago that Compton carried out the experiments that made him famous.

Compton's research interest was in the exact way that X-rays scatter upon impact with matter. His experiments showed that the classical theory of scattered X-rays being of the same wavelength as the primary beam was incorrect. Instead, results indicated that the wavelengths of the rays from the target metal varied with the angle at which they emerged. His only explanation was that quantum mechanics provided the best answer. Compton's conclusion was that the interaction of matter and X-rays was a quantum of radiant energy bouncing off an electron. This result is now called the Compton effect. His theory and measurements were confirmed by measurements made in a Wilson cloud chamber. Compton's work helped provide the first hard experimental evidence of the dual nature of electromagnetic radiation. For this discovery and related research, Compton received the 1927 Nobel Prize for Physics.

As important as his scientific work, Compton's leadership in the wartime scientific community was more significant. He was placed in charge of all scientific work on a nuclear chain reaction and the construction of a working atomic reactor. It was his decision to move the research on a reactor to the University of Chicago's Metallurgical Laboratory. His leadership made the Metallurgical Laboratory indispensable in the research leading to the building of the atomic bomb. He sponsored the series of reports on the postwar future of atomic energy. Although he disagreed with several of the points of the Franck Report, Compton recommended that the report be forwarded to government officials.

After the war, Compton returned to academia. He became the chancellor of the University of Washington, St. Louis, in 1953. Compton retained that post until 1961, when he retired. After an illness of several years, Compton died on March 15, 1962, in Berkeley, California.

Suggested readings: Henry A. Boorse, Lloyd Motz, and Jefferson Hane Weaver, *The Atomic Scientists: A Biographical History* (New York: Wiley Science Editions, 1989); Arthur Holly Compton, *Atomic Quest: A Personal Narrative* (New York: Oxford University Press, 1956); Marjorie Johnston, ed., *The Cosmos of Arthur Holly Compton* (New York: Knopf, 1967); Anthony Serafini, *Legends in Their Own Time: A Century of American Physical Scientists* (New York: Plenum Press, 1993); Roger H. Stuewer, *The Compton Effect: Turning Point in Physics* (New York: Science History Publications, 1975).

Conant, James Bryant

James Conant (1893–1978) was an American chemist and the president of Harvard and a prime participant in obtaining government support for atomic research. He was born on March 26, 1893, in Dorchester, Massachusetts. Conant married the daughter of Nobel laureate Harvard chemist T. W. Richards. His early education was at Roxbury Latin. Conant studied organic and physical chemistry under T. W. Richards at Harvard University obtaining his Ph.D. in 1916. During World War I, Conant served as a major in the Chemical Warfare Service, specializing in poison gas research. After the war, he returned to Harvard University. His entire academic career was at Harvard University, first as a professor of chemistry from 1919 to 1933 and then as president of the university from 1933 to 1953. His scientific research centered on studying organic reaction mechanisms including chlorophyll and hemoglobin. He was also active in studying acids and bases.

Conant's contribution to atomic energy was as an administrator during World War II. He became head of the National Defense Research Committee and deputy head of the Office of Scientific Research and Development (OSRD). His role as chair of the Atomic Committee, or S-1, of the OSRD placed him in the middle of planning for the Manhattan Project. His close working relationship with Vannevar Bush was instrumental in advancing government scientific projects.

After the war, Conant returned to the helm of Harvard University fulltime. As a member of the General Advisory Committee (GAC) of the Atomic Energy Commission (AEC), Conant opposed the building of the hydrogen bomb on political, social, and moral grounds. He retired from Harvard in 1953 and became the U.S. ambassador to West Germany from 1953 to 1957. After retirement from diplomatic activity, Conant spent the rest of his career as an advocate for reform in higher education. He died on February 11, 1978, in Hanover, New Hampshire.

Suggested readings: John Daintith, Sarah Mitchell, Elizabeth Tootill, and Derek Gjertsen, *A Biographical Encyclopedia of Scientists*, 2nd ed., vol. 1 (Bristol, UK: Institute of Physics, 1994); James Hershberg, *James B. Conant: Harvard to Hiroshima and the Making of the Nuclear Age* (New York: Knopf, 1993).

The Conqueror

The Conqueror was an RKO Studio movie whose actors and support staff were exposed to high levels of radioactivity during its filming in 1954. This movie about Genghis Khan starred John Wayne, Susan Hayward, Agnes Moorehead, and Pedro Almendariz and was directed by Dick Powell. It was made during the summer of 1954 in southwestern Utah in Snow's Canyon, which was 12 miles west of St. George, Utah, near the Nevada Test Site where the United States tested atomic weapons. In addition to the cast, 300 extras were hired, including Shivwits Indians, a band of Utah Paiutes. During the shooting, John Wayne and his sons used a Geiger counter around the area and thought the site contained radioactive ore. The cast and crew spent three months on location. Movie critics gave the movie horrid reviews, and it closed early. Soon members of the 220 movie crew began to experience bouts with cancer. All of the stars died of various forms of cancer except Pedro Almendariz, who committed suicide after learning he had developed cancer of the lymphatic system. Within 30 years of the making of the movie, 91 of the crew had developed carcinomas. The Indian extras have also experienced health problems. A radiological expert stated that this case was far beyond statistical probability and that they had a possible court case. No lawsuit was ever filed by any member of the crew.

Suggested reading: Robert L. Miller, *Under the Cloud: The Decades of Nuclear Testing* (New York: Free Press, 1986).

Cosmos 954

The Cosmos 954 was a Soviet Radar Ocean Reconnaissance Satellite (RORSAT) that crashed in 1978 and spread its nuclear reactor debris over northern Canada. This satellite had been launched in September 1977 with a nuclear reactor powered by 100 pounds of uranium-235. It was designed to be boosted into a permanent higher orbit at the end of its operational life, but for some unexplained reason, the Cosmos 954 went out of control. On January 24, 1978, the satellite finally crash-landed in Canada's Northwest Territory. An extensive search was conducted by American Canadian teams to find the remains and assess the radioactive contaminated areas. For the next two weeks these teams found debris over a 100,000-square-mile area. The Soviet Union agreed in December 1980 to pay Canada $3 million in damages. Soviet scientists spent the next two years redesigning the ROR-SAT to prevent future accidents.

Suggested reading: John May, *The Greenpeace Book of the Nuclear Age: The Hidden History, the Human Cost* (New York: Pantheon Books, 1989).

Critical Mass

Critical Mass, a key antinuclear organization in the 1970s, was led by Ralph Nader, a national social and economic critic and consumer advocate. In 1974, Nader made nuclear power one of his priority issues, and he convened a convention of antinuclear activists in November. A steering committee, called Critical Mass, was established after this convention to coordinate antinuclear activities. A monthly antinuclear newsletter, *Critical Mass*, was founded to publicize the activities of the movement. In November 1975 another Critical Mass convention took place. While only 300 people attended the first convention, more than 1,000 participated in the second. It was at the second convention that its leaders began to investigate political strategies to prevent the spread of atomic power. Nader continued to play a role at the national level, but the strategic initiatives were left to others.

Suggested reading: Etahn M. Cohen, *Ideology, Interest Group Formation, and the New Left: The Case of the Clamshell Alliance* (New York: Garland, 1988).

Crossroads Nuclear Tests

After a hiatus of American testing of atomic weapons because of the end of the war, nuclear testing resumed in July 1946 with the Crossroads series of tests. Operation Crossroads came out of the rivalry between the U.S. Navy and the Army Air Force over the atomic bomb and its impact on future budgets, missions, and prestige. The navy wanted to prove that the naval forces could withstand an atomic attack. Admiral William H. Blandy was placed in charge of the tests. Much of the scientific community considered this operation a waste of time and meaningless, so most scientists boycotted the test. Nevertheless, more than 42,000 military and civilian personnel participated in Operation Crossroads. Plans were made for three tests: an airburst (Able), a ground-level or shallow underwater depth (Baker), and a device several thousand feet underwater (Charlie). A target fleet of 95 naval vessels ranging in size from battleships to submarines was assembled at the Bikini Atoll in the Marshall Islands for atomic test Shot Able. A total of 5,664 animals were placed on 22 target ships (200 pigs, 204 goats, 200 mice, 60 guinea pigs and 5,000 rats). On July 1, 1946, an improved version of the atomic bomb (a Mark III plutonium bomb reengineered from the Fat Man prototype) was detonated over the fleet. Despite heavy damage, most of the vessels remained afloat, but radioactive levels were too high for crews to have survived. A second test, Shot Baker, was an atomic device of a new design to be detonated underwater in the Bikini lagoon. Again, a fleet was present for the test on July 25, 1946. This time the fleet was heavily damaged, but the most spectacular part of the blast was the mushroom of pressurized water vapor. This base surge of radioactive water led some scientists to conclude that the low-level vapor cloud might be more deadly than an airburst. Despite some surprises on radiation and battle damage, this series of tests convinced American military leaders, especially those in the U.S. Navy, that atomic weapons had a future as strategic military weapons.

Suggested readings: Robert L. Miller, *Under the Cloud: The Decades of Nuclear Testing* (New York: Free Press, 1986); Jonathan M. Weisgall, *Operation Crossroads: The Atomic Tests at Bikini Atoll* (Annapolis, MD: Naval Institute Press, 1994).

Curie, Marie Sklodowska

Marie Sklodowska Curie (1867–1934) overcame numerous difficulties to become one of the greatest experimental physicists of all time. She was born Marya Sklodowska in Warsaw, Poland, on November 7, 1867, to a family of school teachers. Her father was a mathematics and physics teacher, and her mother the principal of a girls' school. Her marriage in 1895 to Pierre Curie produced two daughters. At an early age she showed outstanding intelligence, but because Polish law did not allow her to go beyond secondary schooling, she had to save money to go abroad. Her first job was as a governess, which lasted from 1885 to 1891, and most of her money went to support her sister Bronya's medical studies in France. Her married sister lived in Paris, so she moved there in 1891 to study physics and mathematics at the University of Paris–Sorbonne. She received her degree in the physical sciences in 1893, in mathematical sciences in 1894, and then a Ph.D. in 1904. Curie became a professor at the Normal Superior School at Sevres, France, in 1900. After a stay there, she returned to the University of Paris–Sorbonne in 1904 and remained until 1934.

Madame Curie was recognized by her contemporaries as a great experimental scientist. She was a methodological worker and began studying the properties of uranium in the fall of 1897. It was at this time that her husband joined her in the study of uranium. After confirming Becquerel's findings, she and her husband Pierre discovered polonium in July 1898. Then in September 1898, they announced the discovery of radium. She received two Nobel Prizes—one for physics in 1903 and another for chemistry in 1911—and became an international scientific star. Despite these discoveries and an international reputation, the Curies lacked a good laboratory for research. This lack had just been solved when Pierre Curie was killed in a freak traffic accident on April 19, 1906. Finally, France built the Radium Institute in the period before World War I, and Madame Curie became the director of the Radioactivity Laboratory until 1934. She became ill in early 1934, probably from radiation sickness, and died on July 4, 1934, in a sanatorium in Sancellemoz, France. Albert Einstein eulogized her by saying, "Marie Curie is of all celebrated beings, the only one whom fame has not corrupted."

Suggested readings: Henry A. Boorse, Lloyd Motz, and Jefferson Hane Weaver, *The Atomic Scientists: A Biographical History* (New York: Wiley Science Editions, 1989); Keith Brandt, *Marie Curie, Brave Scientist* (Mahwah, NJ: Troll Associates, 1983); Eve Curie, *Madame Curie* (Garden City, NY: Doubleday, Doran and Co., 1938); Françoise Giroud, *Madame Curie, a Life* (New York: Holmes and Meier, 1986); Sean M. Grady, *Marie Curie* (San Diego, CA: Lucent Books, 1992); Emilio Segre, *From X-Rays to Quarks: Modern Physicists and Their Discoveries* (San Francisco: W. H. Freeman, 1980).

Curie, Pierre

Pierre Curie (1859–1906) was a French physicist whose early research on radioactivity paved the way for later advances. He was born on May 15, 1859, in Paris. His father was physician who had been a Communard (a supporter of the Commune of Paris in 1871). In 1895, Curie married the brilliant Polish scientist Marie Sklodowska, and they had two daughters. Curie's early education was under a tutor. His college education was at the University of Paris–Sorbonne. After graduation, he became a research assistant at the Sorbonne, beginning in 1878. In 1883, he assumed the position of laboratory chief at the School of Industrial Physics and Chemistry. Curie stayed there until his appointment as professor of physics at the Sorbonne in 1904.

Most of Curie's research before 1895 was with crystals. He discovered piezoelectricity in 1880, with his brother Jacques Curie's assistance. He also experimented with the effects of temperature on the magnetic properties of substances. He discovered that permanent magnets lose their magnetization when heated above a critical temperature, known as the Curie point. After the discovery of radioactivity, Curie and his wife began an extensive research project studying it. They discovered two new elements, radium and polonium, in 1898. Their research led to the Curies' receiving the 1903 Nobel Prize for Physics.

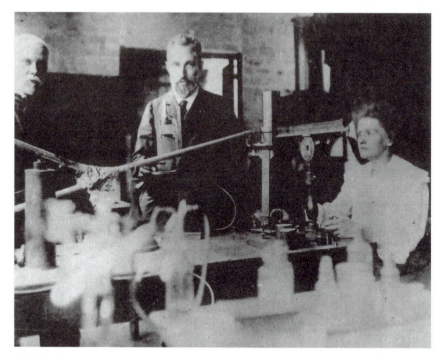

The French physicists Pierre Curie and Marie Sklodowska Curie were pioneers in the field of atomic energy. Here they are in their laboratory surrounded by the type of equipment they used to discover radium.

Despite Curie's successes as a physicist, he received little recognition in France. He was never in favor with the French scientific community, possibly because of his leftward political leanings. His promotion to the Sorbonne was only in reaction to a job offer from Geneva. Curie's failure to find a satisfactory laboratory for research during his lifetime also bothered him. His first candidacy for the Academie Française in 1902 was defeated, but he was finally elected in 1905. Evidence exists that he was suffering from radiation sickness when he fell under a Parisian horse cab and was killed on April 19, 1906.

Suggested readings: Henry A. Boorse, Lloyd Motz, and Jefferson Hane Weaver, *The Atomic Scientists: A Biographical History* (New York: Wiley Science Editions, 1989); Marie Curie, *Pierre Curie* (New York: Dover Publications, 1963); John Daintith, Sarah Mitchell, Elizabeth Tootill, and Derek Gjertsen, *A Biographical Encyclopedia of Scientists*, 2nd ed., vol. 1 (Bristol, UK: Institute of Physics, 1994); (New York: Facts on File, 1981); Emilio Segre, *From X-Rays to Quarks: Modern Physicists and Their Discoveries* (San Francisco: W. H. Freeman, 1980).

One of the key instruments in development of atomic energy was William O. Lawrence's series of cyclotrons at the University of California at Berkeley. This picture shows the 184" cyclotron at the university just before final installation. It shows the size of the apparatus in comparison to its human handlers.

Cyclotron

The cyclotron was an early particle accelerator invented by E. O. Lawrence and M. S. Livingston at the University of California at Berkeley in 1931. Lawrence had been thinking about the problem of accelerating particles when in 1929 he found an article by a Norwegian engineer suggesting that a linear accelerator using timed electrical pulses applied to a series of cylindrical electrodes would work. Once he read the idea, Lawrence thought out the theory behind the cyclotron. Two hollow brass electrodes called "dees," alternatively charged, pushed and pulled ions around in a magnetic field. It took some time for he and Stanley Livingston to build the first apparatus. Lawrence obtained a large electromagnet from U.S. Navy surplus. Once it was working successfully, Lawrence was able to garner enough support to establish the Radiation Laboratory at the University of California at Berkeley in 1932. The university gave him an old wooden building on campus for his cyclotron. Soon Lawrence had a 1-million-volt cyclotron in operation. Lawrence took charge of the cyclotron program and kept busy upgrading its capacity by building ever larger cyclotrons. In 1940 Lawrence obtained from the Rockefeller Foundation a grant of over $1 million to build a super 4,900-ton cyclotron for general research, but soon after its construction it was used to separate quantities of U-235 for the Manhattan Project. On the eve of World War II, the United States led the world with over 30 cyclotrons in operation, many of them built for medical research.

Soon other countries in late 1930s began to build cyclotrons for atomic research. The Soviet Union had a cyclotron in operation at the Radium Institute in Leningrad in 1937. Two additional larger cyclotrons were planned and under construction when the German invasion of the Soviet Union in the summer of 1941 prevented their completion. A small cyclotron was also built in Paris at the College de France under Joliot-Curie's supervision in 1940 and was in operation in 1942, but it was not a powerful enough machine to produce plutonium. The failure of Germany to build a cyclotron until 1944 severely limited its atomic bomb program. Their access to the French cyclotron after 1940 helped a bit, but it was never used to produce weapons-grade uranium.

Suggested readings: Luis W. Alvarez, *Alvarez: Adventures of a Physicist* (New York: Basic Books, 1987); Daniel J. Kevles, *The Physicists: The History of a Scientific Community in Modern America* (Cambridge: Harvard University Press, 1987); Roger H. Stuewer, ed., *Nuclear Physics in Retrospect: Proceedings of a Symposium on the 1930s* (Minneapolis: University of Minnesota Press, 1979).

D

Dalton, John

In the early part of the nineteenth century John Dalton (1766–1844), a British chemist–physicist, firmly established the atomic theory of matter. He was born on September 6, 1786, in Eaglesfield, England. His father was a hand-loom weaver. He was educated at the village school on a scholarship from a fellow Quaker until age 11. At age 12, he began teaching at the local school. In 1781, he moved to Kendal to teach at a school there. Then in 1793 Dalton moved to Manchester, where he remained the rest of his life. In Manchester, he taught at the Presbyterian institute, Manchester New College.

Dalton is famous for a variety of discoveries, but it was his development of atomic theory that was his most important contribution. In his experiments on the solubility of gases in water, he found that the expected same solubility didn't exist. Dalton developed the theory that atoms of different gases had different weights. His atomic theory was first stated in its entirety in a lecture at a Royal Institution lecture in 1802. Although some of his calculations proved to be incorrect, his system on atomic weights was on the right track. His theory remained unconfirmed until more than a decade after his death on July 27, 1844, in Manchester.

Suggested readings: Henry A. Boorse, Lloyd Motz, and Jefferson Hane Weaver, *The Atomic Scientists: A Biographical History* (New York: Wiley Science Editions, 1989); John Daintith, Sarah Mitchell, Elizabeth Tootill, and Derek Gjertsen, *A Biographical Encyclopedia of Scientists*, 2nd ed., vol. 1 (Bristol, UK: Institute of Physics, 1994); Frank Greenaway, *John Dalton and the Atom* (Ithaca, NY: Cornell University Press, 1966).

De Broglie, Louis-Victor Pierre Raymond

Prince de Broglie (1892–1987) was a French physicist noted for his hypothesis of the wave nature of the electron. He was born on August 15, 1892, at Dieppe, France. His family had a long and distinguished history in French politics. His parents died when de Broglie was young, and his older brother Maurice de Broglie raised him. De Broglie remained a lifelong bachelor. His education was at the Sorbonne: licence in history in 1910; licence in science in 1913; and D.Sc. in 1924. He intended to be a historian, but his service with radios in the French Signal Service in World War I turned his interest toward physics. His entire research career was at the Sorbonne, starting in 1926 and lasting until 1962. In 1928, de Broglie also became a professor at the Henri Poincaré Institute.

De Broglie's thesis topic on quantum theory served as the basis for his wave theory. He found that Einstein's special theory of relativity described matter itself as a form of energy. In the course of his research, de Broglie advanced the thesis that just as waves can behave as particles, so particles can behave as waves. This wave-particle duality became a source of controversy, but it also provided answers on the nature of the atom. It was for his work on wave theory that de Broglie received the 1929 Nobel Prize for Physics.

With his reputation as one of France's great scientists, de Broglie spent much of his life working for the cause of peaceful uses for nuclear energy. He was active on the French Commission on Atomic Energy. He died on March 19, 1987, at Louveciennes, Yvelines, France.

Suggested readings: Henry A. Boorse, Lloyd Motz, and Jefferson Hane Weaver, *The Atomic Scientists: A Biographical History* (New York: Wiley Science Editions, 1989); John Daintith, Sarah Mitchell, Elizabeth Tootill, and Derek Gjertsen, *A Biographical Encyclopedia of Scientists*, 2nd ed., vol. 1 (Bristol, UK: Institute of Physics, 1994).

Debye, Peter

Peter Debye (1884–1966) was a prominent Dutch-American physicist and physical chemist who was instrumental in warning Allied scientists of the progress of German scientists' work on a German atomic bomb. He was born March 24, 1884, in Maastricht, Netherlands. His marriage in 1913 to Mathilde Alberer produced a daughter and a son. Debye's undergraduate education was at the Aachen Institute of Technology, Germany, where he studied electrical engineering. After receiving his degree in 1905, he obtained a Ph.D. in physics at the University of Munich in 1908. Debye had worked at Aachen as a scientific researcher from 1905 to 1906. He also held a research position at the University of Munich from 1906 to 1910. Debye obtained a professorship at the University of Zurich and stayed there for two years. He was appointed director of the Kaiser Wilhelm Institute of Physics at Dahlem until the Nazis gave him the choice of adopting German citizenship or being dismissed. Refusing to give up his Dutch citizenship, Debye left Germany. In 1939 Debye traveled to the United States, bringing with him news about a German atomic bomb program under the directorship of Werner Heisenberg. This news confirmed fears among American and British scientists that the Germans were working hard on a possible atomic weapon. Without such information it would have been difficult for these scientists to recommend to their government a crash program to counter the German one.

After leaving Germany, Debye obtained a teaching position at Cornell University. He was appointed chair of the chemistry department there in 1940 and remained until his retirement in 1950. Debye won the 1936 Nobel Prize for Chemistry for his experiments on X-ray diffraction by gases and liquids. He remained active in scientific affairs until his death in 1966.

Suggested reading: Thomas Powers, *Heisenberg's War: The Secret History of the German Bomb* (New York: Knopf, 1993).

Declared Nuclear Weapon States

China, France, Russia, the United Kingdom, and the United States have all declared their nuclear status and are recognized under the Nuclear Non-Proliferation Treaty (NPT) as nuclear weapon states. For purposes of definition, a *nuclear weapon state* is one that manufactured and exploded a nuclear device prior to January 1, 1967. This status has given these countries a role as defenders of nuclear nonproliferation. It has also made them unpopular with the nonnuclear states, as these countries consider their actions as self-serving and unfair because the nonnuclear states have to submit to International Atomic Energy Agency (IAEA) safeguards, but the nuclear states do not. The nonnuclear states have insisted on security guarantees, but the nuclear states have been reluctant to give these guarantees except through the United Nations. Two recent additions to the ranks of countries possessing nuclear weapons are India and Pakistan, but they have not been welcomed by the declared nuclear weapon states. Rather, the nuclear states have imposed economic and political restrictions on both India and Pakistan to punish them for their March 1998 tests.

Suggested reading: Marianne van Leeuwen, ed., *The Future of the International Nuclear Non-Proliferation Regime* (Dordrecht, Holland: Martinus Nijhoff, 1995).

Decommissioning of Nuclear Plants

As the popularity of nuclear energy has waned and maintenance of the plants remains expensive, a movement has begun in both the nuclear industry and governments to decommission nuclear plants. Since nuclear plants have a life expectancy of about 30 years, decommissioning has always been a part of their future. Part of the cost of operations includes building a fund for decommissioning costs, but the final costs have exceeded the amount of funds set aside. When a nuclear plant has outlived its efficiency, the company petitions the Nuclear Regulatory Agency for a permit to close down the facility, with a projected closing day. A contract is issued to a company or companies that specialize in nuclear waste management and the decommissioning procedures. This was the fate of the Shippingport Nuclear Power Plant. Plants that have proven unprofitable or inefficient, however, have been placed in the decommissioning cycle earlier than anticipated. One such plant was the Fermi I fast breeder reactor, which was shut down after only slightly over 6 years of operation in November 1972. Another was the Fort St. Vrain Nuclear Generating Station in Colorado, which only operated from 1979 to 1989. Decommissioning of this unit started in 1990 and ended in 1996 at a cost of $188 million.

The U.S. government has also been active in decommissioning activities. In 1978 the Department of Energy (DOE) established the Surplus Facilities Management Program (SFMP) to handle decommissioning activities of nuclear weapons production facilities. In 1985, the government estimated that it would cost in the vicinity of $1.2 billion to clean up decommissioned government nuclear plants. In 1989, the DOE reorganized and placed decommissioning under the Office of Environmental Management (EM). Then in 1992, the Office of Materials Stabilization and Facility Management (EM-60) was formed under EM to manage its decommissioning activities. Approximately 900 facilities of all types are on schedule for decommissioning from reactors to laboratories in 130 sites in 33 states and Puerto Rico. The biggest task, however, is the decommissioning of the three gaseous diffusion plants at Oak Ridge; Paducah, Kentucky; and Portsmouth, Ohio. It is expected to cost around $17.4 billion over the next few decades, while cost recovery will be only around $7 billion. Decommissioning government nuclear facilities is expensive and will take considerable time to be accomplished.

All nuclear countries face decommissioning activities. Other countries have decommissioned 55 reactors in the last 20 years. The most famous of these, however, is the decommissioning of the three undamaged reactors at the Chernobyl facility. This plan is under consideration, but the problems of nuclear waste and contamination remain serious problems to be solved.

Suggested readings: International Conference on Nuclear Decommissioning, *Nuclear Decommissioning: The Strategic, Practical, and Environmental Considerations* (London: Mechanical Engineering Publications Limited, 1995); *Nuclear Power Reactors in the World* (Vienna, Austria: International Atomic Energy Agency, 1997).

De Gaulle, Charles

Charles de Gaulle (1890–1970) as the head of the French state was interested in nuclear energy only to the extent it ensured France's role on the international scene. Despite his role as leader of the Free French in World War II, de Gaulle learned about work on the atomic bomb from a group of French scientists working for the Anglo-Canadian atomic program in Canada. Jules Gueron, a member of the chemistry division of the Montreal project, informed de Gaulle about the issue in a brief meeting at Ottawa on July 11, 1944. From the beginning, de Gaulle understood the international ramifications of possession of the atomic bomb. He was active in the formation of the Commissariat à l'Énergie Atomique (CEA) in October 1945. Even after de Gaulle left office, he remained interested in the future of French atomic energy. When de Gaulle returned to power, he used the powers of his office to build an independent nuclear weapons strike force.

Suggested reading: Bertrand Goldschmidt, *The Atomic Complex: A Worldwide Political History of Nuclear Energy* (La Grange Park, IL: American Nuclear Society, 1982).

Deuterium. *See* Heavy Water

Diebner, Kurt

Kurt Diebner (1905–1964) was a German physicist who played an administrative role in the German atomic bomb program. He was born in 1905. His education was at the universities of Insbruck and Halle. In 1931, he received his Ph.D. in experimental physics from the University of Halle, studying under Gerhard Hoffman. From 1931 to 1934, he was an assistant professor in physics at the University of Halle. It was during this period that Diebner joined the Nazi Party. He moved to the German Bureau of Standards, where he became head of the research section for nuclear physics and explosives. Diebner was named provisional head of the Kaiser Wilhelm Institute of Physics at Dahlem to replace Peter Debye in 1939, but not without opposition from a group of physicists around Werner Heisenberg. Once the war broke out, Diebner took charge of uranium research for the Army Ordinance Bureau until replaced by Heisenberg in 1942. He continued to conduct reactor research in Berlin at Gottow and later in Stadtilm, Thuringia, until the end of the war. Diebner had a running disagreement with Werner Heisenberg over reactor design and competed with him over the supply of heavy water. Although he was 1 of 10 physicists held by the Allies at the Farm House for six months after the war, he was never popular with his colleagues because of his Nazi contacts. After the war, Diebner returned to scientific work, but this time in the private sector. Much of his postwar work was on the development of nuclear-powered commercial ships. He died in 1964.

Suggested readings: David Irving, *The German Atomic Bomb: The History of Nuclear Research in Nazi Germany* (New York: Simon and Schuster, 1967); Thomas Powers, *Heisenberg's War: The Secret History of the German Bomb* (New York: Knopf, 1993).

Dimona

Dimona is the main center of Israel's nuclear weapons program. It is located about 40 miles from Beersheba in the Negev Desert. Construction work started in 1958 near a tiny Jewish settlement named Dimona. Israel claimed that it was a textile factory, but U.S. Air Force reconnaissance in 1960 proved that Israel was building a nuclear reactor. The Dimona reactor is modeled on France's EL-3 reactor at Brest in Brittany. In its original configuration, the reactor had a capacity of 24 megawatts and could produce enough plutonium for a 20-kiloton atomic bomb a year. Initial stock of uranium came from France. Dimona went into operation in December 1963. It was a heavy water reactor that burned natural uranium fuel. This type of reactor did not depend on enriched uranium, so it meant that Israel was not dependent on the United States for fuel as with the Nahal Soreq reactor. Although Israeli authorities claimed that this reactor was for peaceful purposes, the reactor had the potential to make weapons-grade plutonium. The problem was in finding enough uranium. They smuggled 200 tons of uranium ore from Belgium to develop nuclear weapons. By 1973, Israel had an arsenal of atomic weapons. Information about operations became public in the Vanunu Affair of 1986, when an Israeli citizen gave a story and smuggled pictures to the English newspaper *Sunday Times*. Despite denials from Israeli officials, Dimona remains Israel's source for weapons-grade plutonium for nuclear weapons in the late 1990s.

Suggested reading: Peter Pry, *Israel's Nuclear Arsenal* (Boulder, CO: Westview Press, 1984).

Dirac, Paul Adrien Maurice

Paul Dirac (1902–1984) was a British mathematician–physicist whose contribution to atomic theory made him one of the leading scientists of his day. He was born on August 8, 1902, in Bristol, England. His father was Swiss and his mother, English. Dirac married Margit Wigner in 1937, and they had two daughters. His education was a bachelor of science in electrical engineering and pure mathematics from Bristol University in 1921 and a Ph.D. in mathematical physics from Cambridge University in 1926. Dirac divided his research and teaching career between Cambridge University, where he was from 1932 onward the Lucasian Professor of Mathematics, and Florida State University (1971–1984). His students admired Dirac as a brilliant mathematician–physicist, but those working under him found him uncommunicative.

Dirac was one of the few theoretical physicists at Cambridge University. In 1928, he became famous for his integration of the matrix mechanics of Werner Heisenberg and Max Born, and the wave theory of Louis de Broglie and Erwin Schrödinger, with Albert Einstein's theory of relativity. His book *The Principles of Quantum Mechanics* was influential in advancing the credibility of quantum mechanics with other physicists. Dirac received the 1933 Nobel Prize in Physics for his mathematical treatment of quantum electricity and magnetism, predicting the existence of positrons. He also postulated a theory of the hydrogen atom that predicted it had two possible ground states with equal energies. Willis Lamb later won a Nobel Prize for measuring the minute difference in these energy levels. He died in Tallahassee, Florida, on October 20, 1984.

Suggested readings: Henry A. Boorse, Lloyd Motz, and Jefferson Hane Weaver, *The Atomic Scientists: A Biographical History* (New York: Wiley Science Editions, 1989); John Daintith, Sarah Mitchell, Elizabeth Tootill, and Derek Gjertsen, *A Biographical Encyclopedia of Scientists*, 2nd ed., vol. 1 (Bristol, UK: Institute of Physics, 1994); Helge Kragh, *Dirac: A Scientific Biography* (Cambridge: Cambridge University Press, 1990); C. P. Snow, *The Physicists* (Boston: Little, Brown, 1981).

E

Einstein, Albert

Albert Einstein (1879–1955) is celebrated as the twentieth century's greatest theoretical physicist, but most of his creative powers were devoted to constructing theories outside of atomic theory. He was born in Ulm, Germany, on March 14, 1879. His father was an unsuccessful businessman in the electrical equipment business. He married Mileva Maric in 1902, and they had two sons. After a divorce in 1919, Einstein married his cousin, Elsa Lowenthal. At an early age, his family moved to Munich. Einstein attended a Munich Gymnasium, and his early academic successes turned bitter when he rebelled against rote learning in early adolescence. His rebelliousness caused him to be expelled from the Gymnasium. After failing the entrance examination to the Zurich Polytechnical School, he finished his secondary education at Aarau in Switzerland. Einstein gained admittance to the Zurich Polytechnical School, but he was an unexceptional student. His professors generally ignored him, but he did graduate. Unable to find an academic post, he survived as a private tutor and substitute teacher until 1902, when he started work as a technical expert, third class, in the Swiss Patent Office in Bern. Einstein's rebellion against things German culminated in his request for a revocation of German citizenship, which was granted by the German government in 1896. He obtained Swiss citizenship in 1901.

Einstein gained world fame for a series of papers published in 1905 in the physics journal *Annalen der Physik* in which he outlined his ideas on a variety of problems in physics. The first article was "On the Motion of Small Particles Suspended in a Stationary Liquid According to the Molecular Kinetic Theory of Induction," in which Einstein produced a formula for the average displacement of particles in suspension. His formula was later confirmed in 1908 and was the first direct evidence for the existence of atoms and molecules of a definite size. In the second article entitled "On a Heuristic Point of View About the Creation and Conversion of Light," Einstein was concerned about the nature of electromagnetic radiation. His solution was to suggest that electromagnetic radiation is a flow of discrete particles—quanta. He took Max Planck's idea of quanta from a tool for calculation to that of a physical fact. Einstein's third paper is the most famous, entitled "On the Electrodynamics of Moving Bodies." In this paper, Einstein introduced the special theory of relativity to science. His attempt was to dispense with the ether in the description of how moving bodies interact with electromagnetic fields, by proposing that the speed of light is the same for all frames of reference that are moving uniformly relative to each other. He rejected the ideas of absolute space and absolute time. In his last paper entitled "Does the Inertia of a Body Depend on Its Energy Content?" Einstein posed his famous equation $E = mc^2$, relating to mass and energy.

In 1907, Einstein advanced his general theory of relativity. This theory is based upon the equivalence principle, which says that it is impossible to distinguish between an inertial force and a gravitational one. This was a way to generalize the special theory to cover the case of noninertial frames. During the course of this discussion, Einstein predicted that rays of light in a gravitational field move in a curved path. The confirmation of this effect by Arthur Eddington in 1919 made Einstein an even more significant force in physics. In 1916, Einstein published his seminal paper entitled "The Foundation of the General Theory of Relativity." In this work, he introduced the picture that a body "warps" the space around it so that another nearby free body moves in a curved path. He used various types of mathematics and the help of mathematician Marcel Grossmann to advance his theory of general relativity.

The originality of his theories gained him instant recognition, and in 1908 he received an academic post at the University of Bern. In the next six years, he held posts at the University of Zurich (1909), University of Prague (1911), and Zurich Polytechnical School (1912) before a move to the University of Berlin in 1914. Despite some opposition from German nationalists, Einstein stayed at Berlin until Hitler came to power. His theories, which attacked traditional physics, and his internationalism made him a target for anti-semitic propaganda by the Nazis. Einstein was characterized as the prototype propagator of Jewish Science. He immigrated to the United States in 1933, where he became a fixture at the Institute for Advanced Study in Princeton, New Jersey. As the godfather of physics, he worked there on a quest for a unified field theory without success. With his great work completed by the early 1920s, Einstein played an important role again when Leo Szilard and Eugene Wigner needed to warn President Roosevelt about the danger of the Nazi regime's possibly obtaining an atomic bomb. Einstein's direct contributions to the understanding of atomic theory were minor, but as Arthur Holly Compton put it, "Every physicist who studies atoms makes use of tools of thought that Einstein has supplied." In particular, Einstein's theory of energy and matter in his formula $E = mc^2$ provides the worldview for all research in atomic energy. Einstein died in Princeton, New Jersey, on April 18, 1955. *See also* Einstein-Szilard Letter.

Suggested readings: Jeremy Bernstein, *Einstein* (New York: Viking Press, 1973); Denis Brian, *Einstein: A Life* (New York: Wiley, 1996); Ronald William Clark, *Einstein: The Life and Times* (New York: World Publishing, 1971); Albert Folsing, *Albert Einstein: A Biography* (New York: Viking, 1997); Aylesa Forsee, *Albert Einstein: Theoretical Physicist* (New York: Macmillan, 1993); Andrew Whitaker, *Einstein, Bohr, and the Quantum Dilemma* (Cambridge: Cambridge University Press, 1996); Michael White, *Einstein: A Life in Science* (New York: Dutton, 1994).

Einstein-Szilard Letter

On August 2, 1939, the Einstein-Szilard Letter was sent to President Roosevelt informing him of the possibility of the building of an atomic bomb. Leo Szilard had been working with Enrico Fermi on a possible atomic reactor, and he had informed his good friend Lewis Strauss of the potential of fission as an explosive. In the meantime, Enrico Fermi had given a lecture to a Navy Department technical group on the same topic in March 1939. The Navy Department showed little enthusiasm for either Fermi's or Szilard's schemes for uranium research. Szilard considered approaching Albert Einstein with the idea for a letter to Einstein's friend, Elizabeth, queen of the Belgians, about the danger of the Belgium Congo's uranium ore falling into the hands of the Germans. A meeting with Einstein had been fruitful about a letter, but he was reluctant to send the letter to the queen. Instead, Szilard was advised to take an alternative approach to President Roosevelt through the intermediary of Alexander Sachs, an economist who worked with the Lehman Corporation and who had been an official at the National Recovery Administration (NRA). Sachs had become knowledgeable about atomic energy and had already broached the subject with President Roosevelt. After communicating with Sachs, Szilard and Eugene Wigner decided that this was the best approach. Szilard and Edward Teller traveled again to Einstein's summer home on Long Island to get his signature. Two letters, a long and a short version, both of which had been drafted by Szilard and Wigner, were submitted to Einstein. After some discussion, Einstein signed both letters but recommended the longer version. In the letter, Einstein explains that "the element uranium may be turned into a new and important source of energy in the immediate future." "This new phenomenon would also lead to the construction of bombs." He recommends immediate U.S. government action and concludes by warning that the Germans are working with uranium. Because of the outbreak of the European war, the letter could not be delivered by Sachs until October 11, 1939. Sachs, in the meantime, wrote a cover letter supporting the conclusions of the letter. In the meeting, President Roosevelt decided immediate action was necessary, and the Uranium Committee was formed soon afterward.

Suggested readings: Stanley A. Blumberg and Gwinn Owens, *Energy and Conflict: The Life and Times of Edward Teller* (New York: Putnam's Sons, 1976); Richard Rhodes, *The Making of the Atomic Bomb* (New York: Simon and Schuster, 1986); Robert C. Williams and Philip L. Cantelon, eds., *The American Atom: A Documentary History of Nuclear Policies from the Discovery of Fission to the Present, 1939–1984* (Philadelphia: University of Pennsylvania Press, 1984).

Eldorado Mine

A Canadian prospector, Gilbert A. (Lucky) LaBine, discovered a large deposit of uranium ore in the Great Bear Lake region north of the Arctic Circle in May 1930. He named the mine the Eldorado Mine. While its ore was rich in uranium content, the Shinkolobwe Mine in the Belgian Congo had much richer ore, and it was more economical to mine it. Nevertheless, a small mining community was established of around 200 people, which was called Port Radium. This uranium ore was 1,600 miles from the nearest railroad, so most of the ore had to be moved by water. Despite transportation difficulties, LaBine engaged in a price war with the Belgian Syndicate controlling the Shinkolobwe Mine and reduced the ongoing price for uranium ore by two thirds. It was the demand for radium for medical and scientific research that kept these companies in business. In 1938, the Canadian and Belgian governments established a cartel to control the production of uranium ore and maintain radium's price at $25,000 a unit. Part of the reason for LaBine's success was his Port Hope refinery. When his markets in Europe for radium were cut off by the war, LaBine shut down operations in July 1940. Less than two years later, the Canadian government gave LaBine a huge order for uranium oxide from his stockpile. At its peak, the Eldorado Mine could produce 300 tons of uranium ore, and the United States contracted for 200 tons in April 1942. It was part of this order that provided the uranium for the Chicago Pile One experiment on December 2, 1942. With orders such as these, the mine reopened and continued producing uranium ore for the rest of the war. Since the mine could be operated only three months in the years when the lakes and rivers were unfrozen, the mine could only meet a part of the growing demand for uranium for the Manhattan Project. From 1943 until well into the late 1940s all uranium ore mined at the Eldorado Mine was reserved by the United States in a contract negotiated between the American and Canadian governments. In 1944 the Eldorado Mining Company was nationalized by the Canadian government, but its management team was left in place. This mine remained open until 1960, when the uranium ore finally gave out.

Suggested reading: Earle Gray, *The Great Uranium Cartel* (Toronto, Ontario: McClelland and Stewart, 1982).

Electricité de France

Electricité de France (EDF) is the French electricity company, and in a power struggle with the Commissariat á l'Énergie Atomique (CEA) in the mid-1960s, it won control over French nuclear reactors. From the beginning of its existence, the goal of the EDF was to control all aspects of France's electrical program. The EDF was considered a success story in the postwar era, because it had kept electricity costs down. Its leadership considered nuclear power as an extension of their responsibilities for dams and thermal stations. This company, however, came into conflict with the French Atomic Energy Commission. The battle was fought over nuclear reactor design, with the French Atomic Energy Commission advancing the British-French gas-graphite reactor with a heavy water system as a backup. The EDF, however, advocated adopting the American light water technology. In the ensuing political struggle, neither side had an advantage until Charles de Gaulle left office in 1969. By this time the popularity of the light water reactor was so great that the French government turned over nuclear power plant building to the EDF. Depending in the past on mostly American light water technology, the oil embargo in the early 1970s caused the leadership of the EDF to adopt a more aggressive pro–nuclear energy posture. By 1989 EDF was operating 55 nuclear reactors with 8 additional under construction. It has been estimated that France generates about three quarters of the country's electricity from nuclear power plants.

Suggested reading: Irvin C. Bupp and Jean-Claude Derian, *Light Water: How the Nuclear Dream Dissolved* (New York: Basic Books, 1978).

Emergency Committee of Atomic Scientists

The Emergency Committee of Atomic Scientists (ECAS) was an organization founded in May 1946 to lobby against atomic weapons. Its membership consisted of a number of famous scientists: Leo Szilard, Harold Urey, Victor Weisskopf, Hans Bethe, and Linus Pauling. Albert Einstein was ECAS's first chairperson, but the leading figure was always Leo Szilard. The organization's first action was to build a war chest of $200,000. These fundraising activities were successful, so that by the fall of 1947, a fund of $320,000 was at its disposal. In its short life span of nearly three years, the ECAS campaigned hard for the internationalization of atomic energy and against nuclear weapons. By 1946 none of the ECAS members were engaged in atomic weapons research. As the Cold War intensified, the ECAS lost its public support and ceased as an effective organization.

Suggested reading: Lawrence S. Wittner, *The Struggle Against the Bomb: One World or None: A History of the World Nuclear Disarmament Movement Through 1953*, vol. 1 (Stanford, CA: Stanford University Press, 1993).

Emergency Core Cooling System

The Emergency Core Cooling System (ECCS) system was designed to inject large amounts of water into a nuclear reactor core in case the reactor core threatened to overheat and melt. It was the primary safety feature for the light water reactor developed in the United States in the early 1950s. Members of the Union of Concerned Scientists complained in 1971 that there was a lack of experimental evidence that this system would work in an emergency. These same scientists voiced their views before Atomic Energy Commission (AEC) hearings in 1972. In the meantime, the AEC made test experiments of the ECCS at the AEC's National Reactor Testing Station in Idaho and found serious problems in the system. Six tests were run; the system failed in each of them. These hearings and the failed experiments caused the AEC to tighten standards of performance slightly, but only after the manufacturers had made the case that their emergency cooling systems worked. These new standards also came under fire from critics at an 18-month series of hearings before the AEC between January 1972 and July 1973. A revised "Emergency Core Cooling System Interim Acceptance Criteria" was issued by the government in December 1973. Despite these efforts at revamping the ECCS, it was the failure of this type of system in the Three Mile Island accident that allowed the partial meltdown of the core reactor.

Suggested reading: Irvin C. Bupp and Jean-Claude Derian, *Light Water: How the Nuclear Dream Dissolved* (New York: Basic Books, 1978).

Energy Reorganization Act

In November 1974 Congress passed the Energy Reorganization Act, which abolished the Atomic Energy Commission (AEC). Responding to criticism that the AEC had too close a tie to the nuclear industry, Congress divided the AEC's functions into two agencies—a new department, the Energy Research and Development Administration (ERDA), assumed the promotional role, and the Nuclear Regulatory Commission (NRC) the regulatory role. Initially, the nonregulatory part of the AEC—reactor development, physical research, and military applications—became part of the ERDA. The NRC had administrative responsibility for nuclear reactor regulation, nuclear materials safety and safeguard, and nuclear regulatory research. Existing personnel were simply transferred into the new agencies. This quick fix did little to solve the regulatory problems because the NRC still shared the pro–nuclear industry orientation of the AEC. Later in 1977, the ERDA was transferred into the Department of Energy.

Suggested readings: Frank G. Dawson, *Nuclear Power: Development and Management of a Technology* (Seattle: University of Washington Press, 1976); Daniel F. Ford, *The Cult of the Atom: The Secret Papers of the Atomic Energy Commission* (New York: Simon and Schuster, 1982).

ENIAC

ENIAC was the computer used at Los Alamos for the complex mathematical calculations necessary to work out the theories for the hydrogen bomb. The name ENIAC stands for Electronic Numerical Integrator and Calculator. It was designed by Herman Goldstine, a professor at the University of Pennsylvania's Moore School of Engineering, and engineering colleagues with government funding seeking a calculating machine that used vacuum tubes rather than gears. Goldstine's version contained more than 19,000 vacuum tubes. In a chance meeting with John von Neumann, the mathematician, in the summer of 1944, Goldstine found himself committed to a new use for the ENIAC. Von Neumann redesigned the format of the new machine, turning it into an electronic digital computer. Next, von Neumann convinced the physicists at Los Alamos to try the ENIAC, and for six weeks in December 1945 and January 1946, it crunched numbers. Stanislaw Ulam found the calculations promising, and research on the hydrogen bomb continued. Then in 1950 von Neumann began to design a more advanced computer—the MANIAC (Mathematical Analyzer, Numerical Integrator and Computer)—which was built both at Princeton University and at Los Alamos. Until this new computer was available, the ENIAC was still in operation, and Ulam used it to show that Edward Teller's initial design for a hydrogen bomb was unworkable.

Suggested reading: Richard Rhodes, *Dark Sun: The Making of the Hydrogen Bomb* (New York: Touchstone, 1995).

Eniwetok Atoll Test Site

In 1947 the U.S. Navy took over control of the Eniwetok Atoll from Japan for a series of nuclear tests. It is about 3,000 miles west of Hawaii and is the most northwesternly of the 32 separate atolls in the Marshall Islands. The atoll itself is an oval ring 20 miles long and 10 miles wide, with 40 small islands. After seizing Eniwetok in February 1945, the United States obtained the Marshall Islands as a U.S. Trust Territory. Despite scientists' recommendations that meteorological conditions were unsafe for atomic testing, the U.S. government selected Eniwetok Atoll as a site for such tests. After two tests, testing was moved from Bikini Atoll to Eniwetok because

Eniwetok had a larger land mass and lagoon. In April 1948, nuclear testing began in Operation Sandstone at Eniwetok, and over the next decade, 43 detonations, most of them atmospheric, were exploded at the atoll. The 146 inhabitants had been evacuated earlier to Ujelang, an uninhabited atoll with one third the land surface of Eniwetok and a much smaller lagoon. It was on the small island of Elugelab that the first hydrogen bomb test, Mike, took place on November 1, 1952. The island completely disappeared as the result of this test.

Rehabilitation of this atoll has been slow because of the lethal radiation left behind in the tests. In 1977, a cleanup of radioactive material was undertaken by the U.S. government, and by 1980, former inhabitants began to return to Eniwetok. Fear of contamination, however, has made most of these people reluctant to stay and live at the Eniwetok Atoll. In 1986, the former inhabitants received part of the $150 million for damages caused by the nuclear testing in return for waiving their rights to sue the U.S. government.

Suggested readings: Ronni Alexander, *Putting the Earth First: Alternatives to Nuclear Security in Pacific Island States* (Honolulu: Matsunaga Institute for Peace, University of Hawaii, 1994); Jane Dibblin, *Day of Two Suns: U.S. Nuclear Testing and the Pacific Islanders* (London: Virago Press, 1988); International Physicians for the Prevention of Nuclear War, *Radioactive Heaven and Earth: The Health and Environmental Effects of Nuclear Weapons Testing in, on, and Above the Earth* (New York: Apex Press, 1991); Richard Rhodes, *Dark Sun: The Making of the Hydrogen Bomb* (New York: Touchstone, 1995).

Enola Gay

The *Enola Gay* was the B-29 bomber that delivered the atomic bomb on Hiroshima on August 6, 1945. Colonel Paul Tibbets, the commander of the 509th Composite Group, was the pilot, and Major Thomas Ferebee was the bombardier. The bomber was named after Colonel Tibbets's mother. Major General Curtis LeMay had selected Hiroshima as the target. The flight originated from Tinian on early the morning of August 6, 1945. The bomb had been armed in flight by Captain William Parsons. Over Iwo Jima the two supporting B-29s with a group of scientists joined the *Enola Gay* on its bombing mission. At an earlier briefing the aiming point for the bomb had been the Aioi Bridge in the middle of the city. Forty-three seconds after its release, the bomb detonated at 1,890 feet. A shock wave buffeted the aircraft even though it was already 13 miles from ground zero. This bomber became world famous, and it has been preserved for public display.

Suggested reading: Gordon Thomas and Max Morgan-Witts, *Ruin from the Air: The Atomic Mission to Hiroshima* (London: Hamish Hamilton, 1977).

Col. Paul W. Tibbets is at the controls of the B-26 *Enola Gay* just before takeoff to deliver the atomic bomb on the Japanese city of Hiroshima on August 6, 1945.

ESECOM

ESECOM, or the Environmental, Safety, and Economic Committee on Magnetic Confinement Fusion Reactors, was formed by the U.S. Department of Energy's Office of Fusion Energy in 1985 to study the viability of magnetic fusion reactors. Over the next four years, chairperson John Holdren, a physicist from the University of California at Berkeley, gathered a group of fusion and fission experts to study the environmental, safety, and economic realities of a variety of fusion and fission reactors. The committee examined 10 fusion reactors and four fission reactors designed with these factors in mind. To do this, the committee had to devise new techniques to measure radioactivity, safety, and waste disposal characteristics. Members developed a five-category scale of radioactive materials starting from the most volatile materials, which are most easily released in an accident (Category I), to the least volatile materials, which are unlikely to be released (Category V). Levels of safety assurance proposed ranged from Level 1, the safest, to Level 5, the most dangerous. Finally, a scale of nuclear waste was determined from most hazardous to that qualified for shallow burial. After subjecting the fusion reactors to this analysis, were safer to operate, and had less nuclear waste than fission reactors. It concluded that the cost of electricity from magnetic fusion reactors would be competitive with the high end of costs from fission reactors. This study has led supporters of fusion reactors to propose their use in a new electric power generation system to be built in the near future.

Suggested reading: T. Kenneth Fowler, *The Fusion Quest* (Baltimore, MD: Johns Hopkins University Press, 1997).

Euratom. *See* European Atomic Energy Community

European Atomic Energy Community

The European Atomic Energy Community (Euratom) is an organization formed in 1957 by Belgium, France, West Germany, Italy, Luxembourg, and the Netherlands to promote the speedy growth of nuclear industries in its membership countries. The founders intended Euratom to be an instrument to pool the scientific and industrial potential of the six participating countries toward that goal. Its secondary mission was to control the use of nuclear materials in Europe. This organization was the outcome of almost two years of formal discussions between European representatives. A treaty was signed in Rome in March 1957 that established a council of ministers as its decision-making body. A commission of representatives from each of the six governments implemented policy from the council of ministers. Euratom had two divisions: the Supply Division in Brussels, whose job was to acquire uranium for countries that belonged to the European Economic Community (EEC), and the Safeguards Division in Luxembourg, whose task was to monitor movements of uranium and spot any irregularities. Euratom took over the control of nuclear research centers in Belgium, Italy, and the Netherlands and a plutonium research laboratory in West Germany. These cooperative efforts were promising, but a common policy in the field of industrial promotion of atomic energy was lacking. The organization did serve as a body for international agreements, particularly with the United States. Despite early promise, by the early 1960s French hostility to Euratom had almost wrecked it as an organization. The French showed special dislike for Euratom's monitoring functions.

Suggested reading: Bertrand Goldschmidt, *The Atomic Complex: A Worldwide Political History of Nuclear Energy* (La Grange Park, IL: American Nuclear Society, 1982).

F

Fallout

The term *fallout* refers to the radioactive debris produced by a nuclear explosion that settles to the surface of the earth. A nuclear explosion produces four types of radiation—gamma radiation, beta radiation, alpha radiation, and neutrons. Most of the radiation is spread via soil and debris carried upward in the mushroom cloud, which then falls over a wide area. These radioactive materials can severely damage people, animals, and plant life. Symptoms of radiation sickness are headaches, fever, thirst, dizziness, loss of appetite, lethargy, nausea, diarrhea, vomiting, hair loss, discoloration of fingernails, hemorrhaging, and burns. Long-term effects include various cancers, particularly leukemia. Early tests of nuclear weapons proved that radioactivity was a problem, but scientists and administrators from the Atomic Energy Commission (AEC) were able to short-circuit publicity until incidents from radioactive fallout became too common. Scientists also learned that rain intensified radioactive fallout, so they were careful to schedule tests in areas and times of clear weather. Unfortunately, sometimes rain showers several days later would produce radioactive hot spots around the country.

The magnitude of the problem of fallout surfaced during the Upshot-Knothole tests at the Nevada Testing Site in 1953. Reports reached the AEC about animals suffering side effects of radiation and dying. Veterinarians were called in, and they identified the problem as iodine-131. Iodine-131 is one of the radioactive elements produced during fission and has a half-life of eight days. In both humans and animals, iodine-131 is stored in the thyroid gland. This report was sent to AEC authorities, who had the report rewritten and then ignored it. Moreover, the authorities refused to pay damages for the loss of animals.

The next big fallout news came out of the Bravo Nuclear Test in the Pacific in February 1954. News of radioactive ash contamination of the fish and crew of a Japanese trawler, the *Lucky Dragon*, went international. Some scientists began to speak out about the dangers of radioactive fallout. The AEC responded by conducting a publicity campaign in the national media, telling the American public how safe nuclear testing was and of the benefits of testing nuclear weapons. Fallout reached a peak in 1963 because nuclear testing had been particularly heavy in 1961 and 1962. Scientists found that it takes around two years before radioactive products leave the stratosphere. Once nuclear testing in the atmosphere ended in the 1970s, fallout was reduced to low levels. Only in rare cases, such as the Chernobyl nuclear accident in 1986, has fallout become an international problem.

Suggested readings: Philip L. Fradkin, *Fallout: An American Nuclear Tragedy* (Tucson: University of Arizona Press, 1989); International Physicians for the Prevention of Nuclear War, *Radioactive Heaven and Earth: The Health and Environmental Effects of Nuclear Weapons Testing in, on, and Above the Earth* (New York: Apex Press, 1991); Robert L. Miller, *Under the Cloud: The Decades of Nuclear Testing* (New York: Free Press, 1986).

Fallout Shelters

A fallout shelter craze hit the United States in the early 1960s as the fear of a possible nuclear war with the Soviet Union became a possibility. Any underground facility built to withstand a nuclear blast and supplied with survival food and water constituted a fallout shelter. Most were one- or two-room underground reinforced concrete shelters constructed in the backyard of a homestead. American civil defense specialists had used evidence from nuclear bomb tests to lobby for a national fallout shelter campaign. President John Kennedy had long been a supporter of civil defense, so he was receptive, as were many Republican leaders. In a speech in April 1961 President Kennedy made some remarks that fallout shelters might be needed in a speech on the Berlin Crisis. National media took up the idea, and a national craze over building fallout shelters began. For the next couple of years, companies appeared and banks approved loans for fallout shelters. Once the Kennedy administration realized the uproar, it tried to lay down public fears. While the fallout shelter craze gradually died out, it was indicative of the fear of nuclear war that most Americans had in the 1950s and 1960s.

Suggested reading: John Dowling and Evans M. Harrell, eds., *Civil Defense: A Choice of Disasters* (New York: American Institute of Physics, 1987).

Farm Hall

The Farm Hall was the English country manor in the town of Godmanchester near Cambridge, England, were 10 prominent German physicists were held for six months after Germany's surrender in World War II.

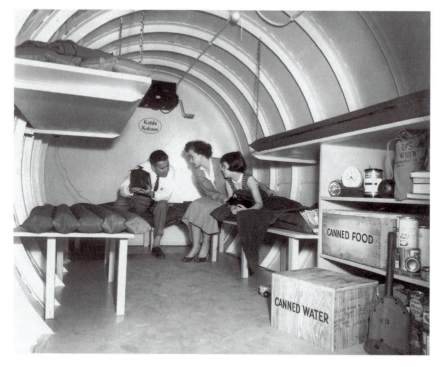

As the Cold War heated up in the mid-1950s, civilians were encouraged to build fallout shelters to protect themselves in case of a nuclear war. This fallout shelter was called a Kidde Kokoon and was manufactured by Walter Kidde Nuclear Laboratories of Garden City, Long Island, New York. (Photo reproduced from the Collections of the Library of Congress)

These physicists had been gathered together as part of Mission Alsos to learn how far Germany's nuclear program had progressed by the end of the war. Unbeknownst to the scientists, the Farm Hall had previously been used as a "safe house" by British Intelligence (MI6), and an elaborate system of hidden microphones was in place. The result was that the conversations between the 10 physicists were recorded over the six-month period. Their conversations confirmed that the Nazi nuclear program lagged far behind the Manhattan Project and that these scientists still had not figured out some of the basic answers. The recordings were kept secret until 1992, but access by certain investigators to the recordings produced a number of publications that have caused a lingering controversy over the role of these German physicists in the Nazi atomic program.

Suggested readings: Jeremy Bernstein, *Hitler's Uranium Club: The Secret Recordings at Farm Hall* (Woodbury, NY: American Institute of Physics, 1996); Samuel A. Goudsmit, *Alsos* (New York: Henry Schuman, 1947).

Fat Man

The atomic bomb that was dropped on Nagasaki on August 9, 1945, was called the Fat Man, in reference to Winston Churchill. It was a fission bomb using 13.5 pounds of plutonium-293, which was exploded by implosion. This bomb was large, measuring 3.5 meters in length by 1.5 meters in diameter and weighing 4.5 tons. It just barely fit into the bomb bay of a B-29 bomber. Fat Man had an explosive punch of 20,000 tons of TNT, or 20 kilotons (20 kt). A study of the Nagasaki blast performed after the war noted that it had been the more efficient of the two bombs, with 17 percent of its plutonium core fissioned to produce energy. Because of the poor aerodynamics of the Fat Man's design, it had to be redesigned after the war. Its replacement was the Mark-4 Fat Man design.

Suggested readings: James N. Gibson, *Nuclear Weapons of the United States: An Illustrated History* (Atglen, PA: Schiffer Publishing, 1996); Ferenc Morton Szasz, *The Day the Sun Rose Twice: The Story of the Trinity Site Nuclear Explosion, July 16, 1945* (Albuquerque: University of New Mexico Press, 1984).

Federal Radiation Council

The Federal Radiation Council (FRC) was formed in August 1959 to advise the president on radiation matters. President Dwight Eisenhower established it in response to the controversy about the effects of nuclear bomb testing fallout in the late 1950s. The FRC followed in the footsteps of the RAND Corporation's report on the ecological impact of radiation fallout in 1953. Its membership consisted of the president's cabinet and scientific advisers. President Richard Nixon transferred most of the organization of the FRC to the Environmental Protection Agency in 1970 as part of a reorganization. The FRC was then disbanded. During its short existence, the FRC accomplished little and was nothing more than a public relations exercise by the Eisenhower administration.

Suggested reading: Howard Ball, *Justice Downwind: America's Atomic Testing Program in the 1950s* (New York: Oxford University Press, 1986).

Federation of American Scientists

The Federation of American Scientists (FAS) became the umbrella organization for a diverse body of opinion on the wisdom of nuclear tests and the dangers of radioactivity. This group was formed on October 31, 1945, in Washington, DC as a forum to advocate civilian control of atomic energy in the congressional debate over the May-Johnson Bill. Membership came from representatives of the Atomic Scientists of Chicago (ASC), the Association of Los Alamos Scientists (ALAS), the Association of Manhattan Project Scientists (AMPS), and the Association of Oak Ridge Scientists at Clinton Laboratory (AORSCL). Its original name was the Federation of Atomic Scientists, but the organization changed its name to broaden its membership and appeal. A national organization was then formed on November 16–17, 1945, and the name changed in a December 7–8 meeting. In the merger of the Chicago groups, the FAS also acquired the *Bulletin of the Atomic Scientists*, which it soon made its official journal. The federation opened an office in Washington, DC to provide a place to educate public officials and the general public on nuclear issues. Its membership has fluctuated from a high of 3,000 members in 1949 to a low of around 1,500 in the early 1960s. The organization stabilized at 2,300 in 1971, and it attracted some prominent scientists—Hans Bethe, George Kistiakowsky, Harold Urey, and Victor Weisskopf, among others. While the membership was ambivalent about the antinuclear positions of other organizations, a majority of the members indicated in a 1975 survey the need for a slowdown on the construction of nuclear power plants until further study on safety problems.

Suggested readings: Milton S. Katz, *Ban the Bomb: A History of SANE, the Committee for a Sane Nuclear Policy, 1957–1985* (Westport, CT: Greenwood Press, 1986); Jerome Price, *The Antinuclear Movement* (Boston, MA: Twayne, 1982); Donald A. Strickland, *Scientists in Politics: The Atomic Scientists Movement, 1945–46* (West Lafayette, IN: Purdue University Studies, 1968).

Federation of Atomic Scientists. *See* Federation of American Scientists

Fermi, Enrico

Enrico Fermi (1901–1954) was an Italian-American physicist whose contributions as an experimental and theoretical physicist directly led to the development of the atomic bomb. He was born on September 29, 1901, in Rome, Italy. His father was a railroad employee who reached a high rank in the civil administration, and his mother was a schoolteacher. His marriage in 1928 to Laura Capon, the daughter of an Italian admiral, produced a son and a daughter. Despite some parental opposition, Fermi won a scholarship to study physics at the Normal Superior School of the University of Pisa. Fermi received his Ph.D. in physics from the university in 1922. After winning a competitive fellowship in 1922, Fermi traveled to the University of Göttingen to work with theoretical physicist Max Born. Soon after his return from Germany, Fermi spent several months at the University of Leyden, where Paul Ehrenfest befriended him. His academic career in Italy was divided between a 2-year stay at the University of Florence and a 12-year term at the University of Rome. His eminence as a theoretical physicist

Italian-American physicist Enrico Fermi
played a leading role in both experimental
and theoretical research on the nature of
the atom. His leadership in building
Chicago Pile One contributed to the
building of the atomic bomb.

allowed him to win a competition for a chair of theoretical physics at the
University of Rome in 1926. Beginning in 1930, Fermi spent several summers in the United States, teaching summer school at the University of
Michigan and learning to like America, especially its well-equipped scientific
laboratories.

Fermi was at home among both experimental and theoretical physicists.
His style was quick and insightful, and because he was rarely wrong, his
friends nicknamed him "the Pope." It was not until the early 1930s that
Fermi turned his research to nuclear physics. His 1933 paper on beta decay
transformed the qualitative hypothesis of Wolfgang Pauli into a quantitative
detailed theory of predictive value, which increased Fermi's reputation as a
theorist. Despite the brilliance of his analysis, the editor of *Nature* rejected
his manuscript, and he had to publish it elsewhere. Soon Fermi and his
colleagues initiated an experiment on artificial radioactivity produced by
neutron bombardment. In the course of their experiment the team discov-

ered the existence of a category of neutrons that were called slow neutrons. They used a retardant to slow the speed of neutrons produced in fission. It was this discovery and the excellence of his theoretical publications that won Fermi the 1939 Nobel Prize for Physics.

Fermi used the occasion of his being awarded the Nobel Prize to remove his family from Fascist Italy to the United States. His wife's Jewish background and Italy's new racial laws made him fearful for her and his children's safety. His fame as a physicist made it easy to find a teaching post in the United States. His first post was at Columbia University, where he stayed from 1939 to 1945. He became an American citizen in July 1944.

From the beginning, Fermi was involved in the building of an atomic reactor. An atomic reactor was a way to produce energy from a controlled chain reaction. The initial stages of the research were conducted at Columbia University. When the research on the atomic reactor was moved from Columbia University to the University of Chicago, Fermi had to commute between New York and Chicago, making it difficult for him to maintain control over the experiment. Moreover, because of Italy's entering the war, Fermi was now considered an enemy alien subject to travel restrictions. A presidential ruling exempting Italians as enemy aliens in 1942 solved this problem. By October 1942, enough graphite and uranium ore were available to set up the reactor (Chicago Pile One) under Fermi's guidance. Fermi monitored the progress of the building of the pile and on December 2, 1942, Fermi conducted a successful test of the reactor, with around 40 scientists and spectators in attendance. After this test, Fermi transferred his reactor research to the Argonne site near Chicago.

Shortly after the reactor experiment, Fermi began making quick visits to Los Alamos Laboratory to serve as a consultant. It was not until 1944 that Fermi and his family moved to Los Alamos to work directly on the atomic bomb project. While there, he developed a close working relationship with Hungarian mathematician John von Neumann. Oppenheimer made Fermi an associate director of the Los Alamos Laboratory and placed him in charge of F Division. This division was to investigate problems that did not fit into the work of other divisions.

Shortly after the end of the war, Fermi accepted a position at the University of Chicago's new Institute for Nuclear Physics under his good friend Samuel Allison. Fermi devoted himself to research and teaching, and the two subjects that interested him the most: high-energy physics and the applications of computers. He also served on the General Advisory Committee of the Atomic Energy Commission (AEC) from 1947 to 1950. Fermi testified at Oppenheimer's Security clearance hearing and supported him without reservations. In the summer of 1954, Fermi began to suffer a serious stomach disease caused by a cancer that had metastasized before the diagnosis. After an unsuccessful operation, Fermi died on November 29, 1954, in Chicago.

Suggested readings: Henry A. Boorse, Lloyd Motz, and Jefferson Hane Weaver, *The Atomic Scientists: A Biographical History* (New York: Wiley Science Editions, 1989); Laura Fermi, *Atoms in the Family: My Life with Enrico Fermi* (Chicago: University of Chicago Press, 1954); Pierre de Latil, *Enrico Fermi: The Man and His Theories* (New York: P. S. Eriksson, 1964); Emilio Segre, *Enrico Fermi: Physicist* (Chicago: University of Chicago Press, 1970).

Fermi I

Fermi I was the name of an experimental commercial breeder reactor that suffered a reactor accident in October 1966. Located at Lagoona Beach, Michigan, about 30 miles from Detroit on the shores of Lake Erie, the Fermi I was owned and operated by Detroit Edison. It was a sodium-cooled fast breeder reactor. Most reactors in production at the time were light water reactors (LWRs), but the fast breeder reactor was designed to produce more plutonium than it consumed. Sodium was chemically unstable but was the best coolant for such a reactor. The fast breeder reactor was considered the reactor of the future, because it promised to produce an inexhaustible supply of energy at a relatively low cost. This plant's construction had been controversial because of its nearness to Detroit and the experimental nature of its reactor, but the Atomic Energy Commission (AEC) had approved its construction over the opposition of its Advisory Committee on Reactor Safety Safeguards (ACRS). Despite court action by labor unions and politicians, the plant was finally completed in early 1966. Sometime on October 5, 1966, a piece of metal became dislodged in the reactor and partially blocked the flow of the core coolant. This blockage caused the core assembly to overheat, threatening the safety of the reactor. At the time, the operators knew something was wrong but not the cause. It took six months to determine the damage and nearly a year before the operators found out the reason for the blockage. In the meantime, the only hope was that the reactor would fix itself and cause no further problems. It was not until 1969 that the AEC gave permission to restart the reactor. Another incident in May 1970 caused the reactor to be shut down again. Despite a restart in July 1970, the cost of running the reactor and the considerable downtime resulted in Detroit Edison's closing the reactor permanently in 1972.

Suggested readings: Frank G. Dawson, *Nuclear Power: Development and Management of a Technology* (Seattle: University of Washington Press, 1976); John Grant Fuller, *We Almost Lost Detroit* (New York: Reader's Digest Press, 1975); John May, *The Greenpeace Book of the Nuclear Age: The Hidden History, the Human Cost* (New York: Pantheon Books, 1989).

Feynman, Richard Phillips

Richard Feynman (1918–1988) was an American physicist who is most famous for his research on quantum electrodynamics (QED). He was born on May 11, 1918, in the Brooklyn borough of New York City, but soon his family moved to Far Rockaway, Long Island. His father was a sales manager of a uniform factory. He married Arlene H. Greenbaum in 1942, Mary Louise Bell in 1952, and Gweneth Howarth in 1960 and has a daughter and a son. His father groomed him to be a scientist at an early age. After graduating from high school in 1935 at Far Rockaway High School, Feynman wanted to go to Columbia University, but he was not admitted. The Massachusetts Institute of Technology (MIT) accepted him, and he started out to study mathematics. In 1939, he obtained his B.S. with honors. His professors at MIT recommended him highly for graduate school at Princeton University. He was finally accepted at Princeton despite an admission quota against Jews. His Ph.D. in physics was obtained from Princeton University in 1942, where his adviser was John Archibald Wheeler. Even before finishing his dissertation, Feynman was called upon to work on designing an automated artillery aiming device at the Frankford Arsenal in Philadelphia.

Feynman was recruited for atomic bomb work at the Los Alamos Scientific Laboratory in 1942. His assignment was in the Theoretical Division under Hans Bethe, and he became a group leader. Some of the older theoretical physicists liked to bounce ideas off him, because he was never afraid to express an opinion. He was also a bit of a prankster. Several times he challenged the security regulations in place at Los Alamos. Despite his fun-loving activities, he was considered a productive member of the theoretical team, exploring ways to build a functional atomic bomb.

After the war, Hans Bethe recruited Feynman for Cornell University. He remained there until 1950, when he moved to the California Institute of Technology (Caltech). The remainder of his academic career was spent at Caltech, where he developed his QED theory. He used this theory to explain the remaining puzzles in electricity and magnetism. QED also predicts the strength of the electron's magnetic field. This highly mathematical model, which was worked out by Feynman and Freeman Dyson, made both of them international figures. Feynman served on the presidential commission to study the space shuttle *Challenger* disaster in 1986 and published his own report criticizing both the space agency and the commission's final report. He died on February 15, 1988, in Los Angeles, California, of abdominal cancer.

Suggested readings: Henry A. Boorse, Lloyd Motz, and Jefferson Hane Weaver, *The Atomic Scientists: A Biographical History* (New York: Wiley Science Editions, 1989); Richard P. Feynman, *Surely You're Joking, Mr. Feynman! Adventures of a Curious Character* (New York: Norton, 1984); James Gleick, *Genius: The Life and*

Science of Richard Feynman (New York: Pantheon Books, 1992); Anthony Serafini, *Legends in Their Own Time: A Century of American Physical Scientists* (New York: Plenum Press, 1993).

First Strike Doctrine

The First Strike Doctrine is the military strategy for a preemptive attack of an enemy before or in the middle of diplomatic negotiations. In one sense, the Japanese attack on Pearl Harbor on December 7, 1941, was a first strike action, although the Japanese planned for a simultaneous declaration of war. This doctrine became more critical in the atomic era. A first strike scenario was adopted by the U.S. Joint Chiefs of Staff in a planning document on September 20, 1945. The intended target was the Soviet Union and its industrial capacity. A committee was formed to study the impact on the Soviet Union of an atomic attack by the United States. This committee reported—the Harmon Report in May 1949—that a planned atomic attack on 70 Soviet cities would not defeat the Soviet Union. Such an attack might cause 2.7 million deaths and destroy Soviet industrial capacity from 30 to 40 percent, but it would probably increase the will of the Soviet people to resist. Although the American military continued to envisage a possible first strike scenario, the U.S. government never formally adopted this doctrine. The idea, however, did pop up from time to time in American military circles as a possibility.

Suggested reading: David Holloway, *Stalin and the Bomb: The Soviet Union and Atomic Energy, 1939–1956* (New Haven, CT: Yale University Press, 1994).

Fission

The discovery of fission, or the splitting of the atom, was one of the most important advances in science in the twentieth century. Fission is the process of splitting a heavy nucleus into two approximately equal parts thus releasing a relatively large amount of energy plus a few neutrons. Depending on the nucleus and the speed of the neutrons the neutrons can sometimes cause other nuclei to split apart, setting up a chain reaction. It is this chain reaction that is controlled in nuclear reactors to produce energy. The chain reaction used in nuclear weapons utilizes a high flux of neutrons to fission the entire fuel in a short period of time. Fissionable isotopes, uranium-233, uranium-235, and plutonium-239 produce the conditions necessary to create a chain reaction. Only uranium-235 occurs naturally; the others are products of nuclear reactors. German experimental physicist Otto Hahn and Austrian theoretical physicist Lise Meitner, first observed fission in an experiment using neutrons to bombard uranium. Later, Meitner and her cousin, Otto Frisch, interpreted further Hahn experiments showing that the uranium nucleus had split into two parts, releasing energy. It was this discovery of fission that led physicists around the world to contemplate the possibility of an atomic bomb.

Suggested readings: Jo Ann Shroyer, *Quarks, Critters, and Chaos: What Science Terms Really Mean* (New York: Prentice-Hall, 1993); Richard Wolfson, *Nuclear Choices: A Citizen's Guide to Nuclear Technology* (Cambridge: MIT Press, 1991).

Flerov, Georgi Nikolaevich

Georgi Flerov (1913–1990) was a Soviet nuclear physicist who was instrumental in convincing Stalin of the feasibility of building an atomic bomb. He was born on March 2, 1913, in Rostov-on-Don, Russia. Flerov graduated from the Leningrad Polytechnical Institute in 1938. His first position was as a member of the staff of the Physico-Technical Institute from 1938 to 1941. It was in 1940 that Flerov and Lev Rusinow conducted experiments that revealed that more than two neutrons are emitted during fission. He also worked with Konstantin S. Petrzhak and discovered the spontaneous fission of heavy nuclei. In the midst of his research, Flerov noted that articles on nuclear energy had disappeared from scientific literature in American, British, and German journals. Soon after the German invasion of the Soviet Union, Flerov was drafted into the Soviet Air Force. It was as a young officer that Flerov wrote a letter to Stalin in April 1942 warning him about a possible atomic bomb. Although there is no evidence that Stalin read the letter, it was passed to top Soviet scientific authorities. In 1943, Flerov left the army and became a member of the staff of the Institute of Atomic Energy. In 1960, after years of research, some at Arzamas-16 (Sarov), Flerov was made the director of the Nuclear Reactions Laboratory of the Joint Institute for Nuclear Research in Dubna. He remained as director until 1988. Flerov died on November 19, 1990.

Suggested readings: Thomas B. Cochran, Robert S. Norris, and Oleg A. Bukharin, *Making the Russian Bomb: From Stalin to Yeltsin* (Boulder, CO: Westview Press, 1995); David Holloway, *Stalin and the Bomb: The Soviet Union and Atomic Energy, 1939–1956* (New Haven, CT: Yale University Press, 1994).

Franck, James

James Franck (1882–1964) was a German-American physicist who is most famous for his research on energy transfer and his opposition to dropping the atomic bomb on Japanese cities. He was born on August 26, 1882, in Hamburg, Germany. His father was a banker in Hamburg. Franck's academic career was at the University of Heidelberg and the University of Berlin. In 1906, he received his doctorate in physics. During World War I, Franck served in the German army and won two Iron Crosses. After the war, he moved to the University of Göttingen, where he held the chair of experimental physics. Franck worked closely with Gustav Hertz, experimenting on the quantized nature of energy transfer. They showed that energy can only be absorbed in quite definite and precise amounts. Their research won them the 1925 Nobel Prize for Physics. By the early 1930s, Franck had become the director of the University of Göttingen's Second Physical Institute. Since he was a decorated veteran of the war, the Jewish law of 1933 did not apply to him, but he resigned his directorship in protest of this law in a glare of publicity. Because of the publicity, he had to leave Germany; he took this opportunity to go to Copenhagen, leaving a considerable fortune behind.

After spending a year in Copenhagen, Franck emigrated to the United States in 1935. Beginning in 1938, Franck held the position of professor of physical chemistry at the University of Chicago until 1949. During the war, he worked in the chemistry section of the Metallurgical Laboratory in the Manhattan Project. His claim to fame in World War II, however, was the Franck Report, authored by Leo Szilard and himself and sent to the secretary of war, arguing against the use of the atomic bomb against Japanese civilian targets—an appeal not heeded. Franck returned to academic pursuits. Most of his postwar research was on photosynthesis. He died in Göttingen, West Germany, on May 21, 1964.

Suggested readings: Alan D. Beyerchen, *Scientists Under Hitler: Politics and the Physics Community in the Third Reich* (New Haven, CT: Yale University Press, 1977); Henry A. Boorse, Lloyd Motz, and Jefferson Hane Weaver, *The Atomic Scientists: A Biographical History* (New York: Wiley Science Editions, 1989); John Daintith, Sarah Mitchell Elizabeth Tootill, and Derek Gjertsen, *A Biographical Encyclopedia of Scientists*, 2nd ed., vol. 1 (Bristol, UK: Institute of Physics, 1994).

Franck Report

This report by a committee of concerned scientists argued against the military use of the atomic bomb against Japan. James Franck was a German emigrant physicist who had worked on the Manhattan Project. He became concerned in the spring of 1945 about the problems that would be unleashed by the use of the atomic bomb. In June 1945 six committees were formed under the auspices of Arthur Compton at the Metallurgical Laboratory in Chicago to recommend to the government policies on the atomic bomb. Franck became the chair of the Committee on Social and Political Implications, which had Donald Hughes, James Nickson, Eugene Rabinowitch, Joyce Stearns, and Leo Szilard as members. They conducted discussions on the University of Chicago campus and issued a report on June 11, 1945, which was sent in secret to the secretary of war. Besides advising against the use of the atomic bomb against Japan, the report warned about postwar nuclear proliferation and recommended international control of atomic energy. They were also concerned about a possible arms race between the Soviet Union and the United States after the war. Compton and Franck tried to meet with Secretary of Defense Henry Stimson in Washington, DC but ended up leaving the report in Stimson's office. There is no evidence that this report was ever brought to the attention of President Harry Truman.

Suggested readings: Martin J. Sherwin, *A World Destroyed: The Atomic Bomb and the Grand Alliance* (New York: Knopf, 1975); Jonathan M. Weisgall, *Operation Crossroads: The Atomic Tests at Bikini Atoll* (Annapolis, MD: Naval Institute Press, 1994).

French Nuclear Program

The French government was slow to enter the nuclear arena despite a considerable history of research on atomic energy by French scientists. Shortly after the end of World War II, Charles de Gaulle, the president of the provisional government, established the French Atomic Energy Commission (Commissariat à l'Énergie Atomique) with the purpose of promoting the development of atomic projects. Behind this was the desire on de Gaulle's part to build a French atomic bomb. French efforts, however, were divided between building reactors to generate cheap and reliable electrical power and weapons research. France built a small heavy water nuclear reactor in December 1948 at Fort de Châtillon, but it was a very low power reactor and only good for research purposes. This reactor was a zero-power uranium oxide pile named Zoe (Zero Oxide eau lourde). French progress was slow in developing atomic energy because of the instability of the French government but also because of the expense of development and uncertainty about a military program. Plans were developed in 1953 for two graphite reactors and a plutonium extraction plant at Marcoule on the Rhone River. French authorities tried to enlist British support for further advances, but the British were afraid of alienating the Americans. In the mid-1950s, the debate over joining the European Atomic Energy Community (Euratom) raged, but France finally joined in 1955. France toyed with the idea of building a nuclear weapons capability, but it was France's isolation after the Suez Canal Crisis in November 1956 that made French leaders opt for a French nuclear arms program.

It was only in the Fifth Republic under Charles de Gaulle that France became aggressive in forming its own independent nuclear policy. In 1957, France tried to buy a large quantity of uranium from Canada, but the deal fell through. Discovery of uranium deposits in the Limousin mountains, however, alleviated part of the problem. Despite some success with the British gas-graphite reactor, the French government decided to enter into agreements in 1958 with the United States for light water reactor technology.

This decision was taken with considerable bitterness, and France's relationship with Euratom deteriorated by the early 1960s. In February 1960, France detonated its first nuclear device at Reggan, Algeria, in the Sahara Desert in North Africa. A total of 14 atomic tests, 4 atmospheric and 10 underground, were conducted in the Sahara before losing the testing to Algerian independence in 1965. Soon President Charles de Gaulle had a nuclear capability to project French policy. Next on President de Gaulle's agenda was the building of a hydrogen bomb. Moreover, he wanted France to develop the technology on its own. It took over five years, but on August 24, 1968, France detonated a hydrogen bomb at its Pacific test center. Between 1968 and 1974, the French conducted 40 tests in the atmosphere at their Mururoa and Fangataufa Test Sites in the Pacific. The last French tests were begun in 1995, with considerable opposition from world opinion. Six underground tests were conducted at the Fangataufa Atoll with the last test on January 27, 1996. Because French nuclear policy depended on the credibility of its nuclear response, nuclear testing had been a central tenet of French government policy. On January 29, 1996. The French government announced the end to French nuclear testing in the South Pacific.

France has the most ambitious nuclear energy program of any country. It plans for almost three fourths of its electricity needs to be supplied by nuclear power plants. The response of the French government to the oil crisis in the mid-1970s was to pursue a policy of "all electricity, all nuclear power." To support this program France has built a strong nuclear industry and depends on the export of nuclear technology. As of December 31, 1996, France had 57 reactors in operation, providing over 77 percent of its electricity needs, with 3 more units under construction.

Suggested readings: Eric Arnett, ed., *Nuclear Weapons After the Comprehensive Test Ban: Implications for Modernization and Proliferation* (Oxford: Oxford University Press, 1996); Helena Flam, ed., *States and Anti-Nuclear Movements* (Edinburgh, Scotland: Edinburgh University Press, 1994); Bertrand Goldschmidt, *The Atomic Complex: A Worldwide Political History of Nuclear Energy* (La Grange Park, IL: American Nuclear Society, 1982); Bertrand Goldschmidt, *Atomic Rivals* (New Brunswick, NJ: Rutgers University Press, 1990); *Nuclear Power Reactors in the World* (Vienna, Austria: International Atomic Energy Agency, 1997).

Friends of the Earth

The Friends of the Earth (FOE) is an environmentalist lobby group of 60 or so branches in the United States and parallel organizations in other industrialized countries that have played an active role in the antinuclear movement. David Brower founded this organization in 1969 in New York City to compete with other environmental groups, but with a specific orientation against nuclear proliferation and a distrust of nuclear energy. Brower had been the executive director of the Sierra Club, but he left this organization after feuding with its executive board over the polices and direction of the club. Later, Brower would also leave the FOE but not before it was established as one of the top environmentalist groups in the world. At one time the Friends of the Earth had a membership in the United States of around 23,000 persons. Among its early missions was to attempt to block passage of the Price-Anderson Act extension bill in 1977.

A spinoff organization to the national groups is the Friends of the Earth International (FOEI). The mission of the FOEI is to coordinate the activities of national Friends of the Earth groups. It opened its headquarters in London in 1970. Within a few years the FOEI had recruited 200 local groups in over 50 countries. Membership in the FOEI worldwide is between 700,000 and 1 million. Its organizational structure resembles a confederation, with local groups being autonomous. A Coordinating Council with representatives from all groups meets once a year and determines general orientation. An executive committee with seven members has general oversight responsibilities. The British groups have been in the forefront of efforts against nuclear projects in the United Kingdom since the mid-1970s. The strategy of local Friends of the Earth has been to concentrate on a small number of specific topics and direct them into single-issue campaigns. One such campaign in 1975 was against British plans for a new reprocessing plant for oxide fuel (the Thermal Oxide Reprocessing Plant—THORP) for the Windscale nuclear facility. Their representatives produced a mass of documentation against the reprocessing plant, but the British government went ahead anyway. The issue was brought before the House of Commons and passed despite considerable opposition.

Suggested readings: Jerome Price, *The Antinuclear Movement* (Boston, MA: Twayne, 1982); Paul Wapner, *Environmental Activism and World Civic Politics* (Albany: State University of New York Press, 1996).

Frisch, Otto Robert

Otto Frisch (1904–1979) was an Austrian-British physicist most famous for helping to work out the problem of the splitting of the atom, with his aunt Lise Meitner, and the Frisch-Peierls Memorandum. He was born on October 1, 1904, in Vienna, Austria. His father was a printer–publisher. Frisch received his education at the University of Vienna and in 1926 obtained his doctorate. After brief employment in private industry, his first position was at the German National Physical Laboratory in Berlin, where he worked in the optics division on a grant to develop a new unit of brightness to replace the candle power from 1927 to 1930. In 1930, he moved to the University of Hamburg. Frisch worked closely with Otto Stern at the University of Hamburg, until the Nazi racial law of 1933 made him leave. He was also forced to give up a Rockefeller Foundation Fellowship when he went to work with Enrico Fermi in Rome. Stern traveled to find positions for his Jewish collaborators, and he founded the Workingmen's College of Birkbeck in London, England, for Frisch. Frisch worked there on a position at Academic Assistance Grant with one of Rutherford's former pupils, Patrick Maynard Stuart Blackett. After his grant ran out, Frisch traveled to Copenhagen, where Niels Bohr arranged for him a five-year grant from a foundation. He worked on building a cyclotron and on various experiments to repeat and extend Enrico Fermi's findings on neutron bombardment. It was in Sweden on Christmas 1938 that Frisch discussed with his aunt, Lisa Meitner, the results of Otto Hahn and Fritz Strassmann's experiment and concluded that fission had taken place. Frisch told Niels Bohr a few days later, who immediately recognized the significance. Frisch and George Placzek confirmed that fission had taken place in an experiment conducted on January 13–14. Letters by Lisa Meitner and Frisch to *Nature* announced their findings. Frisch moved to England, where he collaborated with Rudolf Peierls on the Frisch-Peierls Memorandum, which stated that it was feasible to build an atomic bomb. During World War II, Frisch was recruited with the British team to work on the Manhattan Project at the Los Alamos Scientific Laboratory. After the war, he returned to Great Britain and continued his academic career. He died in 1979.

Suggested readings: Henry A. Boorse, Lloyd Motz, and Jefferson Hane Weaver, *The Atomic Scientists: A Biographical History* (New York: Wiley Science Editions, 1989); Otto Robert Frisch, *What Little I Remember* (Cambridge: Cambridge University Press, 1979).

Frisch-Peierls Memorandum

In 1940, two German physicists worked out the theory behind the atomic bomb in the now-famous Frisch-Peierls Memorandum. Both physicists were refugees from Nazi Germany and knowledgeable about current atomic theory. They took experiments and theories developed in the late 1930s and concluded by mathematical calculations that such a bomb was possible from the energy released from uranium-235. Their estimation that a single kilogram of pure uranium-235 would produce a large bomb was overly optimistic, but the general theory proved to be correct. It meant that enough uranium could be produced to make an air deliverable bomb possible. This memorandum had a limited but effective distribution among American and British scientists. Since both physicists left Germany after 1933, neither scientist had any intention of sharing his views with German scientists still in Germany. They passed their report to Australian physicist Mark Oliphant, who wrote a cover letter for it to the Tizard Committee. Henry Tizard subsequently appointed the MAUD Committee to study the issue. This report convinced physicists, and later the MAUD Committee, that it was practical to build an atomic weapon.

Suggested readings: Lennard Bickel, *The Deadly Element: The Story of Uranium* (New York: Stein and Day, 1979); Ronald William Clark, *The Birth of the Bomb* (New York: Horizon Press, 1961); Richard Rhodes, *The Making of the Atomic Bomb* (New York: Simon and Schuster, 1986).

Fuchs, Emil Julius Klaus

Klaus Fuchs (1911–1988) was a German-British physicist who worked on the Manhattan Project at Los Alamos in World War II and who turned over secret data about the U.S. nuclear program to the Soviet Union. He was born on December 29, 1911, in Russelsheim, Germany. His father was a Lutheran pastor who later became a Quaker. Both his mother and a sister committed suicide. His academic training in physics and mathematics was at the University of Leipzig. After a two-year stay at Leipzig, Fuchs transferred to the University of Kiel. It was at Leipzig where Fuchs joined the Socialist Party, but soon he opposed their policies and joined the Communist Party in 1932. After the Nazis took power in Germany and his life was in danger, Fuchs fled to England, where he continued his studies in mathematical physics, first with Nevill Mott at Bristol University and then with Max Born at the University of Edinburgh in 1937. He was interned in Canada as an enemy alien for over six months in 1940. Soon after Fuchs's return to England in December 1940, he began to work with Rudolf Peierls on the theoretical problems of an atomic bomb. It was sometime in 1941 that his active spying for the Soviet Union began with contacts with Simon Davidovich Kremer (code name Alexander). He basically revealed that a uranium atomic bomb was a possibility, and he had knowledge of Peierls's calculations. On August 7, 1942, he became a naturalized British citizen.

In December 1943, Fuchs moved to the United States with a British team to work on gaseous diffusion research for the Kellex Corporation and Harold Urey at Columbia University. Fuchs reestablished contact with the Soviet in New York City, and his contact was Harry Gold (code name Raymond). A crisis over implosion at Los Alamos made Hans Bethe request assistance from Rudolf Peierls, and Fuchs went with him. Fuchs arrived at Los Alamos in August 1944 and immediately plunged into the implosion problem. His contributions helped make implosion work, and he wrote five papers on implosion theory. Because of his isolation at Los Alamos, he was not able to pass on materials to Soviet spies until February 1945.

From early 1945 until his return to Great Britain in June 1946, Fuchs was inactive as a spy. His new job was head of the division of theoretical physics at the new nuclear research center at Harwell. Part of his position was to advise the British on their atomic program. Sometime in 1947, Fuchs resumed spying for the Soviet Union. Although Fuchs had left Los Alamos without any classified documents, he had a photographic memory. Moreover, he had written the Los Alamos *Handbook on Implosion Technique.* Because of this, he was able to reproduce engineering drawings for both the British and the Soviets. Toward the end of his spying career, Fuchs became less enthusiastic about giving the Soviets atomic secrets because he was beginning to have doubts about how the information was being used by the Soviet scientists. Fuchs was arrested on February 2, 1950, for spying for the Soviet Union, and he spent 9 years of a 14-years sentence in an English prison, first at Wormwood Scrubs and then at Wakefield. In February 1951, he was formally deprived of his British citizenship for reasons of disloyalty. After release from prison in June 1959, Fuchs moved to East Germany, where he became deputy director of the East German Central Institute for Nuclear Physics at Rossendorf, near Dresden. He married his childhood sweetheart, Greta Keilson, soon after his arrival in East Germany. After the director of the institute defected to the West in 1964, Fuchs became its director. Fuchs died on January 28, 1988.

Suggested readings: H. Montgomery Hyde, *The Atom Bomb Spies* (New York: Atheneum, 1980); Norman Moss, *Klaus Fuchs: The Man Who Stole the Atom Bomb* (New York: St. Martin's Press, 1987); Robert Chadwell Williams, *Klaus Fuchs, Atom Spy* (Cambridge: Harvard University Press, 1987).

Fusion

Fusion is the process that fuels the sun and stars and has been considered a rival of fission as a source of energy on earth. The process of fusion involves the formation of a heavier nucleus from two lighter ones accompanied by a release of energy. This thermonuclear reaction yields 10 times more energy than the fission of uranium from the same initial amount of matter. While fusion is a simple nuclear process and produces less radioactive by-products than fission, it is difficult to produce because of the extremely high temperatures and pressures required for ignition. Electrical repulsion must be overcome before the nuclei can get close enough to fuse. Edward Teller learned of the possibility of building a thermonuclear, or hydrogen, bomb based on fusion in the mid-1940s and argued for building one at Los Alamos during World War II. Since such a bomb needed an atomic bomb to detonate it, the building of an atomic bomb took precedence, much to Teller's discontent. After World War II, the American government decided to build a hydrogen bomb. Since fusion is a better source of energy than fission, scientists have searched for ways to unleash its potential for peaceful purposes. The cold fusion fiasco resulted from scientists attempting to take shortcuts. A more promising initiative has been to use lasers to generate fusion.

Suggested readings: T. Kenneth Fowler, *The Fusion Quest* (Baltimore, MD: Johns Hopkins University Press, 1997); Bertrand Goldschmidt, *The Atomic Complex: A*

G

Gas-Cooled Reactors

The chief rival to the light water reactor (LWR) in the design of nuclear power plants in the post–World War II era was the gas-cooled reactor. While the gas-cooled, graphite-moderated reactor is similar to the pressured-water reactor in that both use a fluid to transport heat to the steam generator, these reactors are inherently larger and have a higher operating temperature than water-cooled reactors. They also were more expensive to build and maintain than light water reactors. The technology for gas-cooled reactors was developed most fully by British and French scientists after World War II. The gas-cooled design became the chief competition to the American LWR monopoly in the 1950s and 1960s. This battle pitted American technology against British-French technology with the Americans winning. By 1970, the gas-cooled reactor was no longer a serious contender in this competition.

Suggested reading: Bertrand Goldschmidt, *The Atomic Complex: A Worldwide Political History of Nuclear Energy* (La Grange Park, IL: American Nuclear Society, 1982).

Geiger, Hans Wilhelm

Hans Geiger (1882–1945) was a German physicist who is most famous for his research on alpha particle scattering and the Geiger counter. He was born on September 30, 1882, at Neustadt, Germany. His education was at the University of Munich and the University of Erlangen. He received his doctorate in physics from the University of Erlangen in 1906. Geiger was studying at the University of Manchester in 1907 under Arthur Schuster when Ernest Rutherford arrived, and for the next five years, they worked together to perfect an early version of an electrical device that could detect and count alpha particles. In 1912, Geiger returned to Germany to take a teaching and research position in Berlin at the Physikalische Technische Reichsanstalt. He served in a German artillery unit in World War I. Later, he became a professor of physics at the University of Kiel in 1925. In 1929, he moved to the University of Tübingen. Then in 1936, Geiger became head of physics at the Technical University of Charlottenberg in Berlin. He also was selected to be editor of the prestigious journal *Zeitschrift für Physik*. His health deteriorated during World War II, and he died several months after the war on September 24, 1945, in Potsdam.

Geiger is most famous for his construction of the Geiger counter. During his study of gaseous ionization, he developed in 1908 an instrument for counting alpha particles. He used this counter to study alpha particles, but it proved to have a wide variety of uses. Geiger continued to make improvements on his particle counter until 1928, when in collaboration with Walther Muller the final version of the Geiger-Muller counter appeared. With this counter, Geiger was able to confirm the Compton effect in 1925 and, later, to study cosmic radiation.

Suggested readings: David Abbott, ed., *The Biographical Dictionary of Scientists: Physicists* (London: Blond Educational, 1984); Henry A. Boorse, Lloyd Motz, and Jefferson Hane Weaver, *The Atomic Scientists: A Biographical History* (New York: Wiley Science Editions, 1989); John Daintith, Sarah Mitchell, Elizabeth Tootill, and Derek Gjertsen, *A Biographical Encyclopedia of Scientists*, 2nd ed., vol. 1 (Bristol, UK: Institute of Physics, 1994).

General Advisory Committee

The General Advisory Committee (GAC) was a nine-member body appointed by the president of the United States in 1946 to advise the Atomic Energy Commission (AEC) on scientific and technical issues. This body had a prestigious membership, with J. Robert Oppenheimer as its chair. For several years in the late 1940s, the General Advisory Committee determined AEC policy. Among its early policy decisions was to mute popular enthusiasm for the immediate benefits of nuclear energy. In a July 1947 report the GAC foresaw numerous technical and economic problems that had to be overcome before nuclear energy could fulfill its potential. It was, however, the reluctance of the GAC to advocate the building the hydrogen bomb that found it in trouble with Congress and the group of scientists around Edward Teller. This cautious go-slow approach ultimately caused the political power of the GAC to be replaced by the more aggressive Joint Committee on Atomic Energy of the U.S. Congress.

Suggested reading: Daniel F. Ford, *The Cult of the Atom: The Secret Papers of the Atomic Energy Commission* (New York: Simon and Schuster, 1982).

General Electric

The General Electric (GE) Company has long been a major force in the nuclear energy business. It entered the business shortly after World War II when the company assumed operation of the Hanford facilities, taking over from the Dupont Corporation. GE's primary responsibility was to provide plutonium for bombs. It operated the Hanford facility from 1946 to 1968 under the direct supervision of the Atomic Energy Commission (AEC). Since all three reactors at Hanford were showing signs of disrepair in 1946, General Electric had to reduce plutonium and polonium production until new reactors could be brought on-line in September 1946. General Electric also worked on various nuclear reactor types including navy designs for the light water nuclear reactor for ship propulsion. Its nuclear research center was at the Knolls Atomic Power Laboratory (KAPL) in Schenectady, New York. Efforts to build a successful intermediate breeder reactor ran into serious design problems and were canceled in the spring of 1950. By the mid-1950s, General Electric and Westinghouse were the two biggest players in designing and providing raw materials in the nuclear industry. To stimulate interest among the energy companies, General Electric was active in the turnkey program to build nuclear plants at a fixed price. General Electric continued to be the leading builder of nuclear reactors in the boom time of the 1960s and 1970s. Shortly after the Three Mile Island accident in 1979, the company began to scale back its nuclear operations as demand for nuclear power plants slowed. Much as its competition in the nuclear field in the 1990s, General Electric is no longer active in the nuclear business except for maintaining existing contracts. *See also* Turnkey Nuclear Energy Plants.

Suggested reading: Mark Hertsgaard, *Nuclear Inc.: The Men and Money Behind Nuclear Energy* (New York: Pantheon Books, 1983).

Geneva Conference of 1955

At the 1955 Geneva Conference, scientists from 73 countries gathered together for the first conference on the peaceful uses of atomic energy. American scientists appeared ready to dominate the scientific proceedings, but they soon found out that scientists from other countries were also knowledgeable about nuclear energy. The Americans were especially surprised at the openness of the members of the Soviet delegation. Only 5 countries in 1955 had experience with nuclear reactors—Britain, Canada, France, the Soviet Union, and the United States—but scientists from the nonnuclear states were interested in technical issues of atomic power. In the two-week conference, a variety of scientific papers outlined the potential of nuclear energy. Scientists returned to their respective countries as advocates of the future of peaceful uses for nuclear energy. Within six months of the convention, 29 countries had signed agreements with the United States for help in developing nuclear programs for civilian projects.

Suggested reading: Peter Pringle and James Spigelman, *The Nuclear Barons* (London: Michael Joseph, 1981).

Gen 75

Gen 75 was a committee of senior ministers in the British government who decided issues of atomic energy in the period from 1945 to 1947. It was a quasi-secret group consisting of Clement Attlee, the prime minister, Ernest Bevin, foreign secretary, Stafford Cripps, president of the Board of Trade, Herbert Morrison, lord president of the Council, and later, Hugh Dalton, chancellor of the Exchequer, and Arthur Greenwood, Lord Privy Seal. Gen 75 was in existence for 16 meetings beginning in August 1945 and lasting for 16 months. In these meetings the preliminary discussions were conducted about the launching of a British atomic bomb project, plans for international control of atomic energy, and discussions on the tenuous nature of Anglo-American relations on atomic issues. The actual decision to make a British atomic bomb was made by another committee, Gen 163, a month after the last Gen 75 meeting. To replace Gen 75 the Labor government established a Ministerial Atomic Energy Committee to deal with questions of atomic policy. Most of the actual planning on atomic energy during Gen 75's existence was done by an advisory committee, the Advisory Committee on Atomic Energy. Sir John Anderson was made chairperson of this largely technical committee and was given quasi-ministerial status. This committee was also disbanded in late 1947.

Suggested reading: Margaret Gowing, *Independence and Deterrence: Britain and Atomic Energy, 1945–1952*, 2 vols. (New York: St. Martin's Press, 1974).

Gerlach, Walther

Walther Gerlach (1889–1979) was a German physicist active in Nazi research on an atomic bomb. He was born in 1889. Gerlach studied physics at the University of Tübingen and received his Ph.D. in 1911. Service in the German army during World War I interrupted his academic career. After the war, Gerlach found a teaching position at the University of Frankfurt am Main, where he remained from 1920 to 1924. It was there that his collaboration with Otto Stern on quantum physics experiments made his reputation in physics circles. In 1924, Gerlach moved to the University of Tübingen, and he stayed there until 1929. In 1929, he was appointed professor at the University of Munich. Except for his service in World War II, he remained there until his retirement in 1957.

Gerlach's role in the German atomic bomb program was as an administrator. His contributions to the war effort in the early days of the war were with the German navy's torpedo research project. German scientists and administrators had become unhappy with the performance of Abraham Esau as head of the Nuclear Research Group of the Reich Research Council and looked for a successor. Gerlach replaced Esau as director of the Nuclear Research Group of January 1, 1944. He continued supporting the research efforts of both Kurt Diebner at Gattow and Werner Heisenberg at the Kaiser Wilhelm Institute for Physics in Berlin. Gerlach was 1 of the 10 German physicists held at the Farm Hall for six months after the war.

Suggested readings: Henry A. Boorse, Lloyd Motz, and Jefferson Hane Weaver, *The Atomic Scientists: A Biographical History* (New York: Science Editions, 1989); David Irving, *The German Atomic Bomb: The History of Nuclear Research in Nazi Germany* (New York: Simon and Schuster, 1967); Thomas Powers, *Heisenberg's War: The Secret History of the German Bomb* (New York: Knopf, 1993).

German Atomic Bomb Program. *See* Nazi Atomic Bomb Program

German Nuclear Program

The German nuclear program was late developing because of postwar reconstruction and restrictions placed upon Germans. A fear of German rearmament and their possession of atomic weapons made other western European countries and the United States reluctant to allow Germany access to nuclear technology. The vigorous growth of the German industrial economy only reinforced this fear. In the aftermath of the 1955 Geneva Conference on peaceful uses of atomic energy, the German government began to investigate the potential of nuclear energy for Germany. Two government bodies were formed: the Ministry for Atomic Questions and the German Atomic Commission. The key figure in the promotion of German atomic energy was industrialist Karl Winnacker, chairman of the chemical company Farbwerke Hoechst A. G. But it was only after the formation of the European Atomic Energy Community (Euratom) that other European countries acquiesced to the transmission of nuclear technology to Germany. At the same time German industry began to consider nuclear energy seriously, but the German industrialists insisted on private control of a civil nuclear energy program. A consortium of West German utilities engaged an American firm, Westinghouse, to design and supply materials for a light water reactor at Gundremmingen, Bavaria, in 1961. Siemens was West Germany's primary reactor manufacturer and representatives from this company decided to migrate to light water reactor technology. After negotiations between Siemens and Westinghouse for a consortium fell through, Siemens and AEG-Telefunken combined to provide an industrial base in Germany for the development of nuclear reactors. By the mid-1960s Germany's attention turned to the next generation of reactors and the breeder reactor. Karl Winnacker continued his role as czar of German nuclear energy and promoted the future of the breeder reactor. An experimental breeder reactor project was set up at Karlruhe under Karl Wirtz, a key physicist in the Nazi Atomic Bomb Project during World War II and a leader of the Karlsruhe Nuclear Research Center. A full-scale breeder reactor was constructed in the small town of Kalkar on the Lower Rhine beginning in 1973. As of December 31, 1996, Germany had 20 reactors in operation, providing electricity for slightly over 30 percent of the country. German authorities have also been active in shutting down reactors, with 13 closed in the last two decades. In 1999 Germany still had 20 reactors in operation.

Suggested readings: *Nuclear Power Reactors in the World* (Vienna, Austria: International Atomic Energy Agency, 1997); Peter Pringle and James Spigelman, *The Nuclear Barons* (London: Michael Joseph, 1981).

Glaser, Donald A.

Donald Glaser (1926–   ) is an American physicist who designed the bubble chamber to study the behavior of atomic nuclei. He was born on September 21, 1926, in Cleveland, Ohio. His undergraduate education was at the Case Institute of Technology, where he graduated in 1946. He went to graduate school at the California Institute of Technology and obtained his Ph.D. in 1949. His first position was in the physics department at the University of Michigan. In 1959 he moved to the University of California at Berkeley. In the late 1960s, Glaser turned from physics to work in molecular biology. He was appointed professor of physics and biology in 1964.

In 1952 Glaser decided to design a better version of the Wilson cloud chamber. The weakness of the original cloud chamber was that the gas it used has low density. Glaser proposed using liquid instead of gas. This liquid at its boiling point would permit the observation of bubbles formed on the path of ions when the pressure was suddenly lowered. He experimented with hydrogen and xenon. Luis Alvarez seized on this idea and built a huge bubble chamber using liquid hydrogen at the Lawrence Radiation Laboratory. Later, bubble chambers and computers were tied together to trace particle behavior. Glaser won the 1960 Nobel Prize in Physics for his invention of the bubble chamber. *See also* Wilson, Charles Thomson Rees.

Suggested readings: David Abbott, ed., *The Biographical Dictionary of Scientists: Physicists* (London: Blond Educational, 1984); Emilio Segre, *From X-Rays to Quarks: Modern Physicists and Their Discoveries* (San Francisco: W. H. Freeman, 1980).

Gofman-Tamplin Manifesto

John Gofman and Arthur Tamplin were scientists at the Biomedical Research Division of the Livermore branch of the Lawrence Radiation Laboratory in the late 1960s who challenged the safety of the Atomic Energy Commission's radiation protection standards. The Atomic Energy Commission (AEC) wanted to counter the argument of Ernest Sternglass, who had asserted that 400,000 American babies had died from the aftermath of radioactive fallout from nuclear tests in Nevada. Tamplin was commissioned to counter Sternglass's evidence, and he did so, but in the process he admitted that around 4,000 casualties had ensued from testing. This figure was unacceptable to the AEC, and pressure was placed on him to drop his revised estimate. Tamplin was able to convince John Gofman to support him, and both Tamplin and Gofman suffered from AEC retaliation. Next, the two scientists challenged radiation standards. In a paper delivered at the Nuclear Science Symposium of the Institute of Electrical and Electronic Engineers on October 29, 1969, they maintained the need for a 10-fold reduction in the maximum permissible radiation dose standard to the general population. They based this demand on the basis of their research that previous standards would cause 16,000 cancer cases in the United States annually. Moreover, since young people were more susceptible to cancer from radiation than adults, the incident of cancer would be concentrated in the youngest parts of the population. This manifesto came under severe attack from other AEC scientists both for being scientifically wrong and for being bias on the part of Gofman and Tamplin. They were called before the Joint Committee on Atomic Energy (JCAE) to defend their views. In this forum, they challenged the JCAE to appoint a jury of eminent scientific peers to judge their findings. No such jury was formed. Both Gofman and Tamplin suffered retaliation in the form of censorship attempts and other restrictions including losing their jobs at Livermore and their federal grants, but their findings added fuel to the growing lack of confidence in nuclear safety by antinuclear activists. In 1979, before and after Three Mile Island, Gofman and Tamplin published a book, *Poisoned Power: The Case against Nuclear Power Plants*, continuing their attack on the dangers of radiation. Gofman returned to teaching at the University of California at Berkeley and continued to speak out on the dangers of radiation.

Suggested readings: Corinne Browne and Robert Munroe, *Time Bomb: Understanding the Threat of Nuclear Power* (New York: William Morrow, 1981); John W. Gofman and Arthur R. Tamplin, *Poisoned Power: The Case against Nuclear Power Plants before and after Three Mile Island* (Emmaus, PA: Rodale Press, 1979); Richard S. Lewis, *The Nuclear-Power Rebellion: Citizens vs. the Atomic Industrial Establishment* (New York: Viking Press, 1972).

Gold, Harry. *See* Soviet Atomic Spying

Göttingen Manifesto

Eighteen prominent German physicists issued a manifesto from Göttingen, West Germany, in 1957 to West German Chancellor Konrad Adenauer protesting Adenauer's assertion that nuclear weapons were no different from other military weapons. These scientists realized that the North Atlantic Treaty Organization (NATO) strategy envisaged a war on West German soil that might kill millions of German citizens. This manifesto proposed that West Germany renounce the possession of nuclear weapons. Soon the left wing of the German Social Democratic Party adopted the scientists' program and made it a political issue. Although the Social Democratic Party lost the next election, the issue of nuclear weapons on German soil entered the political arena for good. It was one of the major issues that the Green Party adopted in the 1980s.

Suggested reading: Spencer R. Weart, *Nuclear Fear: A History of Images* (Cambridge: Harvard University Press, 1988).

Goudsmit, Samuel

Samuel Goudsmit (1902–1978) was a Dutch-American physicist whose job at the end of World War II was to find out the nature of German atomic bomb research. Goudsmit had achieved some fame in the physics community because of his collaboration with George Uhlenbeck in introducing the concept of electron spin to the structure of the atom. In 1927, he became a professor at the University of Michigan. During the early part of World War II, Goudsmit kept busy conducting secret research on radar at the Radiation Laboratory at the Massachusetts Institute of Technology (MIT). He was then sent to Great Britain to work on a radar project there. Goudsmit was strongly anti-Nazi because his parents had remained in the Netherlands and perished in a Nazi concentration camp. Goudsmit was recruited for the Alsos Mission because he was not working on the Manhattan Project, knew several European languages, and was knowledgeable about physics. For these reasons he was appointed the scientific adviser to the Alsos Mission under Lieutenant-Colonel Boris T. Pash in the spring of 1943. He also personally knew most of the leading German scientists, especially Werner Heisenberg. It was his analysis of captured German documents confirmed by other scientists that the Germans were nowhere close to building an atomic bomb. He selected the 10 physicists to be sent to Farm Hall. After the war, he wrote a book describing the Alsos Mission, *Alsos*, pointing out the limitations of the German scientists working on the German bomb project. His book was highly controversial, and debate over some of his charges is still relevant. After the war, Goudsmit returned to the University of Michigan, and he died in 1978.

Suggested readings: Henry A. Boorse, Lloyd Motz, and Jefferson Hane Weaver, *The Atomic Scientists: A Biographical History* (New York: Wiley Science Editions, 1989); Samuel Goudsmit, *Alsos* (New York: Henry Schuman, 1947); David Irving, *The German Atomic Bomb: The History of Nuclear Research in Nazi Germany* (New

York: Simon and Schuster, 1967); Thomas Powers, *Heisenberg's War: The Secret History of the German Bomb* (New York: Knopf, 1993).

Graphite

The availability of purified graphite was a major problem in building atomic reactors in the World War II era. Heavy water was preferable as a moderator for a chain reaction, but Enrico Fermi believed that purified graphite would perform the same function. Industrial-grade graphite had too many impurities, especially boron, so purified graphite was necessary. Fortunately, American plants could produce enough graphite for reactors. The National Carbon Company, a subsidiary of Union Carbide, was the only company in the United States capable of producing large quantities of the purified graphite needed for a reactor. Purified graphite had to be processed by petroleum coke instead of the higher boron content mineral coke. Union Carbide officials rejected Physicist Leo Szilard's 1939 request for $2,000 worth of graphite for an experiment on an atomic chain reaction experiment at Columbia University. Norman Hilberry approached the National Carbide Company in 1942, on behalf of the American government, to see if they could produce large quantities of purified graphite. He wanted 250 tons by the following week. Despite this impossible demand, the National Carbon Company provided most of the purified graphite needed for the nuclear reactors in the Manhattan Project. It was this graphite that was used in Chicago's experimental reactor in 1942. German scientists' failure to realize the importance of graphite hurt their reactor program because they had to depend on Norwegian heavy water.

Suggested reading: Arthur Holly Compton, *Atomic Quest: A Personal Narrative* (New York: Oxford University Press, 1956).

Great Bear Lake Mine. *See* Eldorado Mine

Greenglass, David. *See* Soviet Atomic Spying

Greenhouse Nuclear Tests

Greenhouse was the U.S. governments's code name for its 1951 series of experimental tests on nuclear devices at Eniwetok in the Marshall Islands. Four tests were scheduled to check the functioning of devices for the next series of tests on a new type of bomb—the hydrogen bomb. The first two tests were fission devices, Shot Dog and Shot Easy. Shot George and Shot Item were experiments with fusion devices based on deuterium-tritium designs by Edward Teller. On May 8, 1951, Shot George was detonated with a yield of 225 kilotons. Shot Item tested deuterium-tritium as a booster, and its detonation on May 24 produced a 45.5-kiloton yield. This test was the crucial thermonuclear experiment. The results of these experiments gave Edward Teller the information he needed to present his theory for a hydrogen bomb to the scientific community.

Suggested readings: Stanley A. Blumberg and Gwinn Owens, *Energy and Conflict: The Life and Times of Edward Teller* (New York: Putnam's Sons, 1976); Robert L. Miller, *Under the Cloud: The Decades of Nuclear Testing* (New York: Free Press, 1986).

Greenpeace International

The environmental activist organization Greenpeace International coordinates the activities of the national Greenpeace groups. Greenpeace was formed in 1971 in Vancouver, Canada, by a small group of environmentalists and peace activists protesting nuclear testing by the United States on Amchitka Island, Alaska. At first the leaders organized themselves in a Don't Make a Wave Committee (DMWC), which evolved into the Greenpeace organization in 1972. Members of the DMWC decided on tactics to disrupt the nuclear tests on Amchitka in November 1971 by sailing a ship, the *Phyllis Cormack*, into the testing area. The original attempt was unsuccessful, but the effort gained notoriety, and soon the United States announced cessation of testing at Amchitka.

This success made Greenpeace one of the most popular of the antinuclear organizations. By 1994, it had 6 million members worldwide and a staff of over 1,000 full-time employees. The organization also owns a fleet of eight ships, a helicopter, and a hot-air balloon. Its headquarters is located in Amsterdam, Netherlands, but the organization has offices in over 30 countries. David McTaggart, a Canadian, has been a longtime chairperson. Greenpeace International tried to be nonpolitical so as to work with all governments and political parties. Its philosophy is to use direct, nonviolent action and publicize its actions. Each country, however, has a national organization, and sometimes these organizations pursue independent policies. One such organization has been Greenpeace New Zealand, who has formed close links with the Nuclear Free and Independent Pacific Movement (Hawaii).

Suggested reading: Paul Wapner, *Environmental Activism and World Civic Politics* (Albany: State University of New York Press, 1996).

Groves, Leslie R.

Brigadier General Leslie Groves (1896–1970) was the military administrator of the Manhattan Project and the driving force behind the building of the atomic bomb in World War II. He was born on August 17, 1896, in Albany, New York. He married Grace Wilson in February 1921, and they had a daughter and a son. His father, a former lawyer, was an army chaplain of very strict habits. Because of frequent moves and a poor financial situation, Groves grew up without youthful friendships and was forced to work at an early age. He attended Queen Anne High School in Seattle, Washington. After some educational preparation at the University of Washington and at the Massachusetts Institute of Technology (MIT), Groves entered West Point in 1914 via the annual presidential competition for an at-large appointment. He graduated fourth in his class in 1918 as a respected but unpopular cadet with the nickname of Greasy. Both his class ranking and his inclinations drove him into the Corps of Engineers. Among his experiences in the Corps of Engineers was work at Galveston opening the Port Isabel harbor and in Nicaragua investigating the possibility of a new Panama-type canal. On the eve of World War II, Groves became the special assistant to Quartermaster General E. B. Gregory. In this capacity Groves had the oversight responsibility for the $8 billion construction program for the army. His biggest headache was the political controversy about the construction of the Pentagon. He remained in charge of construction until the Pentagon was ready for occupancy in January 1943.

On September 23, 1942, Colonel Groves was appointed commander of the Manhattan Project and given the temporary rank of brigadier general. Among the reasons for his appointment was that he had considerable construction experience and a reputation for getting things done. His lack of tact and toughness had made him many enemies in the military and soon made him unpopular with many of his subordinates in the Manhattan Project. In particular, his relationship with the scientists was always tense. His emphasis on secrecy was especially resented by them. Unlike the scientists, Groves had as his goal obtaining a decisive weapon at the earliest possible date regardless of obstacles. It was this desire that made Groves appoint J. Robert Oppenheimer as head of Los Alamos operations despite his questionable political past. His accomplishments in gathering support from American industry and the military services was unrivaled. He was able to obtain the services of the du Pont Corporation to build the X-10 site at Oak Ridge by an appeal to the patriotism of the corporation's executives. It was his maneuvering that gave the Manhattan Project top priority in equipment, supplies, and raw materials. While his abrasive personality helped drive the scientists to produce the atomic bombs in record time, in the postwar debate over control of atomic energy, his scientific enemies used his tactics against military control of atomic energy.

After the war, Groves reluctantly turned over the Manhattan Project to the Atomic Energy Commission (AEC). He had been instrumental in the drafting of the May-Johnson Bill and fought for its passage. The rival McMahon Bill was unacceptable to Groves, and he used his considerable political influence to amend it more to his satisfaction. David Lilienthal's appointment as chair of the AEC was another blow, since they detested each other. Groves was appointed to the Military Liaison Committee of the AEC, but Lilienthal ignored both the committee and Groves. Groves initiated several political attacks against Lilienthal, but they were unsuccessful except to annoy Lilienthal. In January 1948 Groves retired from the army. He accepted a job from Remington-Rand as vice president in charge of research and development of the computer mainframe UNIVAC. He remained at this job until his retirement in 1961. Groves died of heart failure on July 12, 1970, at Walter Reed Army Hospital.

Suggested readings: Stephane Groueff, *Manhattan Project: The Untold Story of the Making of the Atomic Bomb* (London: Collins, 1967); Leslie R. Groves, *Now It Can Be Told: The Story of the Manhattan Project* (New York: Harper, 1962): William Lawrence, *The General and the Bomb: A Biography of General Leslie R. Groves, Director of the Manhattan Project* (New York: Dodd, Mead, 1988).

Groves-Anderson Memorandum

The Groves-Anderson Memorandum was negotiated during the American, Canadian, and British heads of states talks in Washington, DC in November 1945. It was a renegotiation of the wartime Quebec Agreement. Talks were held between Major General Leslie R. Groves, the American representative, and Sir John Anderson, the British representative. In this agreement the Quebec Agreement would be superseded by new provisions. The American, British, and Canadian governments would not use atomic weapons against other parties without prior consultation with each other. This was a watering down of the consent provision between the United States and Great Britain in the earlier agreement. None of the governments would disclose any information or enter into negotiations concerning atomic energy without prior consultation with each other. Special efforts were to be made to acquire uranium and thorium supplies. The Combined Policy Committee was authorized to continue to work together on common atomic energy concerns. This memorandum was considered an executive document and not a treaty, so its contents were kept secret for a number of years. Later, after the United States had repudiated the provisions of the memorandum, President Harry Truman admitted that he had been unaware of the existence of the Groves-Anderson Memorandum. It was the restrictive provision of the 1946 Atomic Energy Act, however, that curtailed the Groves-Anderson Memorandum and ended its functioning.

Suggested reading: Margaret Gowing, *Independence and Deterrence: Britain and Atomic Energy, 1945–1952*, vol. 2 (New York: St. Martin's Press, 1974).

H

Hahn, Otto

Otto Hahn (1879–1968) was a German physical chemist who is famous for his experiment that proved the atom could be split. He was born on March 8, 1879, in Frankfurt, Germany. His father was a well-to-do tradesman. He took his degrees in organic chemistry from the University of Marburg. His intent was to work in one of the German chemical firms, but his adviser, Theodor Zincke, sent him to England to study English. It was on a visit to London in 1904 that famous English chemist Sir William Ramsay introduced Hahn to research in the field of radioactivity. His research led him to the discovery of radiothorium, which was one of thorium's 12 isotopes. The next year Hahn traveled to Montreal and worked with Ernest Rutherford on thorium radiation. After a year at McGill University, Hahn returned to Germany for a position as an assistant in the Institute of Organic Chemistry of Emil Fischer in Berlin. At the time in 1906, Hahn was the only chemist in Berlin conducting research on radioactivity. It was shortly afterward in 1907 that he started his 30-year collaboration with Austrian physicist Lise Meitner. In 1912, he moved with Meitner to the Kaiser Wilhelm Institute for Physical Chemistry and Electrochemistry, where he stayed until leaving for military service in 1914. Most of his work in World War I was with the development and military uses of poison gas.

After the war, Hahn returned to the Kaiser Wilhelm Institute, where he conducted chemical experiments on radioactive materials. Since Hahn was anti-Nazi, his position at the Kaiser Wilhelm Institute became precarious after the Nazis took power. When Lise Meitner had to leave Germany in 1938, he helped her escape to Sweden. He continued his experiments with another chemist, Fritz Strassmann but consulted with Meitner on the physics side of the results. In December 1937 and January 1938, Hahn communicated to Meitner results of an experiment that puzzled Strassmann and him. Meitner and her nephew, Otto Frisch, interpreted the evidence to be fission of the atom, and so they reported it for publication. Hahn became unhappy about this and later claimed that Meitner played no role in the experiments. In 1944, Hahn received the Nobel Prize for Physics, but because of the Nazi regime's policy forbidding the acceptance of Nobel awards, he was unable to accept it until after the war.

Hahn became a leader in the West German nuclear energy industry in the postwar world. His interest in nuclear affairs made him active in the campaign to warn the world about the dangers of nuclear weapons. He died on July 28, 1968, in Göttingen, Germany.

Suggested readings: John Daintith, Sarah Mitchell, Elizabeth Tootill, and Derek Gjertsen, *A Biographical Encyclopedia of Scientists*, 2nd ed., vol. 1 (Bristol, UK: Institute of Physics, 1994); William R. Shea, ed., *Otto Hahn and the Rise of Nuclear Physics* (Dordrecht, Holland: D. Reidel, 1983); Ruth Lewin Sime, *Lise Meitner: A Life in Physics* (Berkeley: University of California Press, 1996).

Halban Affair

The Halban Affair was an effort by the American government to isolate and control French atomic scientists working on the Manhattan Project and, after the end of World War II, to restrict information to the French government. Hans van Halban, an Austrian expatriate physicist, and Lew Kowarski, a Polish expatriate physicist, had worked closely with Frederick Joliot-Curie on heavy water experiments in Paris on the eve of World War II. They had also applied and been granted patents on some of the processes. After the defeat of France in May 1940, Halban and Kowarski immigrated to England and took with them the French supply of heavy water. They established, with the help of the British government, a heavy water research project at the Cavendish Laboratory. Their experiments indicated that a chain reaction was possible and could be controlled to produce energy. It was their report that was instrumental in the MAUD Committee's recommendation to proceed further on building an atomic bomb. Both Halban and Kowarski were sent to the United States to conduct further research. They were joined by several other French atomic scientists, Pierre Auger, Bertrand Goldschmidt, and Jules Gueron, and for awhile, they were accepted as coequals with the American and British scientists. American authorities, especially General Leslie Groves, soon became uncomfortable with the idea of French scientists learning atomic secrets, so they were reassigned to a research project in Montreal, Canada. In Canada, the French scientists worked on heavy water experiments and slowly fell behind current research in the United States. After the use of the atomic bombs on Japanese cities, the French scientists wanted to return to France and continue basic research

The Hanford Plant near Richland, Washington, produced the plutonium for the "Fat Man" atomic bomb in World War II and plutonium for postwar nuclear weapons. This plant was one of the war-built nuclear reactors.

on atomic energy. Both the American and British governments, however, were reluctant for them to return to France and share their knowledge of atomic energy with their French colleagues, many of whom were Communists. General Groves and other American officials were fearful that details of the Manhattan Project would be given to the Soviet Union by these scientists. Each of the French scientists had to sign secrecy documents before they were allowed to return to France. This affair did little to improve relations between the French scientists and their former Allies. Most of the French scientists became leaders in the French nuclear industry business, and the restrictions of the Halban Affair were never enforced.

Suggested readings: Bertrand Goldschmidt, *Atomic Rivals* (New Brunswick, NJ: Rutgers University Press, 1990); Martin J. Sherwin, *A World Destroyed: The Atomic Bomb and the Grand Alliance* (New York: Knopf, 1975).

Hanford Plant

The Hanford plutonium plants were built in 1943–1944 as part of the Manhattan Project to build an atomic bomb. General Leslie Groves had ruled out the Oak Ridge Complex for plutonium production, fearing a chain reaction accident at one of the reactors. A team of engineers surveyed a series of possible sites in the Pacific Northwest before picking the Hanford area in western Washington. Hanford was selected for its accessibility to the Columbia River and for its isolation. Six hundred square miles of land was acquired by court order and cost around $5.1 million. Lieutenant Colonel Franklin T. Matthias was placed in charge of building the plants. Initial construction plans were for three production reactors to be spaced at six-mile intervals along the Columbia River and for four chemical separation plants with two plants at each site to be constructed south ten miles behind the Gable Mountain. For safety reasons, a small mountain separated the reactors from the chemical plants. The first reactor (B Reactor) started operations on September 27, 1944. It was a water-cooled, natural uranium reactor with graphite moderators built and maintained by the du Pont Company. Within a day of initial operations, the reactor lost its radioactivity. Enrico Fermi and Archibald Wheeler figured independently that it was due to poisoning by xenon-135. After a few adjustments, the reactor resumed its production. Ultimately, three nuclear reactors (B Reactor, D Reactor, and F Reactor) were built near Hanford, as well as three chemical separation plants (T, U, and B). Du Pont had 5,800 management personnel administering a peak workforce of 45,000. By 1944, Hanford was the third largest city in the state. Several times a congressional investigation of this project was threatened for alleged waste of financial resources, but each time the probe was repulsed in the interest of national security. Due to harsh working conditions, the construction of Hanford had high labor turnover, with 140,000 workers passing through in the period 1944–1945. By February 1945, plutonium shipments began to be made to Los Alamos. Because du Pont requested relief from the responsibility of operating the facilities in peacetime, General Electric operated the Hanford laboratories of the Atomic Energy Commission (AEC) from 1946 to 1964. General Electric, however, drove a hard bargain with the AEC by making the AEC fund a complete nuclear research laboratory for the company at Schenectady, New York.

After the war, Hanford continued to produce weapons-grade plutonium for weapon development. A crisis in plutonium production developed in 1946 when the reactors at Hanford came down with "Wigner's disease," in which graphite bombarded intensely with neutrons absorbs the acquired energy by rearranging its crystal lattices, causing it to swell and block the reactor's fuel-element channels. Consequently, plutonium production at Hanford dropped in 1946–1947 as reactors had to be taken off-line for treatment. This caused the production of atomic weapons to slow at a time when the United States had an atomic monopoly and could use its possession of such weapons as a diplomatic tool. A replacement reactor (DR Reactor) and a new reactor (H Reactor) were built in the late 1940s to fill in the gap. In response to increased international tensions during the Korean

War, three new reactors (C Reactor, KW Reactor, and KE Reactor) were built to increase plutonium production. Then in 1962, a dual-purpose reactor (N Reactor) was authorized by Congress to produce plutonium, but also to provide electricity for the private power sector. Despite considerable political controversy, this reactor was constructed by General Electric and was on-line by April 1964. At its peak, Hanford had nine reactors in operation.

As the need for plutonium became less pressing in the mid-1960s, the AEC began to shut down reactors. Beginning in January 1964, three reactors at Hanford were selected for closure (F, DR, and H Reactors). These closures caused a negative political reaction both in the state of Washington and in Congress. In 1967 another round of closures shut down two more Hanford reactors (B and C Reactors). In 1969 yet another cut shut down three more reactors (C, KE, and KW Reactors). Finally, only N Reactor was left. Based on its inability to meet safety standards, N Reactor was shut down in January 1987. This ended Hanford's career as a weapons production facility.

It was at Hanford that the worst case of worker contamination took place in 1976, Harold McCluskey was on the night shift at the americium-241 unit when an explosion exposed him to massive doses of radiation. Although not expected to live, McCluskey survived but had to live for five months in an isolation ward. Doctors tried new medical techniques for treating radiation illness to save his life. He sued the federal government for the accident, and the settlement was for $275,000 plus lifetime medical expenses.

The U.S. government is slowly closing the Hanford Plant. Staff has been cut to a skeleton crew to maintain buildings and equipment no longer large enough for major projects. Much of the last decade's activities have been directed toward solving the radioactive waste problem from the production of plutonium. Current efforts are to return some of the land to private interests for farming and hunting.

Suggested readings: Rodney P. Carlisle, *Supplying the Nuclear Arsenal: American Production Reactors, 1942–1992* (Baltimore, MD: Johns Hopkins University Press, 1996); Stephane Groueff, *Manhattan Project: The Untold Story of the Making of the Atomic Bomb* (London: Collins, 1967); Paul Loeb, *Nuclear Culture: Living and Working in the World's Largest Atomic Complex* (New York: Coward, McCann and Geoghegan, 1982); Richard Rhodes, *The Making of the Atomic Bomb* (New York: Simon and Schuster, 1986); Harry Thayer, *Management of the Hanford Engineer Works in World War II: How the Corps, DuPont and the Metallurgical Laboratory Fast Tracked the Original Plutonium Works* (New York: ASCE Press, 1996).

Hardtack Nuclear Tests

Hardtack was the code name for a series of hydrogen bomb tests conducted in the spring and fall of 1958. The first series was Hardtack I, which took place in the spring of 1958 at the Eniwetok and Bikini Test Sites in the Pacific. Thirty-three nuclear devices were detonated over the course of the summer. These tests were given names of American tree and shrubs—Yucca, Fir, Nutmeg, Sycamore, and the like. These powerful tests produced several surprises such as a tidal wave from Shot Oak. Another experiment with Army Redstone rockets delivering a hydrogen bomb at high altitude in Shot Orange produced an electromagnetic pulse that could destroy a variety of electronic devices from computers to communication links.

Hardtack II was the second series of nuclear tests at the Nevada Test Site in the fall of 1958. Most of these tests were underground in shafts or tunnels. Of the above-ground blasts the largest was Shot Socorro on October 22, 1958, and it was only six kilotons. Several of the later tests, Shot Humboldt and Shot Santa Fe, had high radioactive fallout that spread over southern California. This fallout produced much negative comment from the media in California. This series ended the day before an international agreement on a ban on atomic testing came into force.

Suggested reading: Richard L. Miller, *Under the Cloud: The Decades of Nuclear Testing* (New York: Free Press, 1986).

Harwell Laboratory

Harwell Laboratory was the site of Great Britain's main research laboratory of the United Kingdom's Atomic Energy Authority. Harwell is 13 miles south of Oxford, and it was the site of a World War II airfield. Among its virtues as a site was its closeness to Oxford University, its access to London, and its location in a sparsely inhabited countryside. The site was given to the British nuclear energy program by the Air Ministry with reluctance in February 1946. Its original mission was to be responsible for the production of radioactive isotopes for research in physics, chemistry, and biology, but the need for weapons-grade uranium and plutonium soon changed its role. Two aircraft hangers were used to house two nuclear reactors of graphite and natural uranium for research. It became the headquarters of the Atomic Energy Research Establishment (AERE) under its first director, John Cockcroft. Harwell was organized along the lines of the university, with no hierarchical chain of command. Most of the basic research for the British atomic program was conducted by scientists at Harwell. To support its research mission, Harwell had a Van de Graaff generator, two cyclic accelerators, and a linear accelerator.

Suggested reading: Margaret Gowing, *Independence and Deterrence: Britain and Atomic Energy, 1945–1952* (New York: St. Martin's Press, 1974).

H-Bomb. *See* Hydrogen Bomb

Health Impact of Nuclear Accidents

The fear of radioactivity has made nuclear accidents a health concern both in the United States and abroad. *Nuclear safety* had long been a byword among advocates of atomic energy. Early on, any possible incidents of health problems involving nuclear energy were vigorously denied by members of the Atomic Energy Commission (AEC) in the United States and in similar types of organizations in other countries possessing nuclear reactors or weapons. The first crack in this defense came in 1976 when Zhores Medvedev, a biochemist refugee from the Soviet Union, made reference to a Soviet nuclear waste accident at Chelyabinsk-40 in the winter of 1957–1958 in the journal *New Scientist*. Despite denials from nuclear authorities in France, Great Britain, and the United States, evidence accumulated that such an accident had occurred and heavy loss of life had taken place. Casualty figures have never been released by the Soviet or Russian governments, but the Chernobyl Power Station accident became too notorious to ignore. It has been estimated that 3.2 million people were affected by the accident, with about of million of them children. The death toll for those engaged in the direct fighting of the accident is over 5,700 by 1990 and still growing. The accident has produced a growing health crisis in the Ukraine and Belarus, the extent of which is difficult to predict. While the Ukrainian Health Ministry's prediction of 125,000 Chernobyl-related deaths may be too high, the final total may be somewhere in that range. In contrast, the health impact of the Three Mile Island accident is not on the same scale of risk. Studies of the local inhabitants around the Three Mile Island plant have displayed more psychological than health problems.

Suggested reading: David R. Marples and Marilyn J. Young, *Nuclear Energy and Security in the Former Soviet Union* (Boulder, CO: Westview Press, 1997).

Heavy Water

Heavy water is chemically similar to ordinary water, except that the hydrogen atoms are replaced with deuterium. Heavy water is particularly suitable for use as a moderator to contain and slow neutrons in the core of a reactor, having nearly the same moderating power as ordinary water, but without its tendency to absorb neutrons. Harold Urey, an American chemist, discovered the deuterium nucleus in 1931, but at the time it had no practical value and was expensive to produce. In ordinary water about 1 molecule out of 4,500 is heavy water, so it takes 50 tons of ordinary water, and about a year of processing, to produce 10 kilograms of heavy water. Despite the cost and scarcity, physicists working on atomic energy soon found that heavy water served as an efficient retarding medium for moderating the speed of neutrons set free in splitting the atom. On the eve of

World War II, only the Norsk-Hydro Plant in Norway produced significant quantities of heavy water as a by-product in its manufacture of chemical fertilizer. Since the German atomic bomb program was designed around heavy water as the moderator for its atomic reactors, the Germans controlled this source of heavy water after their invasion of Norway in 1940. Although the British and Canadians experimented with heavy water in Canada, the Manhattan Project used purified graphite rather than heavy water for its reactors.

Suggested reading: Dan Kurzman, *Blood and Water: Sabotaging Hitler's Bomb* (New York: Henry Holt, 1997).

Heisenberg, Werner Karl

German physicist Werner Heisenberg (1901–1976) is most famous for his work on the theories of quantum mechanics and as one of the leaders of the Nazi nuclear program in World War II. Heisenberg was born on December 5, 1901, in Würzburg, Germany. His father was a professor of ecclesiastical history. He married Elisabeth Schumacher in 1937, and they had three sons and four daughters. His doctorate in physics was from the University of Munich in 1923, and his postdoctoral qualification (Habitilation) was from the University of Göttingen in 1924. His thesis adviser for his Ph.D. was Arnold Sommerfeld. After receiving his degrees, Heisenberg left Germany to work for two years at the University of Copenhagen, Denmark, with Niels Bohr. His friendship with Niels Bohr would outlast subsequent political disagreements. Returning to Germany in 1927, he obtained a professorship at the University of Leipzig. While at Leipzig, Heisenberg became famous for devising quantum mechanics and for the Heisenberg Uncertainty Principle, which states that the exact position and precise velocity of an electron cannot be determined at the same time. For this and other work on quantum mechanics problems, Heisenberg received the 1932 Nobel Prize for Physics. Heisenberg remained at the University of Leipzig until 1942.

Despite his international status, Heisenberg found himself in the middle of political controversy in the 1930s. Attacks from Aryan Physics followers for his support of Einstein's theory of relativity and Planck's quantum mechanics made Heisenberg's position precarious in the uncertain early days of the Nazi regime. A vicious article by Johannes Stark in an SS (Schutzstaffel) weekly in 1937, accusing Heisenberg of being a "white Jew," was dangerous for him. The leadership of the SS and head Heinrich Himmler allowed these attacks to continue. Only the intervention of Heinrich Himmler's mother, who was a friend of Heisenberg's mother, and Himmler's subsequent clearing of Heisenberg saved him further difficulty. Soon the German government began to use him as an ambassador for German science in lengthy trips abroad. During these trips to Great Britain and the United States, his physicist friends tried to persuade him to defect. Heisenberg was too much a German nationalist to defect, so he returned to Germany.

Once war broke out in 1939, Nazi authorities recognized the worth of his research and placed him in the Army Ordinance Research Department, along with other prominent German scientists, to work on nuclear weapon problems. Heisenberg and others recognized the potential of energy release in the splitting of atoms, but how to harness this energy release in a practical military weapon was a puzzle. His two reports in December 1939 and February 1940 outlined the principles behind a chain reaction reactor. Heisenberg believed the best approach was a reactor moderated with heavy water with natural uranium fuel. The German atomic bomb program followed his reactor theory. Heisenberg was a great theorist but was also notorious for his ineptness with numbers. In June 1942, Heisenberg was appointed director of the Kaiser Wilhelm Institute for Physics and professor at the University of Berlin. This meant that he also headed the German atomic program. Heisenberg's inexperience as an experimental physicist made him take several controversial research directions, which experimentalists opposed. He continued to experiment with a reactor at the Institute until 1944, when he moved his research to Hechingen in Swabia. A reactor was built in a cave at the foot of a cliff near Haigerloch. American forces captured Heisenberg at the end of the war and transferred him to the Farm Hall in Great Britain for interrogation. After six months there, he was released and spent the rest of his life defending his working for Hitler to other physicists.

His mission after the war was to rebuild German science. Heisenberg moved to Munich, where he became director of a large physics research institute. He died in 1976.

Suggested readings: Henry A. Boorse, Lloyd Motz, and Jefferson Hane Weaver, *The Atomic Scientists: A Biographical History* (New York: Wiley Science Editions, 1989); David C. Cassidy, *Uncertainty: The Life and Science of Werner Heisenberg* (New York: W. H. Freeman, 1992); Elisabeth Heisenberg, *Inner Exile: Recollections of a Life with Werner Heisenberg* (Boston: Birkhauser, 1984); Thomas Powers, *Heisenberg's War: The Secret History of the German Bomb* (New York: Knopf, 1993).

Hiroshima

On August 6, 1945, the first atomic bomb was dropped by a U.S. Air Force B-29 bomber, the *Enola Gay*, on the Japanese city of Hiroshima. The Target Committee had selected four cities as possible targets. Major General Curtis LeMay selected Hiroshima because it had munition factories, no American prisoner-of-war camps, and no previous bombings. A weather observation plane was sent ahead to scout out the weather over three potential targets. The Hiroshima plane radioed back that the weather was acceptable there. Besides the *Enola*, two support planes, the *Great Artiste* and *No. 91*,

This picture illustrates the extent of damage to Hiroshima caused by the atomic bomb detonated on August 6, 1945. The scene shows the destruction near ground zero, with the Matyosa Bridge as the reference point.

were along with scientists aboard to measure the effects of the blast. The bomb detonated at 1,890 feet over Hiroshima, almost directly above the Shima Hospital, at about 8:15 A.M. Its temperature exceeded 300,000 degrees Centigrade. The combination of heat and shock wave killed everything near its epicenter. A powerful vacuum formed at the epicenter, and an enormous updraft started spiraling toward the sky. The total area reduced to ashes by the blast and by fire was about 13 square kilometers. Shortly after the blast a black rain began to fall, which brought radioactive fallout and increased the exposure of the survivors. Most of the individuals in the vicinity spared by the initial heat and blast died soon thereafter of radiation-induced hemorrhage. The estimated number of casualties a year later on August 10, 1946, was 118,661 dead, 30,524 severely injured, 48,606 slightly injured, and 3,677 missing. In a 1976 report to the United Nations the total number of deaths from this one bomb by the end of 1945 was calculated at 140,000 (plus or minus 10,000). Despite initial reports, the Japanese government was slow to understand that an atomic weapon had been used to destroy Hiroshima. It took a report by the Arisue Mission, a team of scientists and military men headed by Lieutenant General Seizo Arisue, on August 10 to convince the Japanese authorities it had been the atomic bomb.

Suggested readings: Committee for the Compilation of Materials on Damage Caused by the Atomic Bombs in Hiroshima and Nagasaki, *Hiroshima and Nagasaki: The Physical, Medical, and Social Effects of the Atomic Bombings* (New York: Basic Books, 1981); Pacific War Research Society, *The Day Man Lost: Hiroshima, 6 August 1945* (Tokyo: Kodansha International, 1972); Gordon Thomas and Max Morgan-Witts, *Ruin from the Air: The Atomic Mission to Hiroshima* (London: Hamish Hamilton, 1977).

serious leg injury from a stray gunshot and business failure during the depression turned him toward politics. He was a liberal Democrat in a largely
Republican community. After supporting the Upton Sinclair campaign for
California governor, Holifield became active in state politics. Then in 1942,
he won election to California's Nineteenth Congressional District. This election marked a 32-year congressional career. His appointment to the Military
Affairs Committee in late 1944 began his interest in atomic energy when
the May-Johnson Bill appeared before it. His opposition to military control
of atomic energy and his active role in the passage of the Atomic Energy
Act of 1946 marked him as a comer. His reward was an appointment to the
JCAE. Holifield staunchly supported David Lilienthal as chairperson of a
civilian-controlled Atomic Energy Commission (AEC) and was willing to
tangle with some of the top Republican senators over his confirmation. Next
Holifield became a leader in the lobbying campaign to build the hydrogen
bomb. After this success, he next was embroiled in the controversy of private
versus public control of atomic energy. He vigorously defended against Republican attempts to allow private industry to monopolize the benefits of
atomic energy paid for by American taxpayers. In 1956, Holifield and Senator Albert Gore of Tennessee sponsored a bill, the Gore-Holifield Bill, for
the federal government to build large-scale atomic power plants to simulate
efforts of private industry. Both the Eisenhower administration, the AEC,
Republican congressmen, and coal interests combined to defeat the bill.
Holifield was an early supporter of civil defense and a frequent critic of the
Eisenhower administration's lack of a civil defense program. Holifield was a
longtime supporter of Admiral Hyman Rickover and his plan for a nuclear-
powered submarine and surface fleet. His 1957 hearings on radioactivity
proved embarrassing to the Eisenhower administration because of disclo-
sures by scientists that a clean nuclear bomb was impossible. The AEC had
been aware of fallout hazards before the Bravo Test, and fallout from nuclear
testing was heavier in the Northern Temperate Zone. Holifield's 1959 hearings on radioactivity were not successful and critics accused him of a cover-
up. In 1961 Holifield became the chair of the JCAE. He used his position
to lobby the Kennedy administration for a strengthened civil defense program, but he proved no more successful with Kennedy than he had been
with Eisenhower. Holifield opposed the construction of Fermi I near Detroit, because he objected to the construction of nuclear reactors near large
population centers as too dangerous. His opposition to the Price-Anderson
Act was also based on his fears that the nuclear industry would not be as
safe as the federal government in building plants in isolated sites. As the
AEC found shortcuts around licensing and safety issues, Holifield became a
more frequent critic and demanded it reform itself. Soon Holifield found
himself in the middle of the environmental attack on nuclear energy, and
he reacted to what he considered unfounded attacks. As he came close to
retirement in 1974, Holifield became more pessimistic about future energy

supplies caused by delays in licensing of nuclear power plants. Holifield returned from the House of Representatives in 1974 only after working hard on the reform of the AEC. He proved to be the most influential politician in Congress supporting the development of atomic energy projects in the postwar era. He died in 1995.

Suggested reading: Richard Wayne Dyke, *Mr. Atomic Energy: Congressman Chet Holifield and Atomic Energy Affairs, 1945–1974* (Westport, CT: Greenwood Press, 1989).

Hyde Park Aide—Memoir. *See* Roosevelt-Churchill Hyde Park Aide—Memoir

Hydrogen Bomb

The hydrogen bomb was the next logical development after the atomic bomb, but its origin was full of controversy. Edward Teller was an early champion of the idea after discussing its feasibility with Enrico Fermi in 1941. At first Teller's calculations indicated that fusion would not work, but he recalculated his figures, and the issue was considered by the scientists at the Berkeley Summer Colloquium in 1942. It became the source of the disagreement between Teller and J. Robert Oppenheimer during the Manhattan Project. After the war, Teller continued to lobby for the building of a hydrogen bomb, but opposition formed in the General Advisory Committee of the Atomic Energy Commission. This opposition was more than matched by the political skills of Teller and his political allies, Senator Brien McMahon and Lewis Strauss of the Atomic Energy Commission. At this time, all that was known was that, theoretically, an atomic bomb could trigger a thermonuclear explosion from a mixture of deuterium and tritium. Even the supporters of the hydrogen bomb calculated only a 50 percent chance of success. Once the Soviet Union had detonated an atomic bomb in 1949, political pressure on the Truman administration to approve the development of a superbomb was intense. While not all scientists agreed that the United States needed to build a hydrogen bomb, President Truman bowed to pressure from the Joint Committee on Atomic Energy and the circle of scientists around Teller and gave the go-ahead. Teller began recruitment of a team for Los Alamos to construct a hydrogen bomb. Under the code name Greenhouse, this team was to build a thermonuclear device and test it in the Pacific as soon as possible. Soon Teller became dissatisfied with the effort at Los Alamos and turned to a new laboratory in Livermore, California. These two laboratories started competing with each other in the construction and testing of nuclear devices.

The hydrogen bomb itself is a complicated combination of processes. It is best described as a fission-fusion-fission bomb, because it goes through phases before the final detonation. At the core is plutonium-239, but the bomb also has a deuterium-tritium trigger and lithium deuteride fusion fuel. Uranium-238 is also present. While the exact sequence is still classified by the U.S. government, the process probably involves the ignition of the plutonium core, thereby compressing and heating the deuterium-tritium mixture together with the fuel. The surrounding shell of U-238 provides confinement for the heating plasma. As the pressure and temperature rise high enough to ignite the deuterium-tritium fusion reaction, it releases energetic neutrons, which interact with the lithium in the fuel and cause it to fission. The fission then produces more tritium to feed the reaction. Neutrons from the fusion reaction are energetic enough to fission the U-238 around the fusion core, which boosts the energy of the explosion. Finally the vaporized U-238 shell is blown away, and the full energy is released.

Soon after the practicality of building a hydrogen bomb was agreed upon, a series of hydrogen bomb designs were tested in the mid-1950s. Both the Mark-14 and Mark-17 were tested and added to the U.S. nuclear arsenal, but both were so heavy (29,851 pounds for the Mark-14 and 41,400 pounds for the Mark-17) that only the enormous B-36 could deliver them. The Mark-15 was a lighter (7,600 pounds) strategic hydrogen bomb that could be delivered by B-47 and B-52 bombers, but it lasted in the nuclear inventory only about three years. It was replaced by the lighter and more versatile M-39, which could be delivered by the B-58 Hustler light bomber. A series of hydrogen bomb types, the Mark-28, B-43, Mark-53, were adopted in the period of the 1960s and 1970s. All of these types have been replaced by the B-83. This versatile hydrogen bomb is the primary armament of the B-2 bomber, and there are approximately 650 in the U.S. nuclear arsenal in the mid-1990s.

Suggested readings: Stanley A. Blumberg and Gwinn Owens, *Energy and Conflict: The Life and Times of Edward Teller* (New York: Putnam's Sons, 1976); James N. Gibson, *Nuclear Weapons of the United States: An Illustrated History* (Atglen, PA: Schiffer Publishing, 1996); Norman Moss, *Men Who Play God: The Story of the H-Bomb and How the World Came to Live with It* (New York: Harper and Row, 1968); Richard Rhodes, *Dark Sun: The Making of the Hydrogen Bomb* (New York: Touchstone, 1995).

I

Idaho National Engineering Laboratory. *See* National Reactor Testing Station

Implosion

The most carefully guarded secret in the U.S. atomic bomb program at Los Alamos was the principle of implosion used to detonate the fission device known as the Fat Man. Implosion is a way to trigger the plutonium core into an overcritical mass by directing blast waves inwardly. The theory of implosion was first advanced by Seth Neddermeyer, a physicist at Los Alamos who had formerly worked at the Bureau of Standards and was a former student of J. Robert Oppenheimer at Caltech. Hans Bethe and John von Neumann made calculations that proved implosion would work and provide a better detonation device, but it required incredible precision. Neddermeyer's idea was rebuffed at first because it was too complicated, but he was given the responsibility to study the problem. Implosion as a method was soon resurrected after it was found that the gun method would not work with plutonium. Emilio Segre conducted an experiment on plutonium and found PU-240 emitted alpha particles on its own. Trace amounts of isotope PU-240 in the fuel made it too unstable for the use of the gun method, because it might cause a premature detonation. At one point the lack of a workable triggering device threatened the success of a plutonium bomb. A special division was formed at the Los Alamos Laboratory to work out the problem. Still plagued by problems of design, a special Cowpuncher Committee was formed in March 1945 to take over the implosion problem. The engineering problem was to make the entire plutonium core go overcritical at the same time, and the solution was to use a series of lenses to produce the necessary implosion wave. Once this hypothetical solution had been found, it was necessary to test Fat Man to see if it worked. This was the reason for the Trinity Test on July 16, 1945.

Implosion was still top secret when details about it came out in the spy trial of former army sergeant David Greenglass. He was convicted of spying for the Soviet Union and implicated Julius and Ethel Rosenberg. Although the Soviet Union already had information about implosion from Klaus Fuchs, publicity about the process bothered the scientists who had worked on it. British scientists also learned about implosion, but they encountered difficulty in making it work.

Suggest readings: Brian Cathcart, *Test of Greatness: Britain's Struggle for the Atom Bomb* (London: Murray, 1994); Stephane Groueff, *Manhattan Project: The Untold Story of the Making of the Atomic Bomb* (London: Collins, 1967); Richard Rhodes, *The Making of the Atomic Bomb* (New York: Simon and Schuster, 1986).

Indian Nuclear Program

India has developed the most advanced civil nuclear industry program in southeastern Asia during the last two decades and now possesses a large nuclear weapons arsenal. India entered nationhood in 1948 under the influence of Mahatma Gandhi, who opposed nuclear weapons. His policy was articulated as official national policy by India's first prime minister, Jawaharlal Nehru, but Nehru sought to use nuclear energy for peaceful purposes. India's Atomic Energy Commission was formed by authority of Nehru and by legislative approval in 1948. Homi Bhabha, director of the Tata Institute for Fundamental Research in Bombay, assumed control of the new commission. While this commission had three members—Bhabha, K. S. Krishnan, director of the National Physics Laboratory, and S. S. Bhatnagar, director-general of the Council of Scientific and Industrial Research (CSIR)—Bhabha was its most important member. A huge budget and absolute secrecy of its deliberations showed the importance attached to the Atomic Energy Commission by the government. By 1952 Indian scientists had found large deposits of uranium ore in Bihar, so plans moved forward on building a viable atomic energy program. India's first nuclear center was at Trombay, 35 miles north of Bombay. Plans were made for three reactors at this site to be named Apsara, Cirus, and Zerlina. Apsara, a simplified research reactor called a "swimming pool"–type using enriched uranium fuel and producing only one megawatt, was the first reactor to go critical in August 1956. By 1957 there were 300 scientists and engineers working at Trombay. India's nuclear program was growing so fast that the Atomic Energy Commission was reorganized and the agency given increased authority in 1962. In 1963 the United States sold to India two civil power reactors to be located at Tarapur in the Bombay region. Indian scientist used these reactors only for research into civilian projects until India's defeat in the Chinese-Indian War of 1962.

A shift toward a more aggressive attitude toward nuclear weapons came out of this setback at the hands of the Chinese Army. China's subsequent testing of a nuclear weapon at Lop Nur in 1964 reinforced this attitude. India refused to sign the Nuclear Non-Proliferation Treaty (NPT) on the grounds that the treaty discriminated between nuclear haves and have-nots. A pro-bomb lobby began to redirect India toward building a nuclear arsenal. Help came from Canada in constructing a plutonium-producing research reactor. A heavy water nuclear reactor was built at Rajasthan. The United States provided the heavy water May 18, 1974 when until India tested an underground nuclear device at Pokhran in the Rajasthan Desert in northwest India. After the United States would no longer supply heavy water for a nuclear weapons program, India turned to the Soviet Union, which shipped 80 tons of heavy water in late 1976. Once India had demonstrated its possession of a nuclear weapon capability, then Pakistan went on a crash program to build nuclear weapons to counter India's perceived threat. It is estimated that by the early 1990s India had accumulated enough weapons-grade plutonium for 8 nuclear weapons. This estimate is based on the large nuclear infrastructure that India has built. By the mid-1990s, India has eight nuclear power reactors in operation or under commission and another nine reactors under construction or under repair. It also has eight research reactors, two uranium enrichment plants, and eight heavy water production facilities. Only two sites are under safeguards inspection of the International Atomic Energy Agency (IAEA). The Stockholm International Peace Research Institute (SIPRI) estimates that India, by 1996, had enough plutonium to produce between 65 and 105 nuclear weapons. This stockpile was authenticated by the five underground tests on three different bomb designs conducted by India on May 13, 1998, at its test site at Pokharan in Rajasthan, near the Pakistani border. These tests provoked a vigorous negative response from the international community and a commitment by Pakistan to conduct nuclear tests. Despite the negative attitude of the international community and its neighbors, these tests were part of a more aggressive foreign policy by a new Hindu nationalist government. Since these tests and Pakistan's rebutted tests, India has had to ride a tide of hostile international opinion and threat of sanctions. None of these sentiments has seen India renounce its position as a proud nuclear power.

Suggested readings: Itty Abraham, *The Making of the Indian Atomic Bomb: Science, Secrecy and the Postcolonial State* (London: Zed Books, 1998); Eric Arnett, ed., *Nuclear Weapons After the Comprehensive Test Ban: Implications for Modernization and Proliferation* (Oxford: Oxford University Press, 1996); Shyam Bhatia, *India's Nuclear Bomb* (Ghaziabad, India: Vikas House, 1979); David Cortright and Amitabh

Mattoo, eds., *India and the Bomb: Public Opinion and Nuclear Options* (Notre Dame, IN: University of Notre Dame Press, 1996); Marianne van Leeuwen, ed., *The Future of the International Nuclear Non-Proliferation Regime* (Dordrecht, Holland: Martinus Nijhoff, 1995).

Inertial Confinement Fusion. *See* Laser Fusion

Institute for Theoretical Physics

The Institute for Theoretical Physics in Copenhagen, Denmark, rivaled the Cavendish Laboratory for importance as a center for advanced study on atomic structure in the interwar period. Niels Bohr was the director of the institute, and it had been designed for his interests and abilities. It was built with support from the University of Copenhagen and from Danish private industry. The city of Copenhagen donated the land for the Institute on the edge of a prominent park. The Institute opened its doors on January 18, 1921. It was housed in a four-story building. Final plans had the Institute with a lecture hall, a library, laboratories, and offices. Later as Bohr's family became larger, a house was built next door to the Institute for them.

It was a center where theorists could listen to each other and advance theories in an open and free atmosphere. An advantage was that Bohr was from a neutral country and national prejudices were absent at the Institute. Often Bohr invited participants and students to his home. An annual conference on theoretical physics was held at the Institute in the 1920s and 1930s, and theoretical physicists from around the world came to it. In one sense, the participants of the Institute for Theoretical Physics provided the theories, and the researchers at places like the Cavendish Laboratory carried out the experiments.

Suggested reading: C. P. Snow, *The Physicists* (Boston, MA: Little, Brown, 1981).

Institute of Nuclear Power Operations

The Institute of Nuclear Power Operations (INPO) was formed in 1979 as a nuclear industry response to the Three Mile Island accident. Efforts had been made before the accident to question the safety features at the commercial nuclear power plants, but at the hearings, the Atomic Energy Commission always assured the critics that proper safety features were in place. After the combination of human failure and inadequate safety features at the Three Mile Island Nuclear Power Plant, the nuclear industry realized that it had a public relations nightmare. Their response was to form the Institute of Nuclear Power Operations to conduct research to improve safety systems. This organization is located in Atlanta, Georgia, and has an active membership of 101 companies. To serve this membership the INPO has a staff of 440 and a budget of nearly $52 million. Its mission is to provide workshops and training for nuclear energy plant operators.

Suggested reading: Joseph V. Rees, *Hostages of Each Other: The Transformation of Nuclear Safety Since Three Mile Island* (Chicago: University of Chicago Press, 1994).

Interim Committee on the Atomic Bomb

The Interim Committee on the Atomic Bomb was formed on May, 2, 1945, by President Harry Truman to study the implications of dropping an atomic bomb and to report its conclusions to the U.S. government. President Truman had access to a report by General Leslie Groves on the development of the bomb, and he followed a suggestion by Henry Stimson, the secretary of war, to appoint this committee. Among its other tasks, this committee had the responsibility to recommend policies on postwar development of atomic energy and to propose the necessary legislation. Stimson was selected by Truman to be the chair, and other members were Ralph A. Bard, under secretary of the navy, Dr. Vannevar Bush, Director of the Office of Scientific Research and Development, James F. Byrnes, representing the president, William L. Clayton, assistant secretary of state, Dr. Karl T. Compton, chief of the Office of Field Service in the Office of Scientific Research and Development, and Dr. James B. Conant, chair of the National Defense Research Committee. To make sure the scientists working on the atomic bomb project were represented, a scientific advisory panel was appointed, with J. Robert Oppenheimer, Arthur H. Compton, Ernest O. Lawrence, and Enrico Fermi as members. The Interim Committee was heavily weighted toward government officials and may be interpreted as a way to counterbalance growing unease among the scientists of the Manhattan Project over the direction of the atomic program. Meetings were held on May 9, 14, and 18 as the committee learned details about the operations of the Manhattan Project. At the May 31 and June 1 meetings, the committee decided to recommend that the atomic bomb be used against a Japanese city and without warning. Members of the committee hoped such a demonstration would make the Japanese amenable to surrender. James Byrnes reported the conclusions of the report to President Truman soon after the committee concluded its deliberations. On June 16 the scientific panel made a similar-type recommendation to use the atomic bomb. A final decision was made at a joint meeting on July 4 in Washington, DC, in favor of using the bomb. It was with this kind of advice that President Truman approved on July 5 the use of the atomic bombs on Japanese cities, which turned out to be Hiroshima and Nagasaki in August of 1945.

Suggested readings: Martin J. Sherwin, *A World Destroyed: The Atomic Bomb and the Grand Alliance* (New York: Knopf, 1975); Dennis Wainstock, *The Decision to Drop the Atomic Bomb* (Westport, CT: Praeger, 1996).

Intermediate Nuclear Forces Treaty

The Intermediate Nuclear Forces (INF) Treaty was signed between the Soviet Union and the United States on December 8, 1987, and ratified by the U.S. Senate in April 1988. This treaty came out of a concern to control the number of nuclear delivery systems, which included intercontinental ballistic missiles and cruise missiles. Attention had been given to this issue during the Carter administration, but it was not until the Reagan administration reopened talks in 1981 that serious negotiations commenced. Negotiations almost collapsed when the United States deployed new missile systems in West Germany, and it took the Reykjavik Summit in October 1986 between President Ronald Reagan and General Secretary Mikhail Gorbachev to restart the talks. The final treaty eliminated all long-range intermediate nuclear forces (1,000 and 5,000 kilometers) and short-range nuclear forces (500 and 1,000 kilometers). These weapons were to be destroyed, and a verification system to ensure this destruction was established. It ended older missile systems and also prevented the development of new intermediate nuclear systems. Although there was considerable debate over the wisdom of this treaty, it retained enough support to get the treaty through the U.S. Senate.

Suggested reading: James L. George, *The New Nuclear Rules: Strategy and Arms Control after INF and START* (New York: St. Martin's Press, 1990).

International Atomic Energy Agency

The United Nations formed the International Atomic Energy Agency (IAEA) on October 1, 1957, to provide an organization from which to spread the benefits of atomic energy to the world. This organization was an outgrowth of President Dwight Eisenhower's Atoms for Peace initiative in 1953. A successful conference at Geneva in 1955 in which peaceful uses of atomic energy were openly discussed added momentum for further international initiatives. A conference of 12 countries met in February 1956 in Washington, DC to plan the organization for the International Atomic Energy Agency. Differences between the participating representatives showed itself in the debate on the number of members of the Board of Governors, with the original number finally being set at 23. Final negotiations on the IAEA statutes took place in October 1956 in New York City, with representatives of 81 countries present.

With its headquarters in Vienna, Austria, the IAEA had the mission to promote the peaceful spread of nuclear energy and provide safeguards against the development of nuclear weapons. It has a 35-member Board of Governors, which serves as its policymaker. Progress was slow in the spread of nuclear technology to the nonnuclear countries and even slower in establishing effective safeguards. The safeguards system was retarded by opposition from the Soviet Union, so it was 1963 before implementation. Even after implementation the IAEA had difficulties making the safeguards work, because a panel of safeguard specialists meeting in 1970 maintained that a nation diverting material from a plutonium plant would be able to make a bomb within 10 days. This meant that a breeder reactor of a reprocessing plant would have to be inspected at least every 10 days. Since this was an impossible schedule, the IAEA safeguards were more an early warning than a prevention system. One of the IAEA's major crises was the attack on the Iraqi nuclear facility at Tammuz by the Israeli Air Force on June 7, 1981. Israel was suspended from the IAEA by its Board of Directors, and later a vote passed to deny Israel's credentials to attend the next agency meeting. It took the threat of the United States suspending funding for the IAEA before Israel was reinstated. In the mid-1990s, the IAEA was monitoring more than 800 installations worldwide. A series of static budgets in the 1980s and 1990s, however, have hurt the IAEA's inspections.

Suggested readings: Congressional Quarterly, *The Nuclear Age: Power, Proliferation and the Arms Race* (Washington, DC: Author, 1984); David Fischer, *History of the International Atomic Energy Agency: The First Forty Years* (Vienna, Austria: International Atomic Energy Agency, 1997); Bertrand Goldschmidt, *The Atomic Complex: A Worldwide Political History of Nuclear Energy* (La Grange Park, IL: American Nuclear Society, 1982); *International Atomic Energy Agency: Personal Reflections* (Vienna, Austria: International Atomic Energy Agency, 1997); Leonard S. Spector, Mark G. McDonough, and Evan S. Medeiros, *Tracking Nuclear Proliferation: A Guide in Maps and Charts, 1995* (Washington, DC: Carnegie Endowment for International Peace, 1995).

International Commission on Radiological Protection

The International Commission on Radiological Protection (ICRP) is an organization formed as an authority on radiological protection. This organization was a successor to an earlier body, the International X-Ray and Radium Commission, which had been founded in 1928. Representatives from Canada, the United Kingdom, and the United States gathered together in 1950 and established the ICRP. It serves as a clearinghouse of information on the hazards of radiation. Because the ICRP has no governmental status, its advice and warnings on the dangers of radiation are unofficial. By the middle of the 1950s the ICRP became a leader in the advocacy of stricter standards on radiation protection. This organization continues its monitoring role, but it operates behind the scenes without seeking publicity.

Suggested reading: Margaret Gowing, *Independence and Deterrence: Britain and Atomic Energy, 1945–1952*, vol. 2 (New York: St. Martin's Press, 1974).

International Nuclear Fuel Cycle Evaluation Program

The International Nuclear Fuel Cycle Evaluation Program was an effort by the Carter administration to convince other countries to rely on conventional nuclear fuel and avoid the reprocessing process leading toward weapons-grade nuclear materials. This program was launched in October 1977 with the participation of nearly 500 experts from 46 countries. A side issue was to discourage countries from developing breeder reactors. A report from this program came out in February 1980, and it did not endorse the American position. It did give a forum for the United States to present its position without forcing other countries to comply. Other factors such as the expense and difficulty of building and maintaining breeder reactors proved the American position to be correct, but this was only proven in the future.

Suggested reading: Congressional Quarterly, *The Nuclear Age: Power, Proliferation and the Arms Race* (Washington, DC: Author, 1984).

International Nuclear Information System

The International Nuclear Information System (INIS) is an international clearinghouse for information on nuclear energy. This clearinghouse is administered by the United Nation's International Atomic Energy Agency (IAEA), but the information is provided by member countries. At last count data come from 102 states and 16 international or interdisciplinary organizations. In 1905 the proliferation of information on atomic energy prompted the director of the IAEA to invite a representative from the Soviet Union and the United States to serve as consultants in forming an international database. Based on their 1966 report, the INIS was founded in 1968 with the mission to provide a forum for information exchange. Participating member states began reporting their nuclear literature in 1970, and the first INIS publication, *Atomindex*, appeared in printed form and on magnetic tape later the same year. Data input is sent by the member state to the INIS headquarters at the IAEA in Vienna, Austria, where it is merged, checked, and corrected for final copy. Input electronic mail began in 1993. This database provides a place where information on nuclear energy can be found regardless of the source.

Suggested reading: *Worldatom: International Atomic Energy Agency* (http://www.iaea.or.at).

International Thermonuclear Experimental Reactor

The International Thermonuclear Experimental Reactor (ITER) is a joint enterprise by Russia and the United States to promote fusion research by building a large magnetic fusion reactor. Both Soviet and American scientists had become concerned about the future of fusion research following the cutbacks in scientific research in the early 1980s in their respective countries. Alvin Trivelpiece, a former physics professor at the University of Maryland and director of research in the U.S. Department of Energy in the Reagan administration, and Evgenii Velikhov, a Russian fusion scientist and a member of the Supreme Soviet, advised their respective heads of state, President Ronald Reagan and General Secretary Mikhail Gorbachev, to support a cooperative venture in fusion research at the 1985 Reagan-Gorbachev Summit. In the subsequent negotiations, European and Japanese participation was recruited for the ITER project. The project was organized under the auspices of the United Nations' International Atomic Energy Agency (IAEA). By 1988 the ITER project was under way with designs for a super tokamak reactor much larger than previous models. Design activities are under way at three sites—San Diego, United States; Garching, Germany; and Naka, Japan—and plans for the ITER continue beyond 1998. In comparison, the Tokamak Fusion Test Reactor (TFTR) at Princeton University has peaked at 10.7 million watts of fusion power, but the ITER is designed for 1,500 million watts of fusion power. Ultimately, the lessons learned from the ITER will provide information for the use of fusion as an energy source for civilian power. The energy from fusion reactors promises to be much cleaner and more safe than that from fission reactors. The only remaining problem is for political authorities in the United States and elsewhere to provide the funding to build the ITER.

Suggested reading: T. Kenneth Fowler, *The Fusion Quest* (Baltimore, MD: Johns Hopkins University Press, 1997).

Ioffe, Abraham Feodorovich

Abraham Ioffe (1880–1960) was the leading physicist in the Russian physics community in the first half of the twentieth century. He was born on October 29, 1880, in Ronmy in the Poltava region of the Ukraine. His father was from a middle-class Jewish family and worked as a midlevel bank official. After attending secondary school until 1897, Ioffe became a student at the St. Petersburg Technological Institute. Although from the beginning his interests leaned toward the study of physics, he graduated in 1902 as an engineer. After a practicum working as an engineer on two bridges near Kharkov, Ioffe had the opportunity to travel to the University of Munich to study physics under the famous German physicist Wilhelm Roentgen. For the next three years, Ioffe participated in advanced physics experiments under the loose supervision of Roentgen. In 1906 Ioffe returned to Russia with a doctorate in physics from the University of Munich. Despite his degree and extensive research experience, Ioffe could only find a job as a laboratory assistant at the St. Petersburg Polytechnical Institute. He slowly advanced up the administrative ladder at the Institute and at the same time made contacts with key physicists. In 1913, Ioffe became Extraordinary Professor at the St. Petersburg Polytechnical Institute. It was at this time that he became close friends with Austrian physicist Paul Ehrenfest. Ioffe defended his doctoral dissertation on the elastic aftereffect in quartz crystals in 1915. By the evening of the Russian Revolution, Ioffe had become one of the leaders in the physics community, partly because of his research accomplishments but more so because of the circle of brilliant, young physicists he had surrounded himself with at the Institute. His evening seminars and his friendly, supportive guidance won him the affection and respect of these young physicists. Among his protégés were Pyotr Kapitsa, Nikolai Semenov, Iakov Frendel, and P. I. Lukinskii.

Ioffe's role after the Bolshevik victory in 1917 was to advance the cause of scientific research regardless of the regime. He assumed a leadership role in the founding of the Russian Association of Physicists in February 1919. In the early 1920s Ioffe traveled to western Europe to acquire scientific journals, books, and equipment for research in the Soviet Union. Slowly Soviet physics improved and began to make contributions. In the early 1930s, Ioffe decided to divert part of the Institute's resources toward nuclear physics research. This effort took courage because Soviet authorities viewed atomic research as impractical and lacking in support of the state. Nevertheless, he persuaded the chair of the Supreme Council of the National Economy, Sergei Ozdzhonikidze, to allocate funds for such research. Ioffe soon made his Leningrad Physico-Technical Institute the leading center of atomic research in the Soviet Union. He made Igor Kurchatov the first head of the nuclear department. To his credit, he also encouraged atomic research at Leningrad's Radium Institute, at Moscow's Lebedev Physical Institute and the Institute of Physical Problems, and at Kharkov's Physico-Technical Institute. Later, Ioffe learned from his friend Frederic Joliot-Curie about the German experiment producing fission. He remained director of Leningrad Physico-Technical Institute until 1951. He died on October 14, 1960.

Suggested readings: David Holloway, *Stalin and the Bomb: The Soviet Union and Atomic Energy, 1939–1956* (New Haven, CT: Yale University Press, 1994); Paul R. Josephson, *Physics and Politics in Revolutionary Russia* (Berkeley: University of California Press, 1991); Arnold Kramish, *Atomic Energy in the Soviet Union* (Stanford, CA: Stanford University Press, 1959).

Iranian Nuclear Program

The Iranian nuclear program has had a sporadic history. Although Iran had ratified the Nuclear Non-Proliferation Treaty (NPT) in 1970 as a non-nuclear weapon state, the shah of Iran was interested in building a nuclear weapons capability. But first Iran needed a civilian nuclear program, so in the early 1970s the Iranian government became active in building a nuclear program, and contacts were made with France for the purchase of a French light water power reactor. On the fringe of the civilian program the shah of Iran was also supporting weapon design, reprocessing technology, and enrichment technology. The shah's close ties with Israel also were beneficial in learning about nuclear technology. While an alliance had been formed by France and Iran to build Iran a nuclear energy program, this alliance was interrupted by the Iranian Revolution in 1979, and the reactor was abandoned in the early stages of construction. In 1984 a new nuclear research center at Isfahan opened, and the Iranian government tried to recruit Iranian nuclear physicists to come back from exile. Iran has a proven supply of uranium ore and in 1988 purchased more uranium ore from South Africa. Iran has the capability of sustaining a major nuclear power program. At present, Iran has two nuclear reactors under construction to produce electrical power, but exactly when these reactors will begin commercial operation is uncertain. At present, Iran does not have the technical infrastructure to produce nuclear weapons and shows little inclination to move in that direction.

Suggested readings: Eric Arnett, ed., *Nuclear Weapons After the Comprehensive Test Ban: Implications for Modernization and Proliferation* (Oxford: Oxford University Press, 1996); Marianne van Leeuwen, ed., *The Future of the International Nuclear Non-Proliferation Regime* (Dordrecht, Holland: Martinus Nijhoff, 1995); *Nuclear Power Reactors in the World* (Vienna, Austria: International Atomic Energy Agency, 1997).

Iraqi Nuclear Program

The Iraqi government became interested in a nuclear program in the early 1970s. Although Iraq was a signer of the Nuclear Non-Proliferation Treaty in 1969, the Iranian government soon began to explore nuclear weapons opportunities. In November 1975 France agreed to supply Iraq with two nuclear reactors—a small research reactor and a larger one with a capacity of 70 megawatts. This project was called the Tammuz Nuclear Center by the Iraqis. The reactors were under secret construction at El-Tuwaitha on the outskirts of Baghdad. On June 7, 1981, Israeli fighter-bombers attacked and destroyed the Iraqi nuclear reactors. France took the opportunity to back out of their agreement with Iraq and left the Iraqis without a nuclear program. Iraqi President Saddam Hussein tried to convince the international community to build a nuclear bomb for Iraq to counter the Israelis, but no such aid was forthcoming. Iraq rebuilt its nuclear infrastructure and, in 1982, began to concentrate on the enrichment process for nuclear weapons. Experts claim that Iraq was within two years of producing highly enriched uranium for weapons when the Gulf War broke out. It was only in the aftermath of the Persian Gulf War in 1991 that the International Atomic Energy Agency (IAEA) inspectors realized that the Iraqi government was secretly building a nuclear weapons capability. This development shocked the world community, and the IAEA declared officially that Iraq had not complied with its nuclear safeguards agreement. Allied inspectors have been watching the Iranians closely to make sure that their nuclear weapons program is not resurrected, but the Iranian government has made monitoring as difficult as possible.

Suggested readings: Richard Kokoski, *Technology and the Proliferation of Nuclear Weapons* (Oxford: Oxford University Press, 1995); Marianne van Leeuwen, ed., *The Future of the International Nuclear Non-Proliferation Regime* (Dordrecht, Holland: Martinus Nijhoff, 1995).

Isotope

An isotope is one of a series of atoms with the same atomic number, but with different atomic weights. All isotopes of a given element have the same number of protons but differing numbers of neutrons. Variations among isotopes caused early scientists considerable confusion, because they produced irregular spacings between the atomic weights of the elements. Experiments proved that a single element can be composed of a mixture of two or more isotopes. For instance, ordinary oxygen contains 99.76 percent of an isotope of weight 16 and smaller amounts of weight 17 and weight 18. Harold Urey discovered heavy water (deuterium) in 1932 by finding that ordinary water (H_2O) contained a small admixture of a heavier molecule in proportion of 1 part in 5,000. This discovery became invaluable, because heavy water is an excellent moderator in chain reactions for reactors. As more knowledge became available through experimentation, an understanding of isotopes was crucial to building a picture of the behavior of the atom.

Suggested readings: Hari Arnikar, *Isotopes in the Atomic Age* (New York: Wiley, 1989); Alfred Romer, *Radiochemistry and the Discovery of Isotopes* (New York: Dover Publications, 1970).

Israeli Nuclear Program

The Israeli nuclear program has always been a closely guarded secret, but Israel has had a nuclear weapons program since the 1970s. Almost from the founding of the modern state of Israel in 1948, the government under Prime Minister David Ben-Gurion considered nuclear energy indispensable to provide electrical power and as a guarantee of the independence of the Israeli state. This drive for nuclear energy started in the early 1950s with the exploration and discovery of low-grade uranium ore in the Negev Desert and the invention of a new way to produce heavy water. Then, the Ben-Gurion government secretly founded the Israel Atomic Energy Commission (IAEC) in 1952, with Ernest David Bergman as its head. A large number of Israeli scientists studied nuclear energy in the United States at the Argonne National Laboratory and at Oak Ridge as part of the Atoms for Peace Program between 1955 and 1960. In 1955, the United States constructed a small light water reactor in Israel at Nahal Soreq, 10 miles south of Tel Aviv, with the proviso that no nuclear weapons were to be made there. Israel's most important step toward nuclear power status, however, was with the construction in 1958 of the Dimona nuclear reactor. Ben-Gurion had signed a secret agreement with the French government in 1957 to supply a nuclear reactor. This reactor had the capability of producing weapons-grade plutonium, and seven of the eight members of the Israeli Atomic Energy Commission resigned in protest. Only its head, Bergman, stayed, and he was placed in charge of the nuclear reactor program. The French traded their technical expertise for Israeli heavy water knowledge and American technology. Most of the original support of equipment came from France in the guise of textile machinery. Initial stocks of uranium also came from France. Heavy water was bought from Norway. The leader in the Israeli nuclear program was physicist Yuval Ne'eman. It was the lack of uranium that hindered further development. Acquisition of 200 tons of uranium ore in 1968 as the result of the Plumbat Affair ended this weakness. Sometime after the 1967 Six Day War, Israel decided to construct nuclear weapons. By 1973, the Israelis had a small arsenal of atomic weapons estimated in the range of three to six devices. Moreover, the Israeli Defense Force (IDF) had a missile system capable of delivering a nuclear warhead about 250 miles in any direction. Although the Israeli government had never acknowledged the existence of this nuclear arsenal, the U.S. government was aware of it several years before the outbreak of the Arab-Israeli War in 1973. It proved to be a bargaining chip in the negotiations for American support to end the war. The official Israeli government policy continues to be of deliberate ambiguity about their possession and possible use of nuclear weapons. But since Israel has never signed the Nuclear Non-Proliferation Treaty, it is not certain just what nuclear weapons Israel possesses. An estimate made by informed sources (probably based on Central Intelligence Agency [CIA] sources) in 1990 indicated that the Israelis had a nuclear arsenal of between 50 and 200 nuclear weapons, including fusion-boosted devices. A more recent figure from the Stockholm International Peace Research Institute (SIPRI)

J

Japanese Atomic Bomb Program

The Japanese atomic bomb program never had the resources or the personnel to be successful, but Japanese authorities did seriously consider building an atomic bomb during World War II. After the Japanese military had surveyed potential uranium deposits, a Japanese atomic bomb program was given a tentative go-ahead in October 1940. Yoshio Nishina, a leading physicist and director of its Physical and Chemical Research Institute, was placed in charge of the program. He had studied with Niels Bohr in Copenhagen and had a reputation for his theoretical work on the Compton effect. He had already built a small cyclotron at his Tokyo Laboratory and was working on constructing a larger cyclotron in 1940. His institute had 110 researchers including Japan's most promising physicists. In 1940 the institute had four distinct research groups: the cyclotron atomic nucleus group, the cosmic ray group, the theoretical group, and a radiation group. In April 1941, the Imperial Army Air Force authorized full-scale research for an atomic bomb. Soon after the Army Air Force's initiative, the Imperial Navy also decided in the spring of 1942 to commit itself to nuclear power, but this time for ship propulsion. They were interested enough in a possible atomic bomb, however, to form a Navy Committee to study its possibility. Nishina was elected chair, but he was far too busy conducting his experiments to participate actively. Rumors surfaced in Japan in August of 1942 that the United States was about to begin building an atomic bomb. This news stimulated the members of the committee to give a go-ahead on research. Nishina asked Tadashi Takeuchi to assist him in the production of an atomic bomb. In a meeting in March 1943, it became apparent that with recent military defeats that Japan needed help now in the form of more aircraft and radar. While an atomic bomb was possible, it might take Japan as long as 10 years to

develop it. This committee also concluded that neither Germany nor the
United States had the spare industrial capacity to produce an atomic bomb
to be of use in the present war. Japanese scientists needed to be released for
more pressing tasks. Despite these conclusions, Nishina and the scientists at
his institute continued working on atomic projects until the end of the war.

Suggested readings: Pacific War Research Society, *The Day Man Lost: Hiroshima,
6 August 1945* (Tokyo: Kodansha International, 1972); Richard Rhodes, *The Making
of the Atomic Bomb* (New York: Simon and Schuster, 1986).

Japanese Atomic Energy Commission

The Japanese Atomic Energy Commission (JAEC) is the planning agency
for Japan's atomic energy program. It was founded on December 19, 1955,
as part of a law passed by the Japanese Parliament for a peaceful atomic
energy program. To ensure political control, the JAEC was placed in the
government as part of the Prime Minister's Office. The gold of the JAEC
has been for an independent and self-sustaining nuclear program, but Japan's limited uranium supply and the inability to enrich uranium slowed
development. After a slow start, Japan made rapid progress in building nuclear power generating plants in the 1970s and 1980s. Its nuclear reactors
have the capability of producing weapons-grade plutonium, but the JAEC
is bound by the Japanese Constitution not to make nuclear weapons.

Suggested reading: John E. Endicott, *Japan's Nuclear Option: Political, Technical, and Strategic Factors* (New York: Praeger, 1975).

Japanese Nuclear Program

The Japanese nuclear program was late forming partly because of its experiences in World War II. While Japanese scientists had worked with atomic
issues, the atomic bomb attacks left a lasting impression. Moreover, the
American occupation government was reluctant to allow the Japanese access
to anything resembling a weapon. Also, the Japanese constitution had pro-
hibitions against atomic weapon. Japan, however, has an almost total lack
of conventional energy resources, from coal to oil. Nuclear energy for civilian purposes began to receive government attention in the late 1950s.
The Japanese government formed the Japanese Atomic Energy Commission
in December 1955 under the industrialist Shoriki Matsutaro's leadership to
distribute government funds, promote research, and train personnel. Japan
has a limited supply of uranium ore, so nearly all of its uranium had to be
purchased from abroad. A British-type natural uranium, graphite-
moderated, gas-cooled reactor was purchased in 1959 for a site at Tokai-
Mura. A French contractor won the award for the construction contract.
Between 1965 and 1974 a further 20 orders were placed for American light

water reactors. Two reasons surfaced for the boom of the building of nuclear reactors in Japan: the need for diversification of Japan's sources of energy away from dependence on foreign oil and the need to control pollution problems. Moreover, the Japanese economy in the early 1970s was second only to the United States, and the Japanese wanted independence from foreign powers. Soon Japanese corporations took over from American companies in the construction and running of these reactors. Soon reactor construction was in the hands of large Japanese corporations (Mitsubishi, Toshiba, Hitachi, among others) with support from the Japanese government. It was over the issue of enriched fuel for these reactors that the Japanese and American governments came into conflict. In the end, the Japanese government refused to accede to American demands for control of nuclear fuel for Japanese reactors. In the mid-1960s, Japanese nuclear energy leaders became interested in the potential of breeder reactors. Japan has remained steadfast in its commitment to nuclear energy. In 1994 Japan had 49 nuclear power reactors in operation, which produced about 30 percent of the electricity generated in Japan. By December 31, 1996, the number had increased to 53 nuclear reactors providing more than 33 percent of the country's electricity.

Suggested readings: John E. Endicott, *Japan's Nuclear Option: Political, Technical, and Strategic Factors* (New York: Praeger, 1975); *Nuclear Power Reactors in the World* (Vienna, Austria: International Atomic Energy Agency, 1997).

Japan Tokamak GO. *See* Tokamak Fusion Reactor

Jefferies Report

The Jefferies Report was a study undertaken in 1944 by a committee at the Metallurgical Laboratory (Met Lab) on the postwar future of atomic energy. This study was initiated by Zay Jefferies, a scientist at General Electric, who proposed a comphrensive look at the prospects for the field of atomic energy. Arthur Compton, the head of the Met Lab, had suggested such a study in June 1943. He gave Jefferies the go-ahead, and a committee was formed by July 1944. Members of this committee included Nobel Prize physicist Enrico Fermi; James Franck, associate director of the Chemistry Division at the Met Lab; Thorfin R. Hogness, director of chemistry for the Metallurgical Project at the Met Lab; Robert S. Stone, head of the Met Lab's Health Division; Charles A. Thomas, research director of Monsanto Chemical Company; and Robert S. Mulliken, a physicist who was also the information director of the Metallurgical Project. This committee and subcommittees produced a report entitled *The Prospectus on Nucleonics,* which reflected the consensus of the Met Lab scientists on future research initiatives. In seven sections, a series of recommendations were made, including one to continue an active program in atomic research and another on the need for international control of atomic energy. This report was sent to Compton on November 18, 1944.

Suggested reading: Martin J. Sherwin, *A World Destroyed: The Atomic Bomb and the Grand Alliance* (New York: Knopf, 1975).

Jewish Physics

The pejorative term *Jewish Physics* originated out of the resentment of conservative German physicists of the theories and popularity of Albert Einstein, but in the course of German politics, it developed a more sinister connotation under the Nazis. In the beginning the charge of Jewish Physics was leveled by physicist Philipp Lenard and his followers at a physics conference in Leipzig, Germany, in the summer of 1922. Feelings of resentment were so high over Einstein and his theories of relativity that Einstein had to cancel his lecture. Soon the new Nazi movement seized the charge and used it as part of their racial program. The result was that when the Nazi government passed a decree in 1933 banning all Jews from government posts, which included university teaching posts, Jewish scientists had no option except to try and leave Germany. Even personal appeals to Hitler and other top Nazi officials that such an action would endanger German scientific preeminence failed to stop the implementation of the decree. This purge of Jewish physicists in the interest of racial purity slowed the progress of German physics precisely at a time when new developments in other countries were to transform research on the energy potential of the atom.

Suggested readings: Alan D. Beyerchen, *Scientists Under Hitler: Politics and the Physics Community in the Third Reich* (New Haven, CT: Yale University Press, 1977); Mark Walker, *Nazi Science: Myth, Truth, and the German Atomic Bomb* (New York: Plenum Press, 1995).

Joachimsthal Mines

The Joachimsthal (Jachymov) Mines on the German-Czechoslovakian border were the chief source of uranium in Europe in the nineteenth and early twentieth centuries. These mines were famous for their silver deposits, but miners also discovered a shiny black ore they called pitchblende. It was from pitchblende mined here that uranium was first isolated in 1789 by German chemist Martin Klaproth. Until 1912, these mines were the only sources of uranium- and radium-bearing ores in existence. It was also the site where the Curies obtained their ore for radium research. The closing of this mine by the Germans to outside sales in the late 1930s was taken by American scientists to mean that the Germans were interested in exploiting the uranium ore for atomic research. Much of the uranium for the German atomic bomb project came from the Joachimsthal Mines. The Soviet Union concluded a secret agreement with the Czech government in November 1945, granting the Soviets exclusive rights to this uranium ore. It is estimated that the Joachimsthal Mines provided about 15 percent of the Soviet uranium requirement through 1950. Following intensive mining of the uranium ore, the depleted mines were closed in the mid-1950s.

Suggested reading: Bertrand Goldschmidt, *Atomic Rivals* (New Brunswick, NJ: Rutgers University Press, 1990).

Johnson-May Bill. *See* May-Johnson Bill

Joint Committee on Atomic Energy

The Joint Committee on Atomic Energy (JCAE) was a committee formed by provisions of the Atomic Energy Act of 1946. It was created by statute rather than by congressional rules and had sweeping jurisdiction over all bills, resolutions, and other matters in Congress relating to atomic energy. Brien McMahon sponsored the resolution calling for a special Senate on atomic energy and became its chairperson. Its membership consisted of 18, 9 from the Senate and 9 from the House of Representatives. Until 1952 all the chairs of the JCAE were senators, but a compromise was worked out and sanctified by the Atomic Energy Act of 1954 that the leadership alternate between senators and congressmen. The first meeting of this committee was on August 2, 1946, and it was soon apparent that it needed special staffing to brief the congressmen on atomic energy issues. An early preoccupation of JCAE members was on security and secrecy issues. This committee had an oversight function, but from the beginning, it exercised exceptional powers over all aspects of atomic energy, and its relationship with the Atomic Energy Commission (AEC) was one of sponsorship. Among its first tasks was the selection of the five commissioners of the new AEC, and the selection of David Lilienthal as chairperson was controversial among the Republican members of the JCAE. Lilienthal gained enough support to be appointed to the commission, but not with a unanimous recommendation. Because the Joint Committee was its staunchest ally and protagonist for nuclear energy projects, the leadership of the AEC always kept the committee well informed about their activities, plans, and problems. On occasion, however, differences developed with the JCAE and the AEC, and one such event was on the decision to proceed with the development of a hydrogen bomb. Despite reluctance from the AEC and its General Advisory Committee (GAC), the JCAE had enough political clout to persuade President Truman to approve the building of a hydrogen bomb. Such influence over atomic energy derived from the stature, seniority, and ability of the congressmen on the Joint Committee. Moreover, the stipulation that not more than five of the nine members from each house come from the majority party and the stability of members also contributed to its authority. Members of this committee became powerful advocates for atomic energy. It was members of this committee that drafted and ensured passage of the Atomic Energy Bill of 1954. Because of its proactive participation in favor of nuclear energy and its attempts to intimidate critics, the Joint Committee became unpopular with the rest of Congress in the early 1970s. Consequently, it lost its exclusive jurisdiction over nuclear power in 1974. This committee was disbanded in 1977 as part of a general reform of the atomic energy program of the U.S. government. Seven congressional oversight committees or subcommittees assumed the responsibilities of the Joint Committee. None of these committees had been able to obtain the influence of the JCAE. A study of the JCAE indicates that it had been the most powerful congressional committee ever to exist.

Suggested readings: Corbin Allardice and Edward R. Trapnell, *The Atomic Energy Commission* (New York: Praeger, 1974); Brian Balogh, *Chain Reaction: Expert Debate and Public Participation in American Commercial Nuclear Power, 1945–1975* (Cambridge: Cambridge University Press, 1991); Frank G. Dawson, *Nuclear Power: Development and Management of a Technology* (Seattle: University of Washington Press, 1976); Harold P. Green and Alan Rosenthal, *The Joint Committee on Atomic Energy: A Study in Fusion of Governmental Power* (Washington, DC: George Washington University, 1961); Harold Orlans, *Contracting for Atoms: A Study of Public Policy Issues Posed by the Atomic Energy Commission's Contracting for Research, Development, and Managerial Services* (Washington, DC: Brookings Institution, 1967).

Joint European Torus. *See* Tokamak Fusion Reactor

Joint Institute for Nuclear Research

The Joint Institute for Nuclear Research is a Soviet nuclear research institute whose mission was to coordinate and control Soviet nuclear initiatives abroad. It was founded in March 1956 and was located at Dubna, north of Moscow where the Moscow-Volga Canal meets the Volga River. The original name of the site was Ivankovo, but shortly after the war, the name was changed to Bolshaya Volga. In 1956 the name was changed again to Dubna. Among its facilities were a theoretical physics lab, nuclear problems lab, cyclotron, and other research labs. The costs of the Joint Institute for Nuclear Research were shared between the Soviet Union (47 percent) and the other Socialist countries (53 percent), but the institute's director was always a Soviet scientist. It was at Dubna that scientists and technicians from the Soviet Bloc countries were trained. Three thousand Bloc scientists had been trained there in the first five years of its existence. Soviet scientists were also trained at Dubna, and many of them were sent abroad to build nuclear power plants in countries of the Soviet Bloc and other strategic countries such as India. In this sense, the Joint Institute for Nuclear Research was an extension of Soviet foreign policy. The Soviets, moreover, were always careful to control nuclear materials by demanding that all countries receiving reactors from them—Bulgaria, Czechoslovakia, East Germany, Hungary, Poland, and Rumania—return spent fuel rods to the Soviet Union for processing. This way these countries were not allowed to have access to weapons-grade nuclear materials. Since the downfall of the Soviet regime, the Russian government still uses the institute as a center for advanced physics research and as a way to keep abreast of current research by inviting scientists from around the world to attend its conferences.

Suggested reading: Gloria Duffy, *Soviet Nuclear Energy: Domestic and International Policies* (Santa Monica, CA: Rand, 1979).

Joliot-Curie, Frederic

Frederic Joliot-Curie (1900–1958) was a French nuclear chemist whose research on the fundamentals of radioactivity and nuclear fission made him an international figure in the history of atomic energy. He was born the son of a prosperous calico dealer in Paris on March 19, 1900. His father had been an active member of the Paris Commune and had gone into exile after its suppression. In 1926, Joliot married Irene Curie, the daughter of Marie Curie, and took on the surname Joliot-Curie. They had a daughter and a son. After attending a number of small private schools, Joliot-Curie became a student at the Lycee Lakanal at Sceaux from 1910 to 1917. He was a good student and an outstanding soccer player. His undergraduate degrees in chemistry and physics (1923) and his licence in science (1927) were taken at the School of Industrial Physics and Chemistry (École municipale de physique et de chimie industrielles—EPCI) in Paris. He studied under the famous physicist Paul Langevin. After a short stint as an engineer, Joliot-Curie started in December 1924 as a researcher under Marie Curie at the Radium Institute, where he stayed until 1931. In 1930, he obtained his doctorate from the University of Paris. Then in 1931, Joliot-Curie was transferred to the National Caisse of Sciences. He was appointed in 1937 to a new chair of nuclear chemistry at the University of Paris.

Joliot-Curie and his wife Irene worked in collaboration on their research on radioactivity. In 1931, they began research on bombarding aluminum with alpha rays. In the process, they discovered the first evidence of transmutation when the aluminum absorbed the alpha particles and transmuted it into an unknown isotope with a half-life of 3.5 minutes. Their discovery opened a new discipline of artificial radioactivity. This research won them the 1935 Nobel Prize for Physics. After Joliot-Curie and his wife came close to discovering nuclear fission themselves in 1938, he was one of the earliest

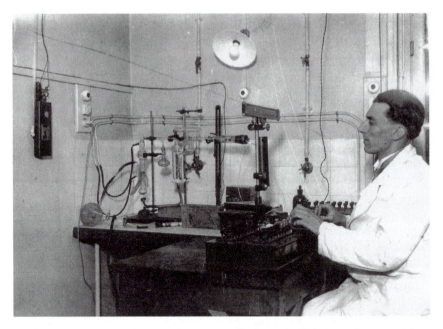

Frederic Joliet-Curie, a French physicist, worked closely with his wife, Irene Joliet-Curie, to discover artificial radioactivity. Although Joliet-Curie was active in the French Resistance, his membership in the Communist Party lead to his dismissal as head of the postwar French atomic energy program.

to recognize the significance of Otto Hahn's discovery of nuclear fission and postulated the probability of a chain reaction. In a series of experiments conducted between March and April of 1939, with coworkers Hans von Halban and Lew Kowarski, it was demonstrated that the splitting of a uranium nucleus by a single neutron produced the emission of several further neutrons on the order of three. They concluded that it was possible to achieve a chain reaction with uranium-235 if secondary neutrons could be slowed down. Joliot-Curie also realized the importance of heavy water as a moderator and persuaded the French government to obtain the entire stock of heavy water from Norway. He made sure that the heavy water stocks reached Great Britain, instead of falling into German hands.

During World War II, Joliot-Curie turned down an opportunity to flee to Great Britain. His lack of English, fear of being unable to continue his research, and wife's illness made him stay in France after May 1940. He continued his research in Paris, but under the supervision of German authorities. In 1934, Joliot-Curie had joined the Socialist Party but had disagreed with the Blum government's nonintervention in the Spanish Civil War. Soon after the German victory over France in October 1940, Joliot-Curie became active in the French Resistance in the College de France. His

German supervisor, former colleague Wolfgang Gentner, became aware of his activities but never turned him in to the Gestapo. It was after the execution of a number of prominent scientists by the Germans in May 1942 that Joliot-Curie formally joined the Communist Party. He continued his resistance activities until he went underground in June 1943.

After the war, Joliot-Curie played an active role in French nuclear politics. His initial concern was to rebuild French science. He was also disappointed in the American use of the atomic bomb, and his concern was to bridge the differences between France and the Soviet Union. Between 1946 and 1950, he was the commissioner of Atomic Energy in the Commissariat à l'Énergie Atomique (CEA). His well-known affiliation with the Communist Party presented the French government with a political problem, and pressure increased from the United States to remove him from the CEA. In April 1950, Joliot-Curie made some intemperate remarks at a Communist Party rally in Gennevilliers, and he was removed as commissioner. Most of his activities in the early 1950s concerned peace and antinuclear movements. In 1956, Joliot-Curie became head of the Radium Institute. He died in Paris on August 14, 1958, of viral hepatitis. Charles de Gaulle gave him a state funeral.

Suggested readings: Henry A. Boorse, Lloyd Motz, and Jefferson Hane Weaver, *The Atomic Scientists: A Biographical History* (New York: Wiley Science Editions, 1989); Maurice Goldsmith, *Frederic Joliot-Curie: A Biography* (London: Lawrence and Wishart, 1976).

Joliot-Curie, Irene

Irene Joliot-Curie (1897–1956) was a French nuclear chemist whose research on radioactivity almost matched those of her famous mother. She was born on September 12, 1897, in Paris. Her parents were famous physical chemists Pierre Curie and Marie Sklodowska Curie. In 1926, Irene Curie married French chemist Frederic Joliot and assumed the name Joliot-Curie. They had a daughter and a son. She received little formal schooling and instead was tutored by an all-star cast of scientists—physics by her mother, mathematics by Paul Langevin, and chemistry by Jean Baptiste Perrin. After two years of this study, she entered the College Sevigne where she was an excellent student. Serving as a radiologist for the French Army in World War I, she worked with her mother at the Front. It was at this time that she had her first serious exposure to radiation, and it weakened her health. In 1921, she began research at Madame Curie's Radium Institute. It was in 1934 that the collaboration of Frederic Joliot-Curie and Irene Curie produced the first experimental evidence of artificial radioactivity. For this discovery Irene shared the 1935 Noble Prize for Physics with her husband. They also conducted experiments that came close to the discovery of nuclear

K

Kaiser-Wilhelm-Society

The Kaiser-Wilhelm-Society (KWS) for the Advancement of Sciences provided the resources and site for research on atomic energy in Germany from 1920 to 1945. It was established in 1911 to coordinate scientific research in imperial Germany. Kaiser Wilhelm II served as the sponsor for this society after his adviser on education, Adolf von Harnack, convinced him of the close association between scientific advances and their military applications. Earlier, a 1905 committee of prominent university and industrial chemists had proposed a chemical institute for advanced research. Two institutes were planned as a result of these initiatives: the Kaiser-Wilhelm-Institute for Chemistry and the Kaiser-Wilhelm-Institute for Physical Chemistry and Electrochemistry. Lands for the institutes were donated from the crown lands around Berlin in Dahlem, and the institutes began opening in 1912. By 1914 there were seven institutes in operation. The Kaiser-Wilhelm-Institute for Physical Chemistry received its funding from a wealthy financier Leopold Koppel, and its original emphasis was on industrial research. Soon, however, the institute began to make a place for researchers of radioactivity and atomic theory. Despite its reputation for physics research, the Kaiser-Wilhelm-Institute for Physics had to wait until 1937 to have a building of its own.

Some of the great names in German chemistry and physics worked at the Kaiser-Wilhelm-Institute. Among these notables were Otto Hahn, Werner von Heisenberg, and Lise Meitner. During the Nazi period, the Kaiser-Wilhelm-Society was partially isolated from pressures from the Nazi government, as a separate corporation, but the regime's anti-Jewish legislation made many of the institute's top names leave. Those that remained had to make personal concessions to the Nazis, and scientific research suffered.

Parts of the institute underwent the Nazi deemphasis on individual research projects to an emphasis on work-community projects. Initial work on the Nazi atomic bomb program began in the Virus House on the grounds of the Kaiser-Wilhelm-Institute in Dahlem. During World War II, most of the buildings of the institute were heavily damaged. After the war, American and British pressure caused the Kaiser-Wilhelm-Society to be dissolved, and in its place was constituted the Max Planck Society.

Suggested readings: Alan D. Beyerchen, *Scientists Under Hilter: Politics and the Physics Community in the Third Reich* (New Haven, CT: Yale University Press, 1977); Richard Rhodes, *The Making of the Atomic Bomb* (New York: Simon and Schuster, 1986); Ruth Lewi Sime, *Lise Meitner: A Life in Physics* (Berkeley: University of California Press, 1996).

Kapitsa, Pyotr Leonidovich

Pyotr Kapitsa (1894–1984) was one of the leading Soviet nuclear scientists and he is considered the architect of the Soviet atomic bomb. He was born on July 9, 1894, at Kronstadt, Russia. His father was a general in the Russian Army. Kapitsa's marriages to Nadezhda Tschernosvitova in 1916 and Anna Alekseyevna Krylova in 1920 produced two sons. Kapitsa received his baccalaureate in 1919 from the Petrograd Polytechnical Institute, USSR, and his Ph.D. in 1926 from Cambridge University, England. His professional career began in 1918 with a professorship at the Petrograd Polytechnical Institute. In 1921, Kapitsa was allowed by Soviet authorities and Abraham Ioffe's approval to go to Cambridge University to study for an advanced degree in physics and to conduct research at the Cavendish Laboratory. For the next 14 years, Kapitsa worked closely with Sir Ernest Rutherford on high-level atomic research projects. On a trip back to the Soviet Union in 1934, he was refused permission to return to the Cavendish Laboratory. Kapitsa became depressed over this, and the separation from his wife, so that he was unable to do research for nearly 2 years.

Less is known about Kapitsa's subsequent career in the Soviet Union except that he soon was in a leadership position in Soviet physics. Once it was apparent that Kapitsa would have to remain in the Soviet Union, Ernest Rutherford sent him his laboratory (the Mond Laboratory), and the Soviet government paid 30,000 English pounds for it. Kapitsa was appointed director of the Institute of Physical of Problems of the USSR Academy of Sciences. The Soviet government set up a laboratory similar to the one that Kapitsa had left at Cambridge, and he continued to conduct the same type of experiments. He served two terms there, from 1935 to 1946 and again from 1955 to 1984. His research on liquid oxygen led to the development of a new method to handle liquid oxygen, which had military application, earning him the reputation of a Soviet hero. Kapitsa had been recruited earlier to work on a Soviet atomic bomb, but his inability to get along with Lavrentii Beria, the head of the committee directing the Soviet bomb project and the head of the NKVD (People's Commissariat of Internal Affairs) over the research on the bomb, caused him to resign as a member of the committee. Kapitsa's role had been more as a senior statesman and adviser in atomic physics than as an active participant, but he wanted to find a cheaper and quicker way to make the bomb. Stalin protected him from Beria for the rest of Stalin's life, possibly because Stalin liked him. Kapitsa was kept under house arrest from 1946 to 1955, but since he lived across the way from the institute, he was able to continue his research. He also held a professorship at the Physiotechnical Institute from 1947 to 1984. The Nobel Prize Committee presented the 1978 Nobel Prize for Physics for his work in low-temperature physics. From 1955 onward Kapitsa headed the Soviet Committee for Interplanetary Flight and was also active in the satellite program. During the 1960s, he became active in the Soviet delegation to the Pugwash Conferences on peaceful uses of atomic energy. He died on April 8, 1984, in Moscow.

Suggested readings: Lawrence Badash, *Kapitza, Rutherford, and the Kremlin* (New Haven, CT: Yale University Press, 1985); Pyotr L. Kapitsa, *Peter Kapitza on Life and Science* (New York: Macmillan, 1968); C. P. Snow, *The Physicists* (Boston, MA: Little, Brown, 1981).

Kasli Disaster. *See* Chelyabinsk-40

Kellex Corporation

Kellex Corporation was the organization of corporate executives and engineers formed in early 1943 to build a gaseous-diffusion project to separate uranium-235 from uranium-238. It was an organization of the best executive and engineering talent available in the United States. In peacetime, no such corporation could be formed, because of the high-price talent. Dobie Keith was recruited by General Leslie Grove to head this organization. He was a Texan and had a reputation as one of the top chemical engineers in the United States. Keith had been the M. S. Kellogg Company's vice president in charge of engineering for a secret project on gaseous diffusion before Groves selected him for the Manhattan Project. While his mission was to build a gaseous-diffusion reactor at the Oak Ridge Laboratory, Tennessee, nobody knew how to build facilities for a process as yet undeveloped. His reputation was as a dynamic, hardworking, and authoritarian leader, and his working relationship with General Leslie Groves always remained good because Keith produced results. Keith organized Kellex into a loose organization of warlords reporting only to him. At its peak, Kellex employed 3,000 people in three floors of offices at the Woolworth Building in downtown

Manhattan. Kellex personnel worked closely with Union Carbide executives in designing and building the gaseous-diffusion plant at Oak Ridge. They designed the specifications and set the standards for an incredibly complex set of machinery and piping. Once the plant, K-25, started production in January 20, 1945, Kellex turned it over to the operating company, Union Carbide.

Suggested readings: Stephane Groueff, *Manhattan Project: The Untold Story of the Making of the Atomic Bomb* (London: Collins, 1967); Kenneth D. Nichols, *The Road to Trinity* (New York: Morrow, 1987).

Kemeny Commission

The Kemeny Commission was a 12-person committee that had been appointed by President Jimmy Carter to investigate the Three Mile Island nuclear accident. Its chair was Dartmouth University president John G. Kemeny, who was by training a mathematician and had worked at the Los Alamos Scientific Laboratory after World War II. Among the members of the committee were Anne Trunk, a Middletown housewife and a representative of the Three Mile Island community. The Kemeny Commission's report was a strong indictment of the nuclear industry in general and the Nuclear Regulatory Commission in particular. Although 8 of the 12 members wanted to impose a moratorium on the construction of new nuclear plants, the final report recommended reexamination of pending reactor licenses, evacuation plans reviewed and approved by the Federal Emergency Response Agency, and the restructure of the Nuclear Regulatory Commission and placement of it under the control of the executive branch. This report made it difficult for the nuclear industry to ignore the problems leading to the Three Mile Island incident.

Suggested reading: Mark Hertsgaard, *Nuclear Inc.: The Men and Money Behind Nuclear Energy* (New York: Pantheon Books, 1983).

Khariton-Zeldovich Letters

Yuli Khariton and Yakov B. Zeldovich were two Soviet physicists at the Institute of Physical Chemistry in Leningrad who wrote a series of scientific letters in the period between the summer of 1939 and March 1940 outlining the processes to follow to build a Soviet atomic bomb. These letters were the products of a report by Khariton and Zeldovich at the Conference on Questions of the Physics of the Atomic Nucleus held at Kharkov, between November 15 and 20, 1939. The first letter in the summer of 1939 described the operational steps necessary to produce an atomic explosion and the magnitude of the damage. In a second paper on October 22, 1939, the two physicists described the problems in building atomic reactors. They recommended the use of heavy water as a moderator for a chain reaction. The third letter, produced in March 1940, advanced procedures to initiate and control a chain reaction in an atomic reactor. Despite some criticism about their theories and safety factors, these letters served as the theoretical basis for the Soviet atomic bomb program. Despite their conclusions, the Soviet government took little action except to accelerate efforts to find uranium ore. It was several years before the Soviet authorities realized the high stakes involved in working with atomic energy.

Suggested readings: David Holloway, *Stalin and the Bomb: The Soviet Union and Atomic Energy, 1939–1956* (New Haven, CT: Yale University Press, 1994); Arnold Kramish, *Atomic Energy in the Soviet Union* (Stanford, CA: Stanford University Press, 1959); Richard Rhodes, *Dark Sun: The Making of the Hydrogen Bomb* (New York: Touchstone, 1995).

King Nuclear Test

King was the code name for the 1952 test of the largest fission bomb detonated in American testing. Much of the activity at the Los Alamos Scientific Laboratory was in designing and building a hydrogen bomb, but in the meantime Ted Taylor designed an atomic bomb that went far beyond the maximum yield limit of 250 kilotons of a U-235-plutonium bomb. He was able to design a device using lower-grade uranium. The test took place on November 11, 1952, at the Nevada Test Site, and the blast yielded 500 kilotons. This was the last of the tests on large-yield fission devices, but it proved that a good design produced high-yielding weapons. Some scientists concluded from this test that the United States should depend on large atomic weapons rather than develop the hydrogen bomb, but their views were soon overtaken by political events.

Suggested reading: Richard L. Miller, *Under the Cloud: The Decades of Nuclear Testing* (New York: Free Press, 1986).

Kistiakowsky, George Bogdan

George Kistiakowsky (1900–1982) was a Russian-American chemist whose work on the atomic bomb in the Manhattan Project helped make it a success. He was born on November 18, 1900, in Kiev, Russia. His father was a professor of sociology, and most of the rest of his family were academics. He started his education at the University of Kiev, but the Russian Revolution intervened. After fighting against the Bolsheviks in both the infantry and the task corps in the White Russian Army, Kistiakowsky migrated to Germany and the University of Berlin. After finishing his education with a doctorate in physics in 1925 he migrated to the United States in 1926. He obtained a fellowship to Princeton University, and later Princeton hired him as a faculty member. In 1930 Kistiakowsky moved to Harvard University, where he stayed until retirement in 1972.

During World War II, Kistiakowsky became head of the Explosives Division at the Los Alamos Scientific Laboratory. He was recognized as a top expert on high explosives and had been the director of the National Defense Research Committee (NDRC) explosives laboratory in Bruceton, Pennsylvania. His close friendship with James Conant, the president of Harvard and a key official in the Office of Scientific Research and Development (OSRD), enabled him to persuade Conant on the feasibility of the atomic bomb. Once it was discovered that the plutonium bomb needed an implosion trigger, Kistiakowsky was recruited in January 1944 to work with his colleague, Seth Neddermeyer, in the Ordnance Division on an implosion device. He found Deak Parson and Neddermeyer in the midst of a fight over implosion and the running of the Ordnance Division. In a reorganization, Kistiakowsky replaced Neddermeyer and headed a large division formed to make implosion work.

After the war, Kistiakowsky became an important scientific adviser. During the Eisenhower presidency, he served as special assistant to the president for science and technology from 1959 to 1961. Later, he became a critic of the use of nuclear weapons. He retired from Harvard University as professor emeritus in 1972 and died in 1982.

Suggested readings: Al Christman, *Target Hiroshima: Deak Parsons and the Creation of the Atomic Bomb* (Annapolis, MD: Naval Institute Press, 1998); Richard Rhodes, *The Making of the Atomic Bomb* (New York: Simon and Schuster, 1986).

KIWI. *See* NERVA

Knapp Report

The Knapp Report was a report commissioned by the Atomic Energy Commission (AEC) in 1960 to respond to the Joint Congressional Committee on Atomic Energy's request to determine the dangers of radioactive fallout. Harold Knapp had a doctorate from the Massachusetts Institute of Technology and worked in the Fallout Studies Branch of the AEC's Division of Biology and Medicine. His assignment was to prepare an assessment of radioactive fallout in the St. Louis area. After a comprehensive study of the various types of radioactive dangers, Knapp came to the conclusion that iodine-131 was a more immediate and greater hazard to public health than strontium-90. He concluded that the AEC had been measuring the wrong things. He turned his attention to the St. George area of southwestern Utah and concluded that the AEC had grossly underestimated the radiation doses to various areas of the United States. Knapp sent his final report to the AEC in 1963, and the reaction there was negative. AEC officials tried to suppress the report, but a panel of scientists recommended that the report be published. The AEC finally released the report, but it censored the most damaging material. Knapp quit the AEC and published the entire report in *Nature* magazine. This report further damaged the credibility of the AEC at a time when confidence by the general public in this organization was beginning to wane.

Suggested readings: Howard Ball, *Justice Downwind: America's Atomic Testing Program in the 1950s* (New York: Oxford University Press, 1986); Richard L. Miller, *Under the Cloud: The Decades of Nuclear Testing* (New York: Free Press, 1986).

K-219

The most startling Soviet nuclear submarine disaster was the disabling and sinking of the Yankee-class ballistic missile nuclear submarine the *K-219* on October 3, 1986, 450 miles northeast of Bermuda. This nearly 400-feet-long, 10,000-ton submarine had two VM-4 nuclear reactors and a crew of 119. Its mission was to cruise off the eastern coast of the United States and provide a credible nuclear threat against the eastern seaboard. A small leak in a ballistic missile silo caused an explosion after saltwater mixed with the missile fuel, producing nitric acid. A crewman, Sergei Preminin, lost his life sealing off the two nuclear reactors before the submarine sank in 18,000 feet of water. Altogether four Soviet sailors died of injuries immediately, and two others died later of health complications resulting from the accident. Soviet naval command blamed the submarine's captain, Igor Britanov, for the loss of the vessel and threatened a court-martial. It was recognized by others in the navy that the submarine was in bad shape before deployment. The *K-219* remains undisturbed at the bottom of the Atlantic Ocean off Bermuda, but its nuclear ballistic missiles have been removed by the American navy. *See also* Soviet Nuclear Submarine Accidents.

Suggested reading: Peter Huchthausen, Igor Kurdin, and R. Alan White, *Hostile Waters* (New York: St. Martin's Press, 1997).

Kurchatov, Igor

Igor Kurchatov (1903–1960) was the scientific director of Soviet atomic research in the 1940s and 1950s. He was born on January 12, 1903, in Simskii Zavod in the Chelyabinsk region of the southern Urals. His father was a surveyor and a teacher who moved around for jobs and family health reasons. In 1927, he married Marina Dimitrevna Sinelnikov. Kurchatov attended academic gymnasiums in Ulyanosk and in Simferopol (Crimean Peninsula). His majors were physics and mathematics when he studied at the Crimean State University. Because his university had a laboratory ill-equipped for experimentation, Kurchatov prepared a thesis on a theoretical physics problem. He worked on a degree in shipbuilding at the Polytechnical Institute in Leningrad in 1924 before leaving for the Baku Polytechnical Institute. After a short term at Baku, his friend and future brother-in-law Kirill Sinelnikov recommended him to Abraham Ioffe, the head of the Leningrad Physico-Technical Institute (LFTI), and Kurchatov was offered a position at the Institute. After some research in ferroelectricity, in 1932 Kurchatov switched his field of research to atomic physics. In 1933, Ioffe appointed him to head the Institute's program on nuclear physics. Kurchatov coordinated the research on nuclear physics not only at the LFTI but also at the Institute of Chemical Physics and the Radium Institute. By the late 1930s, Kurchatov had built a small cyclotron, which aside from the one at the University of California at Berkeley, California, was the only one in existence. Kurchatov soon became well known for his investigations in the field of nuclear isomerism.

In November 1942, Stalin entrusted him to become the head of the Soviet atomic weapons program. At the time he had been busy in war work on antimine protection and then conducting research on advanced tank armor. From secret intelligence reports, he was able to organize the main outlines of the Soviet atomic program in March 1943. Laboratory No. 2 was established near Moscow as the site for atomic bomb research. It was his complaints about Vyacheslav Molotov, the Soviet foreign minister and head of the Soviet Nuclear Program, in 1944–1945 that resulted in Molotov's removal as head of the Soviet atomic program. Although many of the Soviet scientists had difficulty dealing with Lavrentii Beria as the new head of the Soviet program in 1945, Kurchatov was an exception. Kurchatov continued to be a leader in nuclear research until his death in Moscow on February 7, 1960.

Suggested readings: Ulrich Albrecht, *The Soviet Armaments Industry* (Chur, Switzerland: Harwood Academic Publishers, 1993); Thomas B. Cochran, Robert S. Norris, and Oleg A. Bukharin, *Making the Russian Bomb: From Stalin to Yeltsin* (Boulder, CO: Westview Press, 1995); David Holloway, *Stalin and the Bomb: The Soviet Union and Atomic Energy, 1939–1956* (New Haven, CT: Yale University Press, 1994); Arnold Kramish, *Atomic Energy in the Soviet Union* (Stanford, CA: Stanford University Press, 1959).

Kyshtym Disaster. *See* Chelyabinsk-40

L

Laboratory No. 2

Laboratory No. 2 of the Academy of Sciences, USSR, was the top-secret laboratory for the development of the Soviet atomic bomb. At first the laboratory was located in the Seismological Institute on Pyzhevskii Lane in Moscow. In the fall of 1942, Igor Kurchatov found an abandoned three-story building on the outskirts of Moscow at Pokrovskoe-Streshnevo and used it to house his laboratory. A nearby artillery range was used to test explosives. Nikolai I. Pavlov was appointed representative of the Central Committee and the Council of Ministers to oversee nuclear weapons development. Although the scientists had no illusions of building an atomic bomb before the defeat of Germany, they began to conduct research with the full support of the Soviet state. By January 1944, the laboratory had a staff of 20 scientists and a support staff of about 30. Kurchatov constructed an atomic reactor in November–December 1946. By December 25, 1946, the Soviets had an operational reactor (F-1). Later in 1946, an experimental radiochemical shop was also constructed. From 1943 to 1946, Laboratory No. 2 was the principal center for atomic bomb research, but it became less important after 1946 as other research centers were opened at remote sites, allowing testing near the centers. The Laboratory was renamed the Kurchatov Institute of Atomic Energy after Kurchatov's death in 1960. Nuclear weapons research was shifted to other facilities, and research at the Kurchatov Institute has been devoted mostly to civilian nuclear power and general nuclear theory. Anatoli P. Aleksandrov replaced Kurchatov as director of the Institute.

Suggested readings: Thomas B. Cochran, Robert S. Norris, and Oleg A. Bukharin, *Making the Russian Bomb: From Stalin to Yeltsin* (Boulder, CO: Westview

Press, 1995); Richard Rhodes; *The Making of the Atomic Bomb* (New York: Simon and Schuster, 1986).

Landau, Lev Davidovich

Lev Landau (1908–1968) was one of the leaders of Soviet theoretical physics. He was born on January 22, 1908, in Baku, Russia. His father was a petroleum engineer in the Baku Oil Fields, and his mother was a physician. In 1939, he married Kondordia Landau, and they had one son. Landau studied at the Universities of Baku (1922–1924) and Leningrad (1924–1927). In 1929, after obtaining a professorship at the University of Kharkov, Landau traveled to Niels Bohr's Institute for Theoretical Physics, where he spent several years studying with Bohr. While in Germany, Landau also worked as a colleague of Hungarian physicist Edward Teller. Returning to the Soviet Union, Landau became head of physics at the University of Kharkov in 1932. In 1934, he received his doctorate in physics from the Kharkov Institute of Mechanical Engineering. He was promoted to professor in 1935. But in 1938 Landau fell into disfavor and was arrested as a German spy. It took the direct intervention of Peter Kapitsa to win Landau's release from prison after a year's stay. In 1937, Landau moved to Moscow, where he was appointed director of Theoretical Physics at the Institute of Physical Problems. In 1943, he became professor of physics at Moscow State University.

By the early 1930s, Landau was the Soviet Union's foremost theoretical physicist. His inquiries into problems of quantum mechanics, nuclear physics, particle physics, astrophysics, and low-temperature physics produced major advances. He was drafted into working on the Soviet atomic bomb program at the Institute of Physical Problems until Stalin's death. After leaving the nuclear project in 1955, he returned to research topics more to his liking. He received the 1962 Nobel Prize for Physics for his research on condensed matter.

At the height of his scientific success, Landau was seriously injured in an automobile wreck in 1962. He was a passenger in a car driven by another Soviet physicist that was hit by a truck unable to stop due to icy conditions. He was sustained on a life-support system for almost six years without recovering his physical powers. He died in Moscow on April 3, 1968.

Suggested readings: Alexander Dorozynski, *The Man They Wouldn't Let Die* (New York: Macmillan, 1965); Anna Livanova, *Landau: A Great Physicist and Teacher* (London: Pergamon Press, 1980).

Large Ship Reactor

The Large Ship Reactor (LSR) Program was started in July 1951 at Westinghouse Corporation to build a propulsion reactor for a surface warship.

Although the consensus of expert opinion was that reactor was intended for the eventual use in a large aircraft carrier, the U.S. government did not acknowledge it to be so. After an extensive consideration of various types of reactors, a consensus was arrived at on the pressurized-water reactor (PWR). Just about the time Westinghouse and Admiral Hyman Rickover's team had come to this conclusion the U.S. government canceled the project. A contract was forthcoming but not until October 15, 1954, when Westinghouse signed a contract for a nuclear propulsion system (A1W Project) for an aircraft carrier, the USS *Enterprise* (CVN 65), and for a cruiser, the USS *Long Beach* (CGN 9).

Suggested reading: John W. Simpson, *Nuclear Power from Underseas to Outer Space* (La Grange Park, IL: American Nuclear Society, 1995).

Laser Fusion

Laser fusion is a process that has been developed to produce energy from fusion by use of high-powered lasers. The chief difficulty researchers encountered in trying to harness the energy source of nuclear fusion was the problem of how to produce the conditions of high pressure and temperature that exist inside the sun within a controlled terrestrial environment. With the advent of the laser it became possible to produce a very hot and dense plasma of deuterium and tritium for a very brief period of time by focusing several laser beams on a very small spot and dropping a frozen deuterium-tritium pellet into the intersection region. The energy of the light pulse heats the vaporized gas to a high temperature, while the pressure from the outer layers ejected first from the pellet compresses the inner region to a very high density, where fusion begins to ignite. Although the nuclear reaction from a single pellet is very short-lived, the energy released from a steady sequence of laser pulses and dropping pellets could potentially exceed the energy input and provide a useful source of energy.

The laser was first proposed by two American scientists, Arthur Schawlow and Charles Townes, in 1958, and by 1960, a working laser had been built by Theodore Maiman. Interest in laser physics led to experiments to produce fusion by bombarding deuterium-tritium (DT) pellets with high-powered lasers. These experiments proved that energy could be produced by this method. Scientists at the Lawrence Livermore National Laboratory began full-scale experiments on laser fusion in 1962. A series of large lasers were built there, with the giant laser Nova finished in December 1984. This research soon became part of a larger effort to find new ways to use laser fusion, the Inertial Confinement Fusion (ICF) program. Research continues on laser fusion, with the chances increasing as the lasers become more powerful and DT pellets more compact. Much of the research initiative for laser fusion has been turned over to the National Ignition Facility (NIF). Funding for further advances in laser fusion has been restricted by recent budget cutbacks on scientific research.

Suggested reading: T. Kenneth Fowler, *The Fusion Quest* (Baltimore, MD: Johns Hopkins University Press, 1997).

Lawrence, Ernest Orlando

Ernest Lawrence (1901–1958) was an American physicist who specialized in building atom-smashing cyclotrons for nuclear research. He was born on August 8, 1901, in Canton, South Dakota. His parents were Norwegian immigrants who had become schoolteachers. His father taught history in a Lutheran academy and was twice elected superintendent of public instruction in the state of Wisconsin. His mother was a mathematics teacher. He married Mary Kimberly Blumer, and they had six children, five daughters and a son. Initially Lawrence wanted to be a physician and entered Saint Olaf College in Minnesota. After a brief stay there, Lawrence transferred to the University of South Dakota, where he received his bachelor of science with honors in 1922. He helped pay his way through college by selling aluminum potware to farmers' wives. His work on a master's in physics at the University of Minnesota was under W. F. Swann. When Swann moved to the University of Chicago, Lawrence followed him. Despite the galaxy of stars there, he was unhappy. In 1924, he transferred to Yale University, again with Swann. In 1928, Lawrence accepted an appointment as associate professor at the University of California at Berkeley. Within two years Lawrence was promoted to full professor with a large salary increase. He spent the rest of his academic career at the University of California at Berkeley.

Lawrence built the cyclotron as a way to probe the structure of the atom. He decided upon a circular apparatus, because within a uniform magnetic field, particles could be accelerated along a circular orbit over a wide range of particle speeds. Lawrence's first cyclotron in 1931 was only a few inches in diameter, but within five years he built larger 27-inch and 37-inch cyclotrons. By the end of 1936, a worldwide boom in cyclotrons had taken place with at least 20 cyclotrons in action. Lawrence convinced the Rockefeller Foundation in 1940 to give him more than $1 million for the construction and maintenance of a 100-million-volt cyclotron. The value of the cyclotron was in that physicists were able to accelerate the nuclei of virtually any sample in a controlled manner and to discover new elements. As these elements became known, Lawrence lobbied harder for even larger cyclotrons. This time the foundations and the university combined to fund these huge cyclotrons. His winning of the 1939 Nobel Prize for Physics for the building of the cyclotron undoubtedly helped.

Once World War II began, Lawrence became involved in military work. Soon Lawrence was working with the Manhattan Project on ways to use the cyclotron to separate uranium-235 from natural uranium. By using the cyclotron as a spectrometer, Lawrence was able to make the cyclotron separate uranium-235. He built the Calutron, a 184-inch dynamo at the Radiation Laboratory, to accomplish the separation. Subsequently, a plant was also built at Oak Ridge, Tennessee, to do this.

After the war, Lawrence continued to campaign for bigger and better cyclotrons. He lobbied for the Livermore facility for nuclear weapons research. His attention also turned to the use of radiation for medical purposes. Many of the physicists in his laboratory understood nuclear physics better than he did, but Lawrence had leadership skills, enthusiasm, and salesmanship. Lawrence was always extremely demanding, but the young scientists working for him on the cyclotron worshiped him. His generosity in allowing other scientists to use his facilities and in sharing radioactive substances with outsiders made him popular in the physics community. As Lawrence grew older and more conservative, his close friendship with J. Robert Oppenheimer became strained. Their friendship ended in the debate over the hydrogen bomb and with Lawrence's testimony against Oppenheimer during his security clearance trial. Lawrence also used the hydrogen bomb debate to campaign for and win the establishment of the Lawrence Livermore National Laboratory as a rival to the Los Alamos Scientific Laboratory. He died of colitis following surgery on August 27, 1958, in Palo Alto, California.

Suggested readings: Henry A. Boorse, Lloyd Motz, and Jefferson Hane Weaver, *The Atomic Scientists: A Biographical History* (New York: Wiley Science Editions, 1989); Herbert Childs, *An American Genius: The Life of Ernest Lawrence* (New York: Dutton, 1968); Nuel Pharr Davis, *Lawrence and Oppenheimer* (New York: Simon and Schuster, 1968); Anthony Serafini, *Legends in Their Own Time: A Century of American Physical Scientists* (New York: Plenum Press, 1993); Spencer R. Weart, *Scientists in Power* (Cambridge: Harvard University Press, 1979).

Lawrence Livermore National Laboratory

The Lawrence Livermore National Laboratory (LLNL) has been a major source in the United States for the design of nuclear weapons. It was founded in July 1952 by the National Security Council and the Atomic Energy Commission (AEC) to concentrate on H-bomb and fusion technology. Edward Teller had lobbied in Washington, DC for this laboratory after he found the senior scientists at Los Alamos hesitant to redirect all of their activities to build hydrogen bombs. The first director was Herb York, who remained until 1958. In that year, Edward Teller replaced him as director and remained until 1960. He was replaced by Harold Brown (1960–1961). Built on the site of a former naval air station near Livermore, California, the facility was first occupied in September 1952 and operated by the University of California under contract with the AEC. The laboratory had four major programs: designer of thermonuclear weapons, diagnostic weapon experiments, a magnetic fusion program (Project Sherwood), and a basic physics program. Scientists from this laboratory soon became active in testing nuclear weapons at the Nevada Test Site. A rivalry developed between the Livermore facility and Los Alamos over the direction of nuclear weapons research. The scientists at the LLNL were younger and more sympathetic toward the military side of their research than their compatriots at Los Alamos. This rivalry manifested itself in intense lobbying by Edward Teller and the Livermore staff in Washington, DC for a greater role in the testing program. An indicator of the success of this campaign was the more active participation of the LLNL scientists in atomic and hydrogen bomb testing at the Nevada Test Site from the mid-1950s to the 1970s. Toward the end of nuclear testing the Livermore scientists dominated the proceedings.

Much as the other nuclear weapons laboratories, the Lawrence Livermore National Laboratory has been active in finding new avenues for nuclear research in the 1990s. Under a proposed reorganization of the national laboratories by the Department of Energy (DOE), the LLNL is to give up its duplicate plutonium laboratory to the Los Alamos National Laboratory and end its longtime competitive role with that laboratory. In return, the LLNL is to become the laboratory for uranium and tritium research on hydrogen bombs. Another proposed major project is to explore the possibility of laser-produced fusion for peaceful purposes.

Suggested readings: Norman Moss, *Men Who Play God: The Story of the H-Bomb and How the World Came to Live with It* (New York: Harper and Row, 1968); Thomas H. Saffer and Orville E. Kelly, *Countdown Zero* (New York: Putnam's Sons, 1982).

Lawrence Radiation Laboratory

The Lawrence Radiation Laboratory was an experimental physics laboratory run by the University of California where several important atomic energy experiments were conducted. In 1932 the president of the University of California, Robert Sproul, gave Ernest O. Lawrence a wooden building for his exclusive research use. The building was the first home of what Lawrence soon named the Radiation Laboratory. It served as the home for E. O. Lawrence's series of cyclotrons. Since the laboratory was a separate entity from the Physics Department, Lawrence had to raise the finances available to sustain it. This meant that Lawrence had to devote much of his energy toward fund-raising. Consequently, Lawrence left most of the daily operations of the laboratory to the young physicists working there. Lawrence was a master salesman, and soon the Radiation Laboratory had a large staff conducting research using the cyclotron and other equipment.

Experts from here soon were building cyclotrons at other sites. Glenn T. Seaborg used this laboratory to conduct his experiments to isolate and identify plutonium. The plans for the electromagnetic separation for the Oak Ridge plant were also developed here. After the war, experiments continued to isolate elements for scientific study.

Suggested readings: Daniel J. Kevles, *The Physicists: The History of a Scientific Community in Modern America* (Cambridge: Harvard University Press, 1987); Emilio Segre, *From X-Rays to Quarks: Modern Physicists and Their Discoveries* (San Francisco: W. H. Freeman, 1980).

Leipzig Reactor

The Leipzig was a small heavy water–uranium pile at the Leipzig Laboratory in the early 1940s. It was the reactor that Werner Heisenberg and Robert Doepel had built to test chain reaction theory. Considerable progress had been made in experiments in the spring and early summer of 1942 until an accident caused an explosion on June 23, 1942. Somehow the heavy water leaked into a sealed-off shell of uranium. Doepel examined the mishap and tried to determine the extent of the leak, but when a technician unscrewed an inlet, the uranium started burning. Water was dumped on the flame, which put out the fire but caused the uranium to go critical. Heisenberg and Doepel attempted to relieve pressure on the uranium shell when they were forced to evacuate the building just before the laboratory blew up. They were lucky to escape serious injury. Firefighters took two days to put out the fire. The destruction of the Leipzig Reactor set back the German atomic bomb program for many months, as Heisenberg had to build another pile with a new supply of uranium and heavy water.

Suggested readings: David Irving, *The German Atomic Bomb: The History of Nuclear Research in Nazi Germany* (New York: Simon and Schuster, 1967); Dan Kurzman, *Blood and Water: Sabotaging Hitler's Bomb* (New York: Henry Holt, 1997).

Lenard, Philip Edward Anton

Philipp Lenard's (1862–1947) reputation as a physicist has been overshadowed by his leadership of the Aryan Physics movement in Germany from the 1920s onward. He was born on June 7, 1862, in Pressburg, in the Austro-Hungarian Empire. He married Katherine Schlehner in 1897, and they had one son. By birth a Hungarian, he later became a German citizen. His father was a wine merchant. Lenard received his doctorate in physics, with Georg Quincke as his adviser, from the University of Heidelberg in 1886. His first job was as a researcher at the University of Heidelberg. He took a brief and unsatisfactory trip to England. His reputation as an experimental physicist gained him a professorship at the University of

Breslau in 1890 and then at the University of Bonn (1891–1894). While at Bonn, he became an assistant to Heinrich Hertz. He returned to the University of Heidelberg before accepting a position as a professor of experimental physics at the University of Kiel in 1898. He became embroiled in a scientific dispute over whether he or J. J. Thomson had priority over the discovery of the electron. This dispute reinforced his dislike for England and British physics. It was during his stay at the University of Kiel that Lenard received the 1905 Nobel Prize in Physics for his work in connection with cathode rays. Finally, in 1907, Lenard returned to the University of Heidelberg, where he stayed until his retirement in 1931. His work on the absorption of cathode rays by matter made an important contribution to the development of the quantum concept. But his emphasis on experimental over theoretical research and his strong political views made him beginning in the early 1920s a dominant figure in German physicist politics. His distaste for theoretical physics and its mathematical foundations extended to those physicists advancing theories based on this methodology.

As important as his physics research was to Lenard, his career in German politics became more important. The defeat of Germany in World War I, the loss of his son, and the loss of his savings in the postwar inflation came together to unhinge him. Moreover, over the years, Lenard had developed strong anti-Semitic views. Although in the beginning his relationship with Albert Einstein was cordial, over the years Lenard began to attack Einstein and what he called Jewish Physics. Some of this resentment came from Einstein's fame and status in the world of physics, but more of it was directed to his rejection of Einstein's theories on relativity. Lenard tried to redirect German physics away from Einstein through publications. This hostility also extended to Max Planck's quantum theories. His alliance with Johannes Stark and a circle of conservative and nationalistic colleagues started open warfare between the two camps. In 1924, Lenard joined Stark in publishing an open letter supporting Adolf Hitler. This warfare continued unabated until the Nazis consolidated control of Germany in 1933. His publication of the four-volume work *Deutsche Physik* revealed the depth of his feelings that science was racial and conditioned by blood. From this time onward the doctrine of Aryan Physics became the official policy of the Nazi state. Lenard profited little from this official policy change, but the Germany physics community suffered greatly, with many talented Jewish physicists leaving Germany. Lenard retired from academic life in 1931. The triumph of Heisenberg's circle reduced his influence, and he played almost no role in German science in World War II. Lenard's age and health prevented him from having to appear before the de-Nazification courts. He died on May 20, 1947, at Messelhausen, Germany.

Suggested readings: Alan D. Beyerchen, *Scientists Under Hitler: Politics and the Physics Community in the Third Reich* (New Haven, CT: Yale University Press, 1977); Mark Walker, *Nazi Science: Myth, Truth, and the German Atomic Bomb* (New York: Plenum Press, 1995).

Leningrad Physico-Technical Institute

The Leningrad Physico-Technical Institute (LFTI) played a leading role in the development of Soviet nuclear physics in the first half of the twentieth century. The LFTI was formed on November 23, 1921, and its head was the prominent physicist A. F. Ioffe. Ioffe held this post from 1921 to December 1950. Since it had become the national center for physics research, it earned the nickname of "Fiztekh." Because so many of its members had close ties to the Petrograd Polytechnical Institute, LFTI's headquarters were across the street from the Polytechnical Institute. Because of its prestige and the leadership of Ioffe, the LFTI played a central role in the development of Soviet physics. It served as a buffer between the Soviet regime and the physics community in the 1920s and at the same time trained a generation of Soviet physicists. LFTI physicists dominated Russian publications in the period 1921 to 1939. Only toward the middle of the 1930s did the combination of new scientific institutes, criticisms about too much attention to applied physics, and charges of empire building come together to weaken the prestige of the LFTI. In 1931, there was a reorganization, and it was divided into three parts—Leningrad Institute of Chemical Physics, the Leningrad Physico-Technical Institute, and the Leningrad Institute of Electrical Physics. The purge of 1936 had a tragic impact on the LFTI, as it lost several supporters and physicists. After 1936 and World War II, the LFTI had only a shadow of its former dominance.

Suggested readings: David Holloway, *Stalin and the Bomb: The Soviet Union and Atomic Energy, 1939–1956* (New Haven, CT: Yale University Press, 1994); Paul R. Josephson, *Physics and Politics in Revolutionary Russia* (Berkeley: University of California Press, 1991).

Libby, Willard Frank

Willard Libby (1908–1980) was an American chemist who is most famous for developing a radiocarbon dating technique used by a variety of disciplines and as a champion of nuclear energy. He was born on December 17, 1908, in Grand Valley, Colorado. All of his higher education degrees were from the University of California at Berkeley. After receiving his doctorate in 1933, Libby remained at the University of California, teaching until he moved to Columbia University in 1941 to work on the development of the atomic bomb.

After the war, Libby returned to academia. He became a professor of chemistry at the Institute for Nuclear Physics at the University of Chicago. It was there that Libby and his students developed a highly reliable radiocarbon dating technique. Serge Korff had discovered carbon-14 in 1939, but Libby and his crew used this radioactive isotope as a way to date objects from the past. In the early 1950s, Libby consulted on the development of the hydrogen bomb. His enthusiasm for it led to his opposing J. Robert Oppenheimer's security clearance. In 1959, Libby returned to the University of California at Berkeley to be the director of the Institute of Geophysics. He became a prominent spokesperson for the Atomic Energy Commission (AEC) in its battles against critics of radioactive fallout. Libby died on September 8, 1980, in Los Angeles, California.

Suggested readings: Henry A. Boorse, Lloyd Motz, and Jefferson Hane Weaver, *The Atomic Scientists: A Biographical History* (New York: Wiley Science Editions, 1989); John Daintith, Sarah Mitchell, Elizabeth Tootill, and Derek Gjertsen, *A Biographical Encyclopedia of Scientists*, 2nd ed., vol. 1 (Bristol, UK: Institute of Physics, 1994).

Light Water Reactors

The standard design for most American nuclear reactors was the light water reactor (LWR). This reactor uses enriched uranium as fuel and regular water as a coolant. High-pressure water passes around the reactor, which heats the water before it is transferred to a steam generator. It was designed in the late 1940s and became popular because of the simplicity of its design.

Westinghouse Corporation's version was marketed as a pressurized-water reactor, and General Electric's as a boiling-water reactor. Admiral Hyman Rickover focused on the development of this type of reactor, and by March 1950, it had been selected as the design basis for a naval propulsion system to power American submarines. The LWR became such a successful design that it has been used to power surface ships and to generate electricity. Serious accidents in this type of reactor can be traced most often to some interruption in the cooling cycle. A by-product of the LWR is that it produces about 150 kilograms of plutonium per year, which remains unburned in the core, for possible use in nuclear weapons. The popularity of the LWR grew to the point that American companies exported the technology abroad, mostly to western European countries. During the boom of nuclear development in the 1960s, the LWR promised low cost for the generation of electricity, but the promise proved illusory. They operated less efficiently than advertised, and the capital investment for construction was never recovered. Later, safety problems began to surface as the equipment started to fail due to aging.

Suggested reading: Irvin C. Bupp and Jean-Claude Derian, *Light Water: How the Nuclear Dream Dissolved* (New York: Basic Books, 1978).

Lilienthal, David Eli

David Lilienthal (1899–1981) was a high-level administrator who became the first head of the Atomic Energy Commission (AEC). He was born on

David E. Lilienthal was the former head of the Tennessee Valley Authority and the first chair of the Atomic Energy Commission (AEC). His opposition to the hydrogen bomb development and difficulties with Republicans on the Joint Committee on Atomic Energy led to an early resignation from the AEC.

July 8, 1899, in Morton, Illinois. His father was in the dry goods business. In 1905 Lilienthal's family moved to Valparaiso, Indiana. Later in 1913, his family resettled in Michigan City, Indiana, where he graduated from high school. His undergraduate career was at DePauw College, where he graduated in 1920. He entered Harvard Law School in 1921 and was a solid, if unspectacular, student. In 1922 Lilienthal joined the Chicago law firm of Donald Richberg, which specialized in labor law. After working several years on labor cases, he migrated to public utilities law. For several years in the early 1930s Lilienthal served on the Wisconsin Public Service Commission. In June 1933 he was nominated as one of the three directors of the Tennessee Valley Authority (TVA). Lilienthal soon found himself in opposition over the direction of the TVA with the chair of the directors, A. E. Morgan. They fought for nearly five years before President Roosevelt removed Morgan from the TVA. A congressional inquiry followed that lasted over six months and finished with a vindication of Lilienthal. Lilienthal then became the most powerful figure at the TVA, but it was not until September 1941 that he became the chair. He was still at the TVA when he was appointed as chair to a consulting group to advise the Acheson Group on atomic energy. This body was soon called the Lilienthal Group. Lilienthal assumed the chair of the AEC in November 1946. His appointment was delayed because President Truman was uncertain as to whom to appoint to replace

him at the TVA. In his role as head of the TVA, he had made powerful
enemies on Capitol Hill, and his appointment hearing was brutal. Lilienthal's support for civilian control of atomic energy made him many enemies
on Capitol Hill and in the military. All of the former Manhattan Project
assets were turned over to the AEC in 1946, and General Leslie Groves
relinquished control. Lilienthal and Groves had become enemies during the
lengthy and contentious debate over the Atomic Energy Act, and later Lilienthal refusal to have anything to do with the increasingly embittered
Groves. Lilienthal accepted as his mandate to keep nuclear power under
civilian control, and he was successful to a point. Nevertheless, most of the
early activities of the AEC dealt with atomic weapons development and the
military uses of atomic energy. His next major crisis was the intransigence
of his fellow commissioner Lewis Strauss. In 1948, Strauss went into opposition to Lilienthal and used his contacts in the Republican Party to
counter Lilienthal's policies. Lilienthal was reluctant to build the hydrogen
bomb, and just before President Truman approved its construction, he resigned as chair of the AEC. After he left the AEC, Lilienthal continued to
speak out on atomic energy issues, especially excessive government secrecy
and the total government monopoly of information.

Suggested reading: Steven M. Neuse, *David E. Lilienthal: The Journey of an
American Liberal* (Knoxville: University of Tennessee Press, 1996).

Lilienthal Group

The Lilienthal Group was a panel of consultants suggested by Dean Acheson, the undersecretary of state, to advise the Acheson Group and the
government on atomic affairs. It was formed in January 1946 and was composed of David Lilienthal, the former head of the Tennessee Valley Authority and chair of the panel; Robert J. Oppenheimer, atomic physicist and
former scientific head of the Manhattan Project; Chester Barnard, the president of the New Jersey Bell Telephone Company; Harry Winne, a vice
president of the General Electric Company and a Manhattan Project veteran;
and Charles Thomas, a vice president of the Monsanto Chemical Company
and a prominent chemist. The first session began on January 28, 1946, and
the panel immediately started exploring options. Members of the panel traveled around the country visiting atomic facilities and talking to scientists and
engineers. They studied the problem of reconciling military and peaceful
uses of atomic energy. Three themes came to dominate their proceedings:
rejection of a ban on atomic weapons, civilian control of atomic energy, and
a desire for strong international control of atomic energy. Their solution
was an International Atomic Development Authority with the mandate to
control the dangerous parts of atomic energy. In a presentation of their plan
before the Acheson Committee at the Dumbarton Oaks on March 12, 1946,

the idea of immediate international control received a chilly reception from General Leslie Groves, Vannevar Bush, and James Conant. Bush insisted upon implementation stages that would take around 15 years before full international control. This was the plan finally presented to James Byrnes, the secretary of state. Byrnes then turned the revised Lilienthal plan over to Bernard Baruch to present to the United Nations in the form of the Baruch Plan. Byrnes later admitted that this had been a serious mistake, because Baruch's revision made it unacceptable to the Soviet Union.

Suggested readings: Steven M. Neuse, *David E. Lilienthal: The Journey of an American Liberal* (Knoxville: University of Tennessee Press, 1996); Peter Pringle and James Spigelman, *The Nuclear Barons* (London: Michael Joseph, 1981).

Limited Test Ban Treaty. *See* Nuclear Test Ban Treaty

Lindemann, Frederick Alexander

Frederick Lindemann (1886–1957) was a British physicist and scientific administrator who is more famous as Winston Churchill's scientific adviser in World War II and in Churchill's postwar government. He was born on April 5, 1886, at Baden-Baden, Germany. His father was a wealthy engineer from Alsace who became a naturalized British citizen rather than become a German citizen in 1871. His mother was an American. After attending schools in Scotland and Germany (Gymnasium in Darmstadt), Lindemann studied under German physicist Walther Nernst at the University of Berlin. In 1910, he received his Ph.D. in physics. During World War I, Lindemann worked with Mervyn O'Gorman at Farnborough to study problems of flight and aircraft instrumentation and became the director of the RAF (Royal Air Force) Physical Laboratory. He had been turned down for a military commission by the British army because of his German birth and German connections. After the war, his close personal friend, H. T. Tizard, helped him to be elected in 1919 to Dr. Lee's Professor of Experimental Philosophy at Oxford University. This professorship also made him head of the Clarendon Laboratory. It took Lindemann nearly 20 years and the influx of a number of Jewish physicists from Germany to turn the Clarendon Laboratory into first-rate research facility.

Lindemann was always more of an administrator than a scientist. Nevertheless, he had a series of scientific successes—Lindemann melting-point formula, the Nernst-Lindemann theory of specific head, the Lindemann electrometer, Lindemann glass, and the Dobson-Lindemann theory of the upper atmosphere. Most of these advances came from early in his career, but this is not unusual in scientific inquiry. But as the years went by, it became apparent that Lindemann shined as an administrator. He was always a social person, and these contacts helped both the laboratory and his career.

In World War II, Lindemann served as Winston Churchill's scientific adviser. They had been friends since 1921, and Churchill had always had confidence in Lindemann's scientific views. Lindemann served on the Tizard Committee in the middle and late 1930s, which promoted the development of radar. It was during his tenure on this committee that the feud between Lindemann and Henry Tizard started over defense priorities. In 1941, Lindemann was knighted and made Baron Cherwell. By wartime, Lindemann's friendship with Tizard had turned to bitter rivalry, and they frequently clashed over war and scientific priorities. When news of a possible atomic bomb reached Lindemann from his participation on the MAUD Committee, he was able to advise Churchill to support the project. His only deviation from the report was to recommend building the atomic bomb in Great Britain. Lindemann's influence resulted in close cooperation between Churchill and Roosevelt on the planning and gathering of raw materials for the Manhattan Project. In 1942, he was appointed paymaster-general in the Churchill government and held this position until 1945.

After the war, Lindemann briefly returned to Oxford University before becoming the head of the British atomic bomb program. When the Americans closed off atomic secrets in 1945 from British scientists, it became necessary to reconstruct the processes to build a bomb from scratch. Lindemann provided oversight in the building of the bomb and at the same time became active in the founding of the UK Atomic Energy Authority. Again in 1951, Lindemann became the paymaster-general in the Churchill government. In 1956, Lindemann was appointed Viscount Cherwell. He died on July 3, 1957, in Oxford, England.

Suggested readings: David Abbott, ed., *The Biographical Dictionary of Scientists: Physicists* (London: Blond Educational, 1984); Ronald William Clark, *The Birth of the Bomb* (New York: Horizon Press, 1961); Bertrand Goldschmidt, *Atomic Rivals* (New Brunswick, NJ: Rutgers University Press, 1990); Roy Harrod, *The Prof: A Personal Memoir of Lord Cherwell* (London: Macmillan, 1959); Richard Rhodes, *The Making of the Atomic Bomb* (New York: Simon and Schuster, 1986).

Liquid Metal Fast Breeder Reactor. *See* Breeder Reactors

Little Boy

Little Boy (Mark-1) was the nickname given to the gun-type uranium-235 bomb dropped on Hiroshima on August 9, 1945. The original version was Thin Man, which was a reference to President Franklin Delano Roosevelt, but its plutonium gun was unworkable. Its successor, Little Boy, was a long cylindrical device, which measured 3 meters in length by 0.7 meters in diameter and weighed four tons. While the bomb was rugged in construction, it required a lot of nuclear material (around 135 pounds of

Castings of the two atomic weapons—"Little Boy" (foreground) and "Fat Man" (background)—used by the U.S. Air Force on the Japanese cities of Hiroshima and Nagasaki. These models are on display at the Los Alamos Scientific Hall and Museum.

uranium-235). The detonation was caused by a subcritical mass of uranium-235 fired down a cannon barrel into another subcritical mass of uranium-235. Little Boy had an explosive punch of 15,000 tons of TNT, or 15 kilotons (15 kt). This version was the simpler of the two bombs designed by the scientists of the Manhattan Project, so it was never tested before detonation over Hiroshima. There was simply not enough uranium-235 processed for a test. A postwar study of the blast at Hiroshima indicated that because of its gun-type assembly that the bomb had an efficiency of only 1.3 percent; that is, only 1.3 percent of the uranium fissioned to produce energy. This inefficiency spelled the end of this type of bomb development, because it was too expensive and wasteful. *See also* Atomic Bomb.

Suggested readings: James N. Gibson, *Nuclear Weapons in the United States: An Illustrated History* (Atglen, PA: Schiffer Publishing, 1996); Ferenc Morton Szasz, *The Day the Sun Rose Twice: The Story of the Trinity Site Nuclear Explosion, July 16, 1945* (Albuquerque: University of New Mexico Press, 1984).

London Suppliers Guidelines

The London Suppliers Guidelines were a series of semisecret negotiations among the leading nuclear powers and interested parties on methods to control further nuclear weapons proliferation. These meetings were held in London between 1975 and 1978, with France, the Soviet Union, the United Kingdom, and the United States as the major nuclear powers and Canada, Japan, and West Germany as the interested parties. Joining in the discussions at one time or another were Belgium, Czechoslovakia, East Germany, Italy, the Netherlands, Poland, Sweden, and Switzerland. The impetus for the discussions was the failure of nuclear nonproliferation in the case of India's testing of nuclear weapons in 1974. Canada had supplied much of the equipment and technical expertise that permitted the Indian government to develop nuclear weapons. The Canadian government broke relations with the Indian government over India's refusal to accept the International Atomic Energy Agency's safeguards. It was in the context of this controversy and the reluctance of France to accept the provisions of the Non-Proliferation Treaty (NPT) that the London talks started. France agreed to abide by the NPT. Efforts to impose requirements by the supplier of nuclear materials on the recipient country were accepted on an optional basis. Other restrictions to make transfer of sensitive technologies more difficult were agreed upon. The nonnuclear countries wanting to build nuclear infrastructures protested against this series of agreements, accusing the nuclear states of forming a cartel to perpetuate their privileged status. The London Guidelines proved moderately effective in preventing the transfer of nuclear weapons technology. These guidelines provided the impetus for the formation of the Nuclear Suppliers Group (NSG) to enforce the London agreements.

Suggested reading: Bertrand Goldschmidt, *The Atomic Complex: A Worldwide Political History of Nuclear Energy* (La Grange Park, IL: American Nuclear Society, 1982).

Long Range Detection Project

The Long Range Detection Project was formed in 1947 to detect Soviet atomic tests. Admiral Lewis Strauss, a new member of the Atomic Energy Commission, approached James Forrestal, the secretary of defense, about a detection unit to ferret out Soviet atomic tests. At first such detection had been part of the Central Intelligence Group's (CIG) responsibilities, and the project was so secret that scientists were excluded from it. Vannevar Bush decided that excluding the very people that had the expertise to make the project work was counterproductive. The Office of Naval Research decided that the best approach was radiological detection, and its scientists worked on detection techniques. Still lacking a method of detection, the air force, in September 1947, was ordered by Forrestal to start its own detection program. When it became apparent that the techniques to trace our own tests were still lacking, Bush assigned scientist Ellis Johnson to shake the air force up. Soon the necessary mechanisms were set up by a private radiological laboratory, Tracerlab, in Berkeley, California, and tested by the Office of Atomic Testing, AFOAT-1 (Air Force Office of Atomic Testing), for effectiveness.

Despite evidence to the contrary, the military refused to accept the premise that the Soviets were close to testing an atomic bomb. American scientists were less sure, because the basic ideas on building an atomic bomb were relatively simple. A group of American and international scientists made predictions in 1946 about when the Soviets would test an atomic bomb, and these predictions were that the Soviets would have a test sometime between 1947 and 1950.

The air force used three B-29 squadrons to monitor air samples over Japan, Alaska, and the North Pole. On September 3, 1949, on a routine monitoring mission, air force planes flying east of the Kamchatka Peninsula found evidence of an earlier Soviet atomic test on August 29. Air samples near Japan had unexpected amounts of radioactivity. Tests at the Tracerlab measured activity at 300 percent greater than the alert level. The new secretary of defense, Louis Johnson, however, refused to believe the evidence. Consequently, the evidence was presented to a special committee of Vannevar Bush, Robert Oppenheimer, William Penney, Robert Bacher, and Hoyt Vandenberg, and it was determined that a Soviet atomic bomb, nickname "Joe One," had been detonated.

Suggested readings: Richard Pfau, *No Sacrifice Too Great: The Life of Lewis L. Strauss* (Charlottesville: University of Virginia Press, 1984); Richard Rhodes, *Dark Sun: The Making of the Hydrogen Bomb* (New York: Simon and Schuster, 1995); Charles A. Ziegler and David Jacobson, *Spying Without Spies: Origins of America's Secret Nuclear Surveillance System* (Westport, CT: Praeger, 1995).

Lop Nur Nuclear Weapons Test Site

The Lop Nur Nuclear Weapons Test Site in Xinjian Uygur (formerly Sinkiang Province) has been China's main nuclear testing site. It is at this site that the first Chinese atomic bomb was detonated on October 15, 1964. Then on June 17, 1967, China exploded its first hydrogen bomb. Between 1964 and 1986, an estimated 28 more nuclear bombs were detonated at this site. At least 2 of them have been above ground and 7 underground. In 1988, Chinese authorities claimed that China would no longer conduct atmospheric nuclear testing. Underground nuclear testing continued at the Lop Nur Nuclear Weapons Test Site, with the last test on July 29, 1996. A total of 45 nuclear tests were conducted at the Lop Nur site from 1964 to 1996.

Suggested reading: John May, *The Greenpeace Book of the Nuclear Age: The Hidden History, the Human Cost* (New York: Pantheon Books, 1989).

Lorentz, Henrik Anton

Henrik Lorentz (1853–1928) was a Dutch theoretical physicist whose early work on electrons paved the way for later advances in the field of atomic physics. He was born on July 18, 1853, at Arnheim, Netherlands. His marriage to Aletta Kaiser in 1881 resulted in two daughters and a son. Both his bachelor of science degree in 1871 and his Ph.D. in 1875 were from the University of Leiden, Netherlands. Lorentz became a professor of theoretical physics at the University of Leiden at the age of 24. He stayed at Leiden until his retirement in 1912.

Lorentz was always eclectic in his research interests, and many of his theories led to ideas further developed by Albert Einstein. At first most of Lorentz's research was on refining the electromagnetic theory of James Clerk Maxwell. In the course of his research, Lorentz advanced the electron theory of positive and negative charges. It was his analysis of light waves and matter that led to the theory of the Lorentz force on charged particles of matter in an electromagnetic field. He also identified negative electrons. It was for research on these topics that Lorentz won the 1902 Nobel Prize in Physics. Another topic of research was on the transformations of space and time, now called the Lorentz Transformations. Einstein reinterpreted Lorentz's idea for his theory of special relativity.

Lorentz was always interested in science education. He held Monday morning forums on current scientific theories to educate his colleagues and students. Upon retirement from the University of Leiden in 1912, he was appointed the director of the Teyler Laboratory in Haarlem, Netherlands. Lorentz was the first president of the Solvay Congress for physics in Brussels in 1911. He continued in this role until his death. He died on February 4, 1928, in Haarlem.

Suggested readings: Henry A. Boorse, Lloyd Motz, and Jefferson Hane Weaver, *The Atomic Scientists: A Biographical History* (New York: Wiley Science Editions, 1989); John Daintith, Sarah Mitchell, Elizabeth Toothill, and Derek Gjertsen, *A Biographical Encyclopedia of Scientists*, 2nd ed., vol. 2 (Bristol, UK: Institute of Physics, 1984); G. L. De Haas-Lorentz, *H. A. Lorentz: Impressions of His Life and Work* (Amsterdam, Holland: North-Holland, 1957).

Los Alamos Atomic Accidents

In the early stages of handling radioactive materials, two scientists died at the Los Alamos Scientific Laboratory in atomic accidents in the period

1945–1946. Although the dangers of working with uranium had long been common knowledge among scientists, the experiments with chain reactions produced new awareness. Two young scientists from the critical assembly team demonstrated this further in separate accidents.

Harry Daghlian was the first fatality at the Los Alamos facility. He served as an assistant to Otto Frisch on the critical assembly team in the Omega Laboratory in the Los Alamos Canyon. On the evening of August 21, 1945, Daghlian was working on an assembly problem when he accidently dropped a block of tamper material used to reflect neutrons. He received a severe dose of radiation and died less than a month later, on September 15, of acute radiation syndrome. His mother received a check from the U.S. government for $10,000 in compensation for the accident. Despite the outcome of this accident, safety procedures at the Los Alamos Scientific Laboratory remained lax, and it took another fatality in the handling of radioactive material for safety features to be incorporated as standard operating procedure.

The second accident took place nine months later when Louis Slotin, a brilliant, young Canadian physicist, was fatally exposed to radiation. He was a chemistry graduate of the University of Manitoba, and he had received his doctorate from the University of London. After a tour in the Royal Air Force, Slotin was recruited in 1944 to work at the Los Alamos Scientific Laboratory on the atomic bomb project. He became the leader of the critical assemblies group. Harry Daghlian assisted him on projects, but Slotin was on vacation in August 1945 when Daghlian had his accident. Slotin returned in time to watch Daghlian die. In a demonstration of a chain reaction exercise on May 21, 1946, a screwdriver slipped in Slotin's hand, and two plutonium hemispheres came too close together, starting a chain reaction. Slotin separated the reacting mass with his bare hand, absorbing a lethal amount of gamma rays and saving the other seven men in the room. Realizing immediately that he had been exposed to a fatal dose of radiation, Slotin began suffering bouts of nausea. Scientists estimated that he had received four times the dose that had killed Daghlian. Slotin died nine days later of radiation sickness. He was only 33 years old. The seven other men who had been exposed suffered temporary sterility. His experiment, nicknamed "tickling the dragon's tail," was outlawed at Los Alamos. Both Daghlian's and Slotin's deaths were covered up, as no news of either accident was permitted outside of the Los Alamos Scientific Laboratory.

Suggested readings: Lennard Bickel, *The Deadly Element: The Story of Uranium* (New York: Stein and Day, 1979); John May, *The Greenpeace Book of the Nuclear Age: The Hidden History, the Human Cost* (New York: Pantheon Books, 1989); Jonathan M. Weisgall, *Operation Crossroads: The Atomic Tests at Bikini Atoll* (Annapolis, MD: Naval Institute Press, 1994).

A postwar view of the town of Los Alamos (left) and the Los Alamos National Laboratory (right) shows the New Mexico countryside. This site was picked by J. Robert Oppenheimer and General Leslie Groves for its isolation and security.

Los Alamos National Laboratory

The Los Alamos National Laboratory (LANL) was founded in 1942 and was the weapon design part of the Manhattan Project. Soon after the first successful reactor experiment at the University of Chicago, it was decided by the leadership of the Manhattan Project for safety and secrecy reasons to locate a remote site to build an atomic bomb. General Leslie Groves, Robert Oppenheimer, and others examined several possible sites in California and New Mexico before finding Los Alamos, New Mexico. Oppenheimer knew the area well, having spent summers at a nearby ranch. It was isolated on a mesa about 35 miles northwest of Santa Fe. Several existing buildings from a boarding school for boys were acquired and made ready for immediate occupancy. The main laboratory and housing for families had to be built.

Robert Oppenheimer was appointed the first director of the then-called Los Alamos Scientific Laboratory. First plans were to have a scientific population of about a hundred, but before the Manhattan Project was complete, nearly 5,000 scientists and engineers worked in the laboratory. This expansion was partly the result of the Los Alamos Review Committee, or the Lewis Committee, which recommended on May 10, 1943, the go-ahead on activities at Los Alamos. Oppenheimer recruited an outstanding cast of scientific talent for the laboratory with the average age of around 32. A cyclotron from Harvard University, a Cockcroft-Walton accelerator from the University of Illinois, and two Van de Graaff accelerators from the University of Wisconsin were brought in and set up. Physicists, chemists, and metallurgists arrived in droves. Several divisions were formed: theoretical physics, chemistry and metallurgy, experimental physics, and others. So many big names were recruited to work at Los Alamos that it soon earned the nickname "Nobel Prize Winners' Concentration Camp." This was in reference to the fact that every worker there had signed a guarantee not to leave until six months after the end of the war.

The laboratory was operated under contract by the University of California at Berkeley, but the university was only a figurehead. Administrative personnel from the university were only told about the atomic bomb after the war ended. This agreement was signed on January 1, 1943, and formalized in April 1943. Oppenheimer believed it would be easier to recruit scientists under the auspices of a university, and it proved to be as he thought. The university determined salaries and took care of minor housekeeping chores such as basic supplies, but otherwise, Los Alamos was independent of the university's supervision.

Despite the intention to keep the number of scientists working at Los Alamos to a minimum, the population of Los Alamos grew rapidly. The original estimate was for a population of around 1,500, and plans for housing and research facilities reflected this figure. By January 1944, Los Alamos had a population of 3,500 and was still growing. In January, it now had 5,700 inhabitants. So much rapid growth put a strain on construction and on the personnel assigned there.

Although senior scientists remained civilians, many of the younger scientists and technician had been drafted in the army and assigned to Los Alamos. These scientists and technicians served in the 9812th Special Engineer Detachment (SED) of the Manhattan Project. Approximately 42 percent of the scientific personnel at Los Alamos belonged to the SED. Despite their military service, the members of SED acted like civilians, much to the displeasure of the military authorities. There was a constant tug-of-war between the military and the senior civilian scientists to improve the lot of these young soldier–scientists. One of the SEDs, Val Fitch would later win the 1980 Noble Prize in Physics.

The scientists' main preoccupation in the late 1940s and 1950s was to perfect the atomic bomb and stockpile weapons. A rivalry developed between Los Alamos and the new Livermore Laboratory (later named the Lawrence Livermore Laboratory) in the 1950s over the hydrogen bomb and research priorities. Norris Bradbury was the director at Los Alamos, and he had difficulty dealing with the political maneuvering of Edward Teller and Ernest Lawrence at Livermore. Los Alamos had most of the responsibilities for testing the first nuclear weapons at the Nevada Test Site, but by the late 1950s, the Livermore scientists had gained the ascendency.

As the Cold War tension lessened, the Los Alamos National Laboratory had to shift roles. A variety of different work was conducted at Los Alamos, from the NERVA (Nuclear Engine for Rocket Vehicle Applications) space rocket to finding ways for private companies, such as Exxon and Ford, to redesign processes more efficiently. In the last decade, nevertheless, the laboratory has lost around 12 percent of its staff. This facility will remain functional because the Department of Energy has proposed that the Los Alamos National Laboratory become its main site for the manufacture and repair of nuclear bombs.

Suggested readings: Lillian Hoddeson, Paul W. Henriksen, Roger A. Meade, and Catherine Westfall, *Critical Assembly: A Technical History of Los Alamos during the Oppenheimer Years, 1943–1945* (Cambridge: Cambridge University Press, 1993); James W. Kunetka, *City of Fire: Los Alamos and the Birth of the Atomic Age, 1943–1945* (Englewood Cliffs: Prentice-Hall, 1978); Richard Rhodes, *The Making of the Atomic Bomb* (New York: Simon and Schuster, 1986); Robert W. Seidel, *Los Alamos and the Development of the Atomic Bomb* (Los Alamos, NM: Otowi Crossing Press, 1995).

Los Alamos Scientific Laboratory. *See* Los Alamos National Laboratory

Lucky Dragon

The *Lucky Dragon* was a Japanese tuna fishing boat that ventured into a U.S. nuclear testing zone on February 28, 1954. It was about 100 miles east of Bikini when the hydrogen bomb of Bravo Test was detonated. The boat received high-grade radioactive contamination from ash from a mushroom cloud, and one crewman, Aikichi Kuboyama, subsequently died of liver and blood damage. Each member of the crew of 23 suffered from radiation sickness, experiencing skin blisters and hair falling out. The catch of tuna was also radioactive and had to be destroyed. Soon other boats fishing in southern waters also had radioactive fish catches. This accident caused a major international uproar in Japan. Scientists were puzzled about the range and intensity of the fallout, because the radioactivity was much greater than could be expected from a bomb of that size. After surveying the evidence, Ralph Lapp concluded that the explosion had enriched the uranium and caused a higher rate of radioactivity. In 1956 the U.S. government gave $2 million in compensation to the Japanese government.

Suggested readings: Jane Dibblin, *Day of Two Suns: U.S. Nuclear Testing and the Pacific Islanders* (London: Virago Press, 1988); Ralph E. Lapp, *Atoms and People* (New York: Harper, 1956).

M

Manhattan Project

The Manhattan Project was the largest military–civilian program undertaken in World War II, and its goal was the building of an operational atomic bomb. It was a costly project, with the tab at around $2 billion for its first three years. Several Prominent American scientific leaders believed in the summer of 1940 that Germany had a six-month headstart on the development of an atomic bomb. Three factors fed this belief: Germany had access to the only supply of uranium in Europe; Germany had seized Norway's heavy water plant; and Germany had able scientists, especially Werner Heisenberg. Despite this news, most American physicists were busy working on radar projects, so much of the early work on atomic problems was done by European emigrant scientists. Early work at Columbia University under brilliant physicist Enrico Fermi looked promising, and the Uranium Committee provided graphite and uranium for the project. A decision was made to transfer the reactor experiment to the University of Chicago, where on December 2, 1942, a successful demonstration took place under Fermi's supervision. After news of this experiment reached President Roosevelt, he called together key members of the Appropriation Committees of Congress and asked for an ultra-secret project that would cost at least $1 billion. General Leslie R. Groves was selected to head this project on September 23, 1942, and it was to be called the "Manhattan Engineer District" as a code name to confuse possible enemy agents. The name was soon shortened to the Manhattan Project. A Military Advisory Committee had overall supervisory authority and this committee had four members: Vannevar Bush, chair, James Conant, vice chair, Admiral William Purnell, representing the navy, and General Wilhelm Styer, representing the army. An early decision was made to pursue the development of both U-235 and plutonium weap-

In this picture the two principal figures in the building of the atomic bomb meet in November 1945 at a congressional hearing. Alexander Sachs (left) played a key role in gaining President Franklin D. Roosevelt's attention concerning the possibility of an atomic bomb. General Leslie Groves (right) was the administrative head of the Manhattan Project, which perfected the atomic bombs used against the Japanese cities of Hiroshima and Nagasaki.

ons. Consequently, the project leaders were Arthur H. Compton, for plutonium work; Ernest O. Lawrence, for electromagnetic separation work; and J. Robert Oppenheimer, for weapons work.

An early problem in the Manhattan Project was the difference in outlook between the high-level physicists and the engineers. The physicists could build an atomic bomb, but they had no experience with industrial engineering. The military and the engineers had the outlook on building a weapon and then placing it into production. With an engineering background, General Groves tended to favor the engineers' point of view. He had little patience with the idiosyncracies of atomic physics, but he was reluctant to interfere with their work. This incompatability of views would haunt the Manhattan Project throughout its existence.

After General Groves was instructed to coordinate research among the scientists, he gained the approval of the Military Advisory Committee to set up an atomic bomb laboratory. Despite some security questions about J. Robert Oppenheimer, General Groves appointed Oppenheimer to head the new weapons project. The next task was to find a location, and the area around Los Alamos, New Mexico, proved ideal for the bomb laboratory. Oppenheimer traveled around the country, recruiting scientists for Los Alamos. It was only after I. I. Rabi and Robert F. Bacher, recruited to head the Experimental Physics Division at Los Alamos, protested to Oppenheimer that the idea of making the scientists join the army was dropped. At one time Oppenheimer believed that only about 30 scientists would be necessary at Los Alamos to design the bomb. Others estimated that it would take as many as 500 scientists and engineers. In fact, the end of the war, there were 5,000 scientists and technicians working at the Los Alamos site.

The laboratory underwent several changes in organizational structure. The original organization entailed four divisions: a Theoretical Division under Hans Bethe; an Experimental Physics Division under Robert F. Bacher; a Chemical Division under Joe Kenney; and an Ordinance Division under Captain William S. Parsons. The implosion crisis (*see* Implosion) led to a reorganization in early 1945, when an enlarged implosion development division was added. All of this organizational structure began to show its merits in early February 1945, when progress on the building of the bomb accelerated. This progress was necessary because plutonium shipments began to arrive from the Hanford plants.

Perhaps the major scientific problem facing the scientists at Los Alamos was how uranium or plutonium reached a critical mass. To bring all the scientists up to date on current theory, Oppenheimer's assistant, Robert Serber, conducted lectures on both the theoretical and experimental states of knowledge on fission research. These lectures were heavily attended, and the lectures were printed as the *Los Alamos Primer*. The reactors constructed in Chicago used a process in which the graphite moderator slowed down fast neutrons. To create an explosive device the number of slow neutrons had to reach a critical value, but the danger was in the timing. One solution was the gun method, in which half of the fissionable material was fired into the other half, creating a larger mass whose combined neutron flux exceeded the critical value. Another way was implosion in which the fissionable material was compressed inward by an explosion. Experiments showed that the gun method worked with uranium but not with plutonium. The race to find a way to trigger implosion became key for a plutonium bomb.

The authority for building the gas-diffusion reactor at Oak Ridge was given to the Kellex Corporation, headed by Dobie Keith. Designing facilities before the designs were invented made the problem unique. An electric plant was constructed at Oak Ridge, Tennessee, without knowing the type of electricity needed, but the priority for it was so high that it was done anyway.

Because there were several different approaches to producing the uranium-235 and the plutonium-239 for the bombs, the leadership of the Manhattan Project planned for redundancy. If one approach failed, then another might work. When it was found that Philip Abelson's thermal diffusion process at the Naval Research Laboratory in Philadelphia could produce enriched uranium, General Leslie Groves decided to use the process by building a large thermal diffusion plant at Oak Ridge. Slowly, plutonium-239 and uranium-235 became available to construct a test bomb (Trinity) and two operational bombs for military use. High-security couriers with escorts arrived at Los Alamos weekly with lead containers filled with either plutonium-239 or uranium-235.

The use of the atomic bombs on the Japanese cities in August 1945 ended the mission of the Manhattan Project. A decision had been made to use the bombs as they became available. At the time of the decision to drop the bombs on Japanese cities, the United States had two bombs available. Since bomb construction was based on the availability of enough plutonium-239 from Hanford, or uranium-235 from Oak Ridge, the Los Alamos scientists could build only one bomb a month. Unless the Japanese surrendered after the Nagasaki bomb, the next bomb was scheduled to be ready sometime after August 24, 1945. Fortunately, the war ended before this bomb had to be used.

The Manhattan Project was the most successful crash program in the history of the United States. At a cost of around $2 billion and with the labor of hundreds of thousands of government officials, scientists, soldiers, and workers, a figment of an idea about the possibility of a weapon of indeterminant size and power was turned into an atomic bomb that ended World War II. Moreover, the Manhattan Project was the second-best-kept secret of the war. Only the breaking of the German coding machine Ultra was a better-kept secret, but it did not involve building large facilities in three states. Both Allies and enemies did learn about the existence of the Manhattan Project, but there was little they could do with the information. When the Manhattan Project's facilities and personnel were turned over to the new Atomic Energy Commission (AEC) in 1947, the organization was so large and uncontained that the AEC took years to build an administrative structure to handle it.

Suggested readings: Stephane Groueff, *Manhattan Project: The Untold Story of the Making of the Atomic Bomb* (London: Collins, 1967); Kenneth D. Nicholas, *The Road to Trinity* (New York: Morrow, 1987); Richard Rhodes, *The Making of the Atomic Bomb* (New York: Simon and Schuster, 1986); Robert W. Seidel, *Los Alamos and the Development of the Atomic Bomb* (Los Alamos, NM: Otowi Crossing Press, 1995).

Maralinga Proving Grounds

The majority of the British atomic bomb tests in Australia in the mid-1950s was conducted at the Maralinga Proving Grounds. *Maralinga* is an aboriginal term for "field of thunder." This proving ground was in a remote site at Woomera in the South Australian desert, about 65 miles north of the transcontinental railway line. In 1950, after the Monte Bello and Emu Test Sites were proven to be unsatisfactory, Australian and British authorities selected the Maralinga Proving Grounds as the permanent site for atomic weapons testing. It took several years to prepare the site, and the first test occurred on September 27, 1956. A total of seven tests were held at this site before pressure to end atmospheric nuclear testing ended its usefulness. The last atomic test at Maralinga was on October 9, 1957. This site became controversial when the Australian Royal Commission concluded in 1985 that evidence existed that there was radioactive contamination by plutonium-239 at Maralinga. The Royal Commission made the recommendation that the Maralinga Proving Grounds be cleaned up, with the cost charged to the British government. A further study in 1990 placed a price tag of $650 million to clean up the Maralinga Proving Grounds.

Suggested readings: Denys Blakeway and Sue Lloyd-Roberts, *Fields of Thunder: Testing Britain's Bomb* (London: George Allen and Unwin, 1985); International Physicians for the Prevention of Nuclear War, *Radioactive Heaven and Earth: The Health and Environmental Effects of Nuclear Weapons Testing in, on, and Above the Earth* (New York: Apex Press, 1991).

March 4th Movement. *See* Union of Concerned Scientists

MAUD Committee

The MAUD Committee was a body of British scientists studying the feasibility of building an atomic bomb in the early 1940s. At first this committee was called the Thomson Committee, in reference to its chair, G. P. Thomson. The Thomson Committee was formed by Henry Tizard, chair of the Committee on the Scientific Survey of Air Defence, in April 1940 in response to the Frisch-Peierls Memorandum on the practicality of an atomic weapon. Henry Tizard, the scientific adviser to the Chamberlain government, called the committee together after reading the Frisch-Peierls Memorandum but not without considerable skepticism. This skepticism showed itself in the selection of the chair, G. P. Thomson, an Imperial College physicist and J. J. Thomson's son, who was known by Tizard to be cautious about war research on uranium's potential as an explosive. Other members of the original committee were James Chadwick, John Douglas Cockcroft, and P. B. Moon. The committee held its first meeting on April 10, 1940, in London, and Jacques Allier told its members about Germany's interest in heavy water and its implications. After listening to Allier, members of the committee decided to pursue the theoretical outlines of the Frisch-Peierls Memorandum. Members of the committee also had access to two of the French physical chemist Joliot-Curie's researchers, Hans von Halban and

Lew Kowarski, who had conducted chain reaction experiments in France in 1939–1940. In June 1940, Thomson changed the name of the committee to the MAUD Committee (the reference was to Maud Ray, who had been a governess to Niels Bohr's boys and lived in Kent). Thereafter, MAUD served as a code name for the committee's activities. After prolonged study of the problem with the benefit of the Frisch-Peierls Memorandum, consultation with other British scientists, and a heavy water and uranium experiment by Halban and Kowarski at the Cavendish Laboratory proving a chain reaction was possible, the members of the MAUD Committee concluded not only that it was possible to build an atomic bomb but that it should receive the highest priority. Although the committee encouraged collaboration with the United States, it felt development work should continue on both sides of the Atlantic. To encourage further developments the MAUD Committee dissolved itself in the summer of 1940 into two separate committees—the Technical Committee and the Policy Committee. The British government, however, was aware of the lack of resources available in the British Isles, so the MAUD Report was sent to the United States in 1941. There the report confirmed what American scientists were thinking and proved to be one of the important documents in the American decision to launch the Manhattan Project.

Suggested readings: Ronald William Clark, *The Birth of the Bomb* (New York: Horizon Press, 1961); Bertrand Goldschmidt, *Atomic Rivals* (New Brunswick, NJ: Rutgers University Press, 1990); Richard Rhodes, *The Making of the Atomic Bomb* (New York: Simon and Schuster, 1986).

May-Johnson Bill

The May-Johnson Bill was introduced in October 1945 by two congressmen, Representative Andrew J. May of Kentucky and Senator E. C. Johnson of Colorado, to give the military control of postwar atomic affairs. Details of this bill had been prepared by experts from the War Department to continue the military monopoly of atomic energy. Justification was that the primary purpose of this legislation was to promote the national defense since commercial applications of atomic power were far into the future. The bill's most important feature was a commission of nine part-time members with five civilians and four serving military officers to serve for nine years without pay. To implement policy an administrator and a deputy administrator were to be appointed by the commission and responsible to it. Its most controversial provision was the inclusion of a military veto of the committee's decisions. The sponsors scheduled hearings, expecting a quick passage. After a debate of only five hours and with only four witnesses, the House of Representatives' Military Affairs Committee passed the bill. A majority of scientists, however, opposed the bill as a continuation of army dominance, excessive secrecy, and loss of freedom to pursue basic research. A counter-attack on it was led by Samuel K. Allison, one of the directors of the Chicago nuclear center. Additional support to Allison and his supporters at the Chicago laboratory came out of the Oak Ridge National Laboratory. This furor among the scientists led to the formation of the Federation of Atomic Scientists to serve as a pressure group to lobby Congress. While several key scientists, such as Robert Oppenheimer, Enrico Fermi, and Ernest O. Lawrence, supported the May-Johnson Bill, they were ignored. There was significant opposition to the bill in the House of Representatives led by Chet Holifield from California and Melvin Price from Illinois. Lobbying by scientists delayed the rapid adoption of the legislation until a commission of senators from both parties was formed to study future atomic proposals. This commission held public and secret hearings from November 1945 to April 1946. The May-Johnson Bill died from lack of support, however, when President Truman acknowledged scientific opinion and withdrew his backing. In its place was substituted the McMahon Bill, and it passed shortly thereafter. The key provision of the McMahon Bill was the substitution of civilian control of atomic energy for the military control of the previous bill.

Suggested readings: Corbin Allardice and Edward R. Trapnell, *The Atomic Energy Commission* (New York: Praeger, 1974); Brian Balogh, *Chain Reaction: Expert Debate and Public Participation in American Commercial Nuclear Power, 1945–1975* (Cambridge: Cambridge University Press, 1991); Frank G. Dawson, *Nuclear Power: Development and Management of a Technology* (Seattle: University of Washington Press, 1976); Bertrand Goldschmidt, *The Atomic Complex: A Worldwide Political History of Nuclear Energy* (La Grange Park, IL: American Nuclear Society, 1982).

McMahon, Brien

Brien McMahon (1903–1952) was a senator from Connecticut who was instrumental in forming a civilian-controlled Atomic Energy Commission (AEC). He was a lawyer who had served in the Justice Department for six years, specializing in tax and criminal matters, before winning election as a senator. McMahon and General Leslie Groves were both from Darien, Connecticut, and they were political rivals. At the end of World War II, a controversy developed in Washington, DC over whether the control of a new atomic agency should continue to be under the military or placed under civilian control. A military-inspired bill for military control, the May-Johnson Bill, died without President Truman's signature. An intense lobbying campaign broke out, and Senator McMahon became the champion of civilian control and on the opposite side from General Groves. McMahon made himself available to nuclear physicists and learned the issues before calling for public hearings to examine the issues. It was the military insistence on secrecy and security that made it difficult for the scientists to accept military control of atomic energy. Debate over the bill lasted 11 months, and it had to be heard in various congressional committees. The Atomic Energy Act was passed on August 1, 1946, and it was called at the time the McMahon Act. McMahon remained a staunch ally of the AEC and its works until his early death of cancer in 1952.

Suggested reading: Brian Balogh, *Chain Reaction: Expert Debate and Public Participation in American Commercial Nuclear Power, 1945–1975* (Cambridge: Cambridge University Press, 1991).

McMahon Act. *See* Atomic Energy Act of 1946

McMillan, Edwin Mattison

Edwin McMillan (1907–1991) was an American physicist whose research on uranium led to the discovery of several atomic elements. He was born on September 18, 1907, in Redonda Beach, California. His undergraduate work was at the California Institute of Technology. In 1932, McMillan received his doctorate in physics from Princeton University. His entire academic career was at the University of California at Berkeley from his appointment in 1935 until his retirement in 1973. In 1946 he was named professor of physics.

Most of McMillan's experimental work was done with the cyclotron at the Lawrence Radiation Laboratory. It was with experiments with the cyclotron that he and Philip Abelson discovered the first element heavier than uranium. They named this new element neptunium (element 93). McMillan believed that further elements existed, and shortly thereafter, Glenn Seaborg found plutonium (element 94). These twin discoveries resulted in McMillan and Seaborg sharing the 1951 Nobel Prize for Chemistry. McMillan also extended the usefulness of the cyclotron by removing its fixed frequency and inserting instead a variable frequency, which produced a synchrocyclotron. This solution was arrived at in 1945 and allowed much more powerful accelerators than the most advanced cyclotron previously. His synchrocyclotron went into operation on November 1, 1946. In recognition of his work, McMillan was named director of the Lawrence Radiation Laboratory in 1958. He held this post until his retirement from the University of California at Berkeley in 1973. McMillan died on September 7, 1991, in El Cerrito, California.

Suggested reading: John Daintith, Sarah Mitchell, Elizabeth Tootill, and Derek Gjertsen, *A Biographical Encyclopedia of Scientists*, 2nd ed., vol. 2 (Bristol, UK: Institute of Physics, 1994).

Medvedev Case

Grigori Medvedev is a Russian nuclear engineer and a longtime critic of the Russian nuclear energy program. He had extensive experience in the

Soviet nuclear energy program, becoming a departmental chief in the Directorate for Nuclear Energy. His career as a critic began in 1971 when he was treated for radiation sickness in Moscow. At Clinic No. 6 he witnessed other patients dying of radiation exposure. Medvedev decided to expose the dangers of nuclear energy by writing a series of fictionalized versions of dangerous practices based on his personal experiences. These stories in a short novel form—*The Reactor Unit, The Operators, The Hot Chamber*, and *The Expert Opinion*—were submitted to literary publishers but soon ran into censorship. It took almost a decade before some of these stories made it into print. Since one of these stories dealt with a nuclear accident at the Chernobyl Nuclear Power Plant, Medvedev decided to write a book on the 1986 nuclear accident at the plant. He interviewed the participants for his book, *The Truth About Chernobyl*, but again ran into censorship. It took nearly three years before his book was published in Russia. He is now considered to be the leading domestic critic of Russian nuclear policies.

Suggested readings: Grigori Medvedev, *No Breathing Room: The Aftermath of Chernobyl* (New York: Basic Books, 1993); Shores Medvedev, *The Legacy of Chernobyl* (New York: Norton, 1990).

Meitner, Lise

Lise Meitner (1878–1968) was an Austrian-Swedish physicist who is most famous for her interpretations of the fundamentals of atomic fission. She was born on November 7, 1878, in Vienna, Austria. Her father was a lawyer. She never married. Her public education at a girls' school was over before the age of 14, and she studied French at a private school in preparation for a teaching career. Then, she took private lessons in mathematics and physics to pass a leaving examination, necessary to attend a university. Meitner went to the University of Vienna, where she studied under famous theoretical physicist Ludwig Boltzmann. She received her Ph.D. in physics from the University of Vienna in 1906, with the distinction of being only the second woman to obtain such a degree. After a year of conducting experiments on radioactivity in Vienna, she moved to the University of Berlin in 1907 and studied under Max Planck. Soon afterward, she started work with Otto Hahn on radioactivity. He provided the expertise in chemistry, and she used her abilities in mathematics and physics to interpret the results. Despite her obvious brilliance and competence as an experimental and theoretical physicist, she experienced prejudice from her male colleagues and was forced to work in the woodworking shop in the basement of the Fisher Institute, away from contact with them. She moved with Hahn in 1912 to the Kaiser-Wilhelm-Institute for Physical Chemistry and Electrochemistry, and Max Planck gave her an assistant position (her first paying job). In 1915, she returned to Austria to serve as a nurse in the Austrian Army. Meitner re-

Lise Meitner, an Austrian physicist, won fame for her role in the discovery of atomic fission. She had to flee Germany in 1938 to escape political persecution.

turned to Berlin in 1916, and in a short time, she became the head of the Department of Radiation Physics at the Kaiser-Wilhelm-Institute. In 1919, she became a professor in the institute, and in 1922 she qualified as a privatdozent, which allowed her to teach at a university. During the next 20 years, Meitner combined experiments with theory to study the behavior of the nucleus of the atom. Her many research articles made her a respected figure in the international physics community. In 1936 Meitner and Hahn began to collaborate on testing Fermi's experiments on the elements. Hahn handled the chemistry and Meitner, the physics. Because she was an Austrian citizen, Meitner escaped the anti-Jewish decree of 1933 until the annexation of Austria in 1938. Her German colleagues then smuggled her out of Germany to Sweden.

Meitner was able to find a new position at the Nobel Institute in Stockholm, but again not without some opposition. She spent the next 20 years in Sweden, but always as an outsider, since the institute's director, Manne Siegbaln, never accepted her. It was while at Stockholm that Otto Hahn informed her of some puzzling data about his experiments with uranium. Meitner discussed the problem with her nephew, Otto Frisch, and they published a letter to the British scientific journal *Nature*, explaining the results as fission of the uranium nuclei. This news caused a flurry of theories about the possibility of an atomic bomb. Hahn later denied Meitner's role in the discovery of fission, and bad feelings resulted. In 1943 Meitner was invited to join a group of British scientists headed for Los Alamos to work on an atomic bomb, but she refused to use her life's work on the atom for military use. Meitner stayed in Sweden after the war and became a Swedish citizen in 1949. She was tempted to return to Germany when she was offered a position there, but she never felt comfortable in a country where the Nazis had been in control. In 1960, Meitner moved to Cambridge, England, where she remained until her death on October 27, 1968.

Suggested readings: Henry A. Boorse, Lloyd Motz, and Jefferson Hane Weaver, *The Atomic Scientists: A Biographical History* (New York: Wiley Science Editions, 1989); John Daintith, Sarah Mitchell, Elizabeth Tootill, and Derek Gjertsen, *A Biographical Encyclopedia of Scientists*, 2nd ed., vol. 1 (Bristol, UK: Institute of Physics, 1994); Ruth Lewin Sime, *Lise Meitner: A Life in Physics* (Berkeley: University of California Press, 1996).

Meshappen Nuclear Power Plant

The Meshappen Nuclear Power Plant was constructed in the early 1960s in Wyoming County, Pennsylvania, and it was the cause of the first antinuclear campaign in the United States. A controversial scientist at the University of Pittsburgh, Ernest Sternglass, had publicized his research findings by warning of an increase in cancer deaths due to radiation from nuclear power plants. Other scientists began to modify his conclusions but otherwise support his general thesis, so a delegation of residents from Wyoming County testified before a Senate Select Committee to block the building of a nuclear reactor on the grounds of radiation danger. This delegation of Quakers and a citizens' committee of prominent members of the community had a respectful hearing, but the nuclear plant was built anyway. This proved to many antinuclear activists that this legislative approach was not the way to proceed and turned to other approaches.

Suggested reading: Jerome Price, *The Antinuclear Movement* (Boston, MA: Twayne, 1982).

Messmer Plan

This plan was named after French Prime Minister Pierre Messmer. It envisaged the construction by the French electricity company, Electricité de France (EDF), of 13 new nuclear reactors in the period 1974–1975. The intent was to end French dependency on oil and allow France to have 85 percent of its energy needs fulfilled by nuclear energy by the year 2000. This plan called for France to have in operation by that time 170 nuclear reactors. The scope of this plan was challenged by the French scientific community and antinuclear activists. Although this goal was never reached, by 1989 France did have 75 percent of its electricity provided by 55 nuclear reactors. The main outline of the Messmer Plan is still French government policy and France still has the highest ratio of public power dependent on nuclear energy. Today France produces over 78 percent of its electricity from nuclear power plants.

Suggested reading: Helena Flam, ed., *States and Anti-Nuclear Movements* (Edinburgh, Scotland: Edinburgh University Press, 1994).

Metallurgical Laboratory of Chicago

The Metallurgical Laboratory (Met Lab) of Chicago sponsored the initial experiments on atomic reactors to produce plutonium. Although the Laboratory was first set up at Columbia University, it was moved in January 1942 to the University of Chicago campus, despite objections from some of its physicists. Arthur Compton was its head, and he united all of the chain reaction experiments under the Met Lab's auspices. He organized the scientists into teams to specialize on different parts of the problem. Compton was so successful in recruiting talent that there were 1,250 people working for the Laboratory by June 1942. At its peak, the Metallurgical Project employed about 5,000 persons, with about 2,000 at the Metallurgical Laboratory in Chicago. By October 1942, enough graphite and uranium were available to test for a chain-reacting reactor. The Chicago Pile One was constructed under the Staff Section of the Football Field at the University of Chicago. A successful test was conducted on December 2, 1942. A series of reactor tests followed at the new Argonne site over the next several months. With its initial successes, the Metallurgical Laboratory was given a new mission to produce plutonium. Its primary goals were to design the plutonium bomb (in February 1943 this task was transferred to Los Alamos), find a method for producing plutonium by irradiating uranium, design a method for extracting plutonium from the irradiated uranium, and accomplish these tasks in time to affect the conduct of the war. Since the Hanford Plants were where the plutonium production was to take place, the personnel of the Metallurgical Laboratory developed a working relationship with the du Pont Chemical Company, who administered the Hanford facilities. This relationship, however, had rocky moments because the scientists at the Met Lab often were frustrated by the conservative approach to engineering practiced by the du Pont personnel. When it became apparent that the Met Lab had been supplanted by the du Pont Corporation and the army, scientists became even more hostile. General Groves returned this hostility and in October 1944 began reducing personnel at the Met Lab. In May 1945 Arthur Compton resigned as head of the Met Lab to become chancellor of Washington University (St. Louis). It was in this milieu of distrust that Met Lab scientists began to oppose the use of the atomic bomb on Japan. After the war, the Metallurgical Laboratory was merged into the Argonne National Laboratory.

Suggested readings: Rodney P. Carlisle, *Supplying the Nuclear Arsenal: American Production Reactors, 1942–1992* (Baltimore, MD: Johns Hopkins University Press, 1996); Arthur Holly Compton, *Atomic Quest: A Personal Narrative* (New York: Oxford University Press, 1956); William Lawrence *The General and the Bomb: A Biography of General Leslie R. Groves, Director of the Manhattan Project* (New York: Dodd, Mead, 1988).

Mike Nuclear Test

Mike was the code name for the first test of a hydrogen bomb device. It was scheduled for October 1952 on the island of Elugelab in the Eniwetok Atoll of the Marshall Islands in the Pacific Ocean. The Panda Committee designed and built this bomb at the Los Alamos Scientific Laboratory. The device resembled a refrigerator, which when fully loaded weighed in the neighborhood of 65 tons. Its job was to freeze heavy hydrogen down to liquid form and then fuse it into helium, using an atomic bomb as a trigger. The device was detonated on November 1, 1952. Shot Mike's yield was far beyond expectations, with a blast of 10.4 megatons. It vaporized the island of Elugelab, leaving an underwater crater of around 1,500 yards in diameter. An analysis of the blast effects showed traces of two new elements—numbers 99 and 100, which were named einsteinium and fermium, respectively. Fallout from this test was moderate, but it showed that radioactive fallout was to be a big problem in hydrogen bomb testing. While this test demonstrated the validity of Edward Teller's design for a hydrogen bomb, the device was far from a tactical nuclear weapon. The Soviet Union's test of a hydrogen bomb on August 12, 1953, showed that the Soviet scientists were further advanced both in theory and in design for a tactical nuclear weapon than the Americans.

Suggested readings: Barton C. Hacker, *Elements of Controversy: The Atomic Energy Commission and Radiation Safety in Nuclear Weapons Testing, 1947–1974* (Berkeley: University of California Press, 1994); Robert L. Miller, *Under the Cloud: The Decades of Nuclear Testing* (New York: Free Press, 1986); Richard Rhodes, *Dark Sun: The Making of the Hydrogen Bomb* (New York: Touchstone, 1995).

Military Policy Committee

The Military Policy Committee was formed on September 21, 1942, to serve as the military policymaking body for the building of the atomic bomb.

Vannevar Bush, the head of the National Defense Research Committee, conceived of this body as a way to ease priority and procurement problems with the military concerning supplies and personnel. Henry Stimson, secretary of war, wanted a committee of seven to nine members, but General Leslie Groves countered with a recommendation for a committee of three. Groves's version was accepted. The three-member committee consisted of Vannevar Bush as chair and head of the Office of Scientific Research and Development (OSRD) (James B. Conant as an alternate), Admiral W. R. E. Purnell, Office of Chief of Naval Operations, and General Wilhelm Styer, Chief of Staff of Army Services and Supply. General Groves served as the executive officer to carry out the policies of the Military Policy Committee. This committee reported directly to the secretary of war. On major policy decisions, the Military Policy Committee had access to President Roosevelt through either Stimson or Bush. While this body remained the policy-formulating agency for the Manhattan Project throughout World War II, General Groves made the daily decisions to make it function.

Suggested readings: Kenneth D. Nichols, *The Road to Trinity* (New York: Morrow, 1987); Martin J. Sherwin, *A World Destroyed: The Atomic Bomb and the Grand Alliance* (New York: Knopf, 1975).

Millikan, Robert Andrews

Robert Millikan (1868–1953) was an American physicist whose experimental work contributed to the acceptance of quantum mechanics and whose leadership at the California Institute of Technology helped make it a major research institution in physics. He was born on March 22, 1868, in Morrison, Illinois. Millikan grew up in Maquoketa, Iowa, where his father was a minister. A marriage to Greta Irvin Blanchard in 1902 produced three sons. Despite an indifferent scientific education in high school, Millikan received his A.B. in 1891 and A.M. in 1893 from Oberlin College. It helped that his granduncle had helped found the school. Good grades at Oberlin won him a scholarship of $700 to Columbia University. There he received his Ph.D. in physics in 1895, under Michael Pupin. In 1895, Millikan traveled to Germany to study under Hermann von Helmholtz, where he found the debate over cathode rays in full force. He moved around, making contacts with such physicists as Max Planck, Walther Nernst, and French mathematician Jules Poincare. While in Europe, Millikan received an offer for a position at the University of Chicago from Albert Abraham Michelson. Millikan spent the first half of his academic career at the University of Chicago, where he remained from 1896 to 1921.

Millikan's experimental research involved determining the charge of the electron. He used a technique borrowed from Thomson's Cavendish Laboratory to measure the charge in a charged cloud of water vapor. By accident

Millikan found that applying too much electrical power to the field reduced the cloud to a few water drops. After shifting to oil drops as a medium, Millikan found himself able to prove in research from 1909 to 1912 that electrons were unique, identical particles. His results convinced even skeptics that the atomic theory of matter was correct. Then, in 1912, Millikan decided to test Einstein's theory of photoelectric effect. He designed an experiment that proved Einstein's equation and theory were correct. Millikan's research on electrons and on Einstein's theory garnered him the 1923 Nobel Prize for Physics. In the early 1930s, Millikan became entangled in and lost a debate with Arthur Compton over cosmic rays.

The second half of Millikan's academic career was at the brand-new California Institute of Technology. In 1921, he accepted the directorship of the Norman Bridge Laboratory. Then in 1922, Millikan became the president of the California Institute of Technology. He had a long and successful career as a university president. His career was marred only by charges that he fudged the results of some of his experiments, but the consensus remains that his contributions outweighed these questions. He died at San Marino, California, on December 9, 1953.

Suggested readings: Robert A. Millikan, *The Autobiography of Robert A. Millikan* (New York: Prentice-Hall, 1950); Henry A. Boorse, Lloyd Motz, and Jefferson Hane Weaver, *The Atomic Scientists: A Biographical History* (New York: Wiley Science Editions, 1989); Robert H. Kargon, *The Rise of Robert Millikan: Portrait of a Life in American Science* (Ithaca, NY: Cornell University Press, 1982); Daniel J. Kevles, *The Physicists: The History of a Scientific Community in Modern America* (Cambridge: Harvard University Press, 1987); Anthony Serafini, *Legends in Their Own Time: A Century of American Physical Scientists* (New York: Plenum Press, 1993).

Monte Bello Atomic Test Site

Great Britain entered the ranks of nuclear powers by exploding an atomic bomb on the remote Australian islands of Monte Bellos on October 3, 1952. The British government, with Prime Minister Attlee's personal intervention, persuaded Australian Prime Minister Robert Gordon Menzies to allow the testing of an atomic bomb in Australian territory. This bomb was a plutonium device, code-named Hurricane, and it closely resembled the Fat Man bomb of Nagasaki. The exercise was named Operation Hurricane, and the site selected was on a desolate island off Australia's western coast. It cumulated a seven-year process in designing and building a nuclear device. Between February and June 1952, five Royal Navy ships sailed to the Monte Bello Islands. The HMS *Plym* carried the nuclear device and served as the target vessel. Besides naval personnel, the ships also contained about 100 civilian scientists to work on the bomb. The device was detonated at 9:29 A.M. on October 3, 1952. This explosion proved to be excessively radioactive, with widespread fallout present in the Monte Bello area. Later calculations found the yield to be in the 25-kiloton range. With this successful test explosion, Great Britain became the third-ranking nuclear power. Great Britian moved most of its atomic testing to other sites in Australia, but in May and June 1956, testing returned to the Monte Bello Test Site. These devices were much smaller, in the low-kiloton range. After 1956, the British abandoned the Monte Bello Test Site permanently for other locations.

Suggested readings: Denys Blakeway and Sue Lloyd-Roberts, *Fields of Thunder: Testing Britain's Bomb* (London: George Allen and Unwin, 1985); Brian Cathcart, *Test of Greatness: Britain's Struggle for the Atom Bomb* (London: Murray, 1994).

Moscow Treaty. *See* Nuclear Test Ban Treaty

Moseley, Henry G. J.

Henry Moseley (1887–1915) was a British physicist whose promising career was ended by his death at Gallipoli in World War I. He came from a long line of distinguished scientists, with his father a biologist who had worked with Charles Darwin. Moseley's education had been at Eton and, later, Oxford University. He started working for Ernest Rutherford at the University of Manchester in 1910. Moseley used crystals as spectroscopes to study the atom and found regular patterns in the spacings between spectral lines. His experiments verifying Bohr's atom theory at the University of Manchester confirmed this theory for most physicists. He was considered by the international physics community as one of the rising stars of English science. Once the war broke out, Moseley volunteered as a lieutenant in the Royal Engineers and was made a signaling officer in the 38th Brigade, 13th Infantry Division. He suffered a head wound and died at Gallipoli on August 10, 1915. His death caused such an uproar in the science community that British scientists were thenceforth directed toward war work rather than fighting units. His case also received attention in the United States, and in World War II, both the Americans and the British made efforts to direct scientific talent into war industries rather than onto the front lines.

Suggested readings: Henry A. Boorse, Lloyd Motz, and Jefferson Hane Weaver, *The Atomic Scientists: A Biographical History* (New York: Wiley Science Editions, 1989): Richard Rhodes, *The Making of the Atomic Bomb* (New York: Simon and Schuster, 1986).

Mururoa Atoll

Mururoa Atoll in the Pacific Ocean has been the site of France's nuclear testing since 1966. After France lost its nuclear testing site in the Algerian Sahara at Reggan in 1965, the French government investigated two atolls (Mururoa and Fangataufa) in their Polynesian Territory east of Tahiti. Construction of testing facilities at both sites soon followed. France conducted its first atomic test at Mururoa on July 2, 1966. In September 1966, Charles de Gaulle attended a nuclear test at the Mururoa Test Site. On August 24, 1968, the French tested their first hydrogen bomb at the Fangataufa Atoll next to Mururoa. In the subsequent 20 years from 1966 to 1986, the French have conducted 167 nuclear tests, 44 of them atmospheric and the remainder underground. Since June 5, 1975, all of the French tests have been underground. This testing has been in defiance of the 1963 Test Ban Treaty and of the countries in the region. Several times French agents have repulsed efforts of the Greenpeace activists to enter the testing area, with the last incident taking place in August 1995 when the *Rainbow Warrior II* and *MV Greenpeace* were seized off Mururoa during a peace demonstration. The last test was held in September 1996, and France has announced no further scheduled tests.

Suggested reading: Jane Dibblin, *Day of Two Suns: U.S. Nuclear Testing and the Pacific Islanders* (London: Virago Press, 1988).

N

Nagasaki

Nagasaki was the Japanese city where the second atomic bomb was detonated on August 9, 1945. The intended target for the second atomic bomb attack was Kokura, but heavy overcast caused the mission to be diverted to Nagasaki. A weather observation B-29 accompanied the B-29 carrying the plutonium bomb, *Bock's Car*. At about 11:05 A.M. the atomic bomb detonated over Nagasaki. The total area reduced to ashes by blasts and fires was about 6.7 square kilometers. Although the bomb dropped on Nagasaki was more powerful than the one dropped on Hiroshima, differences in topography and distribution of buildings was such that there was less physical damage in Nagasaki. In addition, the bomb was off target by nearly two miles. This error also meant casualties were lower. In a reconstruction of the casualties as of December 31, 1945, the total was 73,884 killed, 74,909 injured, and 120,820 affected. In an updated report to the United Nations in the autumn of 1976, the estimation of total deaths from the blast and aftermath was 70,000 (plus or minus 10,000).

Suggested reading: Committee for the Compilation of Materials on Damage Caused by the Atomic Bombs in Hiroshima and Nagasaki, *Hiroshima and Nagasaki: The Physical, Medical, and Social Effects of the Atomic Bombings* (New York: Basic Books, 1981).

National Defense Research Committee

The National Defense Research Committee (NDRC) was the brainchild of Vannevar Bush, and its intent was to mobilize science for the war effort. Bush, former vice president of Massachusetts Institute of Technology (MIT) and president of the Carnegie Institution, and his associates Frank Jewett, president of the National Academy of Sciences and president of the Bell Telephone Research Laboratories, Karl T. Compton, president of MIT, and James B. Conant, president of Harvard University, were aware that previous efforts to coordinate scientific research for the war effort in World War I had been largely ineffective. Bush had worked on a magnetic submarine detector, but bureaucratic delays never allowed it to be put to use. After a May 1940 discussion about the possibility of a quasi-federal agency with Harry Hopkins, the secretary of commerce, which garnered his approval, Bush took a final proposal to President Roosevelt in early June 1940. He argued that a new committee would form a link between the military services and the members of the National Academy of Sciences, with the mission to correlate research in fields of military importance excluding aeronautics. President Roosevelt gave Bush his verbal approval at that meeting. On June 27, 1940, President Roosevelt approved the official order under the authority of the World War I Council for National Defense and promised money from the president's emergency funds. The already existing Uranium Committee was placed under the authority of the NDRC, and its chair, Lyman J. Briggs, reported to James Bryant Conant. Soon national leaders realized more coordination of scientific research was necessary, so the Office of Scientific Research and Development was founded in June 1941. Vannevar Bush resigned from the NDRC, and he was replaced by James B. Conant. A special review committee of scientists from the National Academy of Sciences met in November 1941 to study the feasibility of an all-out effort to produce an atomic bomb. This committee's report was a recommendation to pursue the development of an atomic bomb and was given to Vannevar Bush on November 6, 1941. Since the NDRC coordinated all civilian and military war-related research, the adherents of building an atomic weapon had to compete with other initiatives, including radar development, for support. Among others, Bush had to be convinced that a bomb could be built in a reasonable time frame. Access to the MAUD Report and the committee's recommendations convinced Bush. He took the report at once to President Roosevelt. The decision to go forward was announced on December 6, 1941, which was the day before the Japanese attack on Pearl Harbor.

Suggested readings: Arthur Holly Compton, *Atomic Quest: A Personal Narrative* (New York: Oxford University Press, 1956); Daniel J. Kevles, *The Physicists: The History of a Scientific Community in Modern America* (Cambridge: Harvard University Press, 1987); Martin J. Sherwin, *A World Destroyed: The Atomic Bomb and the Grand Alliance* (New York: Knopf, 1975).

National Ignition Facility. *See* Laser Fusion

National Reactor Testing Station

In 1949 the U.S. government acquired an 800-square-mile reactor testing site at Arco, 40 miles east of Idaho Falls, Idaho, for advanced research on nuclear systems and called it the National Reactor Testing Station (NRTS). Its first director was Leonard E. Johnson. A nuclear reactor was built to begin research on a breeder reactor. A shortage of uranium made a breeder reactor attractive for both military and civilian purposes. Facilities were also constructed for research on a prototype nuclear propulsion system for submarines. This research allowed Admiral Hyman Rickover to build the type of reactor that he wanted for a fleet of nuclear-powered submarines. It was in 1951 that the nuclear reactor at the NRTS also produced the first electricity generated from a nuclear-powered reactor. In the early 1960s, this site was used to build a small reactor to test emergency cooling systems and to subject the reactor to uncontrolled loss-of-coolant accidents. These tests soon revealed that there were serious deficiencies in industry safety standards, but when these results were reported to the Atomic Energy Commission (AEC), a committee was formed to study them. In meetings between this committee and representatives from the nuclear industry, the reports from the NRTS were rejected as unrepresentative of conditions at civilian nuclear power facilities. Nevertheless, the conclusions of the research on nuclear safety at NRTS were handed to the AEC in a two-volume report, the Brockett Report, in April 1971. In 1975, the name of the NRTS was changed to the Idaho National Engineering Laboratory (INEL).

In total, 52 reactors were built and experiments run on them at this facility. From the beginning this station cooperated closely with other laboratories and the nuclear industry. Among the 52 reactors, 11 were built by the Argonne National Laboratory, 11 by General Electric, 11 by Aerojet General, 11 by Phillips Petroleum, and the rest shared by Westinghouse, General Atomics, and Combustion Engineering. Among its newer projects was the integral fast reactor, which burned plutonium as its fuel, but this project was shut down because of the lack of interest on the part of the U.S. government. In 1995, the navy reactor program closed down. The end of these projects and programs has hurt staff morale at INEL and caused cutbacks in personnel. It appears that the future of this facility is in developing technology for radioactive waste sites.

Suggested readings: Rodney P. Carlisle, *Supplying the Nuclear Arsenal: American Production Reactors, 1942–1992* (Baltimore, MD: Johns Hopkins University Press, 1996); Daniel Ford, *The Cult of the Atom: The Secret Papers of the Atomic Energy Commission* (New York: Simon and Schuster, 1982); Richard G. Hewlett and Francis Duncan, *Nuclear Navy, 1946–1962* (Chicago: University of Chicago Press, 1974).

The USS *Nautilus* (SS-571) was the U.S. Navy's first atomic-powered submarine. Seen here on its initial sea trials in January 1955, this submarine was designed by Admiral Hyman Rickover and his supporters. (Photo reproduced from the Collections of the Library of Congress)

Nautilus

In 1954 the USS *Nautilus* was the first nuclear-powered submarine launched by the U.S. Navy. The submarine was launched on January 21, 1954, in a ceremony presided over by Mamie Eisenhower. Electric Boat of Groton, Connecticut, was the shipbuilder. It was the result of a long polit-ical battle within the U.S. Navy and on the national scene on the feasibility, and then desirability, of a nuclear submarine. Admiral Hyman Rickover pushed both the concept and then the design of the *Nautilus* as a first step toward a nuclear-powered submarine and surface fleet. The first roadblock was the design of a nuclear power plant. Initial research was done at the National Reactor Testing Station (NRTS) in Idaho. By the spring of 1950, a pressurized high-temperature water reactor had been developed for sub-marine use. The next hurdle was to gain political support. Once Rickover gained the backing of Senator Brien McMahon, chair of the Joint Con-gressional Committee on Atomic Energy and the godfather of atomic energy, for a nuclear submarine, the job was accomplished. The actual building of the submarine was anticlimatic after the politics to gain its approval. It took only 18 months after the startup of the prototype at the NRTS in Idaho for the installation of the propulsion plant for the *Nautilus* to be completed and sea trials started. Rickover handpicked the officers and crew. They spent a year in school at General Electric's Bettes Laboratory near Pittsburgh, Pennsylvania, studying science and reactor engineering before reporting to the *Nautilus*. Sea trials confirmed that the *Nautilus* had made conventional submarines obsolete. After a 32-year career, the USS *Nautilus* was retired from active service on July 6, 1985, and sent to Groton, Connecticut, for permanent display at the Nautilus Museum.

Suggested readings: Richard G. Hewlett and Francis Duncan, *Nuclear Navy, 1946–1962* (Chicago: University of Chicago Press, 1974); Theodore Rockwell, *The Rickover Effect: How One Man Made a Difference* (Annapolis, MD: Naval Institute Press, 1992); John W. Simpson, *Nuclear Power from Underseas to Outer Space* (La Grange Park, IL: American Nuclear Society, 1995).

Naval Research Laboratory

The Naval Research Laboratory was the military research organization dealing with the use of nuclear energy for ship propulsion. This research and development laboratory had been founded in 1923, and its home was in the District of Columbia. Rear Admiral Harold G. Bowen assumed command of the laboratory in 1939, just in time to confront the issue of atomic energy. Ross Gunn, a Yale University graduate with a doctorate in physics, worked at the Naval Research Laboratory and soon became intrigued with the potential of nuclear fission. Because of his close association with Merle Tuve of the Carnegie Institution, Gunn was appointed in June 1940 to the National Defense Research Committee. He was able to obtain a $100,000 grant for a study of isotope separation and lured Philip Abelson to Philadelphia to work on thermal diffusion experiments. This experiment proved successful enough that a thermal diffusion plant was built at Oak Ridge, Tennessee, to process uranium-235 for the Manhattan Project.

After the war, the navy's attention turned toward nuclear propulsion and submarines. Despite earlier cooperation, the Naval Research Laboratory found itself kept outside of the Manhattan Project and isolated from the research on atomic energy being done by its scientists. Also, the issue of nuclear propulsion became a political issue within the navy and in particular in the Bureau of Ships. It took the dedication and drive of Captain Hyman Rickover to break the impasse. In a reorganization, Code 390 was formed to pursue nuclear ship propulsion issues.

Suggested reading: Richard G. Hewlett and Francis Duncan, *Nuclear Navy, 1946–1962* (Chicago: University of Chicago Press, 1974).

Nazi Atomic Bomb Program

The Nazi atomic bomb program started out with great promise, but in the end, it turned out to be a failure. At a secret meeting on April 29, 1939, Abraham Esau, the president of the Reich Bureau of Standards and the head of the Reich Research Council, physics section, presided over a select group of German officials and physicists who decided to secure immediately all available uranium stocks in Germany, encourage German physicists to participate in a joint research group, and ban the export of uranium compounds. German authorities wanted to centralize all atomic energy research at the Kaiser-Wilhelm-Institute of Physics at Dahlem, but the scientists refused to leave their laboratories and move to Berlin. This refusal meant that atomic weapons research remained fragmented throughout the war. After some initial hesitation, the Nazi atomic bomb program received a boost in 1940 with the capture of heavy water at the Norsk-Hydro Plant in Norway, the seizure of 3,500 tons of uranium ore in Belgium, and access to a partially complete cyclotron in Paris. In December 1941, General Erich Schumann called a meeting to discuss the feasibility of continued support at a time when the war effort needed more manpower and raw materials. Just when it seemed that Werner Heisenberg was making progress with his atomic pile at his Leipzig Laboratory, an accident destroyed the laboratory and the experiment. This mishap set the Nazi atomic bomb program back at least a year.

German scientists were never able to bridge the gap between theory and the completion of building a working atomic reactor. They knew the theory, but refusal to challenge authority meant that Walther Bothe's faulty research findings on graphite were never questioned. Moreover, the difference in status between a theoretical physicist and an engineer in Germany was so broad that it impacted on research. In the end, less than a hundred German scientists and technicians worked on the German atomic bomb program, and less than 10 million was spent on it. Despite these failings, the Nazi atomic bomb project might have made more progress except for the chronic shortage of heavy water for experiments. Then at the key time in the experimentation with separation techniques for uranium-235, the Allied bombing campaign made transfer of laboratories to remote areas a necessity. This disruption started in the middle of 1943 and lasted until the end of the war. After the transfer of equipment and raw materials, an atomic reactor was constructed by Heisenberg's group at Haigerloch in the spring of 1945, and an experiment, B-VIII, came close to going critical. Only the lack of heavy water prevented a chain reaction. Before a further test took place, American soldiers occupied Haigerloch.

Suggested reading: David Irving, *The German Atomic Bomb: The History of Nuclear Research in Nazi Germany* (New York: Simon and Schuster, 1967).

NERVA

The Nuclear Engine for Rocket Vehicle Applications (NERVA) nuclear rocket program of the mid-1960s was a joint initiative of the Atomic Energy Commission (AEC) and the National Aeronautics and Space Administration (NASA) to build a long-range vehicle for space travel. Such a vehicle was considered by scientists necessary to provide propulsion for a manned expedition to Mars. Besides providing propulsion a nuclear engine would alleviate strain on life-support systems. This project for nuclear space reactors was called the Rover Program, and planning for this program began in 1955. First efforts were devoted to building a functional nuclear rocket engine. A KIWI-A experimental reactor was designed by the Los Alamos National Laboratory and tested by the Nuclear Rocket Development Station at Jackass Flats, Nevada, beginning in July 1959. A successful test of a revised model, KIWI B4-E, took place on August 28, 1964. The NERVA-type reactor was then adapted from the KIWI experiments. Work then started on a more powerful reactor, the Phoebus series. A Phoebus-1A was tested by the National Rocket Development Station on June 25, 1965. The most powerful nuclear rocket reactor ever built, the Phoebus-2A, was tested successfully in June 1968. After a series of successful tests, the NERVA program was renamed the Ground Experimental Engine (XE). A further test of the XE-prime prototype nuclear rocket engine was successful in March 1969. Despite the development of an operational nuclear engine for space travel, this program was allowed to lapse in January 1973 by the Nixon administration, as priorities changed in the national space program away from manned expeditions.

Suggested reading: Joseph A. Angelo and David Buden, *Space Nuclear Power* (Malabar, FL: Orbit Book Company, 1985).

Neutron Bomb

The neutron bomb, or enhanced radiation weapon (ERW), was a hydrogen bomb designed as a tactical nuclear weapon to produce a high burst of radiation, rather than destructive blast and heat. It was designed by the United States to serve as a deterrence against a conventional attack by the Soviet Union and the Warsaw Pact. Samuel T. Cohen, a scientific advise at the Rand Institute, conducted a feasibility study of such a weapon in 1958, which led to the development of a neutron bomb. His study determined that a greater portion of the energy from such a bomb would come from high-energy neutrons rather than from blast, heat, or radioactive fallout. This level of radiation would kill humans but leave buildings and property relatively unscathed.

Because radiation kills people but does not destroy property, the existence of this bomb became a highly controversial topic during the Carter administration. President Dwight Eisenhower was never convinced of its need, but other political leaders lobbied hard for its development in the late 1950s and early 1960s. In 1962 personnel from the Lawrence Livermore National Laboratory began testing a neutron bomb. The next step after development was to deploy this weapon to NATO (North Atlantic Treaty Organization) forces in Europe, and this also became controversial. Opponents of the neutron bomb characterized it as an amoral weapon. Public pressure against the neutron bomb became so intense that despite pressure from NATO, member countries refused to allow this weapon within their countries. Nevertheless, President Ronald Reagan authorized the building of a large stockpile of neutron bombs in case of war in Europe.

Suggested readings: S. T. Cohen, *The Neutron Bomb: Political, Technological and Military Issues* (Cambridge, MA: Institute for Foreign Policy Analysis, 1978); Sherri L. Wasserman, *The Neutron Bomb Controversy: A Study in Alliance Politics* (New York: Praeger, 1983).

Nevada Test Site

The Nevada Test Site was the only area in the United States that was used for testing of nuclear weapons. All of the early testing of atomic and hydrogen bombs, except for Trinity at the Alamogordo Test Site, had been conducted on Pacific islands. For security and international public opinion reasons, government officials decided to find a test site in the continental United States. Also, the Korean War had broken out, and a possibility existed that atomic weapons might have to be used in Korea. A specialized military unit, the Armed Forces Special Weapons Project, sponsored a report, Project Nutmeg, to select a North American nuclear weapons test site in late 1948. This report recommended either the arid Southwest or the Eastern Seaboard in North Carolina in the area from Cape Fear to Cape Hatteras. The Las Vegas Bombing and Gunnery Range, which was 75 miles northwest of Las Vegas, Nevada, was selected in 1950. President Harry Truman approved the use of the Nevada Test Site on December 19, 1950. It was picked over Dugway, Utah, and Alamogordo, New Mexico, because of the site's low population density, good meteorological conditions with a prevailing easterly wind, and a huge area of government-controlled land. Government policy was to restrict testing at this site to atomic weapons, leaving testing of the hydrogen bomb to the Pacific test sites. Most of the testing was conducted by scientists from the Los Alamos Scientific Laboratory. The first detonation at the Nevada Test Site took place on January 27, 1951, as part of Operation Ranger. Between January 1951 and July 1962, there were 105 atmospheric nuclear tests conducted at the Nevada Test Site. A further 19 tests were held underground. The 1,350 square miles of the test site became one of the most radioactive places on earth. Shot Harry on May 19, 1953, produced the most radioactive fallout, but all of the tests produced ground radioactivity.

Members of the 11th Airborne Division in full battle
dress participate in an atomic bomb test at the Nevada
Test Site in November 1951. These exercises were to
experiment with combat operations on the nuclear
battlefield. (Photo reproduced from the Collections of
the Library of Congress)

The biggest problem of the Nevada Test Site was radioactive fallout. Prevailing wind currents sent radioactive fallout toward the northeast and east
over southern and central Utah. First indications of a major problem were
on some of the isolated ranches where cattle and sheep suffered burns and
began to die. Then, people in southern Utah began to experience health
problems, especially leukemia among children. A nationwide scare about
strontium-90 entering the food supply through milk in the 1950s intensified
the debate over the dangers of radioactive fallout. Investigators from the
Atomic Energy Commission (AEC) monitored these health problems, but
the AEC policy was to deny responsibility. Even when actors and crew from
a film company making *The Conqueror* became sick after working at a location near the test site, the AEC evaded responsibility. It took court cases
to document the dangers of radioactive fallout from the Nevada Test Site.

The Nevada Test Site was mothballed by the U.S. government in the early 1990s as part of a voluntary test ban. Only in the Yucca Mountain part of the test site has there been any activity, as it is a proposed nuclear waste disposal location. Both government and military officials, however, have been lobbying in recent years to reopen the test site for limited underground testing of nuclear weapons. In 1997 a series of underground subcritical nuclear tests took place at the Nevada Test Site. This series ended in September 1997 and no further tests have been scheduled.

Suggested readings: Howard Ball, *Justice Downwind: America's Atomic Testing Program in the 1950s* (New York: Oxford University Press, 1986); Philip L. Fradkin, *Fallout: An American Nuclear Tragedy* (Tucson: University of Arizona Press, 1989); Varton C. Hacker, *Elements of Controversy: The Atomic Energy Commission and Radiation Safety in Nuclear Weapons Testing, 1947–1974* (Berkeley: University of California Press, 1994).

Nikitin Affair

Aleksandr K. Nikitin, a former naval captain in the Soviet navy and an environmentalist, was arrested and tried for treason by the Russian government for exposing Soviet-era nuclear accidents and present-day nuclear malpractice. He had been the inspector for nuclear reactor safety for the Northern Fleet stationed in Murmansk before leaving the navy in November 1992. His new position was with the Bellona Foundation, a Norwegian environmentalist organization. The Russian Federal Security Service (FSB), the successor to the notorious KGB, arrested Nikitin in February 1996. His charge was for supplying materials on atomic accidents and unsafe nuclear waste practices by the Russian Navy in the Murmansk area for a 1996 report published by the Bellona Foundation. The report was entitled "The Russian Northern Fleet, Sources of Radioactive Contamination." After being held in prison for 10 months, he was placed under travel restrictions in St. Petersburg before his trial in October 1998. This trial has had a chilling effect on environmentalists; some have charged that the trial would discourage cleanup of the radioactive problem at Murmansk. After the trial was held mostly in secret in October and November 1998, the presiding judge referred the case back for further investigation, since he had doubts about a guilty verdict. While the prosecutor has appealed the judge's decision, Nikitin is still confined to St. Petersburg and remains an employee of the Bellona Foundation. Environmentalists both in Russia and abroad are monitoring the affair closely. They consider this trial to be part of the Russian government's continuing campaign to manage the flow of information damaging to the former Soviet state and Russia, beginning with the failed case for revealing state secrets against Dr. Vil S. Mirzayanov in 1992 and continuing with the Vladivostok trial of Grigory Pasko in late 1998 for disclosing the Russian navy spilling of nuclear waste in the Sea of Japan.

No-First-Use

The No-First-Use strategic policy was based on the premise that North Atlantic Treaty Organization (NATO) countries would respond to attack with nuclear weapons but would not initiate attacks. NATO strategy was based for decades on the use of tactical nuclear weapons to counter the overwhelming size of the conventional forces of the Warsaw Pact. This doctrine became controversial in the late 1970s and 1980s as the deployment of new nuclear weapons systems in Europe were proposed. First, the neutron bomb controversy erupted and then deployment of advanced intermediate range ballistic missiles with nuclear warheads. European political figures and antinuclear activists began to question NATO's strategic doctrine to respond to the conventional attack with nuclear weapons. In May 1983 a pastoral letter from the U.S. National Conference of Catholic Bishops condemned the first use of nuclear weapons regardless of the circumstances. This debate lasted for several years without resolution. The collapse of the Soviet Union and the Warsaw act has ended the retaliatory debate. Both the expansion of NATO and its expanded role as peacekeeper have made the No-First-Use strategic policy obsolete.

Suggested reading: Frank Blackaby, Jozef Goldblat, and Sverre Lodgaard, *No-First-Use* (London: Taylor and Francis, 1984).

Non-Proliferation Treaty. *See* Nuclear Non-Proliferation Treaty

Non-Proliferation Treaty Exporters Committee. *See* Zangger Committee

Norsk-Hydro Plant

The Norsk-Hydro Plant at Vemork near Rjukan, in southern Norway, 75 miles west of Oslo, was the sole source of heavy water production in Europe before World War II. Its stock of heavy water was a by-product of the plant's separation of the hydrogen and oxygen in water. The plant was built in 1934 into a 1,500-foot granite bluff beside a waterfall. Its heavy water was produced for synthetic ammonia production. Between 1934 and 1938, the plant had produced only 40 kilograms of heavy water. Production in 1939 continued this output with only about 10 kilograms per month. Consequently, supply in the winter of 1940 was only 185 kilograms (407 pounds). This supply was sold to the French in March 1940, only after turning down German offers. On May 3, 1940, the Norsk-Hydro Plant was captured by German forces. The plant was turned over to the German chemical firm I. G. Farben. As soon as the plant was returned to full operation and a catalytic conversion installed, the Germans increased heavy water production goals to 1,500 liters a year. As plans for German atomic reactors became more real, the necessity of protecting heavy water supplies and transporting them to Germany assumed major importance.

Allied planners knew the significance of heavy water, and ways to destroy the plant and existing stock were tried. General Leslie Groves, head of the Manhattan Project, requested a raid against the heavy water plant in the summer of 1942. The destruction of the heavy water production was the only sure way to shut down the German atomic bomb program. In November 1942, the first attempt to cripple the plant by British commandos ended in complete failure, with the survivors executed by the Germans. Another attempt in February 1943 by Norwegian Special Forces commandos was more successful, and the plant was destroyed without damage to the power plant itself. Their explosives destroyed the machinery making deuterium and caused the loss of half a ton of heavy water. Despite this setback, the Germans rebuilt the heavy water plant faster than expected. A November 1943 air raid caused enough damage to destroy the power station and damage the electrolysis unit. German authorities decided to transfer heavy water production from Norway to Germany. An effort to move 14 tons of heavy water for further treatment to Germany proved to be a failure. The ferryboat *Hydro* carrying the heavy water was sunk by Norwegian commandos on February 20, 1944.

Suggested readings: Dan Kurzman, *Blood and Water: Sabotaging Hitler's Bomb* (New York: Henry Holt, 1997); Thomas Powers, *Heisenberg's War: The Secret History of the German Bomb* (New York: Knopf, 1993).

North Korean Nuclear Program

The North Korean nuclear program has been one of the greatest dangers to the international scene in the last decade. Soon after the end of the Korean War, the North Korean government began looking at a nuclear program. North Korea has ample supplies of natural uranium and graphite. The government negotiated agreements first with the Soviet Union in 1956 and 1959, then with China in 1959 for a research nuclear reactor. This early research reactor was too small for weapons-grade materials, and it had been placed under International Atomic Energy Agency (IAEA) safeguards in 1977. North Korea signed the Nuclear Non-Proliferation Treaty (NPT) in 1985. Then it became known to the international community that the North Koreans had a large reactor in operation at the nuclear complex at Yongbyon, 70 miles north of Pyongyang, capable of producing weapons-grade plutonium. Then in 1989 the North Koreans announced the building of a large reprocessing plant also at Yongbyon. These factors made the United States and others suspect that North Korea was about to become a nuclear state. Considerable international pressure was exerted against North

Korea, and the government there signed an IAEA safeguards agreement in 1991. The IAEA conducted several inspections, and in 1993 their inspectors became suspicious of two facilities at Yongbyon. The North Koreans refused to allow the inspection of the two buildings and took steps to withdraw from the Nuclear Non-Proliferation Treaty. In further talks with American representatives, some assurances were given by the North Koreans, and they suspended their withdrawal from the NPT. Additional talks broke down, and it appeared that North Korea was going ahead with its nuclear bomb program. Inspections were allowed again by the North Koreans in February 1994, and IAEA inspectors found suspicious nuclear activities. An impasse developed with little hope of resolution when the North Korean leader, Kim II Sung, died. North Korea has weapons-grade plutonium available, but the economic and political dislocation beginning in 1996 has removed much of the threat of North Korea developing a large nuclear weapons stockpile.

Suggested readings: Richard Kokoski, *Technology and the Proliferation of Nuclear Weapons* (Oxford: Oxford University Press, 1995); Marianne van Leeuwen, ed., *The Future of the International Nuclear Non-Proliferation Regime* (Dordrecht, Holland: Martinus Nijhoff, 1995).

Novaya Zemlya

Novaya Zemlya (New Land) was the site of the hydrogen bomb testing program of the Soviet Union in the 1950s and the 1960s. It is an archipelago in the Arctic Ocean between the Barents and Kara Seas and is hundreds of miles away from human habitation. This archipelago consists of two large islands—Severnyj and Yuzhnyi. Novaya Zemlya was selected as a test site in July 1954, mostly because of its isolation, with the nearest village over 200 miles away. The weather is harsh, with snow cover around 242 days a year. A series of hydrogen bomb tests took place at this site between September 1955 and April 1959, but the important hydrogen bomb tests took place between September 9 and October 30, 1961. American spy satellites picked up evidence of at least 12 Soviet detonations at this site. One of the tests on October 23, 1961, was huge, somewhere around 25 megatons. Then on October 30, 1961, the biggest detonation of them all took place at Novaya Zemlya, a device in excess of 50 megatons (perhaps as large as 58 megatons). In August and September of 1962, the Soviets tested another 15 devices, including a 30-megaton weapon. In all, the Soviets tested 132 devices at Novaya Zemlya, with 87 in the atmosphere, 42 underground, and 3 underwater. The test site has always been under the administrative control of the Soviet Navy's Sixth Main Directorate. Most of the military and civilian population live at Belyushy Guba (Whale Bay), and in 1992, 10,000 people lived there. There has been no testing at Novaya Zemlya since the testing moratorium in October 1990. In 1991, President Boris Yeltsin officially closed the nuclear test site.

Suggested readings: Thomas B. Cochran, Robert S. Norris, and Oleg A. Bak-harin, *Making the Russian Bomb: From Stalin to Yeltsin* (Boulder, CO: Westview Press, 1995); Richard L. Miller, *Under the Cloud: The Decades of Nuclear Testing* (New York: Free Press, 1986).

NRU Reactor

The NRU (National Research Universal) Reactor was Canada's mainline production reactor at the Chalk River Nuclear Facility for over a decade from the mid-1950s to the mid-1960s. Canadian authorities had become concerned about the instability of the NRX (National Research X-perimental) Reactor and desired a more dependable successor. Although the NRU was an experimental research reactor, it was also designed to produce plutonium. The Canadian government needed a financial commitment from the United States to purchase its plutonium production, however, before proceeding with development plans. Once this commitment was received, construction started on the reactor in 1951. It was designed to be much bigger than the NRX, at 200 megawatts (thermal), and produce 60 kilo-grams of plutonium annually. The reactor was to be moderated and cooled by heavy water. Although budgeted for $26.6 million with a completion date scheduled for 1954, cost overruns and design problems made both estimates faulty. Despite these initial difficulties, the NRU proved to be a successful reactor. Only one incident marred its safety record when a fuel rod broke apart in May 1958, causing the reactor to be shut down for six months. Despite its success, the NRU was considered a transition reactor to be replaced by a more advanced type—the CANDU (Canadian Deuterium-Uranium Reactor).

Suggested readings: Atomic Energy Limited, *Canada Enters the Nuclear Age: A Technical History of Atomic Energy of Canada Limited* (Montreal, Canada: McGill-Queen's University Press, 1997); Robert Bothwell, *Nucleus: The History of Atomic Energy of Canada Limited* (Toronto, Canada: University of Toronto Press, 1988); John May, *The Greenpeace Book of the Nuclear Age: The Hidden History, the Human Cost* (New York: Pantheon Books, 1989).

NRX Reactor

The NRX (National Research X-perimental) Reactor at the Chalk River Nuclear Facility was Canada's principal production reactor in the late 1940s and early 1950s. Design for this reactor began in 1944, and it was intended to be in the 20-megawatt (thermal) size. It was a water-cooled reactor, with heavy water serving as a moderator. Both the size and difficulty in acquiring materials delayed its construction, so it took nearly three years to complete. The NRX Reactor went operational in July 1947. Almost from the begin-ning the reactor experienced technical difficulties, first with leaky valves and then with operator mistakes. One such operator mistake caused the NRX to be shut down for six weeks in August 1947. On December 13, 1952, an explosion killed an operator and sent four others to the hospital. It was 14 months before the NRX was up and running again. Several safety features were studied and implemented before the reactor was placed back on-line. Difficulties of this nature caused Canadian authorities to consider a more advanced reactor, the NRU (National Research Universal), to replace the trouble-prone NRX.

Suggested readings: Atomic Energy of Canada Limited, *Canada Enters the Nuclear Age: A Technical History of Atomic Energy of Canada Limited* (Montreal, Canada: McGill-Queen's University Press, 1997); Robert Bothwell, *Nucleus: The History of Atomic Energy of Canada Limited* (Toronto, Canada: University of Toronto Press, 1988); John May; *The Greenpeace Book of the Nuclear Age: The Hidden History, the Human Cost* (New York: Pantheon Books, 1989).

Nuclear Accidents. *See* Chernobyl Nuclear Power Station Accident; Three Mile Island

Nuclear Control Institute

The Nuclear Control Institute (NCI) is an American research and advocacy center for the control of nuclear materials. It was founded in 1981 by its current president and executive director, Paul Leventhal, a longtime expert on nuclear energy issues, and its headquarters is in Washington, DC. Leventhal was active in legislation to break up the U.S. Atomic Energy Commission (AEC) and in the investigation of the Three Mile Island accident. The institute has a small staff of six professionals to perform its mission of serving as an unofficial oversight committee on Congress to ensure that the nation's nonproliferations laws are enforced. Special attention has been given to publicizing the dangers of proliferation of plutonium and enriched uranium for weapons making. Other areas of concern have been on the prevention of nuclear terrorism and avoidance of a Latin American nuclear arms race. This institute considers its lobbying of Congress and education of the American public about the dangers of nuclear proliferation to be its most important responsibilities.

Nuclear Freeze Movement

The nuclear freeze movement was a response to the antinuclear movement's failure to pressure national governments to outlaw nuclear weapons. Rather than push for nuclear disarmament and fail, the activists decided that a freeze on the production of nuclear weapons might cool the arms race between the superpowers. Randall Forsberg, a writer and researcher at the

Stockholm International Peace Research Institute, originated the idea of a nuclear freeze. She was able to convince representatives of the American peace movement in a January 1980 meeting at the headquarters of the Fellowship of Reconciliation, in Nyack, New York, to adopt her strategy. Forsberg wanted a policy less drastic than total disarmament for the American public to embrace. Her freeze proposal envisaged a ban on the production of plutonium and highly enriched uranium for weapons, a freeze on further production of nuclear weapons, a comprehensive nuclear test ban, and an end to the deployment of new weapons systems. The freeze movement was an immediate success, with support coming from grassroots constituencies. In December 1980, a national clearinghouse was established in St. Louis, Missouri, under the direction of Randy Kehler. Soon freeze resolutions began to appear in Congress. Such a resolution was defeated in the U.S. House of Representatives in mid-1982. Supporters of the nuclear freeze began to target congressional opponents and had some success in defeating them in elections. It was the Senate, however, that was most hostile to the nuclear freeze movement. The nuclear freeze movement made the Reagan administration more receptive to negotiations with the Soviet Union about arms control.

Suggested readings: J. Michael Hogan, *The Nuclear Freeze Campaign: Rhetoric and Foreign Policy in the Telepolitical Age* (East Lansing: Michigan State University Press, 1994); David S. Meyer, *A Winter of Discontent: The Nuclear Freeze and American Politics* (New York: Praeger, 1990); Keith B. Payne and Colin S. Gray, eds., *The Nuclear Freeze Controversy* (Lanham, MD: University Press of America, 1984).

Nuclear Free Zones

The Nuclear Free Zones (NFZ) movement was an attempt to limit the spread of nuclear weapons by establishing areas where no nuclear weapons would be introduced. Three characteristics of a nuclear free zone are nonpossession, nondeployment, and nonuse of nuclear weapons. The idea of a nuclear free zone first attention as early as 1956 when the Polish foreign secretary, Adam Rapacki, proposed a nuclear free zone for eastern Europe to include Czechoslovakia, Poland, and East and West Germany. North Atlantic Treaty Organization (NATO) countries showed little interest in this idea then or later, when another proposal for a nuclear free zone in the Balkans appeared. The first nuclear free zone was in the Antarctic, with the Antarctic Treaty of 1961, which was followed by Latin America with the Treaty of Tlatelolco in 1967. Parties to the Tlatelolco Treaty agreed to use nuclear energy exclusively for peaceful purposes. All but 4 of the 26 Latin American countries—excluding Argentina, Brazil, Chile, and Cuba—became full members of the nuclear free zone for Latin America. Another nuclear free zone formed in 1985 among the southern Pacific states, but it has been less successful adhering to its nuclear free status.

Suggested reading: David Pitt and Gordon Thompson, eds., *Nuclear-Free Zones* (London: Croom Helm, 1987).

Nuclear Nonproliferation Act

The Nuclear Nonproliferation Act was a piece of legislation passed by the U.S. Congress in early 1978 and promulgated in April 1978. This legislation followed shortly after the London Guidelines meeting on nuclear nonproliferation. It embodied all the principles that the negotiators at London had been unable to persuade the international community to accept. Provisions of the act required nonnuclear states to submit their entire nuclear programs to International Atomic Energy Agency (IAEA) safeguards and placed restrictions on their ability to process nuclear materials by reprocessing or enrichment to make weapons. An embargo was to be placed on countries refusing to accept the Non-Proliferation Treaty (NPT) or safeguards. It also empowered the president to reopen existing agreements to bring them into conformity with the new law. The law did allow an escape provision allowing the president or Congress to make exceptions on a case-by-case basis. President Jimmy Carter's administration adhered to this law, but his successors, particularly President Ronald Reagan's administration, were more pragmatic.

Suggested reading: Bertrand Goldschmidt, *The Atomic Complex: A Worldwide Political History of Nuclear Energy* (La Grange Park, IL: American Nuclear Society, 1982).

Nuclear Non-Proliferation Treaty

The Nuclear Non-Proliferation Treaty (NPT) was an agreement between the Soviet Union and the United States in July 1968 to control the spread of nuclear weapons. Earlier the International Atomic Energy Agency (IAEA) had developed a safeguards system in an effort to control the peaceful uses of nuclear energy, but superpower politics had curtailed much of its effectiveness. American policy had been to promote peaceful development of nuclear energy, but always with controls for its usage. While the Soviets had a more adventurous policy, their China experience in the late 1950s showed that they too had to fear the spread of nuclear technology. A central theme among both the Soviet Union and the United States was the fear that the Federal Republic of West Germany might go nuclear. Consequently, in 1968 the Nuclear Non-Proliferation Treaty was negotiated with IAEA safeguards in place. The treaty defined a nuclear power as one that had made and exploded a nuclear device before January 1, 1967. Only China, France, Soviet Union, the United Kingdom, and the United States qualified as nuclear powers under this definition. Transfer of nuclear devices or control over them to any nation was prohibited. This broad definition included a prohibition of peaceful nuclear explosions by nonnuclear countries. This was one of the terms that many of the nonnuclear countries resented. The Nuclear Non-Proliferation Treaty became effective on March 5, 1970, with signatures from representatives of the Soviet Union, the United Kingdom, and the United States. Forty nonnuclear weapons countries also signed the treaty. The treaty's initial term was 25 years, with a conference to be convened in 1995 to study its future. Eventually 160 countries ratified the treaty. This treaty has always been considered discriminatory by the nonnuclear states, because these states have most of the restrictions and the nuclear states so few of them. India has been the most vocal in their opposition to this treaty. In 1975 the 1968 treaty was extended indefinitely by the 179 states participating in the review.

Suggested readings: Greg Egen, *The Origins of the United States' Non-Proliferation Policy: A Study Project* (Washington, DC: Atomic Industrial Forum, 1978); Bertrand Goldschmidt, *The Atomic Complex: A Worldwide Political History of Nuclear Energy* (La Grange Park, IL: American Nuclear Society, 1982); Richard Kokoski, *Technology and the Proliferation of Nuclear Weapons* (Oxford: Oxford University Press, 1995); Marianne van Leeuwen, ed., *The Future of the International Nuclear Non-Proliferation Regime* (Dordrecht, Holland: Martinus Nijhoff, 1995).

Nuclear Power Division. *See* Code 390–590

Nuclear-Powered Ships. *See* Commercial Nuclear Ships

Nuclear Reactor. *See* Breeder Reactors; Chicago Pile One; Light Water Reactors

Nuclear Regulatory Commission

The Nuclear Regulatory Commission (NRC) is the U.S. government agency regulating the nuclear energy industry. It replaced the Atomic Energy Commission (AEC) on January 19, 1975, in an attempt to separate regulatory functions from advocacy of nuclear energy. The NRC's mission was to create a more open licensing procedure and greater citizen participation in the nuclear power debate. In reality, it was the old regulatory staff of the AEC hiding under a new name. A new set of five commissioners was appointed to head the NRC. William Anders, a former AEC commissioner, was the first chairperson. Among the new organization's first acts was to adopt in toto the safety and licensing regulations of the AEC. Its first job, however, was to deal with the nuclear safety issue of the Rasmussen Report. Eventually, in 1979, it renounced the Rasmussen Report but not before considerable controversy. Nuclear safety issues dominated the agenda of the NRC the late 1970s, and no consensus was developed among the members of the commission. Institutional paralysis was the result. It took the Three Mile Island nuclear accident to break the deadlock, but the oversight committee formed by President Jimmy Carter to monitor the NRC's safety policies was abolished by President Ronald Reagan in 1982. Part of the regulatory problems of the NRC was the overexpansion of the nuclear industry, which resulted in a shortage of competent personnel and constant changes in regulations.

Suggested reading: Daniel F. Ford, *The Cult of the Atom: The Secret Papers of the Atomic Energy Commission* (New York: Simon and Schuster, 1982).

Nuclear Secrecy

The international scientific community from the time of Roentgen's discovery of the X-ray in 1896 onward has been engaged in a gigantic intellectual treasure hunt on the nature of the atom. This hunt for the structure of the atom was particularly fierce in the first half of the twentieth century. Each advance in knowledge was one more step toward an elusive goal. Scientists, regardless of nationality, provided either experimental data or theoretical insight into the next stage of the problem. As long as this question was in the interest of pure science, the various governments had little benefit in interfering. Even during World War I, scientific laboratories were raided for researchers for applied military weapons projects, but scientists were allowed after the war to return to pure science. In the interwar period, scientists communicated openly on their experiments and theories. In particular, Niels Bohr's Institute for Theoretical Physics was a place where physicists could speculate on the nature of the atom free from political concerns. A unique case was that of Soviet scientist Peter Kapitsa, who was allowed to spend 14 years conducting experiments at Rutherford's Cavendish Laboratory before being forced to stay in the Soviet Union in 1934 after visiting his family there. Another example was the Solvay Conferences, which served almost as an Olympics of scientific thinking with its weeklong meetings held every 3 years.

Once political authorities were convinced that atomic research could lead to a military weapon, this era of international free exchange of scientific information withered away. The year 1939 stands out not only as a key date for the outbreak of the war but as the year when physicists concluded that the energy unleashed from the splitting of the atom could create a superweapon. A number of physicists from a variety of countries came to the same conclusion. Leo Szilard became the champion of secrecy because he was so concerned about Germany obtaining enough information to build a bomb. Other scientists, especially Neils Bohr, had built lifetimes of sharing information and were reluctant to resort to secrecy. Szilard continued to lobby both physicists and governmental officials. Finally, the Einstein-Szilard Letter and a subsequent interview convinced President Roosevelt and his scientific advisers about the potential of the atomic bomb. From this time onward, a self-imposed censorship among atomic physicists began to occur, instigated mostly on the initiative of Leo Szilard. In the spring of 1940, a Reference Committee was established within the National Research Council of the National Academy of Sciences to serve as a censor. This committee was the suggestion of physicist Gregory Breit, who was chairman of the subcommittee on uranium fission. Certain subjects suddenly no longer appeared in scholarly journals.

Scientists in other countries were slower to learn about the weapons potential of research on the atom, but most of the major powers had scientists with the expertise to figure out the basics. British authorities were only shortly behind those in the United States in learning about a possible bomb, but the British were in the middle of a fight for survival and soon opted out to the Americans. German authorities were slower, but Werner Heisenberg and Max von Laue were certainly abreast of developments. Scientists in Denmark, France, Italy, and the Soviet Union also knew the fundamentals.

Once nuclear programs were initiated, secrecy replaced the open interchange of ideas. Almost from the beginning of the Manhattan Project, security checks were made on the scientists. At first security was directed against possible Axis spies, but as time went on, it was to include agents from the Soviet Union. Robert Oppenheimer had security problems in 1942–1943 because of his associations with left-wing friends. When Niels Bohr had an interview with Winston Churchill in 1944 to argue for the sharing of atomic information even to the Soviets, Churchill never forgave Bohr and distrusted his motives. General Leslie Groves wanted to place the scientists in the military and subject them to military discipline, but an open revolt ended this idea. A system was put into place, however, that scientists at the Manhattan Project could discuss scientific topics, but only among themselves on a need-to-know basis. This policy was called *compartmentalization*. This dispute between General Groves and the scientists over compartmentalization revolved around the difference in perspective between a military engineer and an academic researcher. It is interesting to note that Leo Szilard became one of the biggest critics of this policy because he believed openness among scientists facilitated their research. The resentment against Groves's secrecy policy centered among the scientists at the Metallurgical Laboratory in Chicago. Soon the Federation of American Scientists was formed to fight against army secrecy. It was over secrecy that the atomic scientists mobilized against military control of atomic energy in the May-Johnson Bill. They objected to the rigid security restrictions with penalties for violators of a fine of up to $100,000 and a 10-year prison term.

Despite security measures, atomic secrets did become available to other governments. In the immediate postwar era, the Americans even froze out their nominal allies, the British. British scientists knew so little about the construction of an atomic bomb that it took them years to figure out the basics. They had to turn to the memory of atomic spy Klaus Fuchs for some of the processes. Soviet agents had more successes learning about the atomic secrets from spies, including the big one of implosion.

As the nuclear stakes became higher so did the demand for secrecy. Many of the scientists resented the security checks and restrictions on their contacts with scientists of other nationalities. By the end of the 1950s, the Atomic Energy Commission (AEC) had investigated the backgrounds of more than 150,000 employees and candidates for employment. This demand for security led to the denial of a security clearance to the former director of the Manhattan Project, J. Robert Oppenheimer. His case infuriated many in the scientific community, and the divisions over this case lasted for decades. It served as a fitting contrast to the dividing line between the early days of open scientific inquiry and the postwar need to protect a military weapon. The isolation came to give American scientists an inflated idea of their originality. It took meetings with Soviet scientists in the 1955 Geneva Conference before the realization sank in that Soviet scientists had the capabilities of major advances independently.

Suggested readings: Arthur Holly Compton, *Atomic Quest: A Personal Narrative* (New York: Oxford University Press, 1956); William Lawrence, *The General and the Bomb: A Biography of General Leslie R. Groves, Director of the Manhattan Project* (New York: Dodd, Mead, 1988); Kenneth D. Nichols, *The Road to Trinity* (New York: Morrow, 1987); Martin J. Sherwin, *A World Destroyed: The Atomic Bomb and the Grand Alliance* (New York: Knopf, 1975); Spencer R. Weart, *Nuclear Fear: A History of Images* (Cambridge: Harvard University Press, 1988).

Nuclear Smuggling

The breakup of the Soviet Union and the dismantling of much of its formerly large nuclear program have resulted in the growth of nuclear smuggling to nonnuclear states. The Soviet supply of military-grade plutonium for weapons peaked in 1986. In October 1989 the Soviet government announced the end of producing highly enriched uranium. The government of Yeltsin has continued this policy, and present government policy is to store all plutonium from dismantled warheads for at least a decade. This dismantling of nuclear warheads and the unsettled economic conditions in Russia have led to several occurrences of nuclear smuggling. It has been estimated that for about $40 million a country could obtain on the open market the materials to construct a nuclear reactor capable of producing enough plutonium to produce a nuclear weapon. In 1994 a German businessman was arrested by German authorities for possession of six grams of plutonium-239, which was traced back to the Russian nuclear weapons industry. Despite other warning signs and alerts from international police agencies, the incidents of nuclear materials smuggling have been few. Nevertheless, severe economic and social dislocation in Russia and the former Soviet states lends itself to the possibility of future smuggling. It has been noted that much of Russia's 1,500 tons of weapons-grade nuclear material is left almost unprotected at remote storage sites guarded by underpaid security personnel. Moreover, there exist in Russia around 200 criminal organizations that have established ties to international smuggling rings. The threat is considered serious enough that a Nuclear Black Market Task Force has been formed by the U.S. government to ascertain the risk and find solutions.

Suggested readings: Center for Strategic and International Studies, *The Nuclear Black Market* (Washington, DC: Author, 1996); Richard Kokoski, *Technology and the Proliferation of Nuclear Weapons* (Oxford: Oxford University Press, 1995).

Nuclear Submarines. *See* Chazhma Bay Submarine Accident; *Nautilus*; Rickover, Hyman George; Soviet Nuclear Submarine Accidents

Nuclear Suppliers Group

On January 27, 1976, the Nuclear Suppliers Group (NSG) formed to establish new guidelines on nuclear exports. Earlier meetings in London in 1975 to formulate guidelines were only partially successful, so a new body was needed to implement guidelines on nuclear exports. Formation of this group was a serious effort to control the flow of nuclear materials to prevent nonnuclear states from obtaining nuclear weapons. Negotiations for such a body started in November 1974, mainly under the initiative of the United States. An agreement on basic principles was agreed upon in late 1975. The original participants were Canada, France, Japan, the Soviet Union, the United Kingdom, the United States, and West Germany, but soon other countries joined. After some initial successes, the NSG stopped functioning effectively in the 1980s. In the early 1990s the Nuclear Suppliers Group revived after it was becoming more apparent that nuclear-related sales were allowing countries to obtain nuclear weapons–grade supplies. In April 1992 a new set of guidelines was adopted by the NSG to stop exports of nuclear technology. Information on the extent to which Iraq had obtained nuclear materials on the black market was the impetus for the new restrictions. By March 1995 the NSG had 30 members. This body continues to function, but its task is difficult as more and more countries have initiated nuclear programs.

Suggested readings: Richard Kokoski, *Technology and the Proliferation of Nuclear Weapons* (Oxford: Oxford University Press, 1995); Marianne van Leeuwen, ed., *The*

Future of the International Nuclear Non-Proliferation Regime (Dordrecht, Holland: Martinus Nijhoff, 1995); Leonard S. Spector, Mark G. McDonough, and Evan S. Medeiros, *Tracking Nuclear Proliferation: A Guide in Maps and Charts, 1995* (Washington, DC: Carnegie Endowment for International Peace, 1995).

Nuclear Terrorism

The fear of terrorists acquiring control of nuclear weapons has long been a nightmare to governments around the world. Modern weapons-grade uranium is relatively easy to explode by a process of joining two halves of uranium together. It is gaining access to U-235 that would be difficult for terrorists. Plutonium is more common, but it is much more difficult to work with and to detonate.

Fear about nuclear terrorism was intensified in the early 1960s when the number of light water reactors increased dramatically, since each of these reactors produced plutonium as a by-product. Most governments strengthened physical security at their nuclear power plants, fuel handling, and transport facilities. Still there were cases in which plutonium disappeared from inventories. As the world supply of plutonium has boomed, authorities have become more concerned about terrorists building a nuclear device. Despite this supply of plutonium, it still takes considerable skill to build an atomic bomb. First, a skilled inorganic chemist would need a first-class laboratory to work the plutonium. Another problem is to design a detonator and make it work. Finally, some type of explosive has to be found for the detonation. None of these steps are easy, and they require sophisticated machinery and laboratory equipment. The greater danger is for a nuclear device to be placed by a nation–state at the disposal of a terrorist group, but as of yet, no state has been willing to do this.

Besides governments, antinuclear activists have also used the threat of nuclear terrorism as a part of their antinuclear campaign. They have cited a BDM Corporation report in the mid-1970s detailing 77 terrorist attacks against nuclear facilities between 1966 and 1975. The most serious of these attacks was in Argentina, when a left-wing band of 15 took over a nuclear power plant under construction at Atucha. Both fuel rods and plutonium have been reported missing at plants in the United States and abroad.

Suggested readings: Jeremy Bernstein, *Hans Bethe, Prophet of Energy* (New York: Basic Books, 1980); Jerome Price, *The Antinuclear Movement* (Boston, MA: Twayne, 1982).

Nuclear Test Ban Treaty

The Nuclear Test Ban Treaty, or the Moscow Treaty, was a treaty between the Soviet Union, the United Kingdom, and the United States to prohibit nuclear weapons tests in the earth's atmosphere, in space, and underwater. This treaty was signed on August 5, 1963, in Moscow. Between the end of World War II and the signing of the treaty, somewhere around 500 nuclear tests had been conducted by the nuclear powers; all but 4 (France) were carried out by the three signatories. The treaty allowed for underground nuclear tests, but a provision stipulated that parties could not participate in unauthorized nuclear tests by a nonsignatory state. Duration of the treaty was unlimited, but each signatory had the right to withdraw after a three-month notice to the other signers. President Kennedy signed the treaty on October 7, 1963, for the United States. The treaty became effective on October 10, 1963, and on that date over 100 countries had signed to adhere to the treaty. Two notable exceptions were France and China. This treaty was the first major step toward better relations between the Soviet Union and the United States, and the treaty had public support around the world. Despite subsequent treaties limiting nuclear weapons, this treaty is still operational, with both France and China adhering to it in 1974 and 1980, respectively.

Suggested readings: Bertrand Goldschmidt, *The Atomic Complex: A Worldwide Political History of Nuclear Energy* (La Grange Park, IL: American Nuclear Society, 1982).

Nuclear Waste Disposal

One of the most serious problems of the nuclear age has been nuclear waste. Since most nuclear waste is radioactive, secure places to store or dispose of it become essential. There are five categories of radioactive waste: uranium mill tailings, commercial spent fuel, low-level wastes, transuranic wastes, and high-level wastes. The most common of the radioactive wastes are the uranium mill tailings, which come from mining uranium ore. Commercial spent fuel is the most radioactive of the categories because the extended period that it spends exposed to neutrons during burn-up allows a high level of activity to accumulate. Low-level waste is the least dangerous and comes from medical and scientific uses. Transuranic wastes refers to those radioactive products coming from man-made elements higher than uranium—neptunium, plutonium, americium and so on. High-level wastes from nuclear reactors are so radioactive that they have to be stored away from human contact for thousands of years. For the first couple of decades that nuclear reactors were in operation in the United States these high-level nuclear wastes were stored in tanks at reprocessing plants run by the Atomic Energy Commission at Richland, Washington, Idaho Falls, Idaho, and Aiken, South Carolina. By the late 1960s, these tanks had developed leakage problems at both Richland and Aiken that threatened to contaminate local water supplies. Other types of radioactive materials, including waste from nuclear power plants, also have exacerbated the problem. One solution was to store waste in underground salt beds. After the demise of the Atomic Energy Commission, the Atlantic Richfield Company oversaw the Richland high-level nuclear waste facility until relieved of this task by Rockwell International in 1977. Du Pont manages the waste facility at the Savannah River site in southeastern South Carolina.

The controversy about nuclear waste disposal revolves around site selection and transportation. Early in the history of nuclear waste disposal, communities were eager to bid for waste facilities as a way to attract business to their towns. This open-arms approach ceased in the 1960s as it became more apparent that nuclear waste disposal was a deadly business. Now localities fight vigorously in the courts and through politicians to avoid efforts to locate nuclear waste disposal facilities in their areas. State attorneys general often represent these communities in bringing suits against the federal government. Transportation of nuclear waste through cities and towns is another controversial practice. Efforts have been made to limit such transportation, but these efforts have been less successful than blocking nuclear waste sites.

Suggested readings: Richard S. Lewis, *The Nuclear-Power Rebellion: Citizens vs. the Atomic Industrial Establishment* (New York: Viking Press, 1972); Fred C. Shapiro, *Radwaste* (New York: Random House, 1981); Richard Wolfson, *Nuclear Choices: A Citizen's Guide to Nuclear Technology* (Cambridge: MIT Press, 1991).

Nuclear Waste Policy Act

The Nuclear Waste Policy Act (NWPA) of 1982 is the most important piece of legislation to date enacted by the U.S. Congress on high-level radioactive waste disposal. This legislation was signed into Law on January 7, 1983. It called for the Department of Energy to conduct environmental assessments of five potential disposal sites and select three of them for more study by January 1, 1985. An assumption was that the president of the United States would pick one of the sites by 1987 and another by 1990. Both houses of Congress could vote to override a veto by a state. Among the sites selected for further study were Hanford, Washington, Yucca Mountain, Nevada, and Deaf Smith, Texas. The Nuclear Waste Policy Amendments Act of 1987 mandated the study of only Yucca Mountain, Nevada. This act also imposed a fee on waste generators of one tenth of a cent per kilowatt hour to pay for nuclear waste disposal. Despite this legislation, no action has been taken by the U.S. government to implement this act. The Nevada state government instituted a lawsuit to stop the government from implementation, and this lawsuit is still in litigation as of 1999. Nuclear waste disposal is still a contentious and expensive issue and disposal sites are still being considered.

Suggested reading: K. S. Shrader-Frechette, *Burying Uncertainty: Risk and the Case Against Geological Disposal of Nuclear Waste* (Berkeley: University of California Press, 1993).

Nuclear Weapons Accidents

The strategy of providing a credible nuclear deterrence posture during the Cold War produced a series of nuclear weapons accidents since 1950. A total of twenty-eight such incidents have been recorded by the U.S. Government in the postwar period. Sixteen of these accidents involved aircraft mishaps with nuclear weapons aboard. Fortunately, none of the nuclear weapons were armed so damage was restricted to conventional explosives and blast effects. Three other accidents were missile or rocket malfunctions. These missile and rocket incidents took place at missile ranges or over the Pacific Ocean. An isolated but heavily publicized incident was the loss of the nuclear powered attack submarine, U.S.S. *Thresher*, with a nuclear warhead on board. This accident has never been totally explained, but the most current thesis is that the submarine had a loss of power from its nuclear reactor which caused the submarine to sink below its stress tolerance. The eight other accidents remain unexplained and will continue to be so until the events are declassified by the U.S. Department of Defense.

Suggested reading: Philip L Cantelon, Richard G. Hewlett, and Robert C. Williams, eds., *The American Atom: A Documentary History of Nuclear Policies from the Discovery of Fission to the Present, 1939–1984*, 2nd ed. (Philadelphia: University of Pennsylvania Press, 1991).

Nuclear Winter

Nuclear winter is the thesis that a nuclear war would transform the earth's weather to produce a long winter lasting the better part of a year. This condition would be caused by a cumulative blast in the range of 5,000 megatons. A cloud of nuclear dust would blot out the sun, producing intense cold for up to nine months. Once the sky cleared, the depleted ozone layer would fail to protect survivors from exposure to deadly ultraviolet rays from the sun.

The idea of nuclear winter came from a 1982 project by the Swedish environmental journal of the Royal Swedish Academy of Sciences, *Ambio*. Their editors commissioned Dutch scientist Paul Crutzen and American scientist John Birks to investigate the effects of nuclear war on the atmosphere. These scientists started out by investigating ultraviolet radiation and also studied the issue of smoke. To their astonishment, their calculations showed that nuclear fireballs could produce enough smoke to blot out nearly all of the sunlight from half the earth for many weeks. This evidence was presented to other scientists for confirmation in 1983, and by 1985, various scientific bodies had concluded that a large-scale nuclear war would be followed by severe climatic changes resembling winter. This thesis gained currency in the mid-1980s, partially from the work of Carl Sagan, an astronomy professor at Cornell University and the Ames Research Center at Iowa State University. Critics of this idea have attacked it for various scientific and political reasons, but the concept of nuclear winter still has a large number of believers. More sophisticated computer models in the 1990s have caused a revision of the nuclear winter thesis to a nuclear fall thesis. Severe climatic changes would occur after a major nuclear war, but the changes would not be as severe or prolonged as earlier models indicated.

Suggested readings: Owen Greene, Ian Percival, and Irene Ridge, *Nuclear Winter: The Evidence and the Risks* (Cambridge: Polity Press, 1985); Richard Wolfson, *Nuclear Choices: A Citizen's Guide to Nuclear Technology* (Cambridge: MIT Press, 1991).

NUMEC

NUMEC (Nuclear Materials and Equipment Corporation) was a company in Apollo, Pennsylvania, that secretly supplied Israel with enriched uranium for its nuclear program. The company had been funded in the mid-1950s by Zalman Shapiro to provide uranium for nuclear reactors in the United States. Shapiro was a chemist by training and an active supporter of Israel because many of his relatives had been victims of the Holocaust. The Atomic Energy Commission (AEC) became concerned about frequent visitors to the company from France and Israel. Consequently, the AEC reprimanded the company for security violations in 1962. In 1965, the AEC found 110 pounds of enriched uranium missing from the company's inventory. A more lengthy investigation turned up a total of 587 pounds of enriched uranium gone, with the assumption by the commission that it was in Israel. This much uranium could supply Israel with as many as 18 nuclear weapons. Shapiro was questioned by the Federal Bureau of Investigation (FBI) and other investigating agencies but never charged with a crime.

Suggested reading: Dan Raviv and Yossi Melman, *Every Spy a Prince: The Complete History of Israel's Intelligence Community* (Boston: Houghton Mifflin, 1990).

○

Oak Ridge National Laboratory

The Oak Ridge National Laboratory (ORNL) was founded in 1943 as site for a nuclear reactor, chemical separation facilities, and a scientific laboratory for atomic research. It was located in Oak Ridge, Tennessee, and a number of plants were built for the separation of U-235 from U-238 for the Manhattan Project. The site was selected by General Leslie Groves for various reasons including electrical power from the Tennessee Valley Authority (TVA), abundant water supply from the Clinch Rivers, sparse population, mild climate, and good transportation potential. Knoxville was 18 miles away to the east and Clinton 8 miles to the north. The government through the Army Corps of Engineers acquired 59,000 acres in early 1942 for future development. In February 1943 construction began on a graphite reactor with the code name X-10. It was a pilot plant for the plutonium facility to be built at Hanford, Washington, and with a workforce of 1,513, it cost $12 million to build. X-10 was also called the Clinton Laboratories. By early 1945, 326.4 grams of plutonium had been separated in the X-10 complex. The large gaseous diffusion plant, K-25, which was built at a cost of $500 million and employed 12,000 workers, was located on the western edge of the reservation. Its job was to begin the slow process of separation of U-235 from U-238, and the first batch was ready on January 20, 1945. It had thousands of diffusion tanks and needed twice as much space as the other plants. Another plant, Y-12, built at a cost of $427 million, used the electromagnetic method to process uranium. Construction of this plant started on February 18, 1943. It was located on the reservation's northern edge near the workers' settlement at Oak Ridge. This process had 22,000 employees working on it. Eventually the Y-12 complex had 268 permanent buildings to handle the calutrons and their support functions. Teething

The K-25 gaseous diffusion plant at Oak Ridge, Tennessee, contained the processes to separate uranium-235 from uranium-238. This procedure was difficult to develop, but once it was accomplished, this plant continued operations in the postwar.

troubles with the machines caused them to be shut down for nearly a month in the winter of 1943–1944. A third plant, S-50, built at a cost of $10 million, was the thermal diffusion installation. General Groves never intended to build this plant, but the success of Philip Abelson's process for the U.S. Navy made him change his mind. Only the gaseous diffusion plant, K-25, continued in production after 1945.

The Oak Ridge Laboratory has had several management teams in its history. Its first contract operator was the University of Chicago, which managed the laboratory from February 1943 to July 1945. Martin Whitaker was its first director and Richard Doan its associate director for research. The army selected Monsanto Chemical Company to succeed the University of Chicago. Monsanto selected James Lum as executive director, replacing the resigned Whitaker. In 1946 Eugene Wigner was recruited to become the research director. The major orientation of the laboratory in 1946 and 1947 was in the designing and building of new types of atomic reactors. Farrington Daniels designed a thermal reactor that used low-energy neutrons in the fission process, but the Atomic Energy Commission (AEC) canceled this project. Monsanto lost the contract to operate the Oak Ridge National Laboratory in May 1947, and both Lum and Wigner soon left their administrative posts. After briefly considering the University of Chicago, the AEC asked Union Carbide Corporation to assume the contract for Oak Ridge

National Laboratory. It was at this time that the Clinton Laboratories was upgraded to a national laboratory with a name change, but the price was the transfer of its high-flux materials testing reactor to the Argonne National Laboratory near Chicago. Nelson Rucker became the head of the laboratory, and Alvin Weinberg, a biophysicist and former director of the laboratory's Physics Division, became the head of research.

Despite efforts to turn over nuclear reactor research to the Argonne National Laboratory, the ORNL became active in the design and building of a variety of different nuclear reactors in the 1950s and 1960s. Among the designs were for a homogeneous reactor, a package reactor for the army, a gas-cooled reactor, and a molten-salt reactor. In the late 1960s federal budget cuts curtailed many of the activities of the ORNL. In a severe budget crisis in 1973 Weinberg left the Oak Ridge National Laboratory. In the next 20 years the ORNL has continued to conduct basic research on nuclear energy–related subjects, from nuclear safety to nuclear waste disposal. In 1982 Union Carbide relinquished its contract to manage the ORNL and was succeeded in 1984 by Martin Marietta Corporation. By the mid-1980s the Oak Ridge National Laboratory was no longer a nuclear laboratory, and its research was directed more to issues of technology transfer.

Suggested readings: Leland Johnson and Daniel Schaffer, *Oak Ridge National Laboratory: The First Fifty Years* (Knoxville: University of Tennessee Press, 1994); Kenneth D. Nichols, *The Road to Trinity* (New York: Morrow, 1987); Richard Rhodes, *The Making of the Atomic Bomb* (New York: Simon and Schuster, 1986).

Obninsk Atomic Energy Station

The Obninsk Atomic Energy Station (AES) is Russia's center for research on nuclear reactors. Obninsk is located about 75 miles south of Moscow. Igor Kurchatov, the scientific head of the Soviet nuclear energy program, decided in 1949 to create a research center to build and test nuclear power reactors. After receiving government permission, work on the center began in 1950. For almost a decade this center had no name and merely a post office box. In the beginning Soviet scientists and German physicists, who were still prisoners-of-war, worked together to design a power reactor. Kurchatov appointed D. I. Blokhintsev to head the project, but Kurchatov selected the graphite-moderated, water-cooled type for development, since it could produce weapons-grade plutonium. Research culminated in the first nuclear power station to begin generating electrical power on June 27, 1954. The reactor was so small that it had no commercial value, but it served as a prototype for more advanced reactors. The prison research center was disbanded in 1955 when the German scientists were repatriated to Germany, but Soviet scientists at the Obninsk Physico-Power Institute remained working on reactor designs. Its personnel was also active in the research and design of nuclear reactors for ships and submarines. Obninsk finally received its name in 1958 and it appeared on Soviet maps. It was at this site that experimental research was conducted on both the RBMK (Reactors High-Power Boiling Channel Type) and the VVER (Water-Water Power Reactors). Research has also been conducted on the fast breeder-type reactors. The Obninsk Physico-Power Institute remains the largest nuclear research center in Russia with a staff of over 5,000.

Suggested reading: Zhores Medvedev, *The Legacy of Chernobyl* (New York: Norton, 1990).

Office of Scientific Research and Development

By executive order on June 28, 1941, President Roosevelt established the Office of Scientific Research and Development (OSRD). The OSRD's mission was to serve as a center for mobilizing the scientific resources of the United States and applying them to defense problems. Its head was also to have direct access to President Roosevelt. Vannevar Bush resigned from the National Defense Research Committee and became the director of OSRD. Perhaps the most important of its committees was the Atomic Committee, or S-1. Its head was James Bryant Conant, president of Harvard University, and included as members Lyman J. Briggs, director of the National Bureau of Standards, Arthur H. Compton, professor of physics at the University of Chicago, Ernest O. Lawrence, professor of physics at the University of California at Berkeley, Harold C. Urey, professor of physics at Columbia University, and Eger V. Murphree, director of research for the Standard Oil Development Company of New Jersey. The mission of this committee was to determine whether atomic bombs could be made and report back to the OSRD within six months. This committee received the Berkeley Summer Colloquium Report in August 1942, which recommended a major scientific and technical effort to build an atomic bomb. It was then sent to Bush with the committee's strong support. In May 1942 the OSRD S-1 Section was replaced by the OSRD S-1 Executive Committee with a membership of Conant, chair, Briggs, Compton, Lawrence, Murphree, and Urey. In the early stages of the Manhattan Project, General Groves frequently met with the Atomic Committee for guidance, but once the project was operating at a high level, this interaction ceased and the committee was inactive from the middle of 1943 onward. Its functions had been replaced by the Military Policy Committee.

Suggested readings: Arthur Holly Compton, *Atomic Quest: A Personal Narrative* (New York: Oxford University Press, 1956); Kenneth D. Nichols, *The Road to Trinity* (New York: Morrow, 1987).

Oklo Mine

The Oklo Mine in Gabon, East Africa, contains uranium that sometime in prehistory produced a natural chain reaction. French scientists discovered this phenomenon in 1972 soon after they discovered uranium deposits at Oklo. A worker noticed that samples of uranium from Oklo had lower levels of uranium-235 than expected. Further studies confirmed this and the existence of isotopes from a chain reaction. The French scientists theorized that about 2 billion years ago natural uranium was more enriched than it is at present. It had five to six times more uranium-235 in the mineral deposits at Oklo. Each time there was an incursion of water a chain reaction was set off, but then the water boiled away and the reaction slowed. Several natural piles are located within a few miles of each other in the vicinity of Oklo. These natural piles resemble the mechanism that the light water nuclear power reactors developed in the United States in the 1950s.

Suggested readings: Bertrand Goldschmidt, *The Atomic Complex: A Worldwide Political History of Nuclear Energy* (La Grange Park, IL: American Nuclear Society, 1982); Richard Wolfson, *Nuclear Choices: A Citizen's Guide to Nuclear Technology* (Cambridge: MIT Press, 1991).

Operation Crossroads. *See* Crossroads Nuclear Tests

Operation Hurricane. *See* Monte Bello Atomic Test Site

Oppenheimer, J. Robert

Robert Oppenheimer (1904–1967) was an American physicist who was the director of the Los Alamos Scientific Laboratory during the building of the atomic bomb in World War II. He was born on April 22, 1904, in New York City. His father was a wealthy businessman originally from Germany. He married Kitty Puening, and they had a daughter and a son. Oppenheimer graduated from Felix Adler's Ethical Culture High School in 1921. He received a strong scientific education at Harvard University. After leaving with a summa cum laude degree in chemistry from Harvard University in only three years, Oppenheimer spent an unsuccessful year as an experimental physicist at the Cavendish Laboratory at Cambridge University. Then he studied theoretical physics more successfully under Max Born at the University of Göttingen and in 1927 earned a Ph.D. in physics. During his stay in Germany, Oppenheimer published 16 papers on aspects of quantum mechanics and earned an international reputation as a theoretical physicist. Oppenheimer returned from Europe and accepted dual jobs at the California

J. Robert Oppenheimer was the head of the Los Alamos Scientific Laboratory. His administrative skills created the atmosphere where the building of the atomic bomb was possible.

Institute of Technology (Caltech) and the University of California at Berkeley. Before residing in California, Oppenheimer went back to Europe to work on his mathematics with Paul Ehrenfest at the University of Leyden and with Wolfgang Pauli at the University of Zurich. Oppenheimer remained teaching at the two California schools until 1942. In 1942, he was asked to serve as director of the Los Alamos Scientific Laboratory in Los Alamos, New Mexico. His claim to fame before leaving for Los Alamos was for developing at Caltech and Berkeley the strongest schools of theoretical physics in the United States.

It was as director of the Los Alamos Laboratory that Oppenheimer directed the scientific efforts to build the atomic bomb. Arthur Compton recommended him highly to the government, and the office of Scientific Research and Development (OSRD) named him director of the physics research needed to develop an atomic bomb. Despite differences in personality and scientific perspective, he and General Leslie Groves worked well together. Together they selected Los Alamos as the site for the construction of the atomic bomb. Oppenheimer devoted much of his attention to recruiting a top team of physicists and chemists to move to Los Alamos. Although his fame was as a theoretical physicist, he knew most of the top scientists in the United States, Canada, and Great Britain, and this helped in recruiting them to the Manhattan Project. He was famous for his insight into complex physics problems and his ability to work with others. Both traits proved invaluable as an administrator at Los Alamos. He was always well informed on technical details, so much so that he often appeared arrogant, but the scientists respected him for his abilities and for his concern for their welfare. His success in building a team out of several hundred of the top scientists in the world ensured that the atomic bomb would be built. He received much of the credit for the construction of the atomic bomb and deserved it.

Because of his leftward political leanings and his associations with friends with Communist affiliations, Oppenheimer always had difficulty with security clearances. His wife Kitty had been a member of the Communist Party in the 1930s, and her first husband, Joe Dallet, had been an official in the American Communist Party and had died fighting in the Spanish Civil War. His brother, Frank Oppenheimer, had also been active in the Communist Party. Once in 1943 Oppenheimer had been quizzed about a contact with a Soviet messenger, and he had to divulge his friend's name. A Soviet agent reported in 1943 to his (NKVD) (People's Commissariat of Internal Affairs) contact, however, that neither of the Oppenheimers were susceptible to being recruited for espionage. Later after the war, some scientists accused him of disloyalty because of his reservations about the hydrogen bomb project. In a controversial hearing, Oppenheimer lost his security clearance and his post as chair of the General Advisory Committee of the Atomic Energy Commission. This rejection was a blow from which Oppenheimer never recovered, because government service had been important to him.

After this rejection, Oppenheimer returned to a directorship at the Institute for Advanced Studies at Princeton University. He had been appointed to this post in 1947. Despite an attempt by Admiral Lewis Strauss to have him fired at the institute, Oppenheimer stayed at Princeton until his death from cancer of the throat on February 18, 1967. He received honors for his scientific contributions from the U.S. government, but never again did he serve as a scientific adviser.

Suggested readings: Nuel Pharr Davis, *Lawrence and Oppenheimer* (New York: Simon and Schuster, 1968); Peter Goodchild, *J. Robert Oppenheimer: Shatterer of Worlds* (Boston: Houghton Mifflin, 1980); James W. Kunetka, *Oppenheimer: The Years of Risk* (Englewood Cliffs, NJ: Prentice-Hall, 1982); Richard Rhodes, *The Making of the Atomic Bomb* (New York: Simon and Schuster, 1986); Herbert F. York, *The Advisors: Oppenheimer, Teller, and the Superbomb* (San Francisco: W. H. Freeman, 1976).

Oppenheimer Affair

The Oppenheimer Affair was the denial of the security clearance of J. Robert Oppenheimer by a government hearing in the spring of 1954. Oppenheimer had always been lukewarm about the development of the hydrogen bomb. As a member of the scientific advisory committee of the Atomic Agency Commission (AEC), he had advised concentrating on refining atomic weapons rather than building a superbomb. His opponent on this issue, Edward Teller, attacked him on several occasions. It was not until Lewis L. Strauss became head of the AEC that an attempt to neutralize Oppenheimer over the hydrogen bomb became apparent. William Liscum Borden had been staff director of the Joint Committee on Atomic Energy, and on November 7, 1953, he wrote a letter to Federal Bureau of Investigation (FBI) director J. Edgar Hoover charging that Oppenheimer was probably an agent of the Soviet Union. Borden was a Democrat looking for a job, and he was aware of Strauss's suspicion that Oppenheimer was a Soviet spy. Hoover had long wanted to move against Oppenheimer, and he was a good friend of Strauss's. Adding to the political environment was Senator McCarthy and the House Committee on Un-American Activities actively looking for Communists in government service and ready to seize on any evidence of security lapses. The Eisenhower administration had responded to Senator McCarthy's charges about Communists in government by issuing order 10450 in April 1953, stating that just being a security risk was the basis for not granting security clearance. Strauss discussed Oppenheimer with Eisenhower and received his blessing to proceed against Oppenheimer. Since Strauss was a strong partisan of the hydrogen bomb, he took advantage of his new position to inform Oppenheimer that he could no longer have access to secret documents. Oppenheimer requested that the matter be brought before a committee of inquiry. Strauss constituted a committee of inquiry that was to be named the Gray Committee with three members—Gordon Gray, president of the University of North Carolina and former assistant secretary of the army, Thomas Alfred Morgan, retired board chair of the Sperry Corporation, and Ward V. Evans, a chemistry professor from Loyola University. This committee of inquiry called for hearings. These hearings served as a forum for government officials and scientists to attack and defend Oppenheimer. Oppenheimer's past associations were scrutinized in detail. AEC lawyer Roger Robb had the advantage of surveillance tapes of Oppenheimer's discussion with his lawyer. The committee of inquiry voted two to one to deprive Oppenheimer of his security clearance. This decision was confirmed by the AEC by four votes to one. Oppenheimer lost his position as a government adviser shortly thereafter. This decision was controversial among American scientists, and some friendships ended over this affair. The enmity built up over the hearing later came back to haunt

Lewis Strauss when the Senate refused to approve his nomination as secretary of commerce in the Eisenhower administration.

Suggested readings: Stanley A. Blumberg and Gwinn Owens, *Energy and Conflict: The Life and Times of Edward Teller* (New York: Putnam's Sons, 1976); Charles P. Curtis, *The Oppenheimer Case: The Trial of a Security System* (New York: Simon and Schuster, 1955); Rachel L. Holloway, *In the Matter of J. Robert Oppenheimer: Politics, Rhetoric, and Self-Defense* (Westport, CT: Praeger, 1993); John Major, *The Oppenheimer Hearing* (New York: Stein and Day, 1971); Kenneth D. Nichols, *The Road to Trinity* (New York: Morrow, 1987); Richard Pfau, *No Sacrifice Too Great: The Life of Lewis L. Strauss* (Charlottesville: University Press of Virginia, 1984); Philip M. Stern, *The Oppenheimer Case: Security on Trial* (New York: Harper and Row, 1969); Cushing Strout, *Conscience, Science and Security: The Case of Dr. J. Robert Oppenheimer* (Chicago: Rand McNally, 1963).

P

Pakistani Nuclear Program

The Pakistani nuclear program is a product of the success of India in building a nuclear weapons capability. As news reached Pakistan that India was deeply into atomic energy development, the Pakistani Atomic Energy Commission (PAEC) was established by statutory law in 1955. Almost as soon as Prime Minister Ali Bhutto took office in 1971 he negotiated a contract with France for a reprocessing plant as the first step toward Pakistan's possessing nuclear weapons. In 1972, Pakistan opened its first nuclear power station, the Karachi Nuclear Power Plant. When India detonated an atomic device in 1974, the Pakistani government decided to give higher priority to building a nuclear weapons system to counteract India's. A French company had been constructing an irradiated fuel reprocessing plant in Pakistan in 1976 until international pressure was placed on both the French and the Pakistanis to suspend further construction. Consequently, in 1978 the French canceled the 1976 reprocessing plant deal. This plant near Islamabad was finally finished by the Pakistanis. Pakistani scientists traveled around the world gathering as much information on nuclear weapons systems as possible to fill the void of the loss of French contacts. According to the head of the Pakistani nuclear program, Dr. Abdul Qadeer Khan, Pakistan had the capability to manufacture nuclear weapons in 1984. Since Pakistan is not a signatory of the Nuclear Non-Proliferation Treaty, there are no safeguard inspections of Pakistani nuclear facilities to check on the processing of nuclear fuels except for two sites mandated by agreements from Canada and the United States. In 1985, Pakistan began construction of a 300-megawatt nuclear power station at Chashma in Punjab province with technology supplied by China. Pakistan's acquisition of a uranium conversion plant from Germany in a secret deal with a German businessman increased the probability of Pakistan having, by 1994, 10 to 15 nuclear devices. A recent 1996 study of Pakistan's plutonium production by the Stockholm International Peace Research Institute (SIPRI) suggested that Pakistan had a nuclear arsenal of between 6 and 10 nuclear weapons. Moreover, a delivery system, the Ghauri missile, was developed and tested in April 1998. Government policy had been based, much like Israel's, on nuclear ambiguity, but this changed in May 1998 when India conducted five underground nuclear tests near the border of Pakistan. Almost immediately, the Pakistani government authorized atomic bomb tests. Six underground nuclear tests were conducted at Chagai on May 28–29, 1998. By joining the ranks of the nuclear states uninvited, Pakistan now has to face the displeasure of those states and face economic and political sanctions. The Pakistani government, however, appears willing to accept these sanctions as the price to pay to show India the dangers of an aggressive foreign policy.

Suggested readings: Eric Arnett, ed., *Nuclear Weapons After the Comprehensive Test Ban: Implications for Modernization and Proliferation* (Oxford: Oxford University Press, 1996); Ashok Kapur, *Pakistan's Nuclear Development* (London: Croom Helm, 1987); Marianne van Leeuwen, ed., *The Future of the International Nuclear Non-Proliferation Regime* (Dordrecht, Holland: Martinus Nijhoff, 1995).

Panda Committee

The Panda Committee, or the Theoretical Megaton Group, was the committee charged by the Atomic Energy Commission (AEC) with the task to build the first American hydrogen bomb at the Los Alamos Scientific Laboratory. After Stanislaw Ulam, a mathematician working at Los Alamos, and Edward Teller's breakthrough on the theoretical nature of a thermonuclear explosion, Marshall Holloway, an assistant to Los Alamos director Norris Bradbury, was given the assignment to form a committee to design and build a hydrogen bomb. This committee met for the first time on October 5, 1951, with the mission to complete a bomb within a year. Teller wanted a July 1952 test, but Holloway insisted on a late October 1952 test. Since Teller had resigned from the Los Alamos Scientific Laboratory and was unable to get along with Holloway, the test was scheduled for late October or early November 1952. This test on October 31, 1952, code name Mike, was detonated at Eniwetok Atoll, and the yield was much higher than expected at 10.4 megatons. Subsequent research revolved around ways to make the hydrogen bomb a tactical weapon by making it smaller and deliverable by aircraft, but the fact that such a weapon was possible transformed the international scene with the two superpowers in an arms race to build an ever growing stockpile of these bombs.

Suggested reading: Richard Rhodes, *Dark Sun: The Making of the Hydrogen Bomb* (New York: Touchstone, 1995).

Pantex

Pantex is a nuclear weapons facility administered by the Department of Energy (DOE), which is located just outside of Amarillo Texas. Its original mission was to assemble nuclear weapons from parts manufactured elsewhere. As this job became less essential for national defense, the federal government proposed closing the facility, but this proposal ran into strong opposition from local, state, and congressional sources. DOE officials then changed the work at Pantex to decommissioning nuclear weapons. Since this job has a limited future, the DOE proposed cutting the 3,000-employee workforce. This information again mobilized politicians and Panhandle 2000, a group of local businessmen, was formed to lobby for more nuclear work for Pantex. Mason & Hanger Corporation operates the plant on a contract with the DOE, and this contract has been extended through 2001. A national plutonium resource center had been proposed for Pantex and a modest annual budget of $9 million has been allocated, but the DOE decided toward the center to the Savannah River site. In the meantime, Pantex employees dismantle 5 nuclear weapons a day, or 2,000 a year. This production will continue for the indefinite future as the United States downsizes its nuclear arsenal. The fate of Pantex remains uncertain as the DOE continues to consolidate operations nationwide.

Suggested reading: A. G. Mojtabai, *Blessed Assurance: At Home with the Bomb in Amarillo, Texas* (Syracuse, NY: Syracuse University Press, 1997).

Parsons, William S. (Deak)

Naval Captain Deak Parsons (1901–1953) was the head of the Ordnance Division of the Los Alamos Scientific Laboratory in the Manhattan Project, and after World War II he was promoted to admiral and recognized as the navy's foremost expert on nuclear weapons. He was born on November 26, 1901, in Evanston, Illinois, but at the age of eight his family moved to Fort Sumner, New Mexico. His father was a lawyer who married Martha Cluverius, the daughter of an admiral, and they had three daughters. Parsons's education was at the U.S. Naval Academy where he graduated along with Admiral Hyman Rickover in 1922. It was at the academy where he was given the nickname "Deak," short for deacon. Although he had sea duty on both the U.S.S. *Idaho* and the U.S.S. *Texas* battleships, most of his naval experience had been in ordnance and working on specialized projects on radar and the proximity fuse. Early in the Manhattan Project, Vannevar Bush, the head of the Office of Scientific Research and Development (OSRD), recommended Parsons to General Groves as an ordnance expert. Groves had met Parsons briefly in the 1930s, and with Bush's recommendation in hand he appointed Parsons the associate director at the Los Alamos Scientific Laboratory.

Captain William (Deak) Parsons was a naval
officer active in the ordnance section at the
Los Alamos Scientific Laboratory. In flight,
he armed the atomic bomb that was
detonated over Hiroshima.

With the blessings of the head of the Los Alamos Scientific Laboratory,
J. Robert Oppenheimer, Parsons assumed command of the critical Ordnance
Division in May 1943. He used his contacts in the scientific community to
recruit an outstanding team of scientists to work on ordnance development.
While his military manner and technical expertise won him respect from the
scientists, his reserve and exacting manner intimidated many of them. Parsons was initially opposed to working on implosion; he believed that the
gun method was the simplest and most reliable way to produce a nuclear
reaction. He also was not pleased with the leadership of Seth Neddermeyer
and his progress on implosion. In consultation with the mathematician
Johnny von Neumann, Parsons decided to investigate fast implosion, or the
theory of using lenses to cause a simultaneous explosion of plutonium. Parsons then recruited the explosives expert George Kistiakowsky from the Explosives Research Laboratory at Bruceton, Pennsylvania, to assume
responsibility to solve the implosion problem. In August 1944, Oppenheimer appointed Parsons to be an associate director at Los Alamos for ord-

nance, engineering, assembly, and delivery of the atomic bomb. Then in March 1945, Parsons assumed complete responsibility for Project Alberta, the code name for the delivery of the atomic bomb in combat. After the successful testing of the plutonium bomb, Parsons traveled to Tinian in the south Pacific to oversee the arming of the uranium atomic bomb "Little Boy." His mission was to fly on the Hiroshima mission and arm the atomic bomb. The successful conclusion of this mission and the Nagasaki attack several days later helped end World War II and earned Parsons promotion to the rank of commodore.

Parsons became the leading postwar authority on atomic weapons in the navy. His first assignment after the war was the study of atomic bombs against surface ships. As assistant chief of special weapons in the Special Weapons Office, Parsons moved to Washington, D.C., to serve under Vice-Admiral Spike Blandy. Parsons was an early advocate for using atomic energy to power both surface ships and submarines even before Admiral Hyman Rickover. His professional competency led to his promotion to rear admiral in January 1946, despite the fact that he had never commanded a ship at sea. Parsons was then appointed deputy commander for technical direction under Admiral Spike Blandy for Operation Crossroads only a few days after his promotion. This test was to prove the destructiveness of atomic weapons on surface ships and took place at Bikini in the Pacific Ocean. Despite problems of errant bomb drops and sometimes inclusive results, the tests were considered a success. Parsons returned to Washington, D.C., to find the Special Weapons Office disbanded. He then became director of atomic defense, Navy Department, and in this capacity served as the navy's representative on all nuclear issues. Among his duties was service on the Military Liaison Committee (MLG) to the Atomic Energy Commission (AEC). In 1949 he was assigned to the Weapons Systems Evaluation Group (WSEG), whose mission was to examine present and future weapons from a scientific point of view. Fulfilling his nickname as the "Atomic Admiral," Parsons continued to be active in atomic politics. His relationship with Oppenheimer had always been close, and he became upset over the investigation into Oppenheimer's security status. Shortly after Parsons heard the news, he suffered a heart attack and died on December 5, 1953, at Bethesda Naval Hospital.

Suggested readings: Al Christman, *Target Hiroshima: Deak Parsons and the Creation of the Atomic Bomb* (Annapolis, MD: Naval Institute Press, 1998); William Lawrence, *The General and the Bomb: A Biography of General Leslie R. Groves, Director of the Manhattan Project* (New York: Dodd, Mead, 1988); Richard Rhodes, *The Making of the Atomic Bomb* (New York: Simon and Schuster, 1986).

Pauling, Linus Carl

Linus Pauling (1901–1994) was an American chemist who besides being a two-time Nobel Prize winner also became an early critic of atomic testing due to his fear of global fallout. He was born on February 28, 1901, in Portland, Oregon. His father was a pharmacist. Pauling graduated from Oregon State University with a degree in chemical engineering in 1922. His doctorate in physical chemistry was obtained in 1925 from the California Institute of Technology (Caltech). After a two-year tour in Europe studying under Niels Bohr, William Henry Bragg, Erwin Schrödinger, and Arnold Sommerfeld, he became an associate professor of chemistry at the California Institute of Technology in 1927. In this period, Pauling became deeply involved in research on complex chemical reactions among organic molecules. By applying quantum mechanics to chemical bonding, Pauling concluded that the strength of the bond between two atoms is determined by the relative binding energies of the outermost electrons in the atoms. In 1931 he was promoted to full professor. The remainder of his academic career was spent at Caltech.

Soon after receiving his first Nobel Prize in 1954 for his work on molecular structure, Pauling turned his attention to ending testing of atomic weapons. In 1957 he organized an international petition for scientists for a test ban, which included signatures from more than 11,000 scientists from 48 countries. Pauling asserted that thousands would die from leukemia from continuing nuclear testing. His claim that there was no safe dose of radiation was controversial, but he based it on his genetic research. His campaign to ban testing led to his having to defend his views before a congressional inquiry in 1960. When Edward Teller began to tout the neutron bomb as a clean bomb, Pauling responded by pointing out the dangers of carbon-14. His arguments also attracted the attention of Andrei Sakharov, who from that time onward started questioning for the problem of radioactive contamination from nuclear explosions. Pauling received a second Nobel Prize for his peace activities. Among his other interests, he became the vocal champion of the health benefits of vitamin C. Pauling died on August 9, 1994, in Big Sur, California.

Suggested readings: Ted George Goertzel, *Linus Pauling: A Life in Science and Politics* (New York: Basic Books, 1995); Thomas Hager, *Force of Nature: The Life of Linus Pauling* (New York: Simon and Schuster, 1995); Anthony Serafini, *Linus Pauling: A Man and His Science* (New York: Paragon House, 1989).

Peierls, Rudolf Ernst

Rudolf Peierls (1907–1995) was a German-British physicist who was prominent in the development of the theory behind the atomic bomb and who was the senior British physicist at Los Alamos during World War II. He was born on June 5, 1907, in Berlin, Germany. He married Russian physicist Genia Kannegiesser. His education was at the Universities of Berlin, Munich, Leipzig, and the Federal Institute of Technology, Zurich. Among the physicists Peierls studied with at these universities were Arnold Sommerfeld, Werner Heisenberg, and Wolfgang Pauli. Peierls traveled to Great Britain on a Rockefeller Fellowship to Cambridge University in 1933 and never returned to Germany. He became a professor of mathematical physics at the University of Birmingham in 1937. He stayed at Birmingham until 1963, when he moved to a professorship of physics at Oxford University. It was at Oxford that he remained until retirement in 1974. From 1974 to 1977, he taught at the University of Washington, Seattle. Peierls had become a naturalized British citizen in February 1940.

Peierls was a respected theoretical physicist. Much of his early work was on solid-state theory. In the 1930s his attention turned to quantum theory and theories of atomic reactions. Peierls's most famous contribution to atomic theory, however, was his collaborative report with Otto Frisch on their calculations on the size and the explosive power of U-235. In 1940 their report convinced both scientists and military authorities on the practicality of an atomic weapon. He was part of the British team that joined the Manhattan Project at Los Alamos. Peierls died in 1995.

Suggested readings: David Abbott, ed., *The Biographical Dictionary of Scientists* (London: Blond Educational, 1984); Ronald William Clark, *The Birth of the Bomb* (New York: Horizon Press, 1961); Rudolf Ernst Peierls, *Bird of Passage: Recollections of a Physicist* (Princeton, NJ: Princeton University Press, 1985).

Penney, William

William Penney (1909–1991) was a British mathematician-physicist who was sent to America to work on the atomic bomb at Los Alamos and who later became head of the British atomic bomb program. He was born on June 24, 1909, at Gibraltar. His father was a sergeant-major in the Ordnance Corps of the British Army. His early education was at the Junior Technical School in Sheerness, Kent. His excellence in mathematics won him a scholarship in 1927 to the Royal College of Science of the Imperial College, London. Penney was such a prodigy that he received a first-class bachelor's of science in only two years in 1929. His first job was conducting research on quantum mechanics with visiting Dutch professor R. L. De Kronig, first in London and then in Groningen, Netherlands. Penney became an expert in the mathematics of quantum mechanics. In 1931, he received the first of three doctorates awarded to him. He was a prodigious author of scientific and mathematics papers. After two years at the University of Wisconsin in the early 1930s, Penney received a scholarship in 1933 to Cambridge University. He received his second Ph.D. from Cambridge. Finally, Penney accepted a professorship of mathematics at the Imperial College, London.

During World War II, Penney was called to government service. His first duty was on the Physics of Explosives Committee to study bomb damage.

A study of blast waves under water was Penney's first task. His work on blast effects made him a candidate to be part of the British mission to Los Alamos to work on the atomic bomb. While at Los Alamos, Penney studied both blast problems and implosion problems. He was also placed on the Target Committee, and his job was to estimate bomb size, blast damage, and casualties. His expertise in explosives was especially valuable in the building of the Fat Man bomb. Penney was on another aircraft over Nagasaki when the atomic bomb Fat Man was detonated on August 9, 1945. Less than two weeks later he visited Nagasaki to assess damage.

After the war, Penney became the leading figure in developing the British atomic bomb. Despite the banning of British scientists from information about atomic weapons after the Atomic Energy Act of 1946, Penney was invited to the American Bikini atomic tests in 1947 because of his blast expertise. His title was "Co-ordinator of Blast Measurement." He became the leader of the British nuclear weapons program when Lord Portal appointed him to head the Armament Research Division as Chief Superintendent of Armament Research (CSAR). It was under his administration that the successful test of the British atomic bomb was conducted at Monte Bello, Australia, in October 1952. He was knighted for his achievements and then placed in charge of Britain's hydrogen bomb project. Later in his career Penney became chair of the United Kingdom Atomic Energy Authority. Still later he was appointed rector of Imperial College. Penney died on March 3, 1991.

Suggested readings: Brian Cathcart, *Test of Greatness: Britain's Struggle for the Atom Bomb* (London: Murray, 1994); John Daintith, Sarah Mitchell, Elizabeth Tootill, and Derek Gjertsen, *A Biographical Encyclopedia of Scientists*, 2nd ed., vol. 2 (Bristol, UK: Institute of Physics, 1994); Margaret Gowing, *Independence and Deterrence: Britain and Atomic Energy, 1945–1952*, vol. 2 (New York: St. Martin's Press, 1974).

PEON Commission

The Production d'Electricité d'Origine Nucléaire (PEON) Commission was an attempt by the French government to coordinate the French program for civil nuclear power. Its membership included the two major players in French nuclear energy, the Electricité de France (EDF) and the Commissariat à l'Énergie Atomique (CEA), and representatives from French industry. This commission began operations in 1955 and by the summer of that year had launched a program to develop of series of prototype power stations around France. Each power station would be more powerful than the former, and an interval of 18 months was planned between each construction. The goal was to have an 800-MW(e) (Megawatts) nuclear capacity by 1965, but this goal was not reached until 1969. Overall responsibility for these nuclear power plants was in the hands of the EDF, with the CEA providing the basic plans and supplying the graphite and uranium.

Suggested reading: Bertrand Goldschmidt, *The Atomic Complex: A Worldwide Political History of Nuclear Energy* (La Grange Park, IL: American Nuclear Society, 1982).

Phoebus Reactors. See NERVA

The Physical Review

Over the course of the twentieth century the journal *The Physical Review* was the most influential journal in the field of atomic energy in the United States. Edward L. Nichols founded the journal at Cornell University in 1893. At first the journal was little more than a Cornell University Physics Department organ, because so many articles by Cornell physicists appeared in this publication. Moreover, the journal was slow to reflect the increased interest in the advanced European theories of Max Planck on quantum theory and Albert Einstein's theory of relativity. By 1908, however, increased research on these topics began to show up in articles submitted to the journal. In 1913 the American Physical Society assumed editorial control of *The Physical Review* in response to a demand for more nationwide coverage and a more prestigious journal. The reputation of the journal increased as the quality of research among American physicists improved. Soon by the mid-1920s it took almost six months for an article to be in print from submission. Since priority of a discovery meant fame and possibly a Nobel Prize, *The Physical Review* followed the lead of its English rival *Nature* and began in 1929 to publish biweekly letters to the editor. By the mid-1930s, the journal reflected the high state of American physics, and even physicists in Europe awaited its publication eagerly. This review became so prestigious that its worldwide circulation made it a target of Leo Szilard's campaign to restrict publications on atomic energy topics to prevent the Germans from learning too much. Consequently, advanced research on atomic energy disappeared from *The Physical Review* during World War II.

In the postwar world the journal regained its status as one of the world's leading physics journals. Soon it became a biweekly, and the number of pages of each issue expanded. Two subjects were especially popular in the late 1940s and 1950s: quantum theory and nuclear studies. Later, in the 1960s high-energy physics became the most popular field of research and publication. It became the largest periodical covering the whole spectrum of physics.

Suggested reading: Daniel J. Kevles, *The Physicists: The History of a Scientific Community in Modern America* (Cambridge: Harvard University Press, 1987).

Planck, Max Karl Ernst Ludwig

Max Planck (1858–1947) was perhaps the most influential German nuclear physicist in the first half of the twentieth century. He was born in Kiel, Germany, on April 23, 1858. His father was a professor of civil law at the University of Kiel. Planck was married twice: Mari Merck in 1885 and Marga von Hosslin in 1911. Three sons and two daughters were products of these marriages. In 1867 the Planck family moved to Munich, where Planck attended a Gymnasium. After training at the Universities of Berlin and Munich, Planck received his doctorate in physics at the University of Munich in 1879. While at Berlin, he took courses under two famous physicists, G. R. Kirchhoff and Hermann von Helmholtz. His first teaching position was at the University of Munich in 1880. After five years, Planck moved to the University of Kiel. Four years at Kiel were followed by his advancement to a professorship of theoretical physics at the University of Berlin, where Planck remained for the rest of his academic career. He was especially beloved by his colleagues and friends because of his rectitude, good disposition, and a tolerance for ideas foreign to him. Besides physics, music was his greatest love, and he played piano in a trio that included Albert Einstein on the violin.

Planck's fame as a physicist resides in his research and formulation of quantum theory. His earliest research was in the thermodynamics of irreversible processes. Then he turned his attention to research on electromagnetic radiation and also worked on deciphering the so-called ultraviolet catastrophe of a blackbody for three years, from 1897 to 1900. The blackbody problem, or black radiation, had been a major problem in physics because it appeared to expose an inconsistency between statistical mechanics and Maxwell's successful theory of light. It was while conducting research in this area that Planck made his greatest contribution to atomic theory by his hypothesis that a body emits radiation in the form of discrete quanta of energy equal to the frequency times a constant, or h. This constant was soon called Planck's constant. For the hypothesis of the quantum nature of light and the constant, Planck received the 1918 Nobel Prize for Physics. His role in the development of the quantum theory was cited by the selection committee for special note.

Now recognized as a world figure in atomic physics, Planck spent much of the remainder of his career working with and defending Einstein's theory of relativity and the new quantum mechanics. This stance landed him in the middle of the German controversy between opponents and proponents of the new physics. While Planck never looked for controversy, he spent considerable energy fending off detractors. Planck was never comfortable with the Nazi government, and in 1933 he even argued with Adolf Hitler over the wisdom of the racial laws driving Jews from public office. Planck's argument that such laws would hurt Germany's scientific preeminence was rejected by Hitler. Due to advanced age, Planck played little role in nuclear weapons research in Germany during World War II. His eldest son was executed for his role in the July 1944 assassination plot against Hitler. He died at Göttingen, Germany, on October 3, 1947. His gravestone has the equation for Planck's constant engraved on it.

Suggested readings: Henry A. Boorse, Lloyd Motz, and Jefferson Hane Weaver, *The Atomic Scientists: A Biographical History* (New York: Wiley Science Editions, 1989); Thomas S. Kuhn, *Black-Body Theory and the Quantum Discontinuity, 1894–1912* (Oxford, United Kingdom: Clarendon Press, 1978); Emilio Segre, *From X-Rays to Quarks: Modern Physicists and Their Discoveries* (San Francisco: W. H. Freeman, 1980).

Plowshare

Plowshare was a program advanced by the Atomic Energy Commission (AEC) for research and development on the peaceful uses of nuclear explosives. In the years between 1957 and 1962, various schemes were considered, from Project Chariot for a harbor and ship turning basin on the northwest coast of Alaska to Project Carryall for a railroad pass in the Bristol Mountains near Amboy, California. Each plan floundered on the twin problems of too heavy a population in the explosion area and a weak cost-benefit justification. Another serious problem was radioactive fallout, which showed up in the first Plowshare blast at Gnome, near Carlsbad, New Mexico, when a 160-foot hole was blasted into a salt bed. Radioactive fallout drifted as far as Omaha, Nebraska. The most promising private project was the one to use a nuclear explosion to create a harbor at Cape Kerauden in northwestern Australia, but it also developed liabilities and was dropped. Perhaps the most serious of the Plowshare schemes was the creation of a nuclear canal in Central America. In 1964, the U.S. Congress approved funds for a feasibility and environmental impact study for a possible nuclear excavation on a Panamanian and a Colombian route. Later, President Lyndon Johnson became enamored about the possibility of a new canal, and a five-year study was commissioned with a price tag of $22.1 million. By the late 1960s, it became apparent that both the high cost and uncertainty about the technology made it impossible to build a canal or any other project of such magnitude. Smaller experiments were conducted in Operation Gasbuggy for a nuclear explosion of 29 kilotons to release natural gas in New Mexico on December 10, 1967, and Project Rulison, a 40-kiloton device, for stimulation of gas in western Colorado on September 10, 1969. The last of these big projects was Operation Rio Blanco, which was the detonation of three 33-kiloton devices near the western Colorado town of Rifle on May 17, 1973. Around 30 nuclear blasts took place in Operation Plowshare between 1957 and 1973. Operation Plowshare died a natural death during the Nixon administration

erful nuclear device, and pigs were selected to determine radiation and blast effects. Priscilla was a balloon shot, and a yield of 37 kilotons was registered. Shot Hood was another Livermore Radiation Laboratory imitative, and the shot of July 5, 1957, proved to be the largest bomb ever tested in the continental United States, with 74 kilotons of yield. The soldiers taking part were scared by the extent of the blast, and some off them shook and vomited. Later, it was learned that it was a hydrogen bomb that had been exploded. Shot Diablo turned out to be the most exciting test, because the original detonation was a dud. The scientists had to climb a tower and find the malfunction in a live atomic bomb. It turned out to be a damaged electrical cable, and the test was rescheduled. On July 15, the original test of Civil Defense procedures was completed with the detonation of a 17-kiloton device. Shot John was the testing of the air force's air-to-air missile, the MB-1 Genie, and the blast was at high altitude. A F89D Scorpion jet fighter fired the missile on July 19, 1957, and its 2-kiloton yield exploded at 20,000 feet. Six volunteers had stayed at ground zero during the test. The next test, Shot Kepler, was a tower shot, and it proved to be highly radioactive. It was only a 10-kiloton blast with troop maneuvers, but the radiation from the July 24, 1957, blast forced the maneuvers to be curtailed. Shot Owens was a device similar to the previous one, but it was designed by the Livermore Radiation Laboratory. Despite the forecast of rain, the balloon shot was detonated on July 25, 1957. A new dimension of testing occurred with Shot Stokes with the arrival of the first antinuclear protesters at the Nevada Test Site. This test on August 7, 1957, was an experiment with navy airships. The 19-kiloton explosion proved that blimps were worthless in a nuclear blast. In the meantime, the 11 demonstrators were arrested. The next two tests, Shot Shasta and Shot Doppler, were medium-range bombs. Shot Franklin Prime took place on August 30, 1957, with such disappointing results that troops stationed at the blast site were unimpressed. Shot Smoky was designed to test the army's new pentomic concept with a demonstration of a 44-kiloton device constructed by Edward Teller and the Livermore Radiation Laboratory. It was detonated from a 700-foot tower on August 31, 1957, and it proved to be so hot from radioactive fallout that the soldiers rapidly became concerned about the radiation levels. Troops were quickly moved in and out of the blast area. Because of growing psychological problems of soldiers operating in close proximity to atomic weapons, Shot Galileo was detonated on September 2, 1957, to allow behavioral scientists to observe soldiers in action. The device was in the 11-kiloton range, and the behavioral scientists concluded that soldiers could perform satisfactorily in the presence of a nuclear explosion. Two tests were scheduled for September 6, 1957, Shot Wheeler and Shot Coulomb B. Wheeler was a small Livermore device with a yield of less than a kiloton. Coulomb B was a safety experiment to see if a nuclear device could be detonated by means other than a trigger. A test on January 18, 1956, has shown that it was possible. Coulomb B confirmed the results of the earlier test. Shot LaPlace was a small nuclear device tested as a prototype neutron bomb. Shot Fizeau was a weapons vulnerability test. Scientists wanted to know if a nuclear explosion might set off a sympathy detonation of another nuclear weapon. The blast on September 14, 1957, yielded 11 kilotons and produced no sympathetic explosions. Shot Newton was a balloon experiment with a device suspended 1,500 feet over the target. The next test was Shot Rainier, and it was an underground test. Both the Atomic Energy Commission (AEC) and the military wanted to know how well an underground test could be detected and the fallout levels. A tunnel was dug 900 feet underground, and the device detonated on September 19, 1957, with a yield of 1.7 kilotons. Seismologists reported the blast registered a 4.6 on the Richter scale. Since no radiation was detected on the surface, it was surmised that no fallout was released. The last three tests—Shot Whitney, Shot Charleston, and Shot Morgan—took place between September 23 and October 7, 1957—ending this series of tests.

Suggested reading: Richard L. Miller, *Under the Cloud: The Decades of Nuclear Testing* (New York: Free Press, 1986).

Plutonium

The discovery of plutonium helped the scientists of the Manhattan Project make an atomic bomb to end World War II, but the production of plutonium has become the key ingredient for a country to become a nuclear power. Plutonium is a man-made metallic element with the atomic number 94. It is highly radioactive, and its most important isotope is fissionable plutonium-239. Plutonium is produced by bombarding uranium-238 with neutrons until it changes form. This phenomenon is the reason that plutonium is a transuranium, or beyond uranium. Glenn Seaborg, assisted by Joseph W. Kennedy and Arthur C. Wahl, first discovered plutonium by isolating it in the University of California–Berkeley's cyclotron in 1940. Because of the difficulty of isolating plutonium, it was not until August 20, 1942, that chemists were able to isolate pure elements of plutonium. Subsequent experiments at the Metallurgical Laboratory of Chicago proved that sufficient quantities of plutonium could be gathered for an atomic bomb. It was a slow process, however, as the conversion of uranium to plutonium by neutron capture requires that the uranium be first converted to neptunium and then plutonium as a second step. Nevertheless, the potential was so high that the Hanford Plants in Washington were built during World War II to isolate plutonium for the Manhattan Project. In further experiments with plutonium at Los Alamos it was found that plutonium has some unusual physical properties. Its density changes according to different temperatures, and it can assume five different phases. Also, plutonium's melting point is lower than was expected. It was in the delta phase that plutonium

Transportation of uranium-325 and plutonium-239 to Los Alamos Scientific Laboratory was a problem solved by development of handy carrying boxes lined with lead. These ships traveled by courier and with a dozen heavily armed escorts.

could be stabilized at room temperature by alloying it with gallium. By experimentation, it was soon learned that plutonium could not be detonated by a conventional trigger like uranium-235. A process of implosion was developed to trigger plutonium in a bomb. Plutonium was used in the Fat Man bomb detonated over Nagasaki, Japan. Because of its greater explosive potential, plutonium became preferred over uranium-235 for bombs.

Plutonium has become a controversial subject because of its disposal problems. Since it is highly radioactive for 24,000 years and is vital in the functioning of nuclear weapons, governments find it imperative to find safe storage or acceptable disposal. It is estimated that in excess of 1,000 tons of plutonium is available in the late 1990s. Moreover, plutonium is dangerous if mishandled because it emits alpha rays. Even dust or small fragments of plutonium, if inhaled or ingested can cause cancer, brain tumors, and other health risks. The problem of plutonium supplies is an international issue, but so far no government or international agency has produced a solution.

Suggested readings: Richard Rhodes, *The Making of the Atomic Bomb* (New York: Simon and Schuster, 1986); Glenn Theodore Seaborg, *The Plutonium Story: The Journals of Professor Glenn T. Seaborg, 1939–1946* (Columbus, OH: Battelle Press, 1994).

Pontecorvo, Bruno

Bruno Pontecorvo (1913–1993) was an Italian physicist who is most famous as a spy for the Soviet Union. He was born on August 22, 1913, in Pisa, Italy. His father was a businessman in the textile industry. Showing early academic promise, Pontecorvo entered the University of Pisa at age 16. He graduated with a certificate in physics and mathematics. His doctorate in physics with honors was earned from the University of Rome in 1934. Pontecorvo stayed at the University of Rome after graduation and worked under Enrico Fermi and Enrico Amaldi. In 1936 he was awarded a fellowship to the College de France in Paris. Pontecorvo also worked with Frederic Joliot-Curie at the Radium Institute. In 1940 he and his new Swedish wife, Marianne Nordblom, fled France, making their way to the United States. After a job with the Wells Survey Company in Tulsa, Oklahoma, Fermi invited Pontecorvo to join the Anglo-Canadian atomic research team in Montreal, Canada. For the next six years, he worked at the Chalk River heavy water reactor at Petawawa, Canada. It was during this time around 1943 that Pontecorvo began to pass atomic secrets to Soviet intelligence agents. After rejecting academic job offers in the United States, Pontecorvo accepted a position at the British nuclear research center at Harwell, England, in January 1949. Soon after he arrived at Harwell, the Fuchs Spy Case broke open, and Pontecorvo fell under suspicion by British security. Arrangements were made for him to take an academic position at the University of Liverpool, but Pontecorvo took the opportunity to defect to the Soviet Union while vacationing on the Continent. No news surfaced from the Soviet Union about his fate for several years, but Pontecorvo ended up working at the Institute of Nuclear Physics at Dubna. While Pontecorvo was nowhere as important an atomic spy as Klaus Fuchs, the Soviets did gain from him information about Canadian heavy water research. He died on September 24, 1993, in Dubna, Russia.

Suggested reading: H. Montgomery Hyde, *The Atom Bomb Spies* (New York: Atheneum, 1980).

Portal, Charles, Lord

Lord Portal (1893–1971) had been the wartime Chief of Air Staff for the Royal Air Force (RAF) before becoming head of the British atomic project in March 1946. Portal had served in the Royal Flying Corps in World War I. Then in World War II, he directed Bomber Command, and after 1940 he was the Chief of Air Staff. Portal was reluctant to take the post of Controller of Productions, Atomic Energy, because of a lack of administrative experience outside the RAF. His new position was in the Ministry of Supply, but his line responsibility was to the prime minister. His organization had Michael Perrin, deputy controller (Technical Policy), John Cockcroft, director of the research establishment, and Christopher Hinton, head of design and production plants. Portal set up the Atomic Energy Council within the Ministry of Supply, which held bimonthly meetings to coordinate British atomic policy. Besides being a gifted administrator with extensive contacts in the British government, Portal was bright enough to work with top scientific minds. Portal also had to withstand opposition from Sir Henry Tizard, scientific adviser to the Attlee government, and elements of the armed forces arguing for a reduction in an emphasis on atomic weapons and a buildup of conventional military forces. He confined himself to high policy and left the details to his subordinates. In the administrative warfare at the highest levels, however, Lord Portal was masterful, and he defeated all opposition. In 1951, Lord Portal left as head of the British atomic energy program and joined the board of directors of British Aluminum. He was replaced by General Sir Frederick Morgan. He died in 1971.

Suggested reading: Margaret Gowing, *Independence and Deterrence: Britain and Atomic Energy, 1945–1952*, vol. 2 (New York: St. Martin's Press, 1974).

Power Demonstration Reactor Program

The Power Demonstration Reactor Program (PDRP) was instituted by the Atomic Energy Commission (AEC) on January 10, 1955, in response to the provisions of the Atomic Energy Act of 1954 to expand nuclear power opportunities to the private enterprise sector. It was the brainchild of the chair of the AEC, Lewis L. Strauss. In this program the AEC solicited proposals from companies willing to go into partnership with the AEC in the building and operation of civilian nuclear power plants. The AEC provided a fixed-sum contract in the form of waivers for loans and special nuclear materials. In return the private companies assumed the economic risks of the project. A series of these contracts were negotiated with the AEC, providing a subsidy of about 20 percent per project. Yankee Atomic Electric Company signed the first contract on June 4, 1956, and it resulted in the Yankee Nuclear Power Plant at Rowe, Massachusetts. A second contract was with a consortium, Power Reactor Development Company of Detroit, which built the Fermi I facility outside of Detroit, Michigan. Two other initiatives of the PDRP were a small reactor project program offering in September 1955 and another large reactor proposal in January 1957. This program initiated the first series of nuclear power plants in the late 1950s and early 1960s.

Suggested reading: Frank G. Dawson, *Nuclear Power: Development and Management of a Technology* (Seattle: University of Washington Press, 1976).

Pressurized-Water Reactors. *See* Light Water Reactors

Price-Anderson Bill

The Price-Anderson Bill passed Congress in 1957, and it limited corporate liability in case of a reactor accident to $60 million and total liability to $560 million. A preliminary study by the Atomic Energy Commission of the insurance industry in March 1955 indicated that the limit of insurance available for nuclear accidents was $60 million for liability and $60 million for property damage. These totals were clearly deemed inadequate. Moreover, an experimental breeder reactor test in November 1955 at the National Reactor Testing Station in Idaho showed the danger of a core meltdown and revealed the potential for serious damages in a nuclear reactor accident. Insurance companies refused to insure commercial reactors beyond the $60 million liability range, even Lloyds of London. A Brookhaven National Laboratory report (WASH-740) on a worst-case scenario estimated 3,400 killed, 43,000 injured, and $7 billion worth of damage. Before private enterprise corporations would enter into the nuclear field, they wanted financial protection in case of accidents. Senator Clinton Anderson and Congressman Melvin Price responded with a bill for limits to liability in September 1957. In this bill, companies were absolved from all liability, no matter how the facilities were operated. All victim claims would be settled from a $560 million indemnification fund. This legislation added to the boom of nuclear reactors built by General Electric, Westinghouse, and others in the early 1960s. Congress left an out for itself, however, by having the law expire in 1967. In 1965 Congress passed a law extending the life of the Price-Anderson Bill for another 10 years. This time the private insurance pools were increased in coverage and the government indemnification amount was reduced. A no-fault clause for immediate indemnification of injured parties was also included. In 1988 Congress raised the liability limit to $7 billion by using assessments from nuclear utilities to form the financial pool.

Suggested readings: Frank G. Dawson, *Nuclear Power: Development and Management of a Technology* (Seattle: University of Washington Press, 1976); Daniel F. Ford, *The Cult of the Atom: The Secret Papers of the Atomic Energy Commission* (New York: Simon and Schuster, 1982).

Private Ownership of Special Nuclear Materials Act

In 1964 the U.S. Congress passed the Private Ownership of Special Nuclear Materials Act, ending the federal government's monopoly of nuclear materials. Both leaders in Congress and the domestic nuclear energy industry had lobbied for companies to gain control of nuclear materials. President Lyndon Johnson signed the bill in to law on August 26, 1964. This legislation had been proposed by the Atomic Energy Commission (AEC) in an effort to encourage private enterprise in the American nuclear industry. Once proposed, both the Joint Committee on Atomic Energy (CAE) and the nuclear industry backed this legislation. A timetable of steps for implementation was established with July 1, 1973, as the final target date. Only in the areas of defense and security, and public health and safety were regulatory controls retained. This bill also opened domestic and foreign industries access to government diffusion plants for uranium enrichment.

Suggested reading: Frank G. Dawson, *Nuclear Power: Development and Management of a Technology* (Seattle: University of Washington Press, 1976).

Project Chariot

Project Chariot was the brainchild of Edward Teller, and it envisaged a detonation of several thermonuclear devices at Cape Thompson in northwest Alaska to create a deep-water harbor. This project was part of the U.S. government effort to find peaceful uses of atomic energy in its Plowshare program. The creation of a deep-water harbor was intended to promote economic development in northwest Alaska. Alaskan politicians, the media, and scientists were enthusiastic proponents until questions about the economic viability and danger to wildlife were posed. Public meetings were held in Alaska by supporters of the project, but slowly public opinion turned against it. Many of the environmental questions were left unanswered, and the project died a natural death.

Suggesting reading: Dan O'Neill, *The Firecracker Boys* (New York: St. Martin's Press, 1994).

Psychological Impact of Nuclear Accidents

Studies of both the Three Mile Island and Chernobyl nuclear accidents have found severe psychological consequences among the survivors of the accidents. Doctors noticed in the year or so after the Three Mile Island accident that local inhabitants were ready to blame any malady on the accident. They named this phenomenon on "nuclear neurosis." The sufferers show high levels of anxiety, depression, and hostility. Many of the inhabitants left the area around Three Mile Island despite family ties dating back generations. Surveys indicate that these people have little faith in either the nuclear industry or the U.S. government.

The survivors of Chernobyl have experienced the same type of psychological disorders. A complicating factor has been the large number of deaths and injuries produced by the reactor accident. Whereas the Three Mile Island accident has received maximum publicity, the Russian government was slower to release information about the Chernobyl accident, and the change of sovereignty to the Ukrainian government has not improved matters. Whole areas around Chernobyl have been depopulated for health reasons, and the inhabitants are aware of the health problems caused by the accident. Distrust of the government and anxiety about the future characterized the state of mind of the population in the Chernobyl region.

Suggested readings: Peter S. Houts, *The Three Mile Island Crisis: Psychological, Social, and Economic Impacts on the Surrounding Population* (University Park: Pennsylvania State University Press, 1988); Robert Leppzer, ed., *Voices from Three Mile Island: The People Speak Out* (Trumansburg, NY: Crossing Press, 1980).

Pugwash Movement

The Pugwash Movement is the organization of scientists that formed in 1957 to find ways to abolish wars and control nuclear weapons. It started when British philosopher Bertrand Russell drafted a peace proposal in 1955, which after Albert Einstein signed was called the Russell-Einstein Manifesto. In this manifesto an appeal was made to scientists to take the lead in solving the problem of nuclear weapons. His appeal was answered by the Atomic Scientists' Association and its executive vice president Joseph Rotblat. Rotblat and the other officers of the association solicited help from a sponsor. Cyrus S. Eaton, a Canadian-born American industrialist, believed in peaceful coexistence, so he invited representatives from both the Communist and non-Communist blocs to his summer place in Pugwash, Nova Scotia. Twenty-four scientists from 10 countries met at Pugwash in 1957 to discuss ways to avoid nuclear war. This movement was unique because it had representation from both sides of the Iron Curtain. At first the scientists from the two blocs distrusted each other, but both sides soon found it a forum to discuss arms control and peace issues. For the next 30-plus years this organization continued to seek ways to prevent nuclear war. It shared the 1995 Nobel Peace Prize with its head, Joseph Rotblat, for its peace activities.

Suggested readings: Norman Moss, *Men Who Play God: The Story of the H-Bomb and How the World Came to Live with It* (New York: Harper and Row, 1968); Joseph Rotblat, *Pugwash—The First Ten Years: History of the Conferences of Science and World Affairs* (New York: Humanities Press, 1967).

Quantum Mechanics

Quantum mechanics is a mathematical system developed to describe the physics of the atom. Traditional mathematical approaches were unable to cope with aspects of the behavior of the atom. Part of the answer was the quantum theory. This theory developed out of the idea put forward by Max Planck that the atom emits or absorbs the energy of radiation in discrete units, or quanta, whose energy is in direct proportion to its frequency. In May 1925, Werner Heisenberg published a mathematical framework to explain the erratic nature of the atom, which his colleague Max Born recognized as a form of matrix algebra. Heisenberg, Born, and Pascual Jordan worked out the mathematical details over a three-month period in 1925. Because this method corresponded closely to experimental evidence, it soon had its adherents. The early acceptance of this theory by Arnold Sommerfeld at the University of Munich helped establish it in Germany. Most physicists, however, had difficulty understanding quantum mechanics due to the unfamiliar mathematical methods required. Erwin Schrödinger soon made a challenge to quantum mechanics in 1926 by putting forward a wave theory of the atom, which made use of mathematics more familiar to physicists. Soon thereafter it was realized that Heisenberg's matrix mechanics and Schrödinger's wave mechanics were mathematically equivalent expressions of the same theory, which became known collectively as quantum mechanics. Eventually, quantum mechanics was accepted by most nuclear physicists, but Albert Einstein always remained a critic.

Suggested readings: Richard Rhodes, *The Making of the Atomic Bomb* (New York: Simon and Schuster, 1986); Emilio Segre, *From X-Rays to Quarks: Modern Physicists and Their Discoveries* (San Francisco: W. H. Freeman, 1980).

Quantum Theory

Quantum theory was one of the major theoretical advances of the early twentieth century. This theory claims that the atom emits or absorbs the energy of radiation in discrete units or quanta, whose energy is in direct proportion to its frequency. A historian of science maintains that the quantum theory was a product partly by chance and partly by necessity and that acceptance was based on the development of methods of experimental verification. It was formulated by physicists intrigued by its possible ability to explain phenomena that classical physics was unable to account for. While classical physics assumes a strict determination of certain outcomes for given initial conditions, the quantum theory asserts a restriction on the extent to which one can determine measurable quantities. Development of the theory took place between the years 1900 and 1927. Max Planck laid the foundation for the quantum theory when in 1900 he formulated his constant h in an attempt to explain blackbody radiation. A blackbody is a body that absorbs and emits, but does not reflects electromagnetic radiation (i.e., inside of an oven). Numerous scientists had failed to resolve the conflict between Maxwell's theory of radiation and the statistical theory of a blackbody before Planck made his contribution. This quantum hypothesis divided the international physics community for more than two decades in the early twentieth century, but those that used it as a basis for further understanding of the behavior of the atom became the leaders of the next generation of physicists in the quest for atomic energy.

Suggested reading: Friedrich Hund, *The History of Quantum Theory* (New York: Barnes & Noble, 1974).

Quebec Conference

President Roosevelt and Prime Minister Churchill negotiated the Quebec Agreement on August 19, 1943, to ensure collaborative efforts on war resources and plants for atomic weapons. It was also the first international atomic treaty, and because the agreement limited the spread of information about nuclear technology, it was also the first nonproliferation agreement. A key part of the agreement was a provision for agreement on the use of any atomic weapons. Much of the uranium available for the Manhattan Project came from Canada, but American authorities had started a campaign to shut out the British and Canadians on atomic projects because 90 percent of the funding was coming from the United States. Moreover, it was obvious that there were postwar implications to nuclear development, and the Americans wanted to retain the advantage. The provision to retain information about uranium research to the two of them was dictated by both security and commercial reasons. A Combined Policy Committee of representatives of the United States, the United Kingdom, and Canada was set up to administer the cooperation. It delegated to the Combined Development Trust (CDT) the authority to begin to buy rights to control the world's supply of uranium and thorium ore. Geologists were sent all over the world to explore possible ore fields with portable Geiger counters. By 1945, it has been estimated that the United States and Great Britain had control of 90 percent of the known deposits of high-grade uranium ore. On July 4, 1945, the Combined Policy Committee met and gave its approval for the use of an atomic bomb against Japan.

This wartime agreement to share atomic secrets with Great Britain was abrogated in 1947 by the United States under pressure from American politicians—Senator Arthur Vandenberg and Senator Bourke Hickenlooper. American negotiators insisted that all Belgian Congo ore production come to the United States and the entire British stockpile of uranium be turned over to them. In return, the British would be allowed to participate in the Marshall Plan. The actual terms of the new agreement were slightly less severe. The British would give up their veto on the use of the atomic bomb and relinquish two thirds of their uranium ore stockpile, and the United States would receive all Belgian Congo ore for at least two years. This agreement was signed on January 7, 1948, and was kept secret from both of the U.S. Congress and the United Nations.

Suggested reading: Martin J. Sherwin, *A World Destroyed: The Atomic Bomb and the Grand Alliance* (New York: Knopf, 1975).

R

Rabi, Isidor Isaac

I. I. Rabi (1898–1988) was a prominent American nuclear physicist whose work on atomic theory and international contacts made him one of the most influential scientists in the United States in the middle of the twentieth century. He was born on July 29, 1898, at Rymanov, Galicia, which was then in Austria but is now in Poland. Rabi's parents emigrated with him to the United States while he was an infant, and he grew up in a Yiddish-speaking community in New York City. His father worked in clothing sweat-shops until he became a small-time grocer. Rabi's marriage to Helen Newark produced two daughters. At an early age Rabi discovered the local Carnegie Library of the Brooklyn Public Library and began to read science books, especially on astronomy. He attended the Manual Training High School in Brooklyn, where he graduated in 1916. Rabi entered Cornell University with advanced standing, but with little idea on what to major in, so he picked an electrical engineering major. He graduated from Cornell University in 1919 with a degree in chemistry. For the next three years he worked in a variety of unsuccessful jobs, so in 1922 he returned to Cornell University to graduate school. Soon, however, Rabi transferred to Columbia University. In 1924 he became a part-time tutor in physics at the City College of New York (CCNY). Rabi received his doctorate in physics from Columbia University in 1927. After graduation, he traveled extensively in Europe, making contacts with European physicists. Part of his two-year trip was to study in Germany with Otto Stern, who was doing physics experiments on molecular beams. While there, Rabi discovered a new field configuration of molecular beams, which was later called the Rabi field. Toward the end of his European stay, he was offered a research and teaching position at Columbia University, based upon the recommendation of Werner Heisenberg.

Two leaders of the postwar American atomic energy program were I. I. Rabi, a physicist from Columbia University, and Brien McMahon, a Democratic senator from Connecticut. McMahon used his position as chair of the Joint Committee on Atomic Energy to advance the cause of atomic energy.

Rabi remained at Columbia for the rest of his academic career until retirement in 1968.

Rabi spent the 1930s at Columbia, combining theoretical insight with experiments to become one of the stars in American physics. He established a molecular beam laboratory, and its first product was the 1931 Breit-Rabi theory, which postulated how the classic Stern-Gerlach experiment could be modified to reveal properties of the nucleus of the atom. Based on research at his laboratory, Rabi in 1937 published a paper forming the theoretical basis for the magnetic resonance method. This method modified the classical theory of James Clerk Maxwell on the magnetic susceptibility by finding a way to cancel it with a specialized fluid. Rabi won the 1944 Nobel Prize for Physics for his research on atomic and molecular beam work and for his discovery of the resonance method.

During World War II, Rabi worked on a variety of wartime scientific projects. His first contribution was at the MIT (Massachusetts Institute of Technology) Radiation Laboratory with experiments on microwave electronics leading to an operational radar system. By the time the United States entered the war in December 1941, an operational radar system was ready; and Rabi played a leadership role in this. His next contribution was as an adviser to the atomic bomb project at the Los Alamos Laboratory. He was able to convince J. Robert Oppenheimer of the merits of organization, and along with Robert Bacher persuaded General Leslie Groves and Oppenheimer not to militarize the Los Alamos Laboratory. His lifelong friendship with Oppenheimer allowed him to advise him on scientific matters and to serve as a troubleshooter for him.

In the postwar world, Rabi became a senior statesman in the physics community. His condition for returning to Columbia University was to be appointed chair of the physics department, which was soon done. To improve physics research at Columbia he became one of the founders of the Brookhaven National Laboratory in Upton, New York, on Long Island. In December 1946 Rabi was appointed a member of the General Advisory Committee (GAC) of the Atomic Energy Commission (AEC) with his friend Oppenheimer. Rabi agreed with the majority in the GAC's October 1949 meeting to oppose a crash program to build a hydrogen bomb. He defended Oppenheimer and his record during the security clearance hearing in April 1954, and he never forgave Edward Teller for his testimony against Oppenheimer. He was also active in the founding of the International Laboratory for High-Energy Physics in Geneva (CERN—originally Conseil Européen pour la Recherche Nucléaire, now Centre Européenne pour la Recherche Nucléaire) in 1952. Another interest of his was the international atomic energy conference in Geneva in 1955. Much of his later activity dealt with reconciling the sciences and humanities. Rabi died on January 11, 1988, in New York City.

Suggested readings: Henry A. Boorse, Lloyd Motz, and Jefferson Hane Weaver, *The Atomic Scientists: A Biographical History* (New York: Wiley Science Editions, 1989); John Daintith, Sarah Mitchell, Elizabeth Tootill, and Derek Gjertsen, *A Biographical Encyclopedia of Scientists*, 2nd ed., vol. 2 (Bristol, UK: Institute of Physics, 1994); John S. Rigden, *Rabi: Scientist and Citizen* (New York: Basic Books, 1987).

Radiation Exposure Compensation Act

In 1990 the U.S. Congress passed the Radiation Exposure Compensation Act, and President George Bush signed it. For over a decade, legislation had been introduced to compensate radiation victims, but Congress had been reluctant to pass any of these bills. Senator Ted Kennedy introduced such a bill in 1979, followed by Senator Orin Hatch in 1981. Neither bill received presidential support. Hatch reintroduced his bill in 1983, but it died again without congressional support. In 1989 Hatch had Congressman Wayne Owens propose a more modest bill, and this time it passed. Owens had a personal interest in this legislation because his brother-in-law had died from lung cancer caused by uranium mining. The Bush administration decided this bill might be less costly than losing class-action suits.

This legislation authorized compensation for radiation injuries to those people exposed to atomic testing and for uranium miners. Congress had to pass an appropriation bill for this compensation. Congress in 1991 and 1992 increased the total fund to $200 million. Downwind victims were eligible for $50,000, and uranium miners with lung cancer, $100,000. This law was in response to the refusal of the federal courts to approve claims for damages based on the principle of sovereign immunity of federal agents. Bureaucratic red tape and a hundred-page form have made it difficult for miners to gain compensation. Documentation is difficult to obtain since the government and industry have destroyed records. This lack of documentation has been a special problem with the Navaho miners because the Indian Health Service destroyed their records after 25 years. Only a limited number of claims had been approved by the government by the mid-1990s.

Suggested readings: Howard Ball, *Cancer Factories, America's Tragic Quest for Uranium Self-Sufficiency* (Westport, CT: Greenwood, 1993); Marjorie Garber and Rebecca L. Walkowitz, eds., *Secret Agents: The Rosenberg Case, McCarthyism, and Fifties America* (New York: Routledge, 1995).

Radiation Sickness

Radiation sickness is a reaction to exposure to severe doses of ionizing radiation. Symptoms include nausea, vomiting, diarrhea, blood cell changes, hemorrhaging, loss of hair, and possibly death. Almost from the beginning of the discovery of radioactivity, radiation sickness was known to be a problem. Doctors had reported 50 cases of radiation sickness by 1911. Most of these cases came from individuals exposed to X-rays. Some of the early scientists experimenting with radium fell ill and died of leukemia caused by overexposure to radioactive radiation. Pierre Curie, Marie Curie, and Irene Curie suffered radiation sickness from their experiments. A monument to early pioneers of radiology in Hamburg, Germany, records that 169 radiogists died of radiation sickness in the first decade of the twentieth century. Later studies would show that the incidence of leukemia in radiologists was nine times higher than expected in the normal population. Despite the apparent dangers, X-ray businesses abounded in the 1920s. Early scientists believed that gamma rays were the villains, but soon incidences of radiation sickness showed that alpha and beta particles were also dangerous. The big debate among scientists was on how much exposure to radiation was harmful. Postwar experiments indicated that radiation at any level would affect human genes. Mutations are proportional to the dose of radiation. More recent research indicates that most radiation sickness deaths are the result of bone marrow failure that takes weeks, rather than days, for death. While initial reactions of nausea and vomiting are not life-threatening, blood loss and infection are serious complications. Doctors have found ways to combat radiation sickness by placing the injured in sterile compartments to avoid infection and performing bone marrow transfusions. Doses of radiation exceeding 3 or 4 sieverts (unit of dose equivalent to 100 rads) are considered the median acute lethal dose. Much of the research on radiation sickness has come from survivors of the Hiroshima and Nagasaki atomic bomb attacks. Studies from these survivors indicate that leukemia takes about 14 years from exposure to death, with signs of the disease appearing around year 12.

Suggested readings: Edward Pochin, *Nuclear Radiation: Risks and Benefits* (Oxford, United Kingdom: Clarendon Press, 1983); William J. Schull, *Effects of Atomic Radiation: A Half-Century of Studies from Hiroshima and Nagasaki* (New York: Wiley-Liss, 1995).

Radiological Accidents

A series of 19 radiological accidents at nuclear and medical facilities have caused at least 60 fatalities since 1945. Of these deaths, 40 have been of workers at nuclear plants and twenty of whom were members of the public. These fatality figures are low, given that the deaths at the Chernobyl Nuclear Power Station were much higher than the 29 reported to the International Atomic Energy Agency (IAEA).

The first two deaths were Harry Daghlian and Louis Slotin in 1945 and 1946 at the Los Alamos National Laboratory. Subsequent accidents have taken place at facilities located around the world, from Brazil to Israel. The most deadly was the Chernobyl accident in 1986. Five of the most recent incidents took place in Goiania, Brazil; San Salvador, El Salvador; Soreq, Israel; Nesvizh, Belarus; and Tammiku, Estonia.

In the Goiania incident a private radiotherapy institute left a cesium-137 teletherapy unit insecure, and scavengers stole a portion of the machine for scrap metal. A number of people were exposed to the radiation, and 4 people subsequently died of radiation sickness. In the Nesvizh, San Salvador and Soreq accidents, operators became exposed to cobalt-60 when there were mishaps in the production line operations. These operators later died of radiation sickness. The latest incident was in October 1994 when three brothers illegally entered an installation at Tammiku, Estonia, and removed a metal container filled with caesium-137. In its removal, the container spilled and the three men were exposed to radiation. One brother died and the others recovered after lengthy illnesses. The International Atomic Energy Agency (IAEA) monitors these accidents and compiles lengthy reports on them.

Suggested readings: International Atomic Energy Agency, *The Radiological Accident at the Irradiation Facility in Nesvizh* (Vienna, Austria: Author, 1996); International Atomic Energy Agency, *The Radiological Accident in Goiania* (Vienna, Austria: Author, 1988); International Atomic Energy Agency, *The Radiological Accident in San Salvador* (Vienna, Austria: Author, 1990); International Atomic Energy Agency, *The Radiological Accident in Soreq* (Vienna, Austria: Author, 1993); International Atomic Energy Agency, *The Radiological Accident in Tammiku* (Vienna, Austria: Author, 1993).

Radium Institute

The center of French research on atomic energy in the first half of the twentieth century was the Radium Institute. After the prominent French physicists Pierre and Marie Curie had lobbied hard for years for an adequate research laboratory, French authorities considered building such a laboratory shortly after Pierre Curie's untimely death in an accident, but Madame Curie refused to capitalize on her husband's death. It took an offer from the director of the Pasteur Institute to build a research laboratory to resurrect the idea. Both the Sorbonne University and the Pasteur Institute shared expenses to build the Radium Institute and appointed Madame Curie as the director of the Laboratory of Radioactivity. Construction on the Radium Institute was started in Paris in 1910, but its facilities were not complete until July 1914. In its final form the Institute had two independent laboratories—one for radioactivity research and the other for biological research. The facility was located on land near the École Normale, and it was part of a large scientific complex. Madame Curie was the director of the radioactivity laboratory until her death in 1934. Besides the basic research on radiation by Madame Curie, the laboratory became famous for experiments on artificial radioactivity conducted by her son-in-law Frederic Joliot-Curie and her daughter Irene Joliot-Curie.

Suggested readings: Eve Curie, *Madame Curie* (Garden City, NY: Doubleday, Doran, 1938); Bertrand Goldschmidt, *Atomic Rivals* (New Brunswick, NJ: Rutgers University Press, 1990); Maurice Goldsmith, *Frederic Joliot-Curie: A Biography* (London: Lawerence and Wishart, 1976).

Radon Daughters

Radon daughters was the name given to the radioactive products inhaled by miners working in American uranium mines. The danger of contracting lung cancer by working in uranium and pitchblende mines was already understood in Europe. As early as 1879, lung cancer among miners was identified as a major cause of death. Miners at the Joachimsthal Mines lived only about 20 years after entering the mines. Radon gas had been isolated as the cause of the lung cancers by the 1920s. Radon gas has a half-life of 3.8 days and then decays into other radioactive elements called daughters. These daughters are two parts isotopes of polonium, one part 2 isotope of lead, and one part isotope of bismuth. If these particles repeatedly lodge in lung tissue, they can cause lung cancer. Polonium-210-lead, with a half-life of 22 years, is particularly carcinogenic if it builds up in the lungs. At the peak of the uranium mining boom in 1961–1962 there were around 925 uranium mines in operation, with approximately 5,500 miners working in them. Although this information was known in medical circles, no effort was made to protect American miners in the uranium boom of the 1950s and 1960s. Efforts by medical officials in the Atomic Energy Commission (AEC) to bring attention to the problem were thwarted by a legal interpretation from the AEC that state regulations applied to the ore in the mines and AEC authority only began once the ore had been mined. Only when American miners began to die of lung cancer in significant numbers (between 10 and 20 percent of the miners) in the 1970s did the AEC start paying attention. Many of those sick and dying were Navajo Indians who had worked in the mines, and this startled medical researchers because Navajo Indians had no history of lung cancer in their tribe. It was only in 1967 that health standards were implemented, and that only after opposition from the mining industry. The solution was mechanical ventilation of the mines by any type of electrical or power fan.

Suggested readings: Howard Ball, *Cancer Factories: America's Tragic Quest for Uranium Self-Sufficiency* (Westport, CT: Greenwood, 1993); Peter Pringle and James Spigelman, *The Nuclear Barons* (London: Michael Joseph, 1981).

Rainbow Warrior

The *Rainbow Warrior* was a ship owned by Greenpeace that was sunk in 1985 by French Secret Service in New Zealand on the eve of an antinuclear protest in the Pacific. It was a converted research trawler that had been purchased by Greenpeace in 1977. Greenpeace used the vessel as its flagship for its four-ship protest fleet. Most of the *Rainbow Warrior*'s activities before 1985 had been protests against cruelty to animals—commercial whaling operations and the harp seal pups slaughter—or environmental dumping—nuclear waste and chemical dumping. It had several run-ins with authorities—Spain in 1980 and the Soviet Union in 1983. In 1985 at the Greenpeace International annual meeting, it was decided to highlight the Pacific and send the fleet there to protest nuclear activities in the region. The inhabitants of Rongelap Atoll in the Marshall Islands had been exposed to high levels of radioactivity in fallout from a U.S. hydrogen bomb test (Operation Bravo) on March 1, 1954. Birth defects and deaths from long-term radiation caused the natives to decide to leave Rongelap. They asked Greenpeace for assistance with an evacuation. In the space of a week and a half in May 1985, 304 people, their homes, and their possessions were moved from Rongelap Atoll to Majato Island. From Majato Island the *Rainbow Warrior* headed for New Zealand before going to Mururoa to protest French nuclear testing. On the evening of July 10, 1985, in the harbor of Auckland, New Zealand, at Marsden Wharf, two explosions sank the *Rainbow Warrior*. Photographer Fernando Pereira was killed as the result of the explosions. French Secret Service (Direction Generale de la Securite Exterieure [DGSE]) operatives had place two bombs on the *Rainbow Warrior*. New Zealand authorities arrested two of the saboteurs, but the rest escaped. The French government denied complicity and engaged in a high-level cover-up, but in the end, French Minister of Defense Henri Hernu and the director of the DGSE, Admiral Pierre Lacoste, resigned. The new French minister of defense admitted French complicity, and two operatives were sentenced to long prison terms. The *Rainbow Warrior* was too damaged to repair, so it was sunk at sea on July 10, 1986, in Matauri Bay, New Zealand.

Suggested reading: Michael King, *Death of the* Rainbow Warrior (Auckland, New Zealand: Penguin Books, 1986).

Ranger Nuclear Tests

Ranger was the code name for the nuclear testing at the new Nevada Test Site in 1951. The test site was at Frenchman's Flat toward the southern end of the range. Five detonations were scheduled under the names Shot Able, Shot Baker, Shot Easy, Shot Baker-2, and Shot Fox. All detonations were airdrops and took place over a 13-day period from January 27 to February 6. An Air Force Weather Wing provided aircraft to monitor weather data and to track the path of the nuclear cloud. It was discovered that different wind patterns caused radioactive fallout from the tests to reach different parts of the United States. Fall from Shot Able traveled as far as Rochester, New York, where it contaminated film processing at Eastman Kodak Company. Shot Fox was the most powerful, and it detonated at 22 kilotons, which was about the strength of the Nagasaki bomb.

Suggested readings: International Physicians for the Prevention of Nuclear War, *Radioactive Heaven and Earth: The Health and Environmental Effects of Nuclear Weapons Testing in, on, and Above the Earth* (New York: Apex Press, 1991); Richard L. Miller, *Under the Cloud: The Decades of Nuclear Testing* (New York: Free Press, 1986).

Rasmussen Report

The Rasmussen Report (also known as WASH-1400) was an August 1974 report by the American government on nuclear reactor safety. This 3,000-page report (14 volumes) was the result of a three-year, $4 million study documenting possible reactor safety problems and was funded by the Atomic Energy Commission (AEC). Norman C. Rasmussen, a professor of nuclear engineering at the Massachusetts Institute of Technology (MIT) and a staff of 50 prepared the report. Rasmussen had a number of ties to the nuclear industry, serving as a consultant for several firms. He made it plain that the AEC would have control over the final report and its conclusions. Rasmussen took a leave from MIT, and the work on the report was done at the AEC offices in Germantown, Maryland. Saul Levine of the AEC staff was the project staff director and had control of all aspects of the report. The report concluded that in the worst-possible scenario—coolant loss, failure of backup systems, fuel meltdown, and natural disasters—the cost would be in the neighborhood of 3,300 fatalities, $14 billion in property damage, and the evacuation of inhabitants in a 290-square-mile radius. To arrive at the probability of a nuclear accident the government used a statistical technique called *fault-free analysis*. This method had been developed by the aerospace industry to predict failures in complex systems. By use of this technique, the compilers of this report concluded that the chances of a nuclear accident were remote. This report appeared at a time when the renewal of the Price-Anderson Act was at stake. The AEC launched a favorable public relations blitz for the report. Almost from the date of publication, however, scientific critics of nuclear power plant safety criticized the report for its assumptions and its almost flagrant attempt to minimize the chances of a nuclear accident. The most effective criticism came from a review panel of the American Physical Society, which attacked the validity of certain calculations in the Rasmussen Report. After some more revision, the final Report of the Reactor Safety Study came to 2,300 pages divided into 9 volumes. In 1979 the Nuclear Regulatory Commission formally disowned the Rasmussen Report because of its overoptimistic conclusions. This analysis proved to be correct when the Three Mile Island accident showed just how safe a nuclear reactor was.

Suggested readings: Irvin C. Bupp and Jean-Claude Derian, *Light Water: How the Nuclear Dream Dissolved* (New York: Basic Books, 1978); Daniel F. Ford, *The Cult of the Atom: The Secret Papers of the Atomic Energy Commission* (New York: Simon and Schuster, 1982).

Reactor Poisoning

Reactor poisoning was a problem first encountered at the start-up of the large nuclear reactor (B Pile) in Hanford, Washington, on September 27, 1944. The reactor functioned successfully for 12 hours and then failed. It restarted itself 12 hours later but then slowly went into a power decline. Two prominent theoretical physicists, Enrico Fermi and John Wheeler, were on hand during the first operations of the reactor. Wheeler soon figured out that the poisoning came from iodine-135 decaying into xenon-135, which would absorb idle neutrons at a prodigious rate. So many neutrons were absorbed by xenon-135 that not enough free neutrons were left to sustain a chain reaction. The solution was to increase the reactor's ability to recover by adding enough uranium slugs to overcome the poisoning. After the correction was made, the problem ceased. This solution was possible at the Hanford reactors because the du Pont engineers had built a large margin of error into the reactor design.

The reactor poisoning problem came to the attention of the scientists working in the Soviet nuclear program because of a sentence in the original version of the Smyth Report. The Soviets were in the process of building reactors and needed to know about such problems for design specifications. It gained such a high priority that the Soviet Security Police (NKVD—People's Commissariat of Internal Affairs) sent a young scientist to Denmark to interview Niels Bohr about the problem. Bohr was not enough of a reactor expert to answer the basic questions, nor was he willing to give out too much information to help the Soviets.

Suggested reading: Rodney P. Carlisle, *Supplying the Nuclear Arsenal: American Production Reactors, 1942–1992* (Baltimore, MD: Johns Hopkins University Press, 1996).

Reactor Safeguard Committee

The Reactor Safeguard Committee was a product of the Atomic Energy Commission (AEC) and its General Advisory Committee (GAC). Its mission was to study and advise these bodies on nuclear reactor safety. This committee was formed in 1947, and its membership consisted of six part-time advisers. Its most prominent member and chair was Edward Teller. This emphasis on safety brought them frequently into conflict with the AEC, GAC, nuclear contractors, other government agencies, and the military. Almost from the beginning this committee focused on catastrophic possibilities of nuclear reactors. Their first recommendation was to keep reactors away from heavily populated areas. Next, the committee proposed that reactors be designed around the maximum credible accident theory. This theory reversed the earlier assumption that the likelihood of a reactor accident was low, fixing the probability that a reactor accident would take place and estimating its maximum damage. This revised emphasis was one of the reasons for building the National Reactor Testing Station in Idaho. This pressure compelled the nuclear industry to improve reactor safety, and most complied. Other countries followed the American example with the notable exception of the Soviet Union. In 1953 the Reactor Safeguard Committee was expanded and renamed the Advisory Committee on Reactor Safeguards (ACRS).

Suggested readings: Rodney P. Carlisle, *Supplying the Nuclear Arsenal: American Production Reactors, 1942–1992* (Baltimore, MD: Johns Hopkins University Press, 1996); Spencer R. Weart, *Nuclear Fear: A History of Images* (Cambridge: Harvard University Press, 1988).

Red Specialists

The leaders of the Soviet nuclear program were called Red Specialists. They were a group of young Bolsheviks from humble backgrounds who were handpicked for technical training in the 1920s and 1930s. These individuals were the facilitators and the ultimate decision makers in the Soviet nuclear hierarchy. Unquestioned obedience to the official Communist ideology was a trademark of this elite. They migrated out of heavy industry after World War II and into the Soviet atomic weapons program. Their managerial style was authoritarian and arbitrary, but they owed total allegiance to Lavrentii Beria, the head of the postwar Soviet nuclear program and chief of the Soviet Secret Police. M. G. Pervukhin, Boris Vannikov, and Avraami Zavenyagin were all products of this system, and they held high administrative posts under Beria. They were considered extremely capable administrators, but atomic energy was new to them. A considerable education effort was made by Igor Kurchatov and his associates to bring these administrators to a level of basic understanding of the issues. Beria's successor as head of the Soviet nuclear program, Vyacheslav Malyshev, was also a Red Specialist. This managerial elite had prospered under the Stalinist regime but found it more difficult to coexist in the post-Stalin Soviet Union. They reacted negatively toward Nikita Khrushchev's anti-Stalin campaign in 1956 and engaged in a losing battle on the Politburo to overthrow Khrushchev in June 1957. This political defeat and the nuclear waste accident at Chelyabinsk-40 destroyed the influence of the Red Specialists and set back the Soviet nuclear development for nearly a decade.

Suggested readings: David Holloway, *Stalin and the Bomb: The Soviet Union and Atomic Energy, 1939–1956* (New Haven, CT: Yale University Press, 1994); Peter Pringle and James Spigelman, *The Nuclear Barons* (London: Michael Joseph, 1981).

Reich's Research Council

The Reich's Research Council took over the administration of the Nazi nuclear program in 1943. Abraham Esau became the head administrator. At the time of the takeover, experiments were being carried out at a variety of places: Kaiser-Wilhelm-Institute for Physics in Berlin, University of Leipzig, army proving ground near Berlin, and Kaiser Wilhelm Institute in Heidelberg. Esau was replaced by Walter Gerlach on January 1, 1944, because Esau had lost the confidence of the researchers. Despite the beginnings of results, Allied bombing raids made the scientists move their experiments to smaller communities in the south of Germany. Kurt Diebner headed to Thuringia. Werner Heisenberg moved his reactor research to Hechingen and built his reactor in a cave near Haigerloch. Harteck's heavy water experiments ended up in Celle, near Hanover.

Suggested readings: David Irving, *The German Atomic Bomb: The History of Nuclear Research in Nazi Germany* (New York: Simon and Schuster, 1967); Thomas Powers, *Heisenberg's War: The Secret History of the German Bomb* (New York: Knopf, 1993).

Rickover, Hyman George

Admiral Hyman Rickover (1900–1986) made his career in the U.S. Navy by developing the technology for nuclear-powered submarines and then surface ships. He was born on January 27, 1900, in Makow, Poland. His father was a tailor. In 1906, his father decided there was no future for Jews in Russia, so the family emigrated to the United States. Rickover lived first in New York City and then Chicago. His uncle, Congressman Adolph Sabath, offered him an opportunity to go to the U.S. Naval Academy. Although considered an indifferent student, he graduated 107th out of 540 in the class of 1922. After five years of sea duty, Rickover worked on a master's degree in electrical engineering at the Naval Postgraduate School and Columbia University. In 1929, he returned to naval sea duty, but this time in submarines. After three years of submarine duty and an additional two years on surface ships, Rickover found his place as a troubleshooter at the Bureau of Ships. He stayed in the Electric Section of the Bureau of Ships during most of World War II. In 1945, he assumed command of a large repair facility in Okinawa. He was there when Japan surrendered in 1945.

It was not until after the war that the navy decided to send Commander Rickover to Oak Ridge, Tennessee, to study nuclear technology. Once he understood the technology, Rickover began to work on a nuclear reactor capable of use on a submarine. His goal became to build a nuclear navy regardless of the cost. Top naval brass were less than enthusiastic, and Rickover was constantly in conflict with his superiors. Then when he became concerned with the lack of interest in nuclear ship propulsion by the Atomic Energy Commission (AEC), Rickover convinced the members of the AEC to allow the navy to pursue an independent research design with the Westinghouse Corporation. By a combination of willpower, the successful design of a high-temperature water reactor, and good political connections, Rickover was able to push forward the concept of a nuclear reactor system for submarines. The success of the *Nautilus*, launched on January 21, 1954, and commissioned on December 30, 1954, convinced the navy and Congress that nuclear propulsion made diesel-electric submarines obsolete.

Suggested readings: Richard G. Hewlett and Francis Duncan, *Nuclear Navy, 1946–1962* (Chicago: University of Chicago Press, 1974); Theodore Rockwell, *The*

Rickover Effect: How One Man Made a Difference (Annapolis, MD: Naval Institute Press, 1992).

Roentgen, Wilhelm Conrad

German physicist Wilhelm Roentgen's (1845–1923) discovery of X-rays started the search for the structure of the atom. He was born on March 27, 1845, in Lennep, Germany. His marriage to Anna Bertha Ludwig produced no children, but they did adopt Anna's niece. At an early age his family moved to Apeldoorn, Holland. Roentgen's education began with a mechanical engineering degree in 1868 from the Polytechnic Institute in Zurich and a Ph.D. from the University of Zurich in 1869. His academic career started in 1869 as a researcher at the University of Zurich. Roentgen moved in 1870 and became a research assistant to August Kundt at the University of Würzburg, Germany, where he stayed only two years. Next, he obtained a professorship at the University of Strasbourg in 1872. After three years, Roentgen moved to the Agricultural Academy, Hohenheim, Germany, for one academic year. Back to the University of Strasbourg in 1876, he then took his research to the University of Giessen, Germany, in 1879. Roentgen stayed at Giessen for nine years before a move to the University of Würzburg in 1888. In 1900, Roentgen made his last move to the prestigious University of Munich, where he became the director of the Institute for Experimental Physics. Roentgen remained there until his retirement in 1923.

Despite these numerous changes of position, Roentgen continued to undertake extensive research on a variety of physics topics including those in thermodynamics, mechanics, and electricity. His major contribution, however, was in his accidental discovery of X-rays on November 8, 1895. He submitted his findings about his discovery for publication in December 1895, and it caused a sensation in scientific circles. Reprints of his paper were sent out to other scientists within four days after his report to the Würzburg Physico-Medical Society on December 28, 1895. This discovery resulted in Roentgen's receiving the 1901 Nobel Prize in Physics. Roentgen lived for his life in the laboratory, and the notoriety of his discovery caused him much disquiet. He continued to conduct research but contributed only two more papers on X-rays. He died on February 10, 1923, in Munich, Germany.

Suggested readings: Henry A. Boorse, Lloyd Motz, and Jefferson Hane Weaver, *The Atomic Scientists: A Biographical History* (New York: Wiley Science Editions, 1989); John Daintith, Sarah Mitchell, Elizabeth Tootill, and Derek Gjertsen, *A Biographical Encyclopedia of Scientists*, 2nd ed., vol. 2 (Bristol, UK: Institute of Physics, 1994); Robert W. Nitske, *The Life of Wilhelm Conrad Roentgen: Discoverer of the X-ray* (Tucson: University of Arizona Press, 1971); Emilio Segre, *From X-Rays to Quarks: Modern Physicists and Their Discoveries* (San Francisco: W. H. Freeman, 1980).

Roosevelt, Franklin Delano

President Franklin Delano Roosevelt (1882–1945) made the initial decision to launch a crash program to build an atomic weapon. Alexander Sachs, an unofficial adviser to the president, approached him about the possibility of such a weapon, but it was not until the Einstein-Szilard Letter that Roosevelt gave the go-ahead. He was cautious about initial expenditures, but without his blessing little would have been done. Secretary of War Henry Stimson gave him an occasional briefing. His two foremost advisers on atomic matters, however, were his good friend the mathematician-engineer Vannevar Bush and Bush's deputy, Harvard chemist and former president of Harvard James Conant. In a meeting with Vice President Wallace and Bush on October 9, 1941, Roosevelt gave the go-ahead for the development of the atomic bomb. Roosevelt's meetings with Prime Minister Churchill were also instructive, because Churchill kept pushing him for a European component. Roosevelt agreed that the Anglo-American Alliance would share in the development and postwar use of atomic energy. This personal commitment died with the death of President Roosevelt in 1945.

Suggested readings: Bertrand Goldschmidt, *Atomic Rivals* (New Brunswick, NJ: Rutgers University Press, 1990); Kenneth D. Nichols, *The Road to Trinity* (New York: Morrow, 1987); Martin J. Sherwin, *A World Destroyed: The Atomic Bomb and the Grand Alliance* (New York: Knopf, 1975).

Roosevelt-Churchill Hyde Park Aide—Memoir

An agreement was concluded on September 19, 1944, between President Roosevelt and Prime Minister Churchill about keeping secret the development of the atomic bomb and its possible use against Japan. Plans for further postwar collaboration between the United States and the United Kingdom were also included. Finally, a provision was there to investigate the activities of Niels Bohr, the famous Danish physicist, and make sure that he passed no information about atomic matters to other countries, especially to Russia. This last stipulation was a product of Bohr's activities in trying to persuade President Roosevelt and Prime Minister Winston Churchill to inform the Soviet Union about the atomic bomb and to organize international controls on atomic weapons. He met twice with Roosevelt and once with Churchill, but it was Churchill's opposition that led to this provision in the agreement. Churchill wanted Bohr placed in protective custody, but his advisers warned him against this. Because Roosevelt never disclosed the contents of this agreement to his atomic energy advisers, Vannevar Bush and the others only had a general idea of its nature. These advisers were fearful that the agreement included Great Britain as a partner in the postwar economic and political development of atomic energy. The importance of this agreement was that it did envisage a postwar alliance between Great Britain and the United States on atomic energy. From the British point of view, it would ensure Great Britain's status as a world power. The text of this agreement disappeared after Roosevelt's death and only reappeared years later in the archives of Roosevelt's naval attaché.

Suggested readings: Bertrand Goldschmidt, *Atomic Rivals* (New Brunswick, NJ: Rutgers University Press, 1990); Martin J. Sherwin, *A World Destroyed: The Atomic Bomb and the Grand Alliance* (New York: Knopf, 1975).

Rosbaud, Paul

Paul Rosbaud (1898–1963) was an Austrian scientist who spied for the Allies during World War II, exposing deficiencies in the Nazi atomic bomb program. He was born in Graz, Austria, in 1898. His service in the Austrian Army on the Italian Front in World War I found him a prisoner of the British. After the war, Rosbaud obtained his doctorate in chemistry at the Technical University in Berlin. After a brief stint as a scientific researcher, Rosbaud became the scientific and technical adviser to the journal *Metallurgical Economy*. In 1932, he transferred to the Springer-Verlag publishing house. Rosbaud became an anti-Nazi partly because his wife was Jewish but also because he disliked Hitler and his policies. He was active in helping Jews escape from Germany and in the process developed close ties to British intelligence. Rosbaud also found a way for his wife and children to flee to England. As a publisher's agent, he travel around Germany and to neutral countries. Since he had extensive contacts in the German scientific community, Rosbaud was able to warn the British about German research. He was also the first to alert the British about rocket research at Peenemunde. He became close friends with almost all the members of the Uranium Club but especially with Walter Gerlach and Otto Hahn. Only Werner Heisenberg was suspicious of him. These contacts allowed him to advise the British that the German experimental nuclear fission program was at a standstill at the end of 1942. The Allies ignored this appraisal, but he was right. His information was so accurate that for awhile the British suspected him of being a double agent. He used book codes to transmit his information. Despite several narrow escapes from the Gestapo, Rosbaud survived World War II. He was a target of the Soviet effort to kidnap him, but an Alsos team prevented his capture. Recognizing his vulnerability, British intelligence moved him to England in 1945, where he resumed his scientific publishing career. He died in 1963.

Suggested reading: Nikolaus Riehl and Frederick Seitz, *Stalin's Captive: Nikolaus Riehl and the Soviet Race for the Bomb* (Washington, DC: American Chemical Society, 1996).

Rosenberg Case

The spy case of Julius and Ethel Rosenberg is one of the most contro-versial cases in American jurisprudence both because it involved atomic se-crets and because it resulted in the execution of the Rosenbergs for conspiracy to commit espionage. Julius Rosenberg received an electrical en-gineering degree from the City College of New York (CCNY) in 1939. He married Ethel Greenglass in 1939, and they had two sons. After losing his job as a civilian engineer in the Army Signal Corps for failure to disclose his Communist Party affiliations, he operated his own machine shop with several partners including his brother-in-law David Greenglass. This business ex-perienced financial difficulties, and hard feelings developed between Rosen-berg and Greenglass. In 1944, Rosenberg received the assignment from Soviet agents to begin spying on the United States. Until his arrest in 1949, Rosenberg recruited various agents to engage in industrial and atomic spy-ing.

The Rosenberg Case was broken as an outgrowth of the investigations resulting from intelligence from Igor Gouzenko, a Soviet cipher clerk who had defected in Canada in September 1945. After a lengthy investigation, American intelligence agents gathered enough information to arrest Harry Gold, David Greenglass, and others as participants in a major spy ring op-erated by the Soviet Union to gather data on the atomic bomb. They also arrested Julius and Ethel Rosenberg as the leaders of the spy ring. Their criminal charge was conspiracy to commit espionage, but the case was han-dled by federal prosecutors as a treason case. Moreover, the case was tried in the middle of the McCarthy era when conservatives were hunting Com-munists both inside and outside the U.S. government. Harry Gold and Da-vid Greenglass turned state's evidence for reduced prison sentences and testified against the Rosenbergs. American authorities were certain that the spy ring had more members, so they applied pressure on the Rosenbergs to reveal their identities. The Rosenbergs claimed that they were innocent of spying and refused to cooperate. As part of the pressure, the federal pros-ecutors threatened the Rosenbergs with capital punishment. Neither side broke the other, and the Rosenbergs were executed after a controversial trial.

The controversy over the guilt or innocence of the Rosenbergs started soon after the trial ended. Two prominent critics of the sentence were Albert Einstein and Harold Urey. These critics charged that the Rosenbergs were either innocent or the death penalty was too severe a sentence for the crime. It soon became an international cause, with demonstrations in support of the Rosenbergs conducted around the world. Even after the execution of the Rosenbergs the case was widely debated among critics and supporters of the verdict. As late as 1991 the American Bar Association conducted a mock trial of the Rosenbergs, and the mock jury acquitted the Rosenbergs.

After almost 50 years after the event, it is still difficult to be objective about the Rosenberg Case. Certain facts have emerged about the case that cast new light on both sides of the issue. Julius Rosenberg was undoubtedly a spy for the Soviet Union, but it is by no means certain that he was the leader of the spy ring or handled atomic secrets. Ethel Rosenberg was a minor operative, and her complicity was never proven. In retrospect, Klaus Fuchs was the key figure in the Soviet spy ring in the United States, and he received a sentence in British courts of 14 years. Harry Gold played the important role as the courier, and he was given 30 years in a plea bargain. Both David Greenglass and Ruth Greenglass contributed more to the passing of atomic secrets than either of the Rosenbergs, but when they turned state's evidence, they escaped the death sentence.

Suggested readings: Marjorie B. Garber and Rebecca L. Walkowitz, eds., *Secret Agents: The Rosenberg Case, McCarthyism and Fifties America* (New York: Routledge, 1995); H. Montgomery Hyde, *The Atom Bomb Spies* (New York: Atheneum, 1980); Louis Nizer, *The Implosion Conspiracy* (Garden City, NY: Doubleday, 1973); Ronald Radosh, *The Rosenberg File: A Search for the Truth* (New York: Holt, Rinehart and Winston, 1983).

Rotblat, Joseph

Joseph Rotblat (1908–   ) was an early critic of nuclear policies. He was born on November 4, 1908, in Warsaw, Poland. His education was at the Free University of Poland, where he received a master's in 1932. After graduation, Rotblat worked as a researcher at the University of Warsaw and obtained his doctorate in physics in 1938. In 1939, Rotblat emigrated to Great Britain on a year's scholarship. He found a position at Liverpool University working on James Chadwick's cyclotron. He was recruited to be on the British Manhattan Project team to build the atomic bomb, but before it was finished, he resigned in protest against the possible use on human targets. Discouraged by the application of atomic physics for weapons research, Rotblat left his research on these subjects and turned his attention to using physics on medical research. In 1950, he left Liverpool for the University of London. He was a physics professor at St. Bartholomew's Hospital in London when he challenged the accuracy of the American Atomic Energy Commission's version of the fallout from the Bravo Nuclear Test at Bikini in March 1954. His discovery that the nuclear device at Bikini was a fission-fusion-fission explosion, making it a much dirtier bomb than had been advertised, was a key finding. Rotblat published his evidence in both the *Journal of Atomic Scientists* and the *Bulletin of the Atomic Scientists*. Rotblat was an early participant in the Pugwash movement, becoming one of its leaders. His antinuclear activities led to his sharing the 1995 Nobel Peace Prize with the Pugwash Conferences on Science and World Affairs for his opposition to nuclear policies.

Suggested reading: Joseph Rotblat, *Pugwash—The First Ten Years: History of the Conferences of Science and World Affairs* (New York: Humanities Press, 1967).

Rutherford, Ernest

Sir Ernest Rutherford (1871–1937) was one of the giants in the field of nuclear research in the early twentieth century, and his leadership of the Cavendish Laboratory at Cambridge University made it one of the great centers of experimental nuclear physics. He was born on August 30, 1871, in Spring Grove, New Zealand, the fourth of 12 children. His father was a Scottish immigrant to New Zealand, who was a combination farmer, wheelwright, and miller, and his mother was a schoolteacher. His marriage to Mary Georgina Newton in 1900 produced one daughter. Rutherford had his early academic training at Nelson College and Canterbury College, University of New Zealand by winning academic scholarships. He received his B.A. in 1892, M.A. in 1893, and B.Sc. in 1894. In 1895, he won a scholarship to Cambridge for original research on the magnetization of iron by high-frequency discharges. At Cambridge, he worked on conductivity of air ionized by X-rays, and in 1899 he identified and named alpha and beta radiation. An alpha particle is a positively charged helium nucleus made up of two neutrons and two protons bound together, and a beta particle is either an electron or a positron emitted from a nucleus during radioactive decay. The discovery of these particles was a major advance in understanding the nature of radioactivity. Rutherford obtained a teaching and research position at McGill University, Montreal, Canada, when a Canadian tobacco merchant provided the funds for a laboratory and endowed professorships. Also at McGill was chemist Frederick Soddy, and Rutherford benefited from working with him on radioactivity studies. Together they discovered that each radioactive species had a half-life that could be calculated. It was this research on the chemistry of radioactive substances that won him the 1908 Nobel Prize in Chemistry. After nine productive years at McGill, Rutherford left for the University of Manchester, England. It was at Manchester in 1911 that Rutherford analyzed the results of an experiment by Ernest Marsden and Hans Geiger and concluded that the atom had a compact, massive nucleus. He stayed at Manchester until 1919, when he made his last move, this time to Cambridge University. He took over the directorship of the Cavendish Laboratory from the distinguished physicist and fellow Nobel Laureate Sir Joseph John Thomson.

Rutherford's administration and experiments at the Cavendish Laboratory made it the leading experimental research center in the world. C. P. Snow claimed that Rutherford was the only competitor to Michael Faraday as the greatest experimental physicist of all time. Rutherford had an outgoing and enthusiastic personality, but he always had trouble adjusting to the English class structure. His prestige and experiments, however, attracted the cream of European scientific talent. Moreover, his open personality encouraged scientists to take chances. He ended up training no less than 11 Nobel Prize winners. His daily tours through the Laboratory to talk with the researchers conducting experiments helped build morale. One difficulty was that his booming voice sometimes disturbed sensitive experimental equipment, so signs for quiet were displayed prominently in the laboratory. However, these warnings seemed to have little impact on Rutherford. He was such a dedicated experimental scientist that he always appreciated other scientists who conducted first-rate research, especially Madame Curie.

Rutherford's greatness as an experimental scientist was typified in the experiments that proved the existence of the atomic nucleus. In his experiment a sheet of gold foil intercepted a beam of alpha particles, causing some of them to bounce back. Rutherford postulated that these alpha particles were hitting a positively charged nucleus at the center of the atom. He died on October 19, 1937, following surgery to deal with complications from a fall while trimming a tree.

Suggested readings: Edward Andrade, *Rutherford and the Nature of the Atom* (Garden City, NY: Doubleday, 1964); Henry A. Boorse, Lloyd Motz, and Jefferson Hane Weaver, *The Atomic Scientists: A Biographical History* (New York: Wiley Science Editions, 1989); John Rowland, *Ernest Rutherford, Atom Pioneer* (New York: Philosophical Library, 1957); C. P. Snow, *The Physicists* (Boston, MA: Little, Brown, 1981); David Wilson, *Rutherford, Simple Genius* (Cambridge: MIT Press, 1983).

S

Sakharov, Andrei

Andrei Sakharov (1921–1989) was a Soviet physicist who led in developing the hydrogen bomb for the Soviet Union and was later a famous political dissident. He was born on May 21, 1921, in Moscow. His father was a professor of physics. Sakharov was married twice: to Klavdia Alekseyevna Vikhireva (1943), a chemist, and Elena Georgievna Bonner (1972), a pediatrician and political dissident. In 1938 Sakharov started his studies at the Moscow State University. During the fall of 1941, he evacuated with his school to Ashkhabad in Middle Asia, to escape the war. After graduating from Moscow State University in 1942, Sakharov worked for the next 3 years in a war production plant as an engineer. Sakharov wrote several papers on theoretical physics and sent them to Igor Evgenievich Tamm. Tamm, of the Physics Institute of the Academy of Science (FIAN) subsequently recruited Sakharov in 1945 as a postgraduate student. Sakharov's doctorate was delayed because he failed the ideological examination required before the defense of his thesis, but he was able to defend his dissertation in November 1947. In 1948 Tamm invited him to work with a group of young scientists on the hydrogen bomb. Working at a secret laboratory in Turkmeniya, an area close to the Iranian border, he spent the next 18 years designing the Soviet hydrogen bomb. Others were working on aspects of the hydrogen bomb problem, but Sakharov came up with an alternative design and called it "First Idea." He characterized it as a layer cake—alternating layers of light elements with heavy elements. Tamm accepted this idea immediately, as did Zeldovich at Arzamas-16 (Sarov). Beria ordered Sakharov to continue his work at Arzamas-16. In 1957, Sakharov began to question his responsibility for the problem of radioactive contamination from nuclear explosions. The Soviet Union resumed testing of nuclear weapons in 1961, and Sakharov protested to his superiors and later to Prime Minister Khrushchev. As his protests continued, Sakharov became more isolated until in 1968 he was taken off weapons work. In 1969, Tamm asked Sakharov to come back to the Lebedev Physical Institute in Moscow. Sakharov's activities led to his receiving the 1975 Nobel Peace Prize. His political agitation about the Afghanistan War led to the Soviet government exiling him to Gorki in 1980. While at Gorki, Sakharov used hunger strikes four times as a means of drawing world attention to injustices. He finally left Gorki after Mikhail Gorbachev gave him his freedom to return to Moscow in December 1986. He died on December 14, 1989, in Moscow.

Suggested reading: Andrei Sakharov, Sidney D. Drell, and Sergei P. Kapitza, eds., *Sakharov Remembered: A Tribute by Friends and Colleagues* (New York: American Institute of Physics, 1991).

Salt I. *See* Strategic Arms Limitation Treaty I

Salt II. *See* Strategic Arms Limitation Treaty II

Sandia National Laboratories

The Sandia National Laboratories comprise a series of institutes that make this facility the largest nuclear weapons facility in the United States. Their mission has been to take nuclear devices and incorporate them into operational weapons. These institutes are located on a part of Kirkland Air Force Base, near Albuquerque, New Mexico. Sandia began as the ordnance section, or Z Division, of the Los Alamos Scientific Laboratory of the Manhattan Project, but in 1945 it was established as a separate entity. Sandia was under the administrative control of the Los Alamos National Laboratory until President Harry Truman asked in 1949 for American Telephone and Telegraph (AT&T) to assume management control. AT&T took over management functions on November 1, 1949. At first the Atomic Energy Commission, now the Department of Energy (DOE), had administrative control, but AT&T (formerly Bell Laboratories) remained operating contractor until Lockheed Martin won the contract on October 1, 1993. Sandia has two primary facilities, one in Albuquerque, New Mexico, and one in Livermore, California.

Sandia's role in the atomic bomb program was to be in charge of ordnance design, testing, and assembly. The atomic bombs developed at the Los Alamos Laboratory had been handmade and no arrangement for production in series had been considered. Once the three bombs (one for testing and the other two detonated over Japanese cities) had been used, the American arsenal was depleted. The Sandia Laboratories were established to rectify this deficiency by having facilities develop and make advanced atomic weapons in quantity. It was at Sandia where the triggering mechanism for the hydrogen bomb was designed and put into production. While the Sandia Laboratories still have a role in nuclear weapons manufacture, in the last decade much of their activity has been devoted to environmental and energy projects. Among their private enterprise projects have been a smaller passenger airbag for American automobiles and a clear room for burn victims and the microchip industry. By 1995 the Sandia National Laboratories had a budget of $1.4 billion and a staff of 8,500 employees. In 1999, both the size (8,000 people) and budget ($1.2 billion) was below the 1995 totals, but Sandia remains the largest of the national laboratories. A recent initiative on revitalizing the nation's nuclear stockpile by computer and laser testing ensures that Sandia will continue to be a vital part of the national defense infrastructure.

Suggested readings: Necah Stewart Furman, *Sandia National Laboratories: The Postwar Decade* (Albuquerque: University of New Mexico Press, 1990); Debra Rosenthal, *At the Heart of the Bomb: The Dangerous Allure of Weapons Work* (Reading, MA: Addison-Wesley, 1990).

Sandstone Nuclear Tests

Sandstone was the code name for the series of nuclear tests conducted by the United States in the Pacific during 1948 designed to test new bomb designs. Scientists at the Los Alamos National Laboratory lobbied for a series of atomic weapons tests because plutonium was so scarce that they needed more data to make more efficient atomic bombs. Operating under the designation of Joint Task Force 7, the personnel involved in the operation numbered over 10,000. Three detonations took place between April and May 15, 1948, on the islands of Enjebi, Aomon, and Runit. These tests were named Shot X-Ray, Shot Yoke, and Shot Zebra. Only in Shot Zebra did difficulties surface when several radiochemistry group members became overexposed to radiation and suffered beta-burn symptoms that caused extensive plastic surgery to their hands. By the end of these tests, the United States had its first production model nuclear weapon, the MK-IV. This series ended the first phase of Pacific nuclear tests. Political pressure was placed on the administrators to find a less expensive, less distant, and more secure testing area, so most of the routine testing was transferred to the Nevada Test Site. It was three years before testing resumed at the Eniwetok Atoll.

Suggested readings: Barton C. Hacker, *Elements of Controversy: The Atomic Energy Commission and Radiation Safety in Nuclear Weapons Testing, 1947–1974* (Berkeley: University of California Press, 1994); Richard L. Miller, *Under the Cloud: The Decades of Nuclear Testing* (New York: Free Press, 1986).

SANE

SANE, or the Committee for a Sane Nuclear Policy, was founded in June 1957 by 27 prominent American leaders with the intention to protest against further nuclear testing. At this June 21, 1957, meeting at the Overseas Press Club a cross section of American intellectual elite attended—authors, businessmen, churchmen, labor leaders, public figures, and scientists. A broad-based national committee of 17 members was formed, with Trevor Thomas as temporary executive secretary. SANE's first public act was to place a full-length newspaper advertisement in the *New York Times* on November 18, 1957, entitled "We Are Facing a Danger Unlike Any Danger That Has Ever Existed." Reaction from the national media was muted, but the public response was so great that it launched SANE as a national movement. SANE was never envisaged as a mass movement by its founders, but now it found itself one. By the summer of 1958, SANE had around 130 chapters, with a total membership of nearly 25,000. As SANE grew, it developed more political clout and was instrumental in pressuring the U.S. government into a moratorium on nuclear tests. Just when it seemed that SANE would lead a national antinuclear campaign, it became embroiled in Communist infiltration charges from the Dodd Committee of the U.S. Congress. After considerable turmoil and loss of support over the fallout from this issue, SANE survived to return as a national force in the attempt to control nuclear weapons. Dr. Benjamin Spock was recruited to serve as a national sponsor to regain some of its earlier popularity. Leaders of SANE played an active role in the negotiations and the final passage of the Nuclear Test Ban Treaty in 1963. For the next 10 years, however, SANE was active in anti-Vietnam War activities and only returned to antinuclear weapons lobbying in the mid-1970s. Members campaigned against the MX Missile System and the development of the neutron bomb in the late 1970s and early 1980s. During the Reagan administration, SANE policy was to work for a nuclear freeze and nuclear disarmament, and the organization prospered around these issues in the 1980s. In 1985, SANE members organized protest at the Nevada Nuclear Test Site against nuclear weapons testing. Negotiations began between the leadership of SANE and the nuclear weapons Freeze Movement and in late 1985 a merger of the two was concluded. A restructured organization was called SANE/FREEZE. This new organization became active in promoting arms control and disarmament and formed an opposition to the Gulf War. In 1993, SANE/FREEZE changed it name again, this time to Peace Action. Major campaigns of Peace Action in the late 1990s remain to close nuclear weapons laboratories, abolish nuclear weapons, and weapon proliferation, cut the military defense budget, and promote a peace agenda. Peace Action, with its SANE roots, is the most active and long-lasting of the antinuclear organizations.

Suggested reading: Milton S. Katz, *Ban the Bomb: A History of SANE, the Committee for a Sane Nuclear Policy, 1957–1985* (Wesport, CT: Greenwood Press, 1986).

Sarov. *See* Arzamas-16

Savannah River Project

The Savannah River Project is the site of a plutonium-tritium processing facility that is located along the Savannah River in southeastern South Carolina. The necessity of providing a greater supply of plutonium-tritium to build hydrogen bombs prompted the Atomic Energy Commission (AEC) to look for a site to support the Hanford facility in Washington. After examining 17 potential sites, the AEC selected a site on the Savannah River near Aiken, South Carolina. In 1951 President Truman asked the Du Pont Corporation, at a cost of $200 million, to design, construct, and operate a plutonium production complex to support the hydrogen bomb program. Du Pont accepted and operated the complex until 1988. A. E. Church of Du Pont headed the design team. Colonel Curtis Nelson was given the assignment in 1951 to build five heavy water–moderated reactors modeled on the Canadian reactors at Chalk River to produce plutonium for nuclear weapons. Basic design of the reactors, however, came from the heavy water reactor type researched by Walter Zinn at the Argonne National Laboratory. The first reactor (R Reactor) was started in June 1951 and in operation in July 1953. Others (C, K, L, and P Reactors) took longer to come on-line. As the demand for plutonium dropped in the 1960s, reactors L and P shut down in 1964 and 1968, respectively, and K is on standby in case of national emergency. The Savannah site was also a producer of tritium for the hydrogen bomb program. A crisis developed in the late 1980s when the tritium-producing reactors were shut down for safety checks. In 1985 the L Reactor was restarted for plutonium and tritium production to solve this crisis, and it ran until 1988.

Now the Savannah River Project is a site for research on nuclear waste disposal. Westinghouse Government Services was given a contract in the early 1990s to develop, a technology to separate highly radioactive material from less radioactive liquids, but this process has been deemed too dangerous because of the explosive benzene gas. The government pulled the contract from Westinghouse in January 1998, but no substitute method for handling the radioactive waste has been devised yet. An alternative method will have to be designed and the General Accounting Office of the U.S. Congress estimates the development time from eight to ten years with a cost from $1 billion to $3.5 billion.

Suggested readings: Rodney P. Carlisle, *Supplying the Nuclear Arsenal: American Production Reactors, 1942–1992* (Baltimore, MD: John Hopkins University Press,

1996); Richard Wolfson, *Nuclear Choices: A Citizen's Guide to Nuclear Technology* (Cambridge: MIT Press, 1991).

Schaffer Report

The Schaffer Report in 1980 was the admission by a U.S. government task force that some people had died from radiation-induced cancers as a result of fallout from nuclear testing at the Nevada Test Site. William G. Schaffer, the deputy assistant attorney general for the Civil Division in the Carter administration, was the chairperson of the Task Force on Compensation for Radiation-Related Illness. This task force had 13 members and was drawn from various executive agencies, with six from the Departments of Energy and Defense. In a series of meetings in the fall of 1979, the task force wrote a report in which it conceded that Americans had died from radiation-induced cancers from fallout from nuclear testing from 1951 to 1962. Despite this admission, it recommended guidelines that severely limited the number of possible claimants by establishing a minimum probability requirement for comprehension. This 57-page report was never published, nor were its conclusions widely publicized. It was sent to President Carter on February 1, 1980, and had not been acted upon when President Carter left office. The pro-nuclear orientation of the Reagan administration meant that nothing further was done about implementing provisions of the Schaffer Report.

Suggested reading: Philip L. Fradkin, *Fallout: An American Nuclear Tragedy* (Tucson: University of Arizona Press, 1989).

Schrödinger, Erwin

Erwin Schrödinger (1887–1961) was a German physicist who accepted de Broglie's wave hypothesis and constructed the theory that reconciled it with quantum mechanics. He was born in Vienna, Austria, on August 12, 1887. His father was a prosperous factory owner in Austria, and his mother was English. He married Annemarie Bertel in 1920, but they had no children. Schrödinger received his undergraduate training and his Ph.D. in 1910 from the University of Vienna. Friedrich Hasenhörl was his faculty adviser. His first research post was at the University of Vienna in 1910. He left Vienna in 1914 to serve in the Austrian Army as an artillery officer on the Southwest Front. After the war, his next research position was at the University of Stuttgart in 1920. He had short stints at the universities of Jena and Breslau before landing a post at the University of Zurich. In 1927, Schrödinger moved to the University of Berlin, replacing Max Planck as professor of physics.

Schrödinger made his reputation by developing wave mechanics. He had not been satisfied with the quantum theory, but hearing about Prince de Broglie's wave hypothesis changed his mind. Finding de Broglie's wave relations too simplistic, Schrödinger constructed a theory that refined the underlying ideas. These mathematical equations formed the basis of wave mechanics. His version met with immediate and widespread acceptance because his mathematics were of a type familiar to most physicists. Moreover, his methods were more easily applied to obtain predictions for experimental results than was Heisenberg's quantum mechanics.

Schrödinger's opposition to Hitler made it impossible for him to stay in Germany. His scientific reputation enabled him to become a researcher at Oxford University in 1933. Shortly after leaving Germany, he received the news that the Nobel Committee had given him the 1933 Nobel Prize for Physics. He was honored for his research on wave mechanics. Returning to Austria in 1936, Schrödinger obtained a job at the University of Graz. The Nazi coup against Austria in 1938 made it imperative for Schrödinger to leave again. He escaped on foot with nothing but a few foodstuffs and made it to Rome, where he requested help from Enrico Fermi to escape. After one year at the Fondation Francqui, Belgium, and another year at the Royal Irish Academy, Ireland, he became a professor at the Dublin Institute for Advanced Studies, where he remained until 1956. His last two years were spent at the University of Vienna, Austria, before retiring in 1958. He died in Alpbach, Austria, on January 4, 1961.

Suggested readings: Henry A. Boorse, Lloyd Motz, and Jefferson Hane Weaver, *The Atomic Scientists: A Biographical History* (New York: Wiley Science Editions, 1989); John Daintith, Sarah Mitchell, Elizabeth Tootill, and Derek Gjertsen, *A Biographical Encyclopedia of Scientists*, 2nd ed., vol. 2 (Bristol, UK: Institute of Physics, 1994); Emilio Segre, *From X-Rays to Quarks: Modern Physicists and Their Discoveries* (San Francisco: W. H. Freeman, 1980).

Seaborg, Glenn Theodore

Glenn Seaborg (1912–1999) was an American chemist-physicist who is most famous for his discovery of plutonium and for his championing of nuclear energy. He was born on April 19, 1912, at Ishpeming, Michigan. His parents were Swedish immigrants who moved to southern California when he was a boy. He married Helen Griggs in 1942. His undergraduate academic training was in chemistry at the University of California at Los Angeles (UCLA). In the summers he worked at various jobs, including stevedore, apricot picker, and apprentice Linotype machinist. He graduated with a degree in chemistry from UCLA in 1934 and obtained his doctorate in chemistry from UCLA in 1937. He found a job at UCLA as a personal laboratory assistant to chemist G. N. Lewis. While in this position, Seaborg started showing interest in research on the elements beyond uranium, called transuranium elements. In 1945, he became a full professor at Berkeley.

Glenn T. Seaborg played two important roles in the history of atomic energy. He was awarded the 1951 Nobel Prize for Physics for his research contributions in the discovery of plutonium-239. He was also the chair of the Atomic Energy Commission during the key period from 1962 to 1971.

Seaborg worked with Edwin McMillan in 1940 to continue Enrico Fermi's earlier research on transuranium elements. When McMillan left California–Berkeley to conduct war research on radar at the Massachusetts Institute of Technology, Seaborg replaced him. McMillan's team already knew that bombarding uranium nuclei with neutrons would cause its transmutation into neptunium (element 93). Seaborg continued along the same line of research and soon discovered the next element (element 94), which he named plutonium. Since plutonium was as fissionable as uranium, it was a promising material for use in a weapon. Seaborg and McMillan shared the 1951 Nobel Prize in Chemistry for these discoveries. During World War II, Seaborg was on leave to work as head of the University of Chicago's Metallurgical Laboratory's Chemical-Separation Section. He recommended the use of plutonium for one of the atomic bombs. He was also one of the signers of the Franck Report, which recommended that the atomic bomb be demonstrated, rather than detonated over a city.

For the rest of his research career, Seaborg investigated transuranium element and assisted in the discover of several new elements. Later, Seaborg turned to high-level administration. He became the chancellor of the University of California in 1958 and remained until 1962, when he was appointed chair of the Atomic Energy Commission (AEC). Seaborg retained this position until 1971. In this capacity, Seaborg championed the uses of nuclear energy both for weapons development and for peaceful uses. He considered himself the number-one salesperson for the benefits of nuclear energy. Seaborg died on February 25, 1999.

Suggested readings: John Daintith, Sarah Mitchell, Elizabeth Tootill, and Derek Gjertsen, *A Biographical Encyclopedia of Scientists*, 2nd ed., vol. 2 (Bristol, UK: Institute of Physics, 1994); Daniel F. Ford, *The Cult of the Atom: The Secret Papers of the Atomic Energy Commission* (New York: Simon and Schuster, 1982).

Seaborg Report

The Seaborg Report was an economic status report prepared by Glenn Seaborg, the head of the Atomic Energy Commission (AEC), on atomic energy's future prospects, requested by President John Kennedy in March 1962. President Kennedy had become nervous about the high cost of power generation from nuclear energy facilities and wanted reassurance about the future. Seaborg and the AEC staff presented the report, which was entitled *Civilian Nuclear Power: A Report to the President 1962*. In this report, Seaborg assured the president that breakthroughs in technology and larger nuclear plants would make nuclear energy more affordable and be competitive with other forms of energy. He argued to continue the federal program subsidizing nuclear power development to enable this to happen. The report made the prediction that half of the United States' electrical power would come from nuclear plants by the year 2000. This optimistic report served as the basic philosophy of the AEC for the rest of its organizational life.

Suggested reading: Daniel F. Ford, *The Cult of the Atom: The Secret Papers of the Atomic Energy Commission* (New York: Simon and Schuster, 1982).

Seabrook Nuclear Power Plant

The Seabrook Nuclear Power Plant was the scene of some of the largest antinuclear demonstrations in American history during the 1970s. In 1969, the Public Service Company of New Hampshire bought land near the town of Seabrook in southeastern New Hampshire to construct a nuclear power plant with twin nuclear reactors. In the hearings before the New Hampshire Site Evaluation Committee in 1972, significant opposition from environmental groups was heard. After a year-long hearing, the committee gave the Public Service Company certification to build the plant. Objections from environmentalists continued during the Atomic Energy Commission hearings in 1975. On July 9, 1976, the Nuclear Regulatory Commission (NRC) issued a permit for construction. Efforts to raise utility rates to pay for construction costs produced political controversy and helped defeat New Hampshire Governor Meldrim Thomson in a 1978 governor's race. Protests escalated until in July 1976 the Clamshell Alliance was formed to fight against Seabrook Nuclear Power Plant. Despite these and other protests, the Seabrook Nuclear Power Plant is still in operation in the late 1990s.

Suggested readings: Henry Bedford, *Seabrook Station: Citizen Politics and Nuclear Power* (Amherst: University of Massachusetts Press, 1990); Etahn Cohen, *Ideology, Interest Group Formation, and the New Left: The Case of the Clamshell Alliance* (New York: Garland Publishing, 1988); Donald Stever, *Seabrook and the Nuclear Regulatory Commission: The Licensing of a Nuclear Power Plant* (Hanover, NH: University Press of New England, 1980).

Semipalatinsk-21 Test Site

The Semipalatinsk-21 Test Site was where most of the Soviet nuclear tests took place. In August 1947, Igor Kurchatov, the scientific director of the Soviet Nuclear Program, selected a site about 100 miles west of the city of Semipalatinsk, in Kazakhstan, to test atomic weapons. Facilities were soon built, and by August 10, 1949, the test site was ready for testing to begin. A state committee with Lavrentii Beria as chair was formed to monitor the test. On August 29 the first Soviet atomic device was detonated with a yield of 20 kilotons. A total of 470 nuclear tests were conducted at the Semipalatinsk Test Site—348 tests underground, 215 horizontally emplaced, and 133 vertically emplaced. Most of the tests were at Semipalatinsk-21 near Degelen Mountain. The last test was conducted on October 19, 1989. The president of the newly independent state of Kazakhstan, Nursultan A. Nazarbayev, closed the test site in August 1991. Because of the remoteness of the site and the sparse population in the region, incidences of radioactive fallout affecting the health of the local population have been relatively few. The new government of Kazakhstan, however, has displayed no interest in allowing the Russian government to reclaim the site for further nuclear testing.

Suggested readings: Thomas B. Cochran, Robert S. Norris, and Oleg A. Bukharin, *Making the Russian Bomb: From Stalin to Yeltsin* (Boulder, CO: Westview Press, 1995); David Holloway, *Stalin and the Bomb: The Soviet Union and Atomic Energy, 1939–1956* (New Haven, CT: Yale University Press, 1994); Richard Rhodes, *Dark Sun: The Making of the Hydrogen Bomb* (New York: Touchstone, 1995).

Shinkolobwe Uranium Mine

The Shinkolobwe Mine in the Katanga region of the Belgian Congo was one of only three known sources of uranium on the eve of World War I. Major Richard Sharp, a British citizen, discovered the uranium deposits in August 1915 during World War I when he was prospecting for copper and silver. It proved to be the richest deposit of uranium ever discovered, with an average of 68 percent pure uranium. A Belgian mining company, Union Miniere du Haut Katanga, ran the mine and by 1921 had captured a virtual monopoly on the international market, which it held until the Great Bear Lake Mine became its chief competitor after 1928. Ore was transported to a processing center in Oolen, which was less than 15 miles east of Antwerp, and converted it to radium at a ratio of three tons of uranium ore to a gram of radium. On the eve of World War II the mine was closed down because of a step decline in the price of radium. Edgar Sengier was a director of the Bank of Belgium and chair of the executive committee of the Union Miniere du Haut Katanga, which owned the Shinkolobwe Mine. Sengier received warnings about the potential value of the uranium ore from Sir Henry Tizard, a British scientist. The British government offered to buy the existing ore, but Sengier turned down all offers. When the Germans gained control of Belgium in 1940, they seized available uranium stores in Belgium from the Shinkolobwe Mine. Their efforts to gain control of the Shinkolobwe Mine were unsuccessful, and the mine remained closed down. Sengier, however, had shipped 1,200 tons of uranium ore to the United States in 1940 and stored it in steel drums in the Staten Island Warehouse. Sengier had also moved to the United States and in December 1941 offered the ore to the U.S. government. A representative from the military, Colonel Kenneth Nichols, brought the ore for $1.60 a pound, to be shipped immediately. Arrangements were made to contract for all uranium ore stockpiles above the ground in the Congo. The ore was then sent to the Port Hope refinery in Canada before returning to the United States. This Shinkolobwe ore helped provide enough uranium ore for the Manhattan Project. Uranium ore from this mine was high-grade, some of the richest ever found. After the war, it has been estimated that from 1945 to 1953 85 percent of the total of U.S. supplies of uranium came from the Shinkolobwe Mines. Between 1944 and 1960 the mine supplied the United States and Great Britain with over 30,000 tons of uranium. Increased uranium production and political troubles in Katanga caused the Shinkolobwe Mine to close down in 1961.

Suggested readings: Lennard Bickel, *The Deadly Element: The Story of Uranium* (New York: Stein and Day, 1979); Earle Gray, *The Great Uranium Cartel* (Toronto, Ontario: McClelland and Stewart, 1982); Kenneth D. Nichols, *The Road To Trinity* (New York: Morrow, 1987).

The first American civilian nuclear power plant was the Shippingport Nuclear Power Plant in Shippingport, Pennsylvania. This picture shows the plant at completion in October 1957.

Shippingport Nuclear Power Plant

The first civilian nuclear reactor was built at Shippingport, Pennsylvania. Admiral Hyman Rickover had intended to have a reactor built for a nuclear-powered aircraft carrier engine, but the Eisenhower administration vetoed this project. Rickover then proposed to the Atomic Energy Commission (AEC) and the Joint Committee on Atomic Energy (JCAE) that such a reactor be used for a civil power station, and permission was forthcoming. The AEC advertised for proposals, and the Duquesne Light Company of Pittsburgh, Pennsylvania, had the best offer. This company's bid included the site, a turbine generator plant, and maintenance and operation of the facility at no cost to the government. It also proposed adding $5 million to designing and building the reactor. A contract was signed with Westinghouse Corporation on October 9, 1953, and construction was started on September 6, 1954. The reactor was a light water reactor using reactor fuel elements made of uranium oxide clad in zirconium-alloy tubes. On October 6, 1957, the first reactor core was installed, and by December 18, the first electricity was delivered to the Pittsburgh and surrounding area by the Duquesne Light Company. Although the power plant was budgeted to cost $70 million, its actual cost was about $100 million. In the mid-1980s the Shippingport reactor was decommissioned, and it served as a prototype for decommission of nuclear plants.

Suggested readings: Richard G. Hewlett and Francis Duncan, *Nuclear Navy, 1946–1962* (Chicago: University of Chicago Press, 1974); John W. Simpson, *Nuclear Power from Underseas to Outer Space* (La Grange Park, IL: American Nuclear Society, 1995).

Shot Easy

Shot Easy was an atomic test at the Nevada Test Site on May 7, 1952, that showed for the first time the dangers of radioactive fallout in the United States. The radioactive cloud moved over a small mining community of Tempiute, where 110 civilians were exposed. Then the cloud drifted north over Ely, Nevada, before heading northeast, where it passed over Salt Lake City, Utah. Geiger counters in Salt Lake City registered unusual activity, and the governor of Utah was notified. He wrote an angry letter to the Atomic Energy Commission (AEC) about the incident. The AEC was able to quell any publicity about the radioactive cloud and ignored a rancher's demand for compensation for injured cattle. Despite the outward display of unconcern, key figures in the AEC considered the incident serious enough to commission a study on the suitability of the Nevada Test Site for future atomic testing. Norris Bradbury, the director at Los Alamos, appeared before the committee studying the problem and insisted that there was no alternative to the Nevada Test Site. His argument carried the day.

Suggested reading: Philip L. Fradkin, *Fallout: An American Nuclear Tragedy* (Tucson: University of Arizona Press, 1989).

Shot Harry

Shot Harry was the code name for a test of a hydrogen bomb on May 19, 1953, at the Nevada Test Site that spread radioactive fallout in the southwest Utah region. It was the ninth in a series of 11 tests in Operation Upshot-Knothole. The test produced the worst case of fallout descending on a concentrated population in the history of the United States. Although the Atomic Energy Commission (AEC) assured that those in the fallout area had not been exposed to dangerous levels of radiation, the readings indicated otherwise. No warnings or precautions had been taken, so a large portion of the inhabitants of Alamo, Nevada, and St. George, Utah, and surrounding countryside received a high dose of radiation from fallout. Scientists had learned about the danger of strontium-90, and they wanted to test milk for it, but the study was never performed because the AEC feared alarming the public. In an effort to reassure the public, a movie, an idealized version of the events on the day of Shot Harry, was made in the fall of 1953. This movie made the AEC and its personnel look good, and it was widely distributed. A lawsuit was filed on August 30, 1979, *Allen v. United States of America*, over the high incidence of cancer and birth defects in this region since the bomb tests. After a victory in the local courts for the plaintiffs, higher courts ruled that the government had sovereign immunity despite evidence of misconduct and neglect.

Suggested reading: Philip L. Fradkin, *Fallout: An American Nuclear Tragedy* (Tucson: University of Arizona Press, 1989).

Sierra Club

The Sierra Club is a California-based conservation group that has been active in campaigns against nuclear projects. While the Sierra Club had long been a leader in the American environmental movement, it was not until its board of directors voted in 1974 to oppose the building of new nuclear power plants that the club took a stand on nuclear energy. Most of the concern by members of the Sierra Club was more over technological problems of the nuclear plants than outright opposition to nuclear energy. Membership had long been in opposition to coal burning energy plants, so nuclear energy had some advantages over the older technology. Nevertheless, this conversion to the conservative wing of the antinuclear movement by the Sierra Club was important because it had a large and active membership and access to a large number of financial supporters. After early victories in the antinuclear campaign in California, a schism over tactics developed in the Sierra Club. A leadership change led to the Sierra Club becoming less active in the antinuclear movement and returning to animal conservation issues. While the Sierra Club was active fighting against nuclear plants, its clout among California politicians and on the national scene made it a formidable foe.

Suggested reading: Jerome Price, *The Antinuclear Movement* (Boston, MA: Twayne, 1982).

Silkwood Case

Karen Silkwood was union activist at Kerr-McGee's nuclear fuel plant in Oklahoma. After exposure to plutonium on three occasions, she protested health and safety conditions at the Kerr-McGee plant. The company retaliated with a harassment campaign to discredit Silkwood. She died in an automobile accident on November 13, 1974, on the way to deliver incriminating documents to journalists. Some investigators found the accident suspicious, especially after the documents were found missing. A jury levied a fine on Kerr-McGee of $10.5 million in punitive damages for disregard for her safety and for harassment in a trial held in May 1979. This award was dismissed in an appeal in 1981 on the grounds that state law had infringed on federal control of the nuclear industry. The U.S. Supreme Court, however, reinstated the case in 1984. A settlement of the lawsuit between the Silkwood family and Kerr-McGee was concluded on August 22, 1986, for $1.38 million. This case contributed to the growing disillusionment about nuclear energy by a significant segment of the American population. Karen Silkwood's death also brought the National Organization of Women (NOW) into the antinuclear movement, and it gave the antinuclear movement a martyr. A spinoff from NOW was the Supporters of Silkwood group formed in the mid-1970s.

Suggested readings: Howard Kohn, *Who Killed Karen Silkwood?* (New York: Summit Books, 1981); John May, *The Greenpeace Book of the Nuclear Age: The Hidden History, the Hidden Cost* (New York: Pantheon Books, 1989); Mike Nicholas, *Silkwood*, ABC Motion Pictures, Los Angeles, 1984; Jerome Price, *The Antinuclear Movement* (Boston, MA: Twayne, 1982).

SL-1 Reactor Accident

One of the test reactors, SL-1 (Stationary Low-Power), at the National Reactor Testing Station (NRTS) in Idaho Falls, Idaho, had an accident on January 3, 1961, and three nuclear technicians died. The reactor had been acting up, so the three technicians entered the reactor area to do maintenance. One of the standard operations was to lift a control rod four inches, but somehow the rod was lifted too far. In a millionth of a second, the reactor became critical. The reactor building was so radioactive that fire and safety personnel had difficulty entering it. Two of the technicians died immediately, and the third died shortly afterward. He was so radioactive that the doctor had to wear protective clothing to examine him. The three bodies remained so radioactive that parts of their bodies had to be buried with other nuclear waste. This accident was attributed by the Atomic Energy Commission (AEC) to human error and by a design defect. There was even a hint that this explosion was sabotage caused by a mentally unbalanced technician unhappy about his marriage, but this was never proved. It took 18 months to decontaminate the reactor building.

Suggested readings: Daniel F. Ford, *The Cult of the Atom: The Secret Papers of the Atomic Energy Commission* (New York: Simon and Schuster, 1982); John May, *The Greenpeace Book of the Nuclear Age: The Hidden History, the Human Cost* (New York: Pantheon Books, 1989).

Smyth Report

The Smyth Report was the official account of the history of the Manhattan Project. It was written by Henry DeWolf Smyth, a respected physicist at

Princeton University, during the war and published on August 12, 1945, only a week after the denotation of the atomic bomb on Hiroshima. Its official title was *A General Account of the Development of Atomic Energy for Military Purposes.* This report provided the theory and technical details of the building of the bombs except for those areas still too secret to be revealed. It also served as a justification for the $2 billion Manhattan Project. Without this report it would have been impossible to conduct the postwar debates over atomic energy. No such dissemination of technical information was forthcoming from the U.S. government for almost a decade. It also had a side effect of authenticating for Soviet scientists the information acquired from espionage activities during World War II. The British reacted negatively to this publication, because they felt it was too instructive and violated the secrecy pledge between the United Kingdom and the United States. Only with considerable pressure and resentment did the British yield to its publication.

Suggested reading: Henry DeWolf Smyth, *Atomic Energy for Military Purposes: The Official Report on the Development of the Atomic Bomb Under the Auspices of the United States Government, 1940–1945* (Princeton, NJ: Princeton University Press, 1945).

SNAP

SNAP (Special Nuclear Auxiliary Power) is a special nuclear auxiliary power package that is placed in satellites to power them for indefinite periods of time. This device was designed under the auspices of the Atomic Energy Commission (AEC) in the late 1950s for use in telecommunications satellites. Two types were developed: one utilizing heat from radioisotope decay, and the other, heat by thermoelectric methods. It was delayed in use by the State Department because the launch of such a satellite required launch over Cuba to gain a proper orbit. The State Department was fearful that a SNAP device might fall on Cuban soil and cause the Cubans to claim the United States had dropped an atomic bomb. President John Kennedy overruled the State Department after Chet Holifield of the Joint Committee on Atomic Energy protested the ban. SNAP packages powered a series of 25 satellites over the next couple of decades. The power for the initial satellite lasted 14 years. SNAP devices have also been used by the U.S. Central Intelligence Agency (CIA) for spy missions such as those in the Himalayas to monitor Chinese nuclear tests.

Suggested readings: Joseph A. Angelo and David Buden, *Space Nuclear Power* (Malabar, FL: Orbit Book Company, 1985); John May, *The Greenpeace Book of the Nuclear Age: The Hidden History, the Human Cost* (New York: Pantheon Books, 1989).

Soddy, Frederick

Frederick Soddy (1877–1956) was a British chemist who did pathfinding work on radioactivity. He was born on September 22, 1877, in Eastbourne, England. His father was a corn merchant. Soddy was educated at the College of Aberystwyth and Oxford University. After studying and working with Ernest Rutherford at McGill University in Canada and with William Ramsay at the University College in London, Soddy found a job at Glasgow University in 1904. In 1914 he obtained the chair of chemistry at Aberdeen University. Finally, in 1919 he became the Dr. Lee's Professor of Chemistry at Oxford University.

Soddy had two distinct periods in his scientific career. In the first period, which lasted from 1900 to 1919, Soddy was highly productive. He was active with Ernest Rutherford in proving that radioactive elements could change into other elements through a distinct series of stages. Next he was able to overturn the prevailing views of elements in the periodic table. Soddy claimed that the reason it was so difficult to separate some substances was because they were very nearly chemically identical. His suggestion was to call such nearly identical elements with differing atomic weights "isotopes." Later Soddy was to show the transformation of atoms by a displacement law. For his research, Soddy received the 1921 Nobel Prize for Chemistry. He was also active in publicizing radioactivity with numerous lectures and a 1908 book entitled *The Interpretation of Radium*.

In Soddy's second period, which lasted from 1919 until his death, he became estranged from scientific research. Instead, he devoted his energies to economic and social issues. Among the topics that still intrigued him, however, was the future of atomic energy. His positive view of atomic energy lasted the remainder of his lifetime. He died in Brighten, England, on September 22, 1956.

Suggested readings: Henry A. Boorse, Lloyd Motz, and Jefferson Hane Weaver, *The Atomic Scientists: A Biographical History* (New York: Wiley Science Editions, 1989); John Daintith, Sarah Mitchell, Elizabeth Tootill, and Derek Gjertsen, *A Biographical Encyclopedia of Scientists*, 2nd ed., vol. 2 (Bristol, UK: Institute of Physics, 1994).

Solvay Conference, 1927

At the Solvay Conferences, American and European physicists came together to discuss current developments in physics during the interwar period. Ernest Solvay, a Belgian inventor of a process for preparing sodium carbonate, had endowed an international conference to be held in Brussels every three to five years. At this conference, 30 or so invited physicists would gather and discuss a prearranged topic for a week. The first Solvay Conference was held in 1911. From the beginning, these conferences were productive because of the small number of participants and the lively discussions.

The Fifth Solvay Conference in Brussels in 1927 became the site of one of the most famous scientific confrontations of all time. It was here that the Copenhagen School's interpretation of quantum mechanics clashed with Einstein's theory that "God does not throw dice." In a series of personal debates between Niels Bohr and Werner Heisenberg, on one side, and Albert Einstein and a few allies on the other, the objections to the uncertainty principle of particles in the atom were examined in detail. Each morning Einstein proposed an objection, and each evening the Bohr-Heisenberg team provided an answer. This interchange continued for several days before Einstein began to come around. Einstein left the conference still unhappy about the theory but generally convinced. This debate between the champions of quantum mechanics and the proponent of a theory of general relativity reflected a growing unease among physicists over theories that only a few of them could understand. Quantum mathematics was especially difficult to unravel. This intellectual confrontation at the Solvay Conference has become a part of the folklore of atomic physics. More important in the long run was that after this conference the acceptance of quantum mechanics became widespread.

Suggested reading: Emilio Segre, *From X-Rays to Quarks: Modern Physicists and Their Discoveries* (San Francisco, W. H. Freeman, 1980).

Sommerfeld, Arnold

Arnold Johannes Wilhelm Sommerfeld (1868–1951) was a German theoretical physicist who is less important for his discoveries than for the number of great theoretical physicists to study under him. He was born on December 5, 1868, in Königsberg, Germany. His father was a physician. Most of his education was at the University of Königsberg, where he received his Ph.D. in 1891. Sommerfeld started out as a mathematician and worked with Felix Klein, who converted him to applied mathematics. After teaching stints at the Universities of Göttingen, Clausthal, and Aachen, Sommerfeld became chair of theoretical physics at the University of Munich in 1906. It was at Munich that Sommerfeld mentored a generation of theoretical physicists, including Hans Bethe, Peter Debye, Werner Heisenberg, Wolfgang Pauli, I. I. Rabi, Rudolf Peierls, and Max von Laue. It has been estimated that during his stay at the University of Munich, Sommerfeld trained a third of the theoretical physicists in the German-speaking world. Sommerfeld's weekly seminar was a place where he shone as a teacher. Because he was prominent, he received scientific papers in advance of their publication and shared them with his seminar.

Sommerfeld made an early contribution to the understanding of the Bohr model of the atom. He suggested that some of the electrons moved in elliptical, rather than circular, orbits. Besides this correction, Sommerfeld's book *Atomic Structure and Spectral Lines* (*Atombau und Spektrallinien*) and its number of editions in the 1920s served as the textbook for a generation of physicists working on atomic problems. It has been described as the "bible of atomic physics." A good student could read the book and then, after investigating current literature, could start original research.

Sommerfeld is also important for having been one of the early targets of Aryan Physics. He was Jewish and, in the early 1920s, became embroiled in a feud with Johannes Stark over the direction of German physics. When the 1933 law against Jews holding public office went into effect, Sommerfeld spent the summer term finding jobs for Jewish physicists abroad. When Sommerfeld retired at the University of Munich, he wanted his Prize student Werner Heisenberg to replace him. Political opposition blocked Heisenberg's candidacy. Sommerfeld was a beloved figure among most of his scientific colleagues, and he survived the Nazi regime. After the war, he returned to the Institute of Theoretical Physics in Munich. He died after being struck by a car on March 26, 1951, in Munich.

Suggested readings: David Abbott, ed., *The Biographical Dictionary of Scientists: Physicists* (London: Blond Educational, 1984); John Daintith, Sarah Mitchell, Elizabeth Tootill, and Derek Gjertsan, *A Biographical Encyclopedia of Scientists*, 2nd ed., vol. 2 (Bristol, UK: Institute of Physics, 1994); Richard Rhodes, *The Making of the Atomic Bomb* (New York: Simon and Schuster, 1986).

S-1 Committee. *See* Office of Scientific Research and Development

South African Nuclear Program

The South African nuclear program has always been a mystery because of secrecy by the South African government. A plentiful supply of uranium and the technical personnel to build nuclear weapons existed from the 1950s onward. With the addition of German-supplied equipment, the South Africans built an enrichment plant in the 1960s. Moreover, the South African government had close scientific ties with Israel and the United States. A secret complex was built in 1974 at Pelindaba, about 20 miles outside of Pretoria, to conduct nuclear research. It was run by South Africa's Atomic Energy Corporation and at its peak employed around 1,000 native-born South Africans studying ways to build atomic weapons. In August 1977, the Soviet Union warned the international community that its satellites had noted preparations for a nuclear test in the Kalahari Desert. This news mobilized the nuclear powers to apply successful pressure on the South African government to forego the test. In September 1979, however, an American spy satellite detected a bomb test in the southern hemisphere. Evidence pointed to South Africa conducting a small nuclear test on a remote outpost in the Indian Ocean, but it has never been proven beyond doubt that this test took place. By 1989 the South African nuclear program had produced six atomic bombs, including the prototype named "Melba."

The apartheid government of South Africa built its atomic bomb program as a way to insure its continued existence. Attacked from all sides and subject to sanctions from the international community, the government turned to a program to produce small, dirty bombs of the Hiroshima size and type. The strategy of the government after the production of the first bomb in 1979 was first to keep the world guessing about the possibility of South Africa having an atomic bomb. Then, if South Africa was threatened by a foreign power or its neighbors, the government would formally announce its ownership of an atomic weapon. Finally, as a last resort, the government would test an atomic devise as a warning. This strategy ended when F. W. De Klerk became president of South Africa in 1989. He decided to end the South African nuclear program and to dismantle its existing arsenal. By 1991, all the atomic bombs had been disassembled and South Africa signed the Nuclear Non-Proliferation Treaty. By these actions, South Africa became a unique example of a nuclear-weapons state reversing itself and becoming a nonnuclear state.

Suggested readings: Richard Kokoski, *Technology and the Proliferation of Nuclear Weapons* (Oxford: Oxford University Press, 1995); Roland W. Walters, *South Africa and the Bomb: Responsibility and Deterrence* (Lexington, MA: Lexington Books, 1987).

South American Nuclear Program

Two South American countries, Argentina and Brazil, have been active in nuclear energy. Argentina started first with a program to build nuclear power plants fueled by natural uranium. Two such plants were constructed in the 1970s with international help. A German company built one and the Atomic Energy of Canada the other. These reactors are under international safeguards, but a number of smaller research reactors are not. In the 1980s the Argentinean government financed a program to build an indigenous uranium enrichment technology. In 1985 a heavy water processing plant bought from Switzerland was completed.

Brazil has been less active in nuclear energy than Argentina, but it also contracted from foreign suppliers to build nuclear power plants. In the mid-1970s the Brazilian government tried to buy nuclear technology from the American engineering firm of Bechtel, but the Ford administration refused to approve the sale. Instead the German firm concluded a deal to supply Brazil with eight nuclear power plants, a reprocessing facility, and enrichment technology. Despite this deal, progress on building the Brazilian nuclear industry has been slow, and several of the plants were never built.

Both Argentina and Brazil have long had a natural rivalry as the superpowers on the South American continent. Their nuclear energy programs continue this rivalry. Both countries have acquired enough technical knowledge and nuclear fuel to make nuclear weapons, but there is no information to conclude that either has a nuclear weapons arsenal. Moreover, both Argentina and Brazil are signatories of the Treaty of Tlatelolco, barring the introduction of nuclear weapons into South America, but they have refused to ratify the treaty, as has Chile. This refusal to ratify means that these countries feel themselves free to violate parts of the agreement without totally repudiating the treaty.

Suggested reading: Congressional Quarterly, *The Nuclear Age: Power, Proliferation and the Arms Race* (Washington, DC: Author, 1984).

Southeast Asia Nuclear Weapon Free Zone. *See* Treaty of Bangkok

South Pacific Nuclear Free Zone Treaty. *See* Treaty of Rarotonga

Soviet Atomic Bomb Program. *See* Soviet Nuclear Program

Soviet Atomic Spying

Most of the Soviet scientific spying was conducted through the Soviet Secret Police. In World War II, the People's Commissariat of Internal Affairs (NKVD) was in charge, and Leonid Kvasnikov, the head of the Science and Technology Department, conducted its operations. In the 1930s most of the spying had been industrial espionage—industrial processes and formulas—but in 1940, Kvasnikov became interested in atomic research. While initially most attention was directed toward German research, information was gathered from any source. First indications of British interest in an atomic bomb came from British diplomat and Soviet spy Donald Maclean, who had access to the British Uranium Committee in September 1941. Another British spy, John Cairncross, was private secretary to Lord Hankey, who handled the MAUD Report in August 1941. Hundreds of Soviet agents penetrated Canada and the United States, but most were engaged in non–atomic bomb espionage. The most significant atomic spies were Klaus Fuchs, a German physicist, and the American spies David Greenglass, Harry Gold, and Julius and Ethel Rosenberg. Two other British spies of importance were Bruno Pontecorvo, an Italian physicist working for the British, and Alan Nunn May, a British physicist. May had even been able to give a Soviet agent a sample of enriched uranium-235. Of these spies, Fuchs was by far the most important because he knew more about the construction of the atomic bomb than the others, and he understood the theories behind it. The other physicists had access only to bits of the puzzle. Greenglass, Gold, and Rosenberg transmitted information about the atomic bomb without understanding the processes.

These spying activities were unknown by American, British, and Canadian authorities until a Soviet cipher clerk at the Soviet embassy in Ottawa, Canada, Igor Gouzenko, defected on September 5, 1945, with documents. Almost reluctantly, Canadian security began learning names. Gouzenko's biggest secret was his exposure of Allan Nunn May, a British physicist working at the atomic laboratories in Montreal under John Cockroft. Once the existence of an atomic bomb spy ring was revealed with names, security agents from Canada, Great Britain, and the United States launched investigations that eventually broke the ring. These investigations lasted for years until the early 1950s, but these investigations and arrests only temporarily halted Soviet efforts to learn nuclear secrets from the United States.

British and Canadian courts were more lenient in their sentencing than American courts. Eighteen members of the Canadian spy ring were brought to trial, but only 8 were found guilty. The sentences for those convicted varied considerably from 2- to 6-year terms in prison. Fuchs received a 14-year prison sentence and May a term of 10 years. In contrast, American courts were much harsher. Julius and Ethel Rosenberg received death sentences, and even those who turned state's evidence were given 30-year prison terms. These harsh sentences were to serve as deterrents against further Soviet spying.

Besides the Rosenberg Case, the most controversial aspect of Soviet atomic spying was its role in the Soviet Union's development of atomic weapons. Scientists and politicians have differing views of this issue. Most scientists believed that it was only a matter of time before Soviet scientists learned the basic principles of building an atomic bomb, and information gathered by spies only slightly accelerated the process. Politicians, however, believe that spying significantly advanced the ability of the Soviets to build first an atomic bomb and then a hydrogen bomb. The truth is probably somewhere in the middle. Information on the progress of the Manhattan Project's activities at the Metallurgical Laboratory in Chicago and the Los Alamos Scientific Laboratory helped the Soviet scientists in the early days, but it was the publication of official history of the building of the atomic bomb by Henry DeWolf Smyth—*General Account of the Development of Atomic Energy for Military Purposes*—in 1945 that filled in the gaps. Once Soviet scientists knew the basic procedures, then it was only a matter of time and resources. Joseph Stalin, the Soviet head of state, provided the resources and Lavrentii Beria, the chief of the secret police and the head of the Soviet nuclear program, the incentive to succeed. Conflict developed between Soviet scientists and Soviet leadership when the scientists found better ways to solve problems, but Beria and other leaders wanted an exact copy of the American bombs. Most experts now believe that the Soviet scientists developed a functional hydrogen bomb before the Americans built one, so spying played a limited role. In summary, spying helped the Soviet Union provide basic information about building an atomic bomb, but most of the credit goes to the Soviet scientists and engineers who actually built the bomb.

Suggested readings: Thomas B. Cochran, Robert S. Norris, and Oleg A. Bukharin, *Making the Russian Bomb: From Stalin to Yeltsin* (Boulder, CO: Westview Press, 1995); H. Montgomery Hyde, *The Atom Bomb Spies* (New York: Atheneum, 1980).

Soviet Nuclear Program

The Soviet nuclear program was launched without fanfare by Joseph Stalin in the early 1930s in response to progress on atomic theory abroad. A prestigious research committee of the USSR Academy of Sciences, the Special Committee for the Problem of Uranium, was formed in 1940 to study atomic energy, especially the possibility and consequences of a chain reaction. Soviet physicists Georgi Flerov and L. Rusinov had concluded as early as 1939 that a chain reaction was possible and enough energy from nuclear fission could produce a bomb. After an intense search, a supply of uranium ore was found in the Fergana Valley in Uzbekistan. The decision to go ahead slowly was made by the committee because of the tremendous allocation of resources necessary for such a project and the fear that Stalin would not look favorably on such a long-range plan on the eve of a possible war with Germany. Nevertheless, by 1941 Soviet scientists had constructed a cyclotron, and two others were under construction for basic research. On the eve of war the Soviet Union and the United States were advancing on parallel paths, and progress was about even. When references to atomic research began to disappear from Western scientific literature in 1940, Soviet scientists realized that serious advances on atomic subjects were taking place among the Western Allies. Only the French continued to publish significant research results, and Frederic Joliot-Curie's conclusions about an atomic chain reaction producing energy reinforced the findings of Soviet scientists on the subject. Soviet scientists also believed that the Germans had a serious atomic program under way.

The German invasion of Russia in June 1941 changed this parallel development between the Soviet Union and the United States. An almost total halt to the Soviet atomic program resulted as all energies were turned toward the military problem of stopping the German invasion. Key atomic program personnel and equipment were transferred to the city of Kazan. Intelligence sources gathered from spies in October 1941 indicated that the British and the Americans were developing a bomb based on the explosive reaction of uranium-235. Soviet spies gained access to the MAUD Report in the summer of 1941, revealing that an atomic bomb was possible. Because of this and other information, Lavrentii Beria, the head of the NKVD (People's Commissar of Internal Affairs), became so convinced of foreign progress that he sent a request to Stalin in March 1942 to launch a Soviet atomic bomb program. Stalin was reluctant until he consulted with key Soviet physicists who convinced him of the necessity. After Abraham Ioffe, the leader of the Soviet physicists, turned down the opportunity, Stalin gave Igor V. Kurchatov the go-ahead for a full-scale atomic weapons program. A call was sent out to all intelligence agents for information on atomic bomb projects. For political control, Stalin placed Viacheslav Molotov, the Soviet foreign minister, as the head of the Soviet atomic bomb program. Although the project had the blessings of Stalin, if found difficulty in obtaining personnel and materials from the bureaucratic domains of other wartime projects. Because of timely intelligence reports, Stalin knew of the possibility of an atomic weapon when President Truman told him about it at Potsdam in 1945. Nevertheless, Stalin was so furious about the failure to have an atomic bomb to counter the Americans that he appointed Beria to take charge of the nuclear program. Stalin was so impressed by the American bombing of the Japanese cities that he ordered his scientists to catch up with the United States, regardless of the cost. A Special Committee on the Atomic Bomb (Spetskom) of eight high-level administrators was formed on August 20, 1945, under the chairmanship of Beria to solve the nuclear problem. The First Main Directorate of the USSR Council of Ministers assumed control of administration and coordination of the atomic program, with B. L Vannikov as chair and with oversight to Beria. A shortage of uranium ore was partially solved by the seizure of German uranium ore from a tannery factory west of Berlin in April 1945. An agreement for Czech uranium ore and new exploration for ore in the Soviet Union gave the Soviet program enough ore for military and research purposes. The scarcity of equipment in the Soviet Union was partly alleviated by the dismantling and sending to Russia of the laboratories of Germany's Kaiser-Wilhelm-Institute for Physics. German scientists under Nikolaus Riehl housed at Elektrostal began by the fall of 1946 to deliver uranium metal for the Soviet atomic program.

Because the Soviet Union lacked the industrial experience to build a nuclear program, Beria looked to the industrial expertise of the existing group of industrial leaders, the so-called Red Specialists. Consequently, Beria appointed Boris Vannikov to head the atomic bomb project and to report to him directly. Mikhail Pervukhin and Igor Kurchatov became deputies to Vannikov. The Soviet scientists decided to build a duplicate of America's Fat Man because they had much of its design from Klaus Fuchs. They came to this decision as much for political reasons as scientific because they needed to please Beria. The first Soviet nuclear device was exploded near Semipalatinsk in Soviet central Asia in August 1949. Igor Tamm had recruited Andrei Sakharov at the Physics Institute of the Soviet Academy (FIAN) to work on a hydrogen bomb. Beria continued to head the nuclear program and ruled it with an iron hand until he was executed after Stalin's death in 1953.

Beria was replaced as head of the Soviet nuclear program by Vyacheslav Malyshev in June 1953. Little was known about him except that he was an ex-locomotive driver from Syktyvkar and he was a veteran of the Battle of Stalingrad. Beria's removal did not slow the testing of the first Soviet hydrogen bomb at the Semipalatinsk-21 Test Site on August 12, 1953. Its yield was in the 400-kiloton range. Much as in the United States, the Soviet military wanted tactical atomic weapons to use on the battlefield, and considerable effort was made to design such weapons. Malyshev died in 1957 of cancer. In the 1950s and 1960s most of the Soviet nuclear testing of hydrogen bomb weapons was at the Novaya Zemlya (New Land) Test Site near the Arctic Circle. Testing continued there until a testing moratorium was announced by the Russian government in October 1990. Most of the recent Russian activity has been dismantling their nuclear arsenal in accordance with agreements concluded with the United States.

The Soviet civilian nuclear program was slower to develop than the military program. Although the first civilian nuclear power station opened on June 27, 1954, at Obninsk, near Moscow, it was not until the early 1960s that other power-generating stations were built. It developed two major reactor types: the RBMK (Reactors High-Power Boiling Channel Type), a graphite-moderated reactor, and the VVER (Water-Water Power Reactors), a water-pressurized reactor designed for both domestic and export use. RBMK reactors were the original prototype, and the first three were constructed at Leningrad, Kursk, and Chernobyl. This type of reactor was the most common built, with 14 in operation in 1986 at the time of the Chernobyl accident, providing half of the Soviet Union's nuclear-electricity–generating capacity. From the beginning, design flaws were soon noted with trouble from leaky fittings and the need to handle radioactive water. Finally, this reactor type became unstable if operated at low power. It took the Chernobyl reactor accident to show how deadly these design flaws were. The VVER reactors were intended to replace the RBMK reactors, and by the late 1980s this type became the principal Soviet nuclear-reactor-type. A difficulty with the VVER is that it needs constant recharging, while one-third of the fuel elements need to be replaced annually. Moreover, the cost of producing electricity from VVER reactors is greater than that of the RBMK. Attempts have been made to develop a fast breeder reactor, namely at the Shevchnko plant near the Caspian Sea and at Beloyarsk, but neither plant has been successful because of the high cost of its electricity and constant shutdowns for reloading and breakdowns.

The Soviet nuclear program concentrated on weapons design and testing until around 1988. Since that time, the new Russian state has been downsizing the nuclear facilities and finding alternative uses for the facilities and people. Concern over this process has led to international agreements to help with the disposal of these weapons and nuclear wastes.

Suggested readings: Ulrich Albrecht, *The Soviet Armaments Industry* (Chur, Switzerland: Harwood Academic Publishers, 1993); Thomas B. Cochran, Robert S. Norris, and Oleg A. Bukharin, *Making the Russian Bomb: From Stalin to Yeltsin* (Boulder, CO: Westview Press, 1995); David Holloway, *Stalin and the Bomb: The Soviet Union and Atomic Energy, 1939–1956* (New Haven, CT: Yale University Press, 1994); Amy Knight, *Beria: Stalin's First Lieutenant* (Princeton, NJ: Princeton University Press, 1993); Arnold Kramish, *Atomic Energy in the Soviet Union* (Stanford, CA: Stanford University Press, 1959); David R. Marples and Marilyn J. Young, eds., *Nuclear Energy and Security in the Former Soviet Union* (Boulder, CO: Westview Press, 1997); Grigori Medvedev, *No Breathing Room: The Aftermath of Chernobyl* (New York: Basic Books, 1993); Zhores Medvedev, *The Legacy of Chernobyl* (New York: Norton, 1990).

Soviet Nuclear Submarine Accidents

The Soviet Navy experienced a series of accidents involving its submarines in the period from 1956 to 1994. Part of the problem was that Soviet nuclear submarines were technically inferior to their American counterparts and subject to serious nuclear reactor breakdowns. The most serious were: the K-8 nuclear submarine accident in the Bay of Biscay with the loss of 50 crew members in 1970; the Echo II-class K-431 nuclear submarine refueling accident in Chazhma Bay on August 10, 1985; the Yankee class K-219 sinking on October 3, 1986; and the loss of the Komsomolets Mike-class nuclear-powered attack submarine on April 7, 1989, in the Barents and Norwegian Seas. These accidents resulted in serious loss of life and potential environmental damage. An additional 110 accidents involving nuclear-powered submarines have taken place, with 14 of the accidents so serious that the submarines were decommissioned permanently. Four of the submarines sank following the accidents, and only 1 of these have been recovered. Most of the serious accidents were caused by reactor meltdowns and explosions and nuclear weapons malfunctions.

Suggested readings: Thomas B. Cochran, Robert S. Norris, and Oleg A. Bukharin, *Making the Russian Bomb: From Stalin to Yeltsin* (Boulder, CO: Westview Press, 1995); Peter Huchthausen, Igor Kurdin, and R. Alan White, *Hostile Waters* (New York: St. Martin's Press, 1997).

Special Committee on the Atomic Bomb

The Soviet Union's Special Committee on the Atomic Bomb was formed by Joseph Stalin on August 20, 1945, to oversee the building of a Soviet atomic bomb. While Stalin knew about the existence of the atomic bomb before Hiroshima, the destructive power of the atomic bomb convinced him of the necessity for the Soviet Union to possess such a weapon. Lavrentii Beria, the head of the Soviet Secret Police, was appointed chair of the committee. Other prominent members were Georgi Malenkov, secretary of the Central Committee, Nikolai Voznesenskii, head of the State Planning Committee, Boris Vannikov, an industrial manager, Avraamii Zaveniagin, an industrial manager, Mikhail Pervukhin, an industrial manager, Igor Kurchatov, scientific director, Pytr Kapitsa, world-famous physicist, and General V. A. Makhnev, member of the NKVD (People's Commissar of Internal Affairs). Although this committee was formally the decision-making body, Beria reported to Stalin weekly on the progress of the atomic bomb program. Special organizations were set up to report to the Special Committee on the Atomic Bomb: First Chief Directorate of the Council of People's Commissars to oversee mining, industrial plants, and research facilities and the Scientific-Technical Council to advise on scientific and technical questions. Beria also used the NKVD to report on activities at the atomic research installations.

Suggested reading: David Holloway, *Stalin and the Bomb: The Soviet Union and Atomic Energy, 1939–1956* (New Haven, CT: Yale University Press, 1994).

Special Subcommittee on the Hydrogen Bomb

The Special Subcommittee on the Hydrogen Bomb was formed on November 18, 1949, by President Truman to study hydrogen bomb development. This subcommittee was part of the National Security Council (NSC), which advised the president on national security policies. The three members of this subcommittee were Dean Acheson, secretary of state, Louis Johnson, secretary of defense, and David Lilienthal, chair of the Atomic Energy Commission (AEC). President Truman had received conflicting recommendations from two atomic energy sources. Senator Brien McMahon, the chair of the Joint Committee on Atomic Energy, had presented a 5,000-word letter advancing the building of a hydrogen bomb. The General Advisory Committee of the AEC advised against the hydrogen bomb. Because of the irascible nature of Johnson and the inability of the members of the committee to function smoothly, only two meetings were held in December 1949 and January 1950. In the meantime, intensive lobbying in favor of the hydrogen bomb development was taking place in congressional and military circles. By the January 31 meeting, the recommendation to proceed on development of the hydrogen bomb was agreed upon by a two to one vote (Lilienthal against). President Truman took the recommendation of the subcommittee and adopted it. He announced the findings of the subcommittee and his decision to the press shortly afterward.

Suggested readings: Richard Wayne Dyke, *Mr. Atomic Energy: Congressman Chet Holifield and Atomic Energy Affairs, 1945–1974* (Westport, CT: Greenwood Press, 1989); Norman Moss, *Men Who Play God: The Story of the H-Bomb and How the World Came to Live with It* (New York: Harper and Row, 1968); Richard Rhodes, *Dark Sun: The Making of the Hydrogen Bomb* (New York: Touchstone, 1995).

Speer, Albert

Toward the end of World War II, Albert Speer (1905–1981) was the industrial czar of Nazi Germany, and his support for the German atomic bomb program allowed it to make some progress. Soon after Adolf Hitler appointed him to his post in 1944, Speer learned about the possibility of an atomic bomb from General Friedrich Fromm. Speer tried to placate Herman Goering, head of the German Air Force and longtime friend of Hitler, who was jealous of Speer's new power, by making him head of the Reich Research Council and at the same time give the Council a higher priority on research for the atomic program. He attended a conference on June 4, 1942, where Werner Heisenberg gave a lecture on the atomic bomb. Speer asked Heisenberg about the likelihood of a bomb in the near future and became convinced of the long-range potential of nuclear energy. From this time onward, Speer backed the atomic program, providing raw materials and official support, but he was disappointed by the smallness of the scientists' requests. Perhaps Speer's greatest contribution was to transfer scientists out of military units and give them military exemptions for scientific research. Speer did brief Hitler on the progress of the German atomic bomb process, but they both concluded that such a bomb was in the distant future. After discussion with the physicists, Speer curtailed bomb research but continued to fund research on a heavy water reactor.

Suggested readings: David Irving, *The German Atomic Bomb: The History of Nuclear Research in Nazi Germany* (New York: Simon and Schuster, 1967); Thomas Powers, *Heisenberg's War: The Secret History of the German Bomb* (New York: Knopf, 1993); Richard Rhodes, *The Making of the Atomic Bomb* (New York: Simon and Schuster, 1986).

Spies. *See* Rosenberg Case; Soviet Atomic Spying

Stalin, Joseph

Joseph Stalin (1879–1953) had little knowledge about atomic energy, but he understood the international political consequences of possession of such a weapon as the atomic bomb. Soviet scientists had the basic knowledge to construct an atomic device in the early 1940s, but before this could happen, the commitment to devote the necessary resources to build it had to come from Stalin. He allowed basic research to begin in World War II, but the war effort against Germany was paramount. Soviet intelligence kept Stalin aware of the progress of the atomic bomb being developed by the United States in the Manhattan Project, but the detonations of atomic bombs on Japanese cities in August 1945 convinced him of their power. From this time onward, Stalin gave Soviet scientists a blank check to build atomic weapons. Lavrentii Beria, the head of the NKVD (People's Commissariat of Internal Affairs), might be in charge, but he made weekly reports to Stalin on progress. Stalin desperately wanted an atomic arsenal to counter the one possessed by the United States. He wanted it for strategic reasons, because his view was that the atomic bomb was only useful against rear-echelon forces. This explains Stalin's pressing the issue against Berlin and the Korean War. He made it plain through Beria that success would be rewarded, but failure meant ruin and a prison camp or worse. Once the Soviet Union had an atomic bomb, Stalin pushed even harder for a hydrogen bomb. Stalin died just before the final testing of the Soviet's first hydrogen bomb.

Suggested readings: Ulrich Albrecht, *The Soviet Armaments Industry* (Chur, Switzerland: Harwood Academic Publishers, 1993); David Holloway, *Stalin and the Bomb: The Soviet Union and Atomic Energy, 1939–1956* (New Haven, CT: Yale University Press, 1994).

Stark, Johannes

One of the leaders of the Aryan Physics movement in Germany from World War I to the end of World War II was prominent physicist Johannes Stark (1874–1957). Stark was born on April 15, 1874, in Schickenhof, Bavaria. His father was a landed proprietor. After schooling at Bayreuth and then Regensburg, he studied physics and received his doctorate from the University of Munich in 1897. His first teaching position was at the University of Munich in 1897. After holding various professorships at Göttingen, Hanover, and Aachen between 1900 and 1917, Stark moved to a small, conservative university at Greifswald. He was an outstanding experimental physicist and published more than 300 papers in his academic career. Stark had been an early defender of both the theory of relativity and quantum theory, but in 1913 he suddenly turned against both theories. At the same time, professional disagreements caused his relationship with Arnold Sommerfeld, a theoretical physicist at Munich, to deteriorate into a feud that intensified when Sommerfeld's student, Peter Debye, obtained a professorship in Göttingen. He was in Greifswald when the trauma of the end of the war hit Germany. It hit Stark particularly hard, because he had been a staunch supporter of the German war effort. Stark received the 1919 Nobel Prize in Physics for his discovery of the Stark effect—splitting of spectral lines in an electric field. In 1920, he received a professorship at the University Würzburg. His academic career ended in 1922 at Würzburg with his resignation after numerous fights with his academic colleagues. Shortly afterward, he changed his mind and began to look for another academic position, but none was forthcoming until 1933. He then used his Nobel Prize money to finance an unsuccessful venture into the porcelain industry.

Soon after both academic and business setbacks, Stark began his attempts in 1922 to control German physics. He used his friendship and agreement with Philipp Lenard to attack new trends in German physicist research. In his book *The Contemporary Crisis in German Physics* (1922), Stark attacked current physics as dogmatic, especially quantum mechanics and relativity. His first effort was to break the hold of the German Physics Society by forming a new organization, the Professional Community of University Physicists. When it became apparent that his attempt would fail, he turned to the National Socialist German Workers Party (NSDAP) of Adolf Hitler. An early supporter of Hitler, Stark joined the NSDAP in 1930. In 1933, Stark was rewarded by Wilhelm Frick, the minister of interior, with the presidency of the prestigious Imperial Physical-Technical Institute. Then in 1934, he was appointed president of the German Research Foundation (DFG), which was the clearinghouse for most government funding of scientific research. Stark redirected German physics research by stopping funding of theoretical work and funding only certain types of experimental research. Almost at the top of the physics world, Stark launched an attack on misconduct of local Nazi administrators and found himself almost dismissed from the NSDAP. In the aftermath of this controversy, Stark suffered the loss of his position with the German Research Foundation, but he counterattacked with a campaign against Werner Heisenberg.

This war against Heisenberg proved to be a Pyrrhic victory for Stark. In July 1937, an article appeared in the SS (Schutzstaffel) weekly *The Black Corps* (Das Schwarze Korps) accusing Heisenberg of being a "white Jew." This term was used by the Deutsche Physik crowd to describe Aryans who had been contaminated by the Jewish spirit. It soon became apparent that this campaign against Heisenberg was directed by Stark to prevent Heisenberg's appointment to the chair of theoretical physics at the University of Munich. This campaign succeeded because a follower of the Deutsche Physik, Wilhelm Muller, received the appointment. Stark's constant feuds became less tolerated by scientific authorities, and in 1939, he retired from the Imperial Physical-Technical Institute. The war and the acceptance of relativity and quantum mechanics passed Stark by without his active participation. He stayed on his estate at Traunstein in Bavaria during the war.

After the war, Stark found himself before a de-Nazification court at Traunstein, and he was convicted as a major offender and sentenced to four years of hard labor. In an appeal, Stark had his sentence reversed. He returned to his estate, where he died on June 21, 1957.

Suggested readings: Alan D. Beyerchen, *Scientists Under Hitler: Politics and the Physics Community in the Third Reich* (New Haven, CT: Yale University Press, 1977); Mark Walker, *Nazi Science: Myth, Truth, and the German Atomic Bomb* (New York: Plenum Press, 1995).

START

In 1991 the United States and Russia concluded a treaty, the Strategic Arms Reduction Treaty (START), beginning balanced reductions in their strategic nuclear weapons arsenals. After the withdrawal before a vote of the Strategic Arms Limitation Treaty II (SALT II) in the U.S. Senate in 1979, The Reagan administration proposed in May 1982 major cuts in the strategic forces of both the Soviet Union and the United States. From 1982 to the final agreement in 1991, negotiations consumed time because of the proposed deep cuts on the order of 50 percent or more. This treaty, START I, was the first serious step in reducing nuclear weapons arsenals, but the fragmentation of Russia into independent states complicated the situation. Russia could no longer require Belarus, Kazakhstan, or the Ukraine to comply with the terms of the treaty. Subsequent negotiations have brought these states in line, and work is now proceeding for a START II. This new round concluded in early 1993 would have the United States and Russia reduce nuclear arsenals of no more than 3,500 strategic nuclear warheads slightly after the year 2000. While both countries still keep the size of their nuclear arsenals as classified information, this size reduction is about 70 percent.

Suggested readings: Eric Arnett, ed., *Nuclear Weapons After the Comprehensive Test Ban: Implications for Modernization and Proliferation* (Oxford: Oxford University Press, 1996); Paul R. Bennett, *Russian Negotiating Strategy: Analytic Case Studies from SALT to START* (Commack, NY: Nova Science Publishers, 1997); James L. George, *The New Nuclear Rules: Strategy and Arms Control After INF and START* (New York: St. Martin's Press, 1990); Marianne van Leeuwen, ed., *The Future of the International Nuclear Non-Proliferation Regime* (Dordrecht, Holland: Martinus Nijhoff, 1995).

Sternglass, Ernest J.

Ernest Sternglass (1923–) is an American physicist who challenged the U.S. government's standards on low-level radiation dangers. He was born in 1923 in Berlin, Germany. Both his parents were physicians with medical practices in Berlin. His family left Germany in 1938 when Sternglass was only 14 because of Adolf Hitler. They migrated to the United States. He went to Cornell University to study engineering. During World War II, Sternglass joined the navy too late to see military action. After leaving the navy, he became a civilian employee at the Naval Ordnance Laboratory in Washington, DC, where he worked on imaging devices for military applications. In 1953 he completed his doctorate in engineering physics at Cornell University. He had taken a job at Westinghouse Corporation in Pittsburgh in 1952, working on fluoroscopy. From 1952 to 1967 Sternglass was involved in research on nuclear instrumentation. In the early 1960s he ran across the research of Dr. Alice Stewart on radiation and unborn children and was impressed. Sternglass decided to study the effects of nuclear fallout on infants and children. Then in December 1967 he learned of a project to detonate a nuclear device in central Pennsylvania to create a storage space for underground gas. His editorializing against the project helped kill the idea, but he made major enemies in the nuclear business. It was about this time that Sternglass left Westinghouse and moved to a position at the University of Pittsburgh Medical School. His studies on infant mortality established a correlation between radiation and infant deaths. At one time he claimed 40,000 infant deaths as a result of radioactive fallout and exposure to radiation. His research and its conclusions became highly controversial, but his evidence pointed out the dangers of even low levels of radiation on humans. Sternglass became the director of the Radiological Physics and Engineering Laboratory at the University of Pittsburgh School of Medicine until his retirement. He is now emeritus professor of radiological physics in the Department of Radiology, University of Pittsburgh School of Medicine, and lives in New York City.

Suggested reading: Leslie J. Freeman, *Nuclear Witnesses: Insiders Speak Out* (New York: Norton, 1981).

Stockpile Stewardship Program

The Stockpile Stewardship Program is an effort by the U.S. government to retain and maintain its nuclear weapons arsenal in operating condition. The federal government has numerous sites where it stores nuclear weapons and needs a regular inspection program to ensure that these weapons are available. Since the official policy of the U.S. government is not to make new nuclear weapons, or test them, this program is intended to keep its nuclear weapons arsenal in operable condition and to retain a capability to resume nuclear weapons production and testing if the international climate changes dramatically. While the United States supports the Comprehensive Test Ban Treaty, the Department of Energy (DOE) and the Department of Defense (DOD) want insurance. Legislation establishing this program passed Congress in November 1993, and funds have been appropriated for it. Part of this project includes the National Ignition Facility (NIF), which promotes research on Inertial Confinement Fusion (ICF), in an attempt to use high-energy laser beams to cause fusion from fusion fuel pellets.

Suggested reading: Eric Arnett, ed., *Nuclear Weapons After the Comprehensive Test Ban: Implications for Modernization and Proliferation* (Oxford: Oxford University Press, 1996).

Strategic Arms Limitation Treaty I

The Strategic Arms Limitation Treaty I (SALT I) was signed in May 1972 by the Soviet Union and the United States as the first major step in controlling nuclear weapons. This treaty was an agreement with two parts to it, the antiballistic missile (ABM) agreement and an agreement for a freeze on offensive strategic nuclear systems. These agreements restricted the scope and number of antiballistic missiles sites and put a freeze on the number of ground-to-ground rockets. SALT I was a limited treaty, lasting for only five years, and most of its provisions dealt with limitations on defensive rather than offensive weapons; but its signing was considered a success by the international community. SALT I was ratified by the Senate on August 3, 1972, by a vote of 88 to 12.

Prior to SALT I, basic policy by both the Soviet Union and the United States had been to build as many nuclear weapons as possible and base their strategy on massive retaliation for any nuclear attack. President Lyndon Johnson had proposed a reduction in the production of fissionable materials and later advanced the idea of a verified freeze of nuclear offensive and defensive launch vehicles as early as 1964, but these initiatives were rejected by the Soviet Union over inspection and fairness issues. Actual negotiations between the superpowers on SALT I started in 1968, but the first meeting were held in late 1969. Both powers desired to limit the increase of military budgets and find a balance of power between them. Meetings continued until May 26, 1972, when President Richard Nixon and Soviet leader Leonid Brezhnev concluded a series of agreements that became known as SALT I.

Critics of the treaty pointed out that in the Interim Agreement on Offensive Arms of the treaty the Soviet Union was able to keep a larger number of missiles than the United States. SALT I had no provisions, however, against modernization of these weapons. The leader of the assault against SALT I was Senator Henry M. Jackson, Democrat from Washington, who believed that this treaty was a sellout to the Soviets. Because of this type of opposition and mutual mistrust caused by the Arab-Israeli War of 1973, in which the United States supported Israel and the Soviet Union supported the Arab states, progress toward further arms control was retarded. Despite these critics, this treaty allowed both sides to continue to modernize their nuclear arsenals, and little damage was done to either side.

Suggested readings: Congressional Quarterly, *The Nuclear Age: Power, Proliferation and the Arms Race* (Washington, DC: Author, 1984); James L. George, *The New Nuclear Rules: Strategy and Arms Control After INF and START* (New York:

St. Martin's Press, 1990); Mason Willrich and John B. Rhinelander, *SALT: The Moscow Agreements and Beyond* (New York: Free Press, 1974).

Strategic Arms Limitation Treaty II

The Strategic Arms Limitation Treaty II (SALT II) was the next step in the Strategic Arms Limitation Talks, since SALT I was an interim agreement lasting only five years. Negotiations began in 1972 only months after the signing of the SALT I agreement. Some of the difficult questions left over from the earlier agreement—equal numbers of missiles question, bombers, multiple independently targetable reentry vehicles (MIRVs), and cruise missiles—had to be resolved. In the middle of the negotiations President Ford and General-Secretary Brezhnev concluded another interim agreement, the Vladivostok Agreement in 1976. This agreement, to last through 1985, achieved parity between the Soviet Union and the United States on nuclear delivery vehicles and MIRV systems, placed a ban on the construction of new land-based ICBM (intercontinental ballistic missile) launchers, and outlined a verification system. The new Carter administration assumed responsibility for negotiations in 1977 and tried to renegotiate some of the Vladivostok Agreement limits without much success. An agreement on SALT II was finally signed on June 18, 1979, and it provisions were similar to those in the Vladivostok Agreement. It was when President Carter submitted the SALT II agreement to the U.S. Senate that his difficulties began. A coalition of senators formed to defeat SALT II by denying it a two-thirds majority. Two external factors played a key role—the Soviet Union's invasion of Afghanistan and the Iranian crisis. Because of the prospects of defeat, SALT II was never brought forward to a vote in the U.S. Senate. Nevertheless, both the Soviet Union and the United States have abided by the terms of the treaty. SALT II became an issue in the Carter-Reagan 1980 presidential campaign, with Reagan opposed to it.

Suggested readings: Dan Caldwell, *The Dynamics of Domestic Politics and Arms Control: The SALT II Treaty Ratification Debate* (Columbia: University of South Carolina Press, 1991); James L. George, *The New Nuclear Rules: Strategy and Arms Control After INF and START* (New York: St. Martin's Press, 1990).

Strategic Arms Reduction Treaty. *See* START

Strauss, Lewis L.

Admiral Lewis Strauss (1896–1974) was the third chair of the U.S. Atomic Energy Commission (AEC) and a longtime supporter of atomic energy. He was born on January 31, 1896, in Charleston, West Virginia. His father had been in the wholesale shoe business. He married Alice Hanauer, daughter of one of the partners of Kuhn, Loeb, in 1923. When he was four his family moved to Richmond, Virginia. Although he was valedictorian of his high school, John Marshall High, family financial reverses made him skip college to work as a traveling shoe salesman for his father's firm. After three years of such work, young Strauss joined Herbert Hoover's Food Administration during World War I, and he soon became Hoover's private secretary. After the war, he continued to work for Hoover in the American Relief Administration (ARA). Strauss was able to make many contacts, and in April 1919, he was invited to join the investment banking firm of Kuhn, Loeb and Co. Strauss worked his way up in the organization until in January 1929 he was selected to become a partner in the firm.

Strauss had a long and distinguished career in both the public and private sectors. After becoming wealthy, he was able to devote more time to public service. In 1925 he had accepted a reserve commission as a lieutenant commander. Strauss also had a lifetime interest in physics, and this interest was stimulated by his friendship with such physicists as Leo Szilard. During World War II, he worked in the Bureau of Ordnance in Washington, DC, and became a protégé of the Navy Secretary James Forrestal. Because Strauss had alienated President Roosevelt, he had to wait until Truman became president before being promoted to rear admiral. He was both clever and adaptable but also thin-skinned and pompous. An early supporter of atomic research, starting with Ernest Lawrence's development of the cyclotron, he became one of the first civilian members of the AEC. In the late 1940s, Strauss became a strong backer of the program to build the H-bomb. His close association with Edward Teller helped convince him of its necessity. After the decision to build a hydrogen bomb was made, Strauss resigned from the AEC on April 15, 1950. He worked as a financial adviser until President Eisenhower appointed him back on the AEC to replace Gordon Dean as chair. One of the first acts as new chairman of the AEC in the summer of 1953 was to purge security risks. J. Robert Oppenheimer was one of his early targets. Strauss believed Oppenheimer had been sabotaging the hydrogen bomb program and suspected him of being a Soviet agent. He had taken the job with the stipulation that Oppenheimer must go. Strauss made sure that he was involved in the operations of the Gray Committee, which refused to reinstate J. Robert Oppenheimer's security clearance. Soon afterward, Strauss found himself in the middle of a controversy over Dixon-Yates, a deal that involved a private company and the Tennessee Valley Authority (TVA) supplying electricity to Memphis, Tennessee. This deal had the support of the Eisenhower administration, and Strauss supported its implementation. He was able to save the contract, but not before making important enemies among the new Democratic majority. The collapse of the Dixon-Yates contract in 1955 proved to be an embarrassment to Strauss. Strauss was retaliated against in 1957 when Congress revoked the AEC's discretionary budget and replaced it with a specific project budget. These controversies and opposition in Congress led Strauss to resign from the chair of the AEC on June 30, 1958. Eisenhower made him special adviser to the president on the Atoms for Peace program. Then Eisenhower nominated him to be secretary of commerce. In a lengthy two-month confirmation hearing, Strauss failed to win confirmation by a Senate vote of 46 to 49. This vote ended Strauss's government career, but he continued to work behind the scenes to oppose agreements with the Soviet Union. He died on January 21, 1974, of complications from lymphosarcoma.

Suggested reading: Richard Pfau, *No Sacrifice Too Great: The Life of Lewis L. Strauss* (Charlottesville: University Press of Virginia, 1984).

Strontium-90

The isotope strontium-90 became an international public health problem during the period of surface testing of nuclear weapons in the late 1950s and 1960s. Strontium-90 is a complex isotope that is formed by several processes. Krypton-90 is a radioactive gas formed by fission, and when it emits a beta particle the result is rubidium-90. Rubidium-90 has a half-life of 4.28 minutes. Half of rubidium-90 decays into strontium-90. Since strontium-90 is the radioactive form of an element that chemically resembles calcium, the human body is unable to distinguish calcium and strontium-90. Consequently, both are absorbed into the bones. Strontium-90 has a half-life of 28.9 years. Fallout during nuclear testing became a serious health concern in the United States, because even a small amount of strontium-90 produces bone tumors in human beings. Young children are especially susceptible, because strontium-90 concentrates in newly formed bone. Moreover, the bulk of strontium-90 from a fission explosion shoots up into the stratosphere, where it gradually falls to earth months, and even years, later.

The dangers of strontium-90 were known to scientists for several years before news reached the general public. A controversy developed in the United States in the mid-1950s over strontium-90 fallout from nuclear testing. The news became widespread when Adlai Stevenson, the Democratic candidate for president in 1956, attacked strontium-90 as part of his antinuclear testing plant in his campaign. Then in October 1956, a group of scientists from Brookhaven Laboratory backed up Stevenson's claim that strontium-90 was a threat not only to future generations but to the present one. The most obvious place to test for strontium-90 was in the milk supply, but the Atomic Energy Commission (AEC) was slow to begin testing, for political reasons. Once nuclear testing in the atmosphere ceased, the health threat of strontium-90 began to recede. By the late 1970s, strontium-90 was no longer a serious health concern in the United States.

Suggested reading: Edward Pochin, *Nuclear Radiation: Risks and Benefits* (Oxford, United Kingdom: Clarendon Press, 1983).

Superconducting Super Collider

The Superconducting Super Collider (SSC) was a U.S. government–sponsored effort in the early 1990s to expand the frontiers of knowledge about high-energy physics by constructing a collider with a circumference of more than 50 miles in Waxahachie, Texas. With a price tag of $8.2 billion, the federal government intended this collider to allow physicists to conduct high-energy experiments in a 54-mile tunnel, with 20 TeV (trillion electron volts) beam energy. The High Energy Physics Advisory Panel (HEPAP) had recommended building a superconducting super collider after a series of discoveries at the CERN (European Organization for Nuclear Research) proton-antiproton collider and the Fermilab Tevatron in the early 1980s. One of the main goals of the project was to uncover the mechanism by which elementary particles acquire their mass. Many scientists opposed this project because of the immense amount of resources consumed by it. President Ronald Reagan recommended the building of the SSC in 1988, and President Bush reconfirmed its funding in 1989. However, it soon became an issue in national politics, with cost-cutters and states not receiving the super collider allying to kill it. Congress acted to end the project in 1993 after it was only partially completed. The untimely death of this project showed that the era of big science in the United States had come to an end, with scientific research having to compete for resources at the national level and having to prove the economic merits of each project.

Suggested reading: Sidney D. Drell, *In the Shadow of the Bomb: Physics and Arms Control* (New York: American Institute of Physics, 1993).

Swedish Nuclear Power Program

The Swedish nuclear power program was one of the earliest casualties of the European antinuclear movement. Sweden was in a unique position because it had substantial deposits of uranium and a highly developed nuclear technology but no known coal, natural gas reserves, or oil. At the time of the 1975 Organization of Petroleum Exporting Countries (OPEC) oil embargo, Sweden depended for almost 75 percent of its primary energy on imported oil. The Central Power Supply Management (Central Drifts-Ledningen, CDL), which coordinated Sweden's electricity supply, received approval from the national Parliament (the Riksdag) in the early 1970s to build 11 nuclear power generation plants. When the OPEC embargo hit, the CDL had already proposed the construction of 13 more of these plants between 1975 and 1990. These building plans had business and government approval until a political party, the Center Party, interjected with concerns about nuclear safety. This party mounted a populist campaign against nuclear power in 1973 and 1974. Even the majority parties noticed a shift in public opinion and modified their former strong support. Only 2 of the original 13 nuclear power plants passed the Riksdag. Instead of new power plants, the Swedish government launched a national campaign to reduce consumption of energy to lessen dependence on imports of oil.

Suggested readings: Irvin C. Bupp and Jean-Claude Derian, *Light Water: How the Nuclear Dream Dissolved* (New York: Basic Books, 1978); William D. Nordhaus, *The Swedish Nuclear Dilemma: Energy and the Environment* (Washington, DC: Resources for the Future, 1997).

Szilard, Leo

Leo Szilard (1898–1964) was a Hungarian-American physicist–biologist whose fame is based on his early recognition of the destructive potential of the splitting of the atom. He was born on February 11, 1898, in Budapest, Hungary. His father was a civil engineer. Later in life, he married Gertrud Weiss. He graduated from the famous Minta Gymnasium in Budapest in 1916, after winning a prestigious prize for mathematics. Szilard first studied electrical engineering in Budapest at the King Joseph Institute of Technology before being drafted into the Austro-Hungarian Army. Shortly after the collapse of the Bela Kun regime in 1919, he left Hungary and moved to Berlin to study engineering at the Technology Institute, but changed fields in 1921 and turned to physics under Max Planck at the University of Berlin. After obtaining his Ph.D. in physics in 1922 with Max von Laue as his adviser, he continued conducting research in Berlin until 1933. During this period, he applied to the German patent office for 29 parents, some of which he had worked with Albert Einstein to perfect. Szilard left Germany with many other scientists because of the Nazi regime. After several years of working at Oxford's Clarendon Laboratory and at St. Bartholomew's Hospital, London, Szilard emigrated to the United States in 1938. He had an unconventional approach to life and found academic life too restrictive, so he rarely had an official academic position. Despite this background, Szilard had numerous friends in academic, business, and government circles.

Earlier than most physicists, Szilard recognized where discoveries on the nature of the atom were headed. In 1934, he rejected Rutherford's denial of the energy potential of the atom. Instead, he calculated that if present in sufficient numbers, neutrons from nuclear fission could sustain a nuclear chain reaction. He felt so strongly about this possibility that he applied for a patent to the process, but for secrecy reasons, he assigned the patent to the British Admiralty. When Szilard heard about Otto Hahn's research, and Lise Meitner and Otto Frisch's interpretation of it, Szilard knew that an atomic bomb was possible. His greatest fear was that Germany might acquire the bomb first. He approached his friend Albert Einstein and persuaded him to draft a letter with him to President Roosevelt, warning him of the possibility of an atomic bomb and the likelihood that Germany was working on building such a bomb. The Einstein-Szilard Letter started the process that led to the Manhattan Project.

During World War II, Szilard participated in building the atomic bomb, but gradually he started having doubts. His work with Enrico Fermi was key in developing the uranium-graphite pile experiment at the University of Chicago. From the beginning, Szilard opposed the military leadership of the Manhattan Project and its secrecy, and soon he made an enemy of its head, General Leslie Groves. At one time, Groves considered having him arrested as an enemy alien and jailed until the end of the war. It was only the likely hostility of the scientists that prevented Groves from doing this. As the bomb neared completion, Szilard began to question the need for its use against the Japanese. He was active in the drafting of the Franck Report. In 1943, Szilard became a naturalized American citizen.

Szilard's disillusionment with the atomic program led him to abandon nuclear physics. He turned to the study of biology, taking classes at the Cold Spring Laboratories. In 1946, he was appointed to a chair of biophysics at the University of Chicago. Szilard remained at Chicago until his death on May 30, 1964. In the intervening years, Szilard was active in campaigns to ban nuclear testing and nuclear weapons. In 1959, he received the Atoms for Peace Award.

Suggested reading: Richard Rhodes, *The Making of the Atomic Bomb* (New York: Simon and Schuster, 1986).

T

Tactical Nuclear Weapons

Tactical nuclear weapons (TNWs) are those weapons systems designed for use on the battlefield. They are designed to be battlefield nuclear weapons with deployment, ranges, and yields small enough to be confined to a localized military area. Almost from the beginning of atomic testing, military officials in both the Soviet Union and the United States began to lobby for atomic weapons small enough to be used on the battlefield. Much testing was done to make atomic weapons small enough to be delivered by aircraft, missiles, and guns. The problem of radiation in a battlefield situation was never solved and often ignored. Americans soon had the advantage in development of these weapons, and plans were made in 1952 for deployment of tactical nuclear weapons in Europe. It was this deployment that stimulated the antinuclear movements in Europe in the mid- and late 1950s. In October 1953 the Eisenhower administration directed the military to base planning on the battlefield use of nuclear weapons in the event they were militarily necessary. This directive resulted in the creation of the army's divisional structure into a pentomic division organization. *Pentomic organization* meant that military forces would be scattered on the battlefield in flexible formations to enable them first to survive an atomic attack and then to counterattack with tactical nuclear weapons. It was during the Kennedy administration that this policy was changed from dependence on tactical nuclear weapons back to conventional warfare. This policy has remained in the subsequent administrations despite the improvement in the technology of atomic weapons. A reappearance in the interest in tactical nuclear weapons took place in the Carter and then the Reagan administrations in the development of the neutron bomb. Even after the U.S. government finally lost interest in tactical nuclear warfare, the Soviet Union retained an interest in this type of battle planning. The Soviet military continued to train for battlefield scenarios that included tactical nuclear weapons, but this training also ended as the Soviet Union began to dissolve.

Suggested readings: Joseph Rotblat, *Nuclear Radiation in Warfare* (London: Taylor and Francis, 1981); William R. Van Cleave and S. T. Cohen, *Tactical Nuclear Weapons: An Examination of the Issues* (New York: Crane, Russak, 1978); Sherri L. Wasserman, *The Neutron Bomb Controversy: A Study in Alliance Politics* (New York: Praeger, 1983).

Tamm, Igor Evgenevich

Igor Tamm (1895–1971) was a Soviet physicist whose research on radiation and work on the Soviet H-bomb made him one of the top atomic physicists in the Soviet Union. He was born on July 8, 1895, in Vladivostok, Russia. His father was a city engineer. He married Natlie Shuskaia in 1917, and they had a daughter and a son. After study both at the Universities of Edinburgh in 1913 and Moscow in 1914, Tamm received his doctorate from the University of Moscow in 1918. His first position was at the Crimean University as a professor and administrator from 1918 to 1920. In 1920, he moved to the Odessa Polytechnical Institute as a professor until 1922. Another job change took him to the Sverdlov Communist University, where he remained from 1922 to 1924. Finally, in 1924 Tamm returned to the University of Moscow, where he stayed until his death in 1971. He became head of the department of theoretical physics in 1930. In 1934 Tamm organized the theoretical department at the Physics Institute of the Academy of Sciences (FIAN). In 1937 his brother and his boyhood friend were both arrested and shot in the Stalin purges. Tamm was uncertain about his fate for several years.

Much of Tamm's physics career was devoted to the Soviet nuclear program. He was one of the senior physicists in the developing of the Soviet H-bomb. In 1948 Tamm was appointed the head of a special nuclear bomb research group of the FIAN. His role as head of the theoretical department helped him recruit an outstanding group of young scientists. He worked closely with Andrei Sakharov in designing the Soviet hydrogen bomb. Tamm was the teacher and Sakharov the disciple, but they developed a close working relationship based on mutual affection and trust. Their alliance, with help from Vitaly Ginsburg, produced the basic structure of the hydrogen bomb by the end of 1948. He also collaborated with Pavel Cherenkov and Ilya Frank in work on the Cherenkov radiation effect. These three won the 1958 Nobel Prize in Physics for their research on radiation. He died on April 12, 1971, in Moscow.

Suggested readings: John Daintith, Sarah Mitchell, Elizabeth Tootill, and Derek Gjertsen, *A Biographical Encyclopedia of Scientists*, 2nd edition, vol. 2 (Bristol, UK:

Institute of Physics, 1994); David Holloway, *Stalin and the Bomb: The Soviet Union and Atomic Energy, 1939–1956* (New Haven, CT: Yale University Press, 1994).

Tammuz Nuclear Center Attack

The Israeli Air Force attacked and destroyed the two Iraqi reactors at the Tammuz Nuclear Center at El-Tuwitha on the outskirts of Baghdad on June 7, 1981. Israeli intelligence agencies learned that the Iraqi government had concluded an agreement with France to build a nuclear reactor facility capable of producing weapons-grade materials. In April 1979, Israeli agents destroyed two reactor cores in France by planting bombs on their shipping containers, hoping to discourage the French from proceeding further, but this action did not deter the French. These reactors were nearing completion by the middle of 1980 with the help of French nuclear technicians, and it was to be the centerpiece of the Iraqi nuclear program. In late 1980 Iraqi authorities began to behave erratically. All French technicians were expelled temporarily because of the Iran-Iraq War. Then the Iraqi government indicated that it would no longer permit inspections by the International Atomic Energy Agency (IAEA). The possibility of any Arab state possessing atomic weapons was unacceptable to the Israeli government. Several squadrons of Israeli F-15 and F-16 fighters attacked the facility with surgical precision. Besides completely destroying the nuclear facility, casualties included the death of a French technician. This preemptive strike by Israel produced a huge outcry in the international community and threats of sanctions against Israel. Even the United States protested this raid, but nothing of substance happened because both the Soviet Union and the United States were relieved that Iraq no long had access to the reactors. France took this raid as an excuse to cancel the contract to rebuild the reactors.

Suggested reading: Dan Raviv and Yossi Melman, *Every Spy a Prince: The Complete History of Israel's Intelligence Community* (Boston: Houghton Mifflin, 1990).

Target Committee of the Manhattan Project

The Target Committee of the Manhattan Project made the recommendation in early May 1945 on which targets to drop the two available atomic bombs. This committee was appointed by General Leslie Groves and consisted of General Thomas F. Farrell, Colonel Seeman, Dr. J. Robert Oppenheimer, Dr. John von Neumann, Dr. Norman Ramsey, Captain Deak Parsons, Major J. A. Derry, Dr. J. C. Stearns, Dr. Richard Tolman, Dr. Charles C. Lauritsen, Dr. D. M. Dennison, Dr. R. B. Wilson, and Dr. William Penney. Dr. Hans Bethe and Dr. Robert B. Brodie were also present at the meeting. From a list of 17 potential targets, the committee selected 4 targets following Groves's guidelines that each target should be unda-

maged by earlier raids and be a good subject to study the effects of the blast.
Two other factors were also important—the psychological impact on Japan
of the particular target chosen and the target's impact on the international
community, particularly the Soviet Union. Hiroshima, Kokura, Kyoto, and
Nigata were on the original list. Kyoto was the favorite candidate of the
Target Committee and General Groves, but Secretary of Defense Henry
Stimson decided for cultural reasons to take Kyoto off the list Stimson had
visited Kyoto earlier and believed the destruction of a national, treasure
would be a crime. Nagasaki as a potential target site was determined later.

Suggested readings: William Lawrence, *The General and the Bomb: A Biography
of General Leslie R. Groves, Director of the Manhattan Project* (New York: Dodd,
Mead, 1988); Martin J. Sherwin, *A World Destroyed: The Atomic Bomb and the
Grand Alliance* (New York: Knopf, 1975).

Teapot Nuclear Tests

The Teapot series of nuclear tests in 1955, which were conducted at the
Nevada Test Site, consisted of shots designed to develop newer, lighter, and
more efficient nuclear weapons. Most of the scientific expertise for this series
came from the Los Alamos Scientific Laboratory. Military demands for battlefield nuclear weapons caused the scientists to experiment with miniatur-
ization. Atomic Energy Commission (AEC) officials were also concerned
about keeping radioactive fallout down. Shot Wash was the first test, and it
took place by airdrop on February 18, 1955. Since it was only a 1-kiloton
blast, it disappointed the soldier witnesses and produced little blast. Shot
Moth was next, and it was also a low-yield device testing a prototype atomic
missile warhead. This device was detonated on February 22, 1955, and it
produced more radiation than the previous one. Shot Tesla was next, and
this time troop maneuvers were planned with it. In a surprise, the small
device, which was designed by the Livermore Radiation Laboratory, produced over a three times higher yield than proposed. The 7-kiloton blast
caused some problems for the troops, but the objectives at ground zero
were achieved. The next test, also a Livermore project, was a demonstration
for an intercontinental ballistic missile warhead. The device was placed on
a 500-foot tower and detonated on March 7, 1955. Its yield was 43 kilotons, and the radiation levels were too high for troop maneuvers. The radioactive fallout from this test was high, as the upper-level winds distributed
it all over the country. Shot Hornet was next, and its detonation on March
12, 1955, was a further refinement of a tactical nuclear weapon. Because
the scientists had used smog pots, they found ground zero to be hot for
such a small 4-kiloton yield. Shot Bee was another combined operation to
test military tactics on the nuclear battlefield. An 8-kiloton device was det-

onated on March 22, 1955. In Shot ESS the test was to demonstrate an atomic demolition satchel charge for one-person operations. The device was placed in a 67-foot shaft underground and detonated on March 23, 1955. It was found that the 1-kiloton yield produced a 96-foot-deep crater that contained a highly radioactive material. March 29 proved to be a busy day, as both Shot Apple-1 and Shot Wasp Prime were detonated on the same day. The next test was a demonstration of the first atomic antiaircraft weapon. It was named Shot HA (High Altitude). On April 6, 1955, the test was successful, with a 3-kiloton yield and almost no radioactivity measured. Shot Post was detonated on April 9, 1955, for an experiment for the Livermore Radiation Laboratory. The next test was Shot MET (Military Effects Test), and it was designed to measure radiation effects on equipment and clothing. Detonation was from a high tower, and the April 15 blast produced a yield of 22 kilotons. Shot Apple-2 was a media event, and it featured a Civil Defense theme. After several delays, the bomb was detonated on May 5, 1955, and it produced a yield of 29 kilotons. The fallout was huge, and it spread all over the United States within the next week. The final test was Shot Zucchini. It was detonated on May 15 and had a yield of 28 kilotons.

Suggested readings: Howard Ball, *Justice Downwind: America's Atomic Testing Program in the 1950s* (New York: Oxford University Press, 1986); Richard L. Miller, *Under the Cloud: The Decades of Nuclear Testing* (New York: Free Press, 1986).

Telemark Raid

During World War II, Allied planners became so concerned that Nazi experiments with heavy water from Norway would enable the Germans to produce an atomic bomb that a raid was carried out to cripple production. Successive raids against heavy water facilities at the Norsk-Hydro Plant at Rjukan made the Nazi authorities decide to disassemble the plant and move it to Germany. Available supplies of heavy water were to be transported to Germany for atomic weapons research. Allied intelligence informed Norwegian intelligence about the importance of destroying this shipment. Norwegian resistance leaders decided that the best place to accomplish this was to detonate a bomb on the ferryboat carrying the heavy water in the middle of Lake Tinnsjo. Since the middle of the lake was about 1,300 feet deep, the heavy water would be beyond German control. Two members of the Norwegian resistance planted a bomb aboard the ferryboat, the *Hydro*, and the ferry blew up on February 20, 1944, in the middle of the lake. It sank within four minutes, along with 26 passengers and crew. A few of the heavy water drums floated to the surface and were saved by the Germans, but not enough to be significant. This action ensured that the German atomic program would be chronically short of heavy water for experimentation for the rest of the war.

Suggested readings: Thomas Gallagher, *Assault in Norway: The True Story of the Telemark Raid* (London: Purnell Books Services, 1975); David Irving, *The German Atomic Bomb: The History of Nuclear Research in Nazi Germany* (New York: Simon and Schuster, 1967); Richard Rhodes, *The Making of the Atomic Bomb* (New York: Simon and Schuster, 1986).

Teller, Edward

Edward Teller (1908–) is Hungarian-American theoretical physicist who is more important as an advocate of the building of the hydrogen bomb than for his work as a physicist. He was born on January 15, 1908, in Budapest, Hungary. His father was a prosperous lawyer. In 1934, Teller married his childhood sweetheart Mici Harkanyi. Teller's early education was at the Mellinger School and then the prestigious Minta School in Budapest, but his schooling was marred by the turmoil surrounding Bela Kun's Communist Uprising in 1919. Although Teller wanted to study mathematics, his father insisted on a practical education, so a compromise was worked out for him to major in chemical engineering, with a minor in mathematics. Teller began his college education at the Budapest Institute of Technology, but much of his training was at the German Universities of Karlsruhe, Munich, and Leipzig. While at the University of Munich studying under Arnold Sommerfeld, Teller suffered a serious streetcar accident in 1928 that resulted in the amputation of his right foot above the ankle. After recovering from his injury, he decided to move to the University of Leipzig, where he received a Ph.D. in physics in 1930 studying under Werner Heisenberg. Teller's education at these German institutions exposed him to some of the leading atomic scientists of the day. In 1930 he was offered an assistantship at the University of Göttingen, which lasted for two years. Teller was invited to Rome by Enrico Fermi in 1932 to work on specialized physics problems. Teller left Germany in 1933 soon after the Nazi takeover. After short stays in Copenhagen and London, he emigrated to the United States, where he obtained a professorship in physics at George Washington University, Washington, DC. He followed his good friend George Gamow to George Washington. In March 1941, Teller became a naturalized American citizen.

During World War II, Teller was active in the design and building of the atomic bomb. He worked in the Theoretical Physics Division at Los Alamos. His long friendship with Hans Bethe was hurt by Bethe's appointment as head of the theoretical physics unit. At this time, Teller became a strong proponent of the fusion over the fission bomb. Despite efforts of Los Alamos director J. Robert Oppenheimer to give Teller more freedom, they clashed over this issue. Teller became so difficult to work with because of his obstreperous behavior that he was relieved of work on the atomic bomb and left alone to work on hydrogen bomb theory. Toward the end of the Manhattan Project, Teller drafted a letter advancing the case for the hydrogen bomb and forecasting its completion within five years.

After the war, Teller became a powerful political figure in nuclear politics. In 1946, he left Los Alamos and moved to a professorship at the University of Chicago. Teller kept his links with Los Alamos and was appointed in 1946 assistant director of Weapons Development with the task of designing the hydrogen bomb. He remained at this post until the successful design of the hydrogen bomb. In early 1951 Teller and Stanislaw Ulam, a Polish-American mathematician, worked out the theoretical basis for the hydrogen bomb. Ulam conceived the breakthrough idea based on fission compression, but Teller worked out the rest of the problem. Teller also assumed all of the credit for the breakthrough. During this time, however, Teller began to feud with the Los Alamos director Norris Bradbury, and the appointment of Marshall Holloway to head the thermonuclear program at Los Alamos was the last straw. Teller believed that he should have had that appointment. Soon after the theoretical work was done, Teller left Los Alamos and returned to the University of Chicago. He began to lobby for a second laboratory for weapons work at Livermore, California, and his political contacts helped him win approval. Shortly afterward, Teller's testimony against Oppenheimer in his security hearing caused an uproar in the scientific community, and many physicists ostracized Teller for many years for his testimony.

Teller moved to the University of California at Berkeley in 1953. Much of his work at the University of California at Berkeley was at the Lawrence Livermore Laboratory where research was being carried out on designing atomic weapons. He was director of the laboratory from 1958 to 1960. During the controversy over radioactive fallout, Teller was adamant in denying its dangers. Because of his strong commitment to nuclear energy, Teller also began to look for ways to use nuclear explosions for peaceful purposes. Between his hawkish views toward the Soviet Union and his advocacy of peaceful nuclear engineering projects, Teller became the leading spokesperson for nuclear energy in the postwar era. Teller retired from the University of California at Berkeley and the Lawrence Livermore Laboratory in 1975. He continues to advocate the use of nuclear energy.

Suggested readings: Stanley A. Blumberg and Gwinn Owens, *Energy and Conflict: The Life and Times of Edward Teller* (New York: Putnam's Sons, 1976); John Daintith, Sarah Mitchell, Elizabeth Tootill and Derek Gjertsen, *A Biographical Encyclopedia of Scientists*, 2nd ed., vol. 2 (Bristol, UK: Institute of Physics, 1994); Richard Rhodes, *Dark Sun: The Making of the Hydrogen Bomb* (New York: Touchstone, 1995).

Terrorism. *See* Nuclear Terrorism

Test Ban Treaty. *See* Comprehensive Test Ban Treaty

Theoretical Megaton Group. *See* Panda Committee

Thomson, George Paget

George Paget Thomson (1882–1975) was a British physicist respected for his research on atomic subjects but also important for his leadership role as chair of the MAUD Committee. He was born on May 3, 1892, in Cambridge, England. His father was the famous British physicist Joseph John Thomson. He received all of his college education at Cambridge University. Between 1914 and 1922, Thomson held a position in physics at Cambridge University. During World War I, however, he joined the British army in the 1st Queen's and later in the war was transferred to the Royal Air Corps to conduct research on aeronautics. In 1922, he moved to the chair of physics at Aberdeen University. In 1930, Thomson was appointed to the chair of physics at the Imperial College, London, where he remained until 1952. For the next decade until retirement in 1952, he was master of Corpus Christi College, Cambridge University.

Thomson's claim to fame as a physicist was for his investigations of isotope composition using the mass spectrograph method. In one of these experiments in 1927 Thomson found evidence to support the wave-particle duality of the electron. It was this research that had Thomson share the 1937 Nobel Prize for Physics with Clinton J. Davisson.

Almost as important as the results of his scientific research was his role as chair of the MAUD Committee. It was the duty of this committee to advise the British government on the possibility of building an atomic bomb. He had conducted an early experiment in May 1939 with uranium oxide and various moderators to investigate the possibility of starting a continuous chain reaction. His conclusion at that time was that it was not going to be easy to build an atomic weapon, especially in wartime. Although always skeptical, Thomson ended up agreeing with his scientific colleagues that such as bomb was possible and that it should be built.

Suggested readings: Henry A. Boorse, Lloyd Motz, and Jefferson Hane Weaver, *The Atomic Scientists: A Biographical History* (New York: Wiley Science Editions, 1989); John Daintith, Sarah Mitchell, Elizabeth Tootill, and Derek Gjertsen, *A Biographical Encyclopedia of Scientists*, 2nd ed., vol. 2 (Bristol, UK: Institute of Physics, 1994).

Thomson, Joseph John

British physicist Sir J. J. Thomson (1856–1940) was one of the pioneers in the study of atomic structure and longtime head of the Cavendish Laboratory at Cambridge University. He was born in Manchester, England, on December 18, 1856. His father was a bookseller in Manchester. In 1890, he married Rose Elizabeth Paget, and they had a son and a daughter. His father wanted him to be an engineer, so he entered Owens College at the age of 14. He secured a certificate in engineering in 1876, then spent 2 years studying mathematics and physics. His physics instructor was brilliant physicist Balfour Stewart. Thomson received his B.A. in 1880 from Cambridge University. In 1885, Thomson was the surprise choice for the replacement of John Rayleigh as Cavendish Professor of Experimental Physics. His entire academic career was spent at Cambridge University, from 1883 to 1918.

Thomson spent his career building the reputation of the Cavendish Laboratory as a great experimental research center. Thomson received the 1906 Nobel Prize in Physics for his work on the conduction of electricity through gases. As important as this research proved to be, it was his discovery in 1897 that cathode rays were not waves of radiation but minute particles of matter, each carrying an electric charge, that made his reputation in the nuclear research world. The Thomson atom was "a number of negatively-electrified corpuscles in a sphere of uniform positive electrification." This type of research and his able administration of the Cavendish Laboratory made Thomson an important figure in early research on the structure of the atom. Only during the war years from 1914 to 1918 did the Cavendish Laboratory languish and its research experiments diminish. Thomson's appointment as the master of Trinity College in 1918 and his resignation from the Cavendish Chair of Physics permitted his successor, Sir Ernest Rutherford, to carry on his high standards of experimental research. In honor of his scientific contributions, Thomson was knighted in 1908. Thomson died on August 30, 1940, Cambridge, England.

Suggested readings: Henry A. Boorse, Lloyd Motz, and Jefferson Hane Weaver, *The Atomic Scientists: A Biographical History* (New York: Wiley Science Editions, 1989); John Daintith, Sarah Mitchell, Elizabeth Tootill, and Derek Gjerfsen, *A Biographical Encyclopedia of Scientists*, 2nd ed., vol. 2 (Bristol, UK: Institute of Physics, 1994); J. S. Robert, *The Life of J. J. Thomson* (Cambridge: Cambridge University Press, 1943).

Thomson Committee. *See* MAUD Committee

Three Mile Island

The Three Mile Island Nuclear Plant accident on March 28, 1979, was the most serious nuclear incident in American history. Metropolitan Edison decided to build a nuclear facility on Three Mile Island in the middle of the Susquehanna River in eastern Pennsylvania in December 1966. A local engineering firm, Gilbert Associates, was selected to design Unit One, and Babcock and Wilcox to supply the steam circulation system. The Atomic Energy Commission issued Metropolitan Edison a construction permit for Unit One in May 1968. A second unit, Unit Two, was announced in December 1968. It had a different designer, Burns and Roe. The steam turbine designers were also different: General Electric for Unit One and Westinghouse for Unit Two. Both were pressurized-water reactors designed by Babcock and Wilcox.

The two units had a different history. Unit One's construction and loading proceeded smoothly, on time and within budget. It started operation on April 20, 1974. Unit Two had both construction and cost overruns. It finally received its operating license on February 8, 1978, but due to various problems, mostly leaking valves, Unit Two did not go into operation until December 30, 1978.

Less than three months later, on March 28, 1979, an accident in Unit Two resulted in a reactor meltdown. A few seconds after 4:00 A.M., a safety device malfunctioned while the operators were trying to clear resin beads. The main feedwater pumps lost their water supply, causing the turbine to trip and start overheating the reactor. Then the computer system collapsed and started giving out false readings. The operators had no idea what was going on and they reported difficulties to their station manager. Because the emergency core cooling system had been shut off previously, Unit Two had a core meltdown. Temperatures inside the reactor core increased and fuel rods ruptured. Unknown to the operators, radioactive gases were pouring into the skies above the complex.

It was nearly three hours after the initial problem before personnel in Unit Two called for outside help. At 7:24 A.M. a General Emergency was declared on Three Mile Island. However Representatives from the Nuclear Regulatory Commission (NRC) did not arrive until after 10:00 A.M. Immediately after the accident Metropolitan Edison officials were more concerned about the economics of the situation, since the company had to pay at least a half million dollars a day to purchase replacement electricity for Unit Two. At 1:50 P.M. hydrogen exploded in the containment building that surrounded the reactor. By 4:30 P.M. a NRC radiation-detection helicopter found traces of radioactivity in the air near by.

Early the next day, March 29, the consensus among investigators was that the worst was over until they discovered a 1,000-cubic-foot gas bubble in the containment building. When NRC commissioners found out that this bubble was made of hydrogen, they realized that the situation might become worse. Although they recommended evacuation of the area within a 10-mile radius, nothing was done. Finally, Governor Richard Thornburgh ordered the evacuation of pregnant women and preschool children within a one-mile radius. Fortunately, the hydrogen bubble did not burst but slowly disappeared.

Pennsylvania police and plant security guards prevent outsiders from entering the Three Mile Island Nuclear Power Plant shortly after the March 28, 1979 nuclear accident. The reactor meltdown in Unit Two caused the release of radiation into the atmosphere, but the containment building surrounding the reactor prevented further damage. (Photo courtesy of AP/Wide World Photos)

President Jimmy Carter formed the 12-member Kemeny Commission to investigate the causes of the accident. The members concluded that the training of the operators was totally inadequate for an emergency, and that the NRC had ignored previous incidents and warnings. A recommendation to abolish the NRC was also proposed. Steps were taken to improve training of operators, and more stringent safety regulations were instituted by the Nuclear Regulatory Agency. Changes in leadership within the NRC took place, but the proposal to abolish it failed. The accident, however, did destroy public confidence in nuclear power, the nuclear industry, the safety of nuclear reactors, and utility management.

The cleanup did not get to look at the damaged core until the summer of 1982. They found that the meltdown of the reactor was far worse than previously thought, with 53 percent of the reactor's core melted, but the reactor vessel was virtually unscathed. Nuclear engineers concluded that the reactor in Unit 2 had barely escaped a total meltdown. Cleanup costs were estimated in the billion-dollar range, and the federal government, the state of Pennsylvania, private utilities around the country, and customers are paying the bill. Charges of mismanagement, inefficiency, and waste have come from engineers engaged in the cleanup. Despite claims of health damage from citizens of the area, authorities have steadily denied that the accident endangered the health of the local inhabitants or the workers.

Suggested readings: Philip Starr, *Three Mile Island Sourcebook: Annotations of a Disaster* (New York: Garland, 1983); Mark Stephens, *Three Mile Island* (New York: Random House, 1980).

Tibbets, Paul Warfield

Colonel Paul Tibbets (1915–   ) was the pilot of the B-29 bomber, the *Enola Gay*, that dropped the atomic bomb on Hiroshima on August 6, 1945. He was born on February 23, 1915, in Quincy, Illinois. His father was a whole-sale confectioner. At an early age his family moved to the Miami, Florida, area. After an education at a military school, Western Military Academy at Alton, Illinois and college at the Universities of Florida and Cincinnati, Tibbets joined the United States Air Corps in 1937. During World War II, he had become one of America's most successful bomber pilots flying B-17s in combat in Europe. He was a squadron commander of the 340th Bomb Squadron, 97th Bombardment Group. Later in the war, Tibbets had returned to the United States to serve as a test pilot for the new B-29 Superfortress. After a security vetting by Military Security, Tibbets was placed in command of the 393rd Heavy Bombardment Squadron of 15 bomber crews to train for an atomic strike mission. They trained at Wendover Air Base in Wendover, Utah. The code name for their mission was Silverplate. After months of training, other units were brought in to comprise the 509th Composite Group, attached to the 315th Bombardment Wing of the Second Air Force. Colonel Tibbets found himself commanding the 509th Composite Group. After training at Wendover had become regularized, Tibbets ordered an advanced element to Trinian to prepare the site there for the transfer of the 509th. Soon the rest of the 509th followed, and training resumed there. Tibbets decided to fly the first mission himself and renamed his B-29 bomber the *Enola Gay* after his mother. After the successful mission, Tibbets became famous. He has spent the rest of his life trying to live down this fame. After the war, he participated in the Crossroads atomic bomb tests at Bikini, South Pacific, as a technical advisor. Tibbets served in a variety of commands until his retirement from the air force on August 31, 1966, with the rank of general. Tibbets's civilian career continued his contact with aviation. After a tour in Europe operating Lear jets, he joined an all-jet air tax service— Executive Jet Aviation. In 1982 he became chairman of the board of Executive Jet Aviation, and in 1985 he retired from the company. Since then he has toured around the country speaking and conducting book signings.

Suggested reading: Gordon Thomas and Max Morgan-Witts, *Ruin from the Air: The Atomic Mission to Hiroshima* (London: Hamish Hamilton, 1977).

Tokamak Fusion Reactor

A tokamak (from the Russian for toroidal chamber with magnetic coil) fusion reactor is designed to produce energy from the process of fusion. Such a reactor combines a magnetic filed and a driven plasma to allow the high temperature necessary to produce fusion. Within the reactor confinement chamber, hot fuel forms a low-density ionized gas, or plasma, inside a vessel shaped like a doughnut. Two isotopes of hydrogen, deuterium and tritium, are the active components of the fuel. At sufficient temperature and density, significant numbers of these two species fuse together in what is called the DT (deuterium-tritium) reaction, releasing energy in the process.

The idea for a tokamak reactor came out of the Soviet Union. Soviet physicists Igor Tamm and Andrei Sakharov provided the theory in the early 1950s. Under the sponsorship of the head of the Soviet Nuclear Program, Igor Kurchatov, a series of tokamaks were constructed in the 1950s. American scientists learned about them from Kurchatov in 1956, but Soviet claims of high-temperature successes in the early 1960s were distrusted as unreliable. It was only after a group of British scientists verified the Soviet results in August 1968 that American and British scientists reconsidered and accepted that tokamaks were worthy of further study.

The excitement by the international scientific community over the tokamak led to the building of several large tokamak machines for research. Three-billion-dollar tokamaks were built: the Tokamak Fusion Test Reactor (TFTR) in the United States, the Joint European Torus (JET) in Europe, and the Japan Tokamak 60 (JT-60) in Japan. Several smaller tokamak magnetic reactors have been constructed in the United States and elsewhere for research purposes.

The experiment to prove the viability of controlled fusion energy took place on the TFTR at Princeton University on December 9, 1993. For the first time Princeton scientists combined a 50–50 mixture of deuterium and tritium fuel. A peak of 3 million watts of fusion power resulted from the first experiment. Subsequent experiments produced even greater power outputs, with a November 2, 1994, test reaching a peak of 10.7 million watts. These results in an experimental reactor caused scientists to predict even greater power outputs in larger tokamaks. A campaign ensued for the construction of the International Thermonuclear Experimental Reactor (ITER) to prove that magnetic fusion reactors can produce significant energy for civilian purposes.

Suggested reading: T. Kenneth Fowler, *The Fusion Quest* (Baltimore, MD: Johns Hopkins University Press, 1997).

Tokamak Fusion Test Reactor. *See* Tokamak Fusion Reactor

Tomsk-7

Tomsk-7 is a Russian nuclear facility located in Siberia on the Tom River about 10 miles northwest of the city of Tomsk. It was founded in 1949 as a production facility for chemical separation, plutonium processing, and uranium enrichment. The Siberian Chemical Combine administers Tomsk-7. The closed city of Seversk was built to house the workforce and now has a population of over 107,000. At one time there were five graphite-moderated reactors in operation at this facility, but three of the reactors were shut down in the early 1990s. It is also the site of nuclear waste management facilities.

Suggested reading: Thomas B. Cochran, Robert S. Norris, and Oleg A. Bukharin, *Making the Russian Bomb: From Stalin to Yeltsin* (Boulder, CO: Westview Press, 1995).

Top Policy Group

The Top Policy Group was the body designated by President Franklin D. Roosevelt in October 1941 to decide atomic energy policy. When it became apparent that a decision had to be made on whether or not to institute a crash program to build an atomic bomb, President Roosevelt wanted this decision to come from the highest political level. The MAUD Report and American scientific opinion concurred that an atomic bomb could be built but at great cost and with no guarantee of success. At a meeting on October 9, 1941, between President Roosevelt, Vice President Henry Wallace, and Vannevar Bush, head of the National Defense Research Committee, it was decided to form the Top Policy Group. Its membership consisted of President Roosevelt, Vice President Wallace, Secretary of War Henry L. Stimson, Army Chief of Staff George C. Marshall, Vannevar Bush, and James Bryant Conant. Each member of this group reported directly to the president. This decision also isolated Bush and Conant from criticism from the scientific community. Ultimately, President Roosevelt made the decision to proceed with building the atomic bomb and provided funding outside the knowledge and control of the U.S. Congress.

Suggested readings: Richard Rhodes, *The Making of the Atomic Bomb* (New York: Simon and Schuster, 1986); Martin J. Sherwin, *A World Destroyed: The Atomic Bomb and the Grand Alliance* (New York: Knopf, 1975).

Treason for Spying. *See* Rosenberg Case; Soviet Atomic Spying

Treaty of Bangkok

The Treaty of Bangkok was signed on December 15, 1995, by the member states of the Association of Southeast Asian Nations (ASEAN) to establish a nuclear weapon free zone in southeast Asia. This alliance was formed in 1967 by Indonesia, Malaysia, Philippines, Singapore, and Thailand to promote regional stability by cooperation on political and economic issues. In a series of summits, members of ASEAN came together to discuss issues of regional concern. In 1984, between the second and third summit, the subject of a southeast Asia nuclear weapon free zone was proposed for study. Brunei joined ASEAN in 1987 as its sixth member. In the fourth ASEAN summit in Singapore in 1992, a directive was issued to the foreign ministers of the member states to consider the establishment of a nuclear weapon free zone. At the fifth summit in Bangkok in December 1995, the ASEAN members adopted the treaty. The six ASEAN members were joined in signing by four other concerned states: Vietnam, Cambodia, Laos, and Myanmar. In 1996, Vietnam became the seventh member state of ASEAN.

The provisions of this treaty resemble those of other nuclear weapon free zone treaties. The treaty prohibits the development, manufacture, acquisition, possession, control, stationing, transport, testing, or use of nuclear weapons. Peaceful uses for nuclear energy are allowed, but only under the auspices of the International Atomic Energy Agency (IAEA). Dumping of radioactive waste or materials anywhere in the zone is also prohibited. The treaty comes into force with the ratification of seven of the ten signatories. A provision allows for a 10-year review of the treaty, and a state can withdraw from the treaty after a 12-month notice. The treaty is considered to be weaker in its provisions and less enforceable than other nuclear weapon free zone treaties. Nevertheless, the treaty entered into force on March 28, 1997, when Cambodia became the seventh member nation to ratify it. Singapore ratified the treaty the same day. Only Indonesia and the Philippines still remain to ratify the treaty.

Suggested reading: Ramesh Thakur, ed., *Nuclear Weapon-Free Zones* (New York: St. Martin's Press, 1998).

Treaty of Pelindaba

The Treaty of Pelindaba establishes a nuclear weapon free zone in Africa. First steps toward this treaty began when the Organization of African Unity advocated the denuclearization of Africa at its summit in Cairo, Egypt, in July 1964. Further action in the U.N. General Assembly in 1965 reinforced the desire of African states for a nuclear-weapon free zone in Africa. An unstable political situation with superpowers meddling in African affairs and the strong possibility that the apartheid regime in South Africa possessed nuclear weapon capability limited options. Both the collapse of the Soviet Union in 1989, ending the superpower rivalry, and the renunciation of atomic weapons by the government of South Africa in 1989 opened the way for serious negotiations to proceed. The dismantling of its atomic arsenal and the signing of the Nuclear Non-Proliferation Treaty by the Republic of South Africa in 1991 was the last roadblock for consideration of a treaty. After several years of negotiations, the treaty was opened for signature on April 11, 1996, in Cairo, Egypt. All African countries had signed the treaty by November 19, 1998, except for Equatorial Guinea, Madagascar, and Somalia. It takes ratification of 28 out of the 50 states for the treaty to go into force. As of November 1998, only eight African states had ratified the treaty. The name of the treaty is symbolic because Pelindaba was the site of the atomic bomb program of the Republic of South Africa.

The treaty mandates a complete excusion of all nuclear weapons from Africa. It prohibits the research, development, manufacture, stockpiling, acquisition, testing, possession, control, or stationing of nuclear explosive devices in Africa. Dumping of radioactive wastes is also forbidden by any member state. While it allows the peaceful use of nuclear energy, all parties of the treaty must adhere to all safeguards of the International Atomic Energy Agency. Three protocols are attached to the treaty to insure that the major nuclear powers agree not to use or threaten to use a nuclear weapon against an African state: Protocols I and II require the signatures and ratification by the major nuclear powers, and Protocol III only proposes the signatures of France and Spain. As of November 1998, only France has signed and ratified all three protocols. The People's Republic of China has signed and ratified both Protocol I and II. Russia, the United Kingdom, and the United States have signed but not ratified Protocol I and II. Spain has not signed or ratified Protocol III.

Suggested readings: Joseph Rotblat, ed., *Nuclear Weapons: The Road to Zero* (Boulder, CO: Westview Press, 1998); Ramesh Thakur, ed., *Nuclear Weapons-Free Zones* (New York: St. Martin's Press, 1998).

Treaty of Rarotonga

This treaty was signed at Rarotonga on August 6, 1985, in an attempt by Pacific countries to keep nuclear testing out of the Pacific region. The original idea of a nuclear free zone for the South Pacific came out of a regional meeting in 1975 of the South Pacific Forum, but the electoral loss of the Australian Labor Party, a key backer, ended further initiatives. A nuclear free zone proposal reappeared in 1983 with the political success of the Labor Party in Australia and New Zealand. Negotiations in 1984 and 1985 between members of the South Pacific Forum produced a nuclear free zone treaty in August 1985. Australia, Tuvalu, the Cook Islands, Fiji, Kiribati, Aotearoa/New Zealand, Niue, and Western Samoa were the original signers. Papua New Guinea and Nauru joined later. Three South Pacific Forum countries—Tonga, Solomon Islands, and Vanuatu—declined to sign the treaty for a variety of reasons. This treaty was limited in scope, with provisions to prohibit dumping of nuclear waste in the sea, to end nuclear explosions, and to prevent acquisition of nuclear weapons or their basing within signatories' territorial limits. It does allow peaceful nuclear activities, but with severe restrictions. Critics have charged that this treaty leaves out too much and was dictated by the Australian government. Nevertheless, the treaty, entered into force on December 11, 1986 when the necessary majority ratified the treaty. For several years the South Pacific Nuclear Free Zone was been ignored by its two biggest critics, France and the United States. France continued to test nuclear weapons at Mururoa Test Site until 1995. While the United States was always less confrontational, it refused to honor the existence of the South Pacific nuclear weapons free–zone. In March 1996, France, the United Kingdom, and the United States finally signed the protocols in support of the nuclear weapon free–zone. All of the south pacific forum states have now signed the treaty, but Tonga has yet to ratify it.

Suggested readings: Jane Dibblin, *Day of Two Suns: U.S. Nuclear Testing and the Pacific Islanders* (London: Virago Press, 1988); David Pitt and Gordon Thompson, eds., *Nuclear-Free Zones* (London: Croom Helm, 1987); Ramesh Thakur, ed., *Nuclear Weapons-Free Zones* (New York: St. Martin's Press, 1998).

Treaty of Tlatelolco

The Treaty of Tlatelolco was an agreement among Latin American countries to prohibit the introduction of nuclear weapons into Latin America. Several of the countries—Bolivia, Brazil, Chile, Ecuador, and Mexico—became concerned that they had come close to involvement in the nuclear gamesmanship of the Cold War during the Cuban missile crisis of 1962. The possibility of nuclear war in the hemisphere by the introduction of Soviet nuclear weapons into Cuba in 1962 scared these leaders into action. Presidents of these countries decided to create a denuclearized zone in Latin America. The production, importation, storage and testing of nuclear weapons were to be prohibited. Mexico took a leadership role in the negotiations, and the preparatory commission for the Denuclearization of Latin America (COPREDAL) under the leadership of Ambassador Alfonso Garcia Robles was formed to study the problem. This commission worked for two years and produced the Treaty of Tlatelolco, which was signed by 21 Latin American countries on February 12, 1967. A provision of the treaty did allow Latin American states to detonate nuclear devices for peaceful purposes, but only if advance notice were given and the tests carried out under International Atomic Energy Agency (IAEA) surveillance. Only Cuba refused to sign the treaty, but Argentina never ratified it. Both Brazil and Chile invoked a clause allowing them more leeway in adhering to the terms. The creation in 1969 of the Agency for the Prohibition of Nuclear Weapons in Latin

America and the Caribbean (OPANAL) has insured that there is an administrative organization to oversee the continued implementation of the treaty. Argentina in January 1994 and Brazil in May 1994 finally ratified the treaty. In March 1995 Cuba signed the agreement, making all the Latin American countries signatories. In an effort to control the nuclear rivalry between Argentina and Brazil, a new agency, the Argentine-Brazilian Common System of Accounting and Control (ABACC), was established in December 1992 in Rio de Janeiro, Brazil. This agency continues to monitor the nuclear relationship between the two Latin American superpowers.

Suggested readings: Bertrand Goldschmidt, *The Atomic Complex: A Worldwide Political History of Nuclear Energy* (La Grange Park, IL: American Nuclear Society, 1982); David Pitt and Gordon Thompson, eds., *Nuclear-Free Zones* (London: Croom Helm, 1987); Ramesh Thakur, ed., *Nuclear Weapons-Free Zones* (New York: St. Martin's Press, 1998).

Trinity

Trinity was the code name for the atomic bomb tested on the Alamogordo Test Range on July 16, 1945. The bomb was detonated in a desert area called Jornada del Muerto (Journey of Death), about 250 miles south of Los Alamos between Socorro, New Mexico, and El Paso, Texas. It was the plutonium bomb named Fat Man. At 5:29.45 Mountain War Time, the bomb exploded, sending up a huge fireball, which could be seen in three states. The heat at the center of the blast rivaled that of the center of the sun. Within about seven minutes, the cloud reached 38,000 feet into the atmosphere. Most of the scientists participated in a betting pool on the size of the explosion, and except for a few, they underestimated it. The test also gave scientists their first experience with radioactive fallout, and they learned several important lessons on its range and effects on people and animals. It was a much more powerful weapon than the scientists had envisaged, and the radioactive fallout presented health concerns for people and animals over a wide area. President Harry Truman was at the Potsdam Conference when the news of the test reached him. Truman informed Joseph Stalin of the existence of a new weapon the next day.

Suggested reading: Ferenc Morton Szasz, *The Day the Sun Rose Twice: The Story of the Trinity Site Nuclear Explosion, July 16, 1945* (Albuquerque: University of New Mexico Press, 1984).

Tritium

Tritium is an isotope of hydrogen that is used as part of the triggering mechanism for a hydrogen bomb. In the first designs of the hydrogen bomb, scientists considered tritium to be indispensable, and a ready supply of it was needed. Producing tritium takes a special process, so reactors had to be converted or new reactors built for this process. Later designs of the hydrogen bomb lessened the dependence on tritium, but it remains necessary to produce quantities of tritium for the hydrogen bomb arsenal.

A mix of deuterium and tritium ignites fusion and produces the detonation in the hydrogen bomb. Tritium has a half-life of only 12 years, so a steady supply of tritium is necessary for an arsenal of hydrogen bombs. Special reactors are built to bombard lithium with neutrons to create tritium. Most of the U.S. supply came from reactors at the Savannah River Project, but in the late 1980s these reactors were shut down for safety inspections. This action cut tritium production so low that the U.S. government took emergency steps to restart Savannah's reactors. The Reagan administration even made plans to have civilian reactors produce tritium in a national emergency. Today tritium production continues, but at a modest level, because the U.S. nuclear arsenal still needs tritium for its hydrogen bombs.

Suggested readings: Rodney P. Carlisle, *Supplying the Nuclear Arsenal: American Production Reactors, 1942–1992* (Baltimore, MD: Johns Hopkins University Press, 1996); Richard Wolfson, *Nuclear Choices: A Citizen's Guide to Nuclear Technology* (Cambridge: MIT Press, 1991).

Truman, Harry

President Harry Truman (1884–1972) was not a participant in the decision to build the atomic bomb, but after becoming president, he accepted the responsibility for the decision to drop atomic bombs on Japanese cities. As a senator from Missouri, he was chair of a Senate oversight committee on war expenditures. In this capacity, he heard reports about large government expenditures in Hanford, Washington, and Oak Ridge, Tennessee, and hearsay evidence of possible government waste and abuse. It was only after assurances from Secretary of War Henry L. Stimson that an investigation of either site would endanger national security that Truman backed down. His curiosity was only answered after assuming the presidency and finding out about the Manhattan Project. He was kept appraised of the progress of bomb development after he became president. Everything was in motion to use the atomic bomb, and Truman only had the option to veto its use. He appointed the Interim Committee on the Atomic Bomb (*see* Interim Committee on the Atomic Bomb) in early summer 1945 and followed its advice on the use of the atomic bombs on Japanese cities. Although his decision to drop the bombs on Hiroshima and Nagasaki was controversial at the time and more so since, he always maintained that he had made the right decision.

Truman remained a staunch supporter of atomic weapons during the remainder of his presidency. During his presidency, the nation's nuclear arsenal rose from just a few atomic weapons to 1,600 when Eisenhower took office in 1953. Nevertheless, except for approval of test shots, Truman had a negligible role in the development of nuclear weapons. He was isolated from matters of policy except at the highest level. Moreover, he considered the atomic scientists as naive, untrustworthy, and unstable. He depended on advice about atomic energy from the military, although he did lend his support to the civilian-oriented Atomic Energy Commission in 1946. During the Korean War, Truman toyed with the idea of using atomic weapons, but because of opposition from his allies and international public opinion, he decided against using them.

Suggested readings: Thomas B. Allen, *Code-name Downfall: The Secret Plan to Invade Japan and Why Truman Dropped the Bomb* (New York: Simon and Schuster, 1995); Harry S. Truman, *Harry S. Truman and the Bomb: A Documentary History* (Worland, WY: High Plains Publishing, 1996); Dennis D. Wainstock, *The Decision to Drop the Atomic Bomb* (Westport, CT: Praeger, 1996).

Tube Alloys. *See* Uranium

Tuck, James L.

James Tuck (1919–1980) was a British-American physicist who became most famous for his work on fusion research at the Los Alamos National Laboratory. He was born in a middle-class family on January 9, 1910, in Manchester, England. In 1937 he married Elsie Harper and they adopted two children. His general education was at Manchester local schools. Later, he obtained a degree in general science at Victoria University of Manchester and then began work on a doctorate in physical chemistry at the same school. In 1937 Tuck won an appointment as a Salter Research Fellow at the Clarendon Laboratory, Oxford University, under the directorship of Frederick A. Lindemann. His life at Oxford University was marred by his inability to adapt to British social norms. While there, however, Tuck designed and built a particle accelerator.

Tuck spent the war years engaged in war-related research. His first assignment involved constructing radar installations along the British coast. Lindemann then invited him to join Winston Churchill's private staff as Lindemann's assistant. His role was to be the resident expert on all military scientific projects. In this capacity, he experimented and developed a flash X-ray system to record the impact of shaped charges on armor. This experiment gained him a reputation as an ordnance expert. When the scientists at the Los Alamos Scientific Laboratory developed a problem with implosion, Tuck was recommended to go there with the British mission to help solve the problem. Tuck collaborated with both Seth Neddermeyer and Johnny von Neumann in developing the "lens system" of detonation. His breakthrough was a major contribution to the development of a functioning plutonium atomic bomb.

After the war, Tuck stayed at Los Alamos and worked on atomic bomb testing. He and his family liked the more relaxed atmosphere of the American scientific community and life at Los Alamos. He assisted the American scientists at the Operation Crossroads atomic tests in 1946 and then returned to Great Britain in the fall of 1946 to continue research on his doctorate. He became unhappy with postwar austerity in Great Britain's scientific and economic life. When Norris Bradbury, the new director at the Los Alamos Scientific Laboratory, offered him a position there, Tuck decided to emigrate to the United States. Before assuming his position at Los Alamos, Tuck spent the 1949–50 academic year teaching at the Institute for Nuclear Studies at the University of Chicago. Tuck and his family arrived at Los Alamos in 1950, and he soon found himself in the midst of research on the hydrogen bomb with Edward Teller, Stanislaw Ulam, Robert R. Wilson, and Enrico Fermi. His work assignment was in the T-Division with Edward Teller, testing deuterium-tritium-helium combinations. After Teller left Los Alamos, Tuck directed his research orientation away from weapons development to basic research on fusion energy. In 1952 he was placed in charge of a fusion research program called "Project Sherwood." Tuck remained in charge of Project Sherwood until he retired from Los Alamos National Laboratory in 1972. His leadership in fusion research made the Los Alamos National Laboratory the world's leading fusion research establishment. From his retirement until his death in 1980, Tuck served as the leading spokesperson for fusion research in the United States.

Suggested readings: Bruce Hevly and John M. Findlay, *The Atomic West* (Seattle, WA: Center for the Study of the Pacific Northwest, 1998); Richard Rhodes, *Dark Sun: The Making of the Hydrogen Bomb* (New York: Touchstone, 1995); Richard Rhodes, *The Making of the Atomic Bomb* (New York: Simon and Schuster, 1986).

Tumbler-Snapper Nuclear Tests

The Tumbler-Snapper series of nuclear tests took place at the Nevada Test Site in the spring of 1952. A total of eight detonations were planned. Tumbler tests were carried out mainly for the Department of Defense to check nuclear weapons, and Snapper tests were conducted for the Atomic Energy Commission (AEC) and Los Alamos scientists to improve weapons design. The military had put so much pressure on the AEC to station troops close to the bomb blasts that the AEC abdicated responsibility for the safety of the soldiers. Shot Able was the first test and was delivered by a B-50 bomber on April 1, 1952. It was a low-yield device designed to measure waves from a tactical nuclear weapon. Shot Baker was also a low-yield nuclear device, which was airdropped on April 15, 1952. The next test, Shot Charlie, was a much larger bomb, and this time the national media watched the explosion, and army troops conducted maneuvers near the blast area. A 31-kiloton blast resulted from the airborne detonation on April 22, 1952. The army proclaimed their postblast operations a success. Shot Dog was another low-yield device detonated on May 1, 1952, but this time for the benefit of the U.S. Marines conducting maneuvers. The next test, Shot Easy, was a tower shot with no combat troops involved. Scientists wanted to study the blast in a corrugated iron cab. It was detonated on May 7 and surprised scientists by how much radioactive fallout it produced. Shot Fox was a return to the experiment of using combat troops in simulated military operations. Elements of the First Armored Division were brought in with their equipment. The bomb was detonated on May 25, and the blast of 11 kilotons caused a surprising extent of heavy-equipment damage. Shot George was perhaps the most important of the Tumbler-Snapper series of tests, because it was an experiment by Los Alamos physicist Ted Taylor. He substituted beryllium for the heavy uranium reflector surrounding the plutonium core. Los Alamos scientists had nicknamed the new device Scorpion. The tower blast detonated on June 1, 1952, and it vindicated Taylor's design. The final test, Shot How, was a weapons development test. It took place on June 5, 1952, and afterward, the scientists agreed that the last test was unnecessary.

Suggested reading: Richard L. Miller, *Under the Cloud: The Decades of Nuclear Testing* (New York: Free Press, 1986).

Turner Paper

In a paper written in the spring of 1940, Louis Turner, a Princeton physicist, speculated about the possibility of an element heavier than uranium. His conclusion was based on current research and possible future trends. He sent a copy of his proposed letter to *The Physical Review* with a note asking whether or not it should be withheld from publication. His speculations caused Leo Szilard to persuade him not to publish his articles until after the war. The article finally appeared in *The Physical Review* in 1946. In this paper Turner suggested the possibility of transuranic (heavier-than-uranium) elements. Soon research confirmed his theory, and the new element was named-*plutonium*. Without plutonium, Manhattan Project scientists would have had a more difficult time building the atomic bombs.

Suggested reading: Richard Rhodes, *The Making of the Atomic Bomb* (New York: Simon and Schuster, 1986).

Turnkey Nuclear Energy Plants

In the early 1960s the major nuclear plant builders developed a strategy to promote the construction of nuclear energy plants in the early 1960s by offering to build a complete nuclear power–generating facility at a firm price. A special formula was calculated to adjust to monetary inflation, but otherwise all the electric utility had to do was "open the door" to a completed plant. Hence the name *turnkey* was used for this type of arrangement. At first only General Electric and Westinghouse used this approach, but soon the two other major builders of plants, Babcock & Wilcox and Combustion Engineering, joined them. It was this marketing strategy that reinforced the interest of utility companies to build nuclear power–generating facilities. It is estimated that the first turnkey plants caused the constructors to lose between $10 and $20 million for each installation. These first plants were called "loss leaders," because the nuclear companies depended on future contracts for services and nuclear supplies to make their profit. General Electric started the turnkey phase with their Oyster Creek plant and in the end constructed 10 such plants. Westinghouse Corporation followed with another 6. Modification of later contracts reduced some of this loss, but the turnkey plants always lost money. A major contributing cause was the change in the political environment caused by antinuclear activism and the emerging environmental movement. Nevertheless, between 1962 and 1967, U.S. utilities ordered 75 nuclear power plants. Once power companies began to order nuclear plants built without the stimulus of this program, then General Electric, Westinghouse, Babcock & Wilcox, and Combustion Engineering ended the practice of building turnkey plants.

Suggested reading: Irvin C. Bupp and Jean-Claude Derian, *Light Water: How the Nuclear Dream Dissolved* (New York: Basic Books, 1978).

U

Underground Nuclear Testing

The signing of the Nuclear Test Ban Treaty by the Soviet Union, the United Kingdom, and the United States in August 1963 ended nuclear testing above ground, in space, or on the seas, but it allowed underground nuclear testing. This type of testing limited the amount of radioactivity released into the atmosphere and permitted both the Soviet Union and the United States to continue to refine nuclear weapons. From 1963 to 1980, the United States conducted around 400 underground tests, and the Soviets 300. Americans tended to test more refined weapons, and the Soviet more powerful ones. In 1974 an agreement was reached between the Soviet Union and the United States to limit the power of their underground tests to 150 kilotons. Other countries have also renounced atmospheric testing, but they have continued to test underground nuclear devices. The latest countries to conduct underground nuclear testing have been India, with 5 such tests between May 11 and 13, 1998, and Pakistan with 6 tests between May 29 and 30, 1998.

Even the United States decided to reinstate underground nuclear testing to check on the readiness of its nuclear arsenal. This program is called the Science Based Stockpile Stewardship (SBSS) and under the auspices of the Department of Energy (DOE) it is planning for eight "subcritical" nuclear tests underground at the Nevada test site. The first of these tests, or Rebound, took place on July 2, 1997. A series of tests continued until September 1997. Scientists from the Los Alamos National Laboratory, Sandia National Laboratories in Albuquerque, and Lawrence Livermore National Laboratory are engaged in a 10-year, $45 billion program to test nuclear weapons using high-powered lasers and supercomputers. If this project succeeds, underground nuclear testing in the United States will cease.

Suggested reading: Bertrand Goldschmidt, *The Atomic Complex: A Worldwide Political History of Nuclear Energy* (La Grange Park, IL: American Nuclear Society, 1982).

Union of Concerned Scientists

The Union of Concerned Scientists (UCS) is an independent organization of scientists and others concerned about the effects of nuclear technology on society. It was formed in 1969 as an outgrowth of an informal faculty group at the Massachusetts Institute of Technology (MIT), and the national headquarters has remained in Cambridge, Massachusetts. At first the faculty group had contact with a more radical student organization, the March 4th Movement, which had as its goal to stop all government research activities at MIT in an effort to reduce government support for the Vietnam War. The faculty group drifted away from this more violence-oriented student group after a work stoppage was held on March 4, 1969.

The UCS is a body of around 100 scientists, engineers, and other professionals with knowledge of nuclear technology. This expertise allowed it to become one of the most influential of the organizations critical of nuclear technology. In 1971, the UCS turned its attention toward a critique of the emergency core cooling system used in nuclear power plants. Criticism of nuclear power plants by scientists from the UCS had a worldwide audience. A staff of 40 and a budget of over $4 million allow this organization to carry out its advocacy program. Its efforts have made the UCS one of the most influential of the critics of atomic energy.

Suggested reading: Jerome Price, *The Antinuclear Movement* (Boston, MA: Twayne, 1982).

Units of Measurement of Radiation Standards

Scientists have agreed on several basic units for use as standards to measure radiation. The *curie* is the basic unit to describe the amount of radioactivity in a sample. A 1-Curie sample undergoes 37 billion disintegrations per second, or 2.22 trillion disintegration per minute. The curie is based on the number of disintegrations per second undergone by 1 gram of pure radium-226. It is named in honor of Marie and Pierre Curie, who discovered radium in 1898. The *rad* (radiation absorbed dose) is a measure of radiation absorbed by tissue. A 1-rad dose indicates the absorption of 100 ergs of radiation energy per gram of absorbing material. The *gray* is a unit of radiation dose equal to 100 rads. The *roentgen* is a unit of exposure to X- or gamma radiation named after Wilhelm Roentgen, the discoverer of X-rays.

It is the amount of gamma or X-rays required to produce ions carrying 1 electrostatic unit of electrical charge in 1 cubic centimeter of dry air. A roentgen equals to 0.94 rads. The *rem* (roentgen equivalent man) is a unit of dose of any ionizing radiation that produces the equivalent biological effect in humans as 1 rad of X-rays. The *sievert* is a dose of ionizing radiation equal to 100 rems. The *becquerel* is a measure of radioactivity of a sample equaling 1 disintegration per second. This unit of measurement is named after the discoverer of radiation, Antoine Becquerel. These measurements have allowed scientists a uniform way to measure the strength of radiation in a variety of situations. They are used frequently in works on radiation or radioactivity, sometimes without explanation.

Suggested reading: Erik Bergaust, ed., *The Illustrated Nuclear Encyclopedia* (New York: Putnam's Sons, 1971).

Upshot-Knothole Nuclear Tests

Upshot-Knothole was the fourth series of Nevada Test Site nuclear tests that took place in 1953. This series had 11 tests—3 airdrops, 7 tower shots, and 1 launched from an atomic cannon. The first test was Shot Annie, which was a combination military-Civil Defense demonstration held before the national media. In the airborne detonation on March 17, 1953, both homes and and military equipment were destroyed. The next was Shot Nancy, which involved 3,000-troops and nine officers in a trench 2,500 yards from the blast center. Placement of troops near the blast site was to demonstrate the effectiveness of military operations on the atomic battlefield. Plans also called for 53 B-36 bombers to circle over the blast site. After the detonation on May 24, 1953, however, the officers had to evacuate their trenches prematurely when radioactivity levels became too high. Shot Ruth was next, and it was to test a device designed by Edward Teller from the Livermore National Laboratory. Because the device was so small, no military operations took place when it was detonated on a tower on March 31. Shot Dixie was another airborne drop, and it tested on April 6 a much-refined plutonium bomb with a yield of 11 kilotons. Shot Ray was a tower detonation from the Livermore National Laboratory, but it produced less of a yield than the earlier Livermore effort. Shot Badger was next, and it was part of a U.S. Marine Corps operation. Military officials wanted to measure the effectiveness of helicopters transporting troops onto a nuclear battlefield. The nuclear device was detonated on April 18, and the operation proceeded smoothly until the wind direction changed and the radioactive cloud moved over the units. A hasty evacuation of all troops ended the operation. Shot Simon continued the experiment of using troops close to the blast center. This time 3,000 soldiers were committed to attack objectives near ground zero. A device of 43 kilotons was detonated on April 25, and its blast effects

Early exploration for uranium ore was a primitive affair. An individual armed with an eiger counter explores geological formations hoping to strike it rich.

almost ended the military operations before they started. A navy drone Skyraider was destroyed trying to fly through the cloud. The radioactive fallout was so heavy that monitoring teams moved into southwest Utah to check radiation levels. Shot Encore was yet another effort to coordinate military operations with an atomic weapon. A bomb was detonated by airdrop from a B-50 bomber on May 8, and the blast was in the 25-kiloton range. Despite relatively high radioactive fallout, helicopters delivered troops to their objectives. Shot Harry was a relatively small nuclear device detonated on May 19, but it became famous for its high radioactive fallout over southwestern Utah. The fallout problem was becoming more visible, and members of the Atomic Energy Commission were concerned. Shot Grable was special because it was the first test of an atomic cannon. Two 85-ton cannons (280mm 11-inch guns) had been moved to the Nevada Test Site in early May to participate in the test. The test site was at Frenchman's Flat, and the target was six miles away. On May 25, the shell was fired by conventional manner, and the blast had a yield of 15 kilotons. A final unscheduled test (Shot-Climax) on June 4 was conducted because scientists had discovered in Shot Harry a design improvement that more efficiently caused fission. The resulting airdrop from a B-36 of a device produced a blast of 61 kilotons.

Suggested readings: Barton C. Hacker, *Elements of Controversy: The Atomic Energy Commission and Radiation Safety in Nuclear Weapons Testing, 1947–1974* (Berkeley: University of California Press, 1994); Richard L. Miller, *Under the Cloud: The Decades of Nuclear Testing* (New York: Free Press, 1986).

Uranium

Uranium is the indispensable element in nuclear energy. It is a more abundant mineral than mercury, copper, or silver, but ores rich in uranium have always been difficult to find, so most uranium is mined from low-grade ores. Uranium was discovered in 1789 by German chemist M. H. Klaproth. It was difficult to find and refine, so little use was found for uranium except as a color fixative in ceramics. However, Antoine Becquerel's experimental findings in 1896 that uranium generated energy stimulated further scientific research. Uranium is the basic raw material of nuclear energy. Its atomic number is 92, and its two principal natural isotopes are U-235 (0.7 percent of natural uranium) and uranium-238 (99.3 percent of natural uranium). Both isotopes are unstable and eventually decay via spontaneous fission or via neutron-induced fission in the presence of neutrons. In the case of U-238, the neutrons must be energetic in order to induce fission, whereas for U-235 even slow neutrons are sufficient. This latter fact makes U-235 particularly important for generating a self-sustaining chain reaction.

The discovery of fission in a 1938 experiment by Otto Hahn and Fritz Strassmann in Germany led to the development of the atomic bomb. Further research showed that it was uranium-235 that was susceptible to neutron-induced fission. It was uranium-235's relatively low neutron-proton ratio and its odd-even neutron-proton count that allowed a destabilizing release of energy on neutron capture. For the transuranics, plutonium-239 showed the same destabilizing characteristics.

Before 1912, the only supply of uranium was from the Joachimsthal Mines in Czechoslovakia on the German-Czechoslovakian border. A second supply of incredibly rich uranium was discovered in the Katanga province of the Belgian Congo during World War I. Trace amounts of uranium were found in Colorado and Utah in the 1920s. The next big discovery of uranium ore was in the remote Great Bear Lake Region of Canada. All uranium production came from these remote sites before World War II. The scarcity of uranium caused major problems for these countries experimenting with nuclear energy in the postwar world. Responding to this shortage, the United States attempted to establish a monopoly on available uranium ore in the late 1940s, but was only partially successful. Exploration for uranium ore peaked in the 1950s, with major finds in Australia, Canada, South Africa, the Soviet Union, and the United States. It was only in the late 1950s that the uranium supply matched the demand for research and weapons systems. By the 1970s, uranium glut forced the price of uranium far below previous levels. Moreover, plutonium supplanted uranium as a source for weapons systems. Uranium still remains a valuable commodity for nuclear reactor fuel.

Suggested readings: Lennard Bickel, *The Deadly Element: The Story of Uranium* (New York: Stein and Day, 1979); Earle Gray, *The Great Uranium Cartel* (Toronto, Ontario: McClelland and Stewart, 1982).

Uranium Cartel

The glut of uranium in the early 1970s led to the creation of the Uranium Cartel. This semisecret cartel was formed in February 1972 meetings in Paris by government and business representatives from Australia, Canada, France, Great Britain, and South Africa. Its immediate problem was too much uranium and too little demand. The U.S. government had control of the American market and refused to buy foreign uranium ore. The Uranium Club, also known as the Club of Five, held its first formal meeting in July 1972 in Cannes, France. Slowly, curbs established by the cartel began to boost the non-American price for uranium by late 1973. Soon the price rose from $5 a pound in 1972 to $25 a pound in 1975. Among the reasons for this surge were the ban by the Australian government on the opening of new Australian uranium mines, the start of so many new nuclear energy plants in this period, and finally, the Arab-Israeli War of 1973. International instability caused by the war led several European governments to consider replacing dependence on Middle East oil with nuclear energy.

Information about the cartel came from documents stolen from an Australian mining company by an employee. This employee was a member of an antinuclear protest group, and this information found its way to the U.S. Justice Department. Westinghouse Corporation had found itself defaulting on uranium supplies to power companies, and they used this information to accuse the cartel of an international conspiracy. The cartel fell apart in 1975 in the glare of publicity caused by the Westinghouse Corporation's accusations. Legal battles lasted well into the 1980s, with the Antitrust Division of the U.S. Justice Department and congressional hearings complicating the issues. Finally, the legal issues died because of international complications, but the cartel was long gone.

Suggested readings: Earle Gray, *The Great Uranium Cartel* (Toronto, Ontario: McClelland and Stewart, 1982); Norman Moss, *The Politics of Uranium* (London: Deutsch, 1981).

Uranium Club

The Uranium Club was a group of physicists and chemists recruited by German authorities in April 1939 to conduct research on nuclear fission. Abraham Esau, the president of Germany's Bureau of Standards, called the first meeting. Research on uranium was to be a cooperative project, but rivalry between the Army Ordnance Bureau and the Uranium Club soon developed over the supply of uranium. The Army Ordnance Bureau had reserved all of the uranium for its research. Eventually, an agreement was worked out to share the uranium and research findings. Approximately 50 scientists belonged at one time or another to the club. Each scientist, however, worked at his own institution, and little coordination was possible. This lack of coordination proved to be one of the major failings of the German atomic bomb effort.

Suggested reading: Jeremy Bernstein, *Hitler's Uranium Club: The Secret Recordings at Farm Hall* (Woodbury, NY: American Institute of Physics, 1996).

Uranium Committee

The Uranium Committee, or the Advisory Committee on Uranium, was formed by President Franklin D. Roosevelt in 1939 to study the feasibility of building an atomic bomb. It was an outgrowth of the Einstein-Szilard Letter in the fall of 1939 alerting the president to the possibility of an atomic bomb and the likelihood that the Germans might be working on one. General Edwin M. Watson, Roosevelt's aide, set up the committee almost immediately after Roosevelt gave the go-ahead. This committee had three members: Dr. Lyman Briggs, director of the Bureau of Standards; Colonel K. F. Adamson, Army Ordnance; and G. C. Hoover, Navy Ordnance. At its first meeting on October 21, 1939, with the committee members and invited guests—Alexander Sachs, Leo Szilard, Edward Teller, Eugene Wigner, and Richard B. Roberts—several members of the committee were unconvinced of the necessity to proceed further despite the strong views of the physicists present. Alexander Sachs proved supportive, and a modest appropriation was approved. This appropriation enabled the committee to give to Enrico Fermi's team 4 tons of graphite and 50 tons of uranium. The War Department also released $6,000 for atomic bomb research. A committee report on November 1 was sent to President Roosevelt along with a recommendation to explore a controlled chain reaction for possible use in submarines and to test its explosive potential. This report was read by the president and filed away for future reference.

In a subsequent meeting in June 1940, the committee consulted with a group of nuclear scientists, including Enrico Fermi, Harold Urey, Leo Szilard, and Eugene Wigner, among others, and recommended $140,000 for further study on the nuclear properties of uranium. In the autumn of 1940, the Uranium Committee was placed under the jurisdiction of the National Defense Research Committee (NDRC). The committee was revamped with military men and foreign-born scientists dropped off the committee. Military men lacked the scientific expertise. There were fears that in the event of failure of a research effort the foreign-born scientists would cause problems

After a discovery of uranium ore, uranium mining was initiated by mining companies. Conditions were primitive and dangerous with radioactivity as deadly as mine cave-ins.

in a subsequent congressional investigation. Four subsections were formed in the summer of 1941: isotope separation, power production, heavy water, and theoretical aspects. Numerous reports were submitted from the Uranium Committee to the NDRC on the feasibility of atomic energy, some of which were doubtful of whether or not to continue to give atomic research priority, given the uncertainty of its practical application. It was Ernest O. Lawrence's suggestion in a letter on the possibility of the use of plutonium as the material for a fission bomb that revitalized the Uranium Committee. Despite these initiatives, the slowness of the Uranium Committee made government leaders nervous, and it was decided to bypass it for Vannevar Bush's NDRC.

Suggested readings: Richard Rhodes, *The Making of the Atomic Bomb* (New York: Simon and Schuster, 1986); Martin J. Sherwin, *A World Destroyed: The Atomic Bomb and the Grand Alliance* (New York: Knopf, 1975).

Uranium Mining

Finding new sources of uranium ore became a vital concern after it became apparent that uranium has unique properties both for military and for civilian uses. As much as scientists desired ready access to uranium for research, governments intent on building nuclear weapons were even more insistent. Most uranium ore came from just three sites—Joachimsthal (Czechoslovakia), Shinkolobwe (Belgian Congo), and Great Bear Lake (Canada)—during the first half of the twentieth century. Only about 1,000 tons of uranium per year were being extracted in prewar years. A new strike in Canada in the summer of 1946 at Beaverlodge Lake north of Lake Athabasca was encouraging that new supplies could be found. Nevertheless, by 1947 uranium supplies were exceeded by the ever-increasing demand for uranium for research and weapons. The U.S. government responded by creating financial incentives for new uranium discoveries both in the United States and abroad. A series of discoveries in the Colorado plateau area of New Mexico and Utah in 1951 showed the wisdom of this approach. Uranium was also found as a by-product in the South African gold field. The next big strike, however, was in the Algoma basin, Canada, in the summer of 1953. An increased interest in acquiring uranium by the U.S. government came after the Soviet Union detonated a hydrogen bomb in 1953. Most of the uranium mining in the United States in the 1950s took place in the four corners areas of Arizona, Colorado, New Mexico, and Utah, with 500 of the 550 uranium mines in this region. Certain lucky prospectors became rich finding uranium in Utah—examples were Vernon Pick and his Delta Mine, and Charles Steen and his Mi Vida Mine. This boom reached its high point in 1959, and then a glut of uranium ore drove down the price. Then in 1968 a large uranium field was discovered on the land of Gulf Mineral Resources Corporation of Denver near Rabbit Lake in Saskatchewan. It was an exceedingly rich mine. Soon afterward a major deposit of uranium was found in the Nabarlek region of northern Australia. This glut of uranium led to the formation of the Uranium Cartel in 1972.

Until the private ownership of nuclear materials was approved by Congress in 1964, all uranium supplies were held by the Atomic Energy Commission (AEC). The AEC encouraged domestic production of uranium ore but also purchased ore from abroad. Between 1947 and 1970, the AEC acquired 55 percent of its ore from domestic sources, 24 percent from Canada, and 21 percent for other overseas sources. By the early 1970s, the AEC had a surplus of uranium ore on its hands and looked for ways to dispose of it without destabilizing the uranium mining industry. This glut resulted in the reduction of commercial uranium mines in the United States from 550 in the mid-1950s, 343 in 1980, to 11 in 1991. Since the early 1990s, demand for uranium has been such that the number of mines has remained stable with around ten mines supplying uranium ore. Since a large number of nuclear weapons are being dismantled in the 1990s, uranium is plentiful, with little or no demand for new mining.

Suggested readings: Howard Ball, *Cancer Factories: America's Tragic Quest for Uranium Self-Sufficiency* (Westport, CT: Greenwood, 1993); Frank Barnaby, *How*

Nuclear Weapons Spread: Nuclear-Weapon Proliferation in the 1990s (London: Routledge, 1993); Frank G. Dawson, *Nuclear Power: Development and Management of a Technology* (Seattle: University of Washington Press, 1976); Earle Gray, *The Great Uranium Cartel* (Toronto, Ontario: McClelland and Stewart, 1982); Raymond W. Taylor and Samuel W. Taylor, *Uranium Fever, or No Talk Under $1 Million* (New York: Macmillan, 1970).

Urey, Harold Clayton

Harold Urey (1893–1981) was an American physical chemist who discovered deuterium, or heavy water, which is used as a moderator for atomic reactors. He was born on April 29, 1893, on a farm near Walkerton, Indiana. His father was a teacher and lay minister of the Brethren Church, who died when his son was six years old. A marriage to Frieda Daum in 1926 produced three daughters and a son. Urey majored in chemistry at the University of Montana, where he received a B.S. in 1917. During World War I, Urey worked with an industrial chemistry firm in Philadelphia. After the war, he worked on a Ph.D. in chemistry under the famous chemist Gilbert N. Lewis at the University of California at Berkeley and obtained his degree in 1923. After a year of research at the Institute of Theoretical Physics in Copenhagen under Niels Bohr, Urey returned to the United States to take a teaching position at Johns Hopkins University, beginning in 1924. A move to Columbia University took place five years later in 1929. It was at Columbia University that Urey conducted the research leading to the discovery of deuterium in 1931. He received the 1934 Nobel Prize for Chemistry for this discovery.

During World War II, Urey worked on the Manhattan Project. His discovery of deuterium, or heavy water, made him one of the experts on this process. He was the administrative head of the scientists working on the gaseous diffusion reactor to be constructed at Oak Ridge, Tennessee. His gaseous diffusion theory was based on the fact that gaseous U-235 would diffuse through small holes at a rate different from U-238. Methods of separation were both costly and difficult, and Urey became discouraged and almost had a nervous breakdown.

After the war, Urey continued his work on physical chemistry problems. In 1945, he moved to the University of Chicago, where he stayed until 1958. Then, in 1958, he returned to California at the University of California at La Jolla. He remained there until his retirement in 1970. Urey died on January 5, 1981, at La Jolla, California.

Suggested readings: Henry A. Boorse, Lloyd Motz, and Jefferson Hane Weaver, *The Atomic Scientists: A Biographical History* (New York: Wiley Science Editions, 1989); John Daintith, Sarah Mitchell, Elizabeth Tootill, and Derek Gjertsen, *A Biographical Encyclopedia of Scientists*, 2nd ed., vol. 2 (Bristol, UK: Institute of Physics, 1994), Daniel J. Kevles, *The Physicists: The History of a Scientific Community in Modern America* (Cambridge: Harvard University Press, 1987).

U-235

The discovery of U-235, or uranium-235, was the key to developing the atomic bomb. U-235 is one of the two naturally occurring isotopes of uranium (uranium-238 is the other). It comprises only 0.7 percent of natural uranium. Scientists' initial studies on the properties of uranium found only U-238. In 1935, however, Arthur Jeffrey Dempster, a Canadian-American physicist at the University of Chicago, discovered traces of U-235. Then in 1938, Alfred Otto Carl Nier, a Harvard postdoctoral fellow, measured the ratio of U-235 to U-238 as 0.7 percent. Otto Hahn's discovery of the energy potential of fission in 1939 launched a hunt for the source of the energy. After studying the problem, Bohr concluded in the winter of 1939 that U-235 was responsible for slow-neutron fission. Bohr announced this discovery in a paper to the *Physical Review* on February 7, 1939. Further studies confirmed this contention, and attention turned to the explosive potential of U-235. Because uranium-235 is by nature unstable, it is relatively easy to start a chain reaction producing enough energy for an explosion. Once this characteristic was understood, the next problem was to isolate enough uranium-235 for testing. Several methods—gaseous diffusion, magnetic separation, and thermal separation—were attempted at Oak Ridge, Tennessee, before enough uranium-235 was accumulated for testing at the Los Alamos Scientific Laboratory. U-235 provided the explosive power for Little Boy, which was detonated over Hiroshima on August 6, 1945.

Suggested reading: Richard Rhodes, *The Making of the Atomic Bomb* (New York: Simon and Schuster, 1986).

V

Van de Graaff, Robert Jemison

Robert Van de Graaff (1901–1967) was an American physicist who is most famous for his development of a particle accelerator that was named after him. He was born on December 20, 1901, in Tuscaloosa, Alabama. His undergraduate (1922) and master's (1923) degrees were from the University of Alabama at Tuscaloosa. For a short period he worked for the Alabama Power Company before leaving for a European trip. While taking courses at the University of Paris at Sorbonne in 1924, Van de Graaff attended lectures on physics by Madame Marie Curie and decided to study physics. In 1928 he received his doctorate in physics from Oxford University, studying under John Townsend. Part of his research led him to designing an electrostatic generator. In 1929 Van de Graaff returned to the United States and began work as a research fellow at Princeton University. Then in 1931 he moved to the post of research associate at the Massachusetts Institute of Technology (MIT), where he remained until 1960.

It was while at Princeton that Van de Graaff built the Van de Graaff generator which later he developed as an accelerator. This accelerator used the generator to produce voltage high enough to accelerate charged particles. It was used extensively as a tool in the field of atomic energy. An additional benefit was the use of generator as an X-ray machine for treatment of cancer. During World War II, his apparatus was important in the examination of the interior structure of heavy ordnance.

After the war, Van de Graaff and a colleague, John Trump, founded a company, High Voltage Engineering Corporation, to market the accelerator and X-ray generator. The market consisted of academic institutions, hospitals, industry, and scientific institutes. Van de Graaff was director of the corporation and in 1960 left MIT to devote his full efforts to his company. He died on January 16, 1967, in Boston, Massachusetts.

Suggested readings: David Abott, ed., *The Biographical Dictionary of Scientists: Physicists* (London: Blond Educational, 1984); Henry A. Boorse, Lloyd Motz, and Jefferson Hane Weaver, *The Atomic Scientists: A Biographical History* (New York: Wiley Science Editions, 1989); John Daintith, Sarah Mitchell, Elizabeth Toothill, and Derek Gjertsen, *A Biographical Encyclopedia of Scientists*, 2nd ed., vol. 2 (Bristol, UK: Institute of Physics, 1994).

Vanunu Affair

Mordecai Vanunu was an Israeli citizen from Morocco who publicized the existence of an Israeli nuclear arsenal at Dimona. He was born in Marakkech, Morocco, in 1954, but at an early age he emigrated to Israel with his family. After military duty in the engineering corps of the Israeli Defense Force, he found a job in 1977 as a technician in the top-secret nuclear complex at Dimona. Although Vanunu had previously been a Zionist, he changed his political affiliation in the early 1980s and became active in left-wing student politics. Nuclear authorities warned him about his activities and fired him in November 1985. Vanunu moved to Australia and converted to Christianity. He also started negotiations with journalists about his knowledge and smuggled out pictures of the Dimona complex. He had worked in Machon 2, which was an underground bomb factory making hydrogen bombs. Physicists studying the pictures came to the conclusion that Israel had the potential to have made at least a hundred bombs in the decade between 1977 and 1987. Rupert Murdoch's *Sunday Times* bought his story. A team of Israeli security agents, the Mossad, was sent to London to entice him out of the country. Vanunu was kidnapped in Rome and taken to Israel. He was convicted of treason, espionage, and disclosing of state secrets in March 1988 and sentenced to 18 years in prison. His information about Israel's nuclear program, however, was published in the *Sunday Times*.

Suggested reading: Dan Aviv and Yossi Melman, *Every Spy a Prince: The Complete History of Israel's Intelligence Community* (Boston: Houghton Mifflin, 1990).

Villard, Paul

Paul Villard (1860–1934) was a French physicist who is most famous for his discovery of gamma radiation. He was born on September 28, 1860, in Lyon, France. His father was of English origin and was an artist and musician. Villard was a lifelong bachelor. His early education was at the lycee in Lyon. At the age of 21, Villard entered the École Normale Superieure in Paris. After receiving certification, he taught at several lycees before becoming a teacher at the University of Montpellier. Teaching bored him, so he resigned to study chemistry at the École Normale Superieure in Paris. After receiving his doctorate, he stayed there and devoted his life to experimental research.

It was Villard's research with cathode rays and X-rays that led to his discovery of gamma rays. Gamma rays are high-energy, short-wavelength, electromagnetic radiation. These rays accompany alpha and beta emissions and are present during fission. This discovery formed one of the building blocks in atomic energy. He continued to conduct research experiments after his discovery, but his health suffered from his exposure to radiation. His health finally gave out, and he died on January 13, 1934, in Bayonne, France.

Suggested readings: Henry A Boorse, Lloyd Motz, and Jefferson Hane Weaver, *The Atomic Scientists: A Biographical History* (New York: Wiley Science Editions, 1989); John Daintith, Sarah Mitchell, Elizabeth Toothill, and Derek Gjertsen, *A Biographical Encyclopedia of Scientists*, 2nd ed., vol. 2 (Bristol, UK: Institute of Physics, 1994).

Virus House

The Virus House was a wooden structure built on the grounds of the Kaiser-Wilhelm-Institute in the Berlin suburb of Dahlem to serve as the site of Werner Heisenburg and Carl Friedrich von Weizsäker's experiments on a German atomic reactor. Construction, directed by Karl Wirtz, was started in July 1940 and finished in October. Werner Heisenburg and Wirtz named the building the Virus House to keep away spectators and unwanted guests. Besides a laboratory, it housed a special brick-lined pit. Experiments under the control of Heisenberg, von Weizsäcker, and Wirtz started in December 1940, and soon it became apparent that ordinary hydrogen was a poor moderator for a chain reaction. The only alternatives were graphite or heavy water. Hans Bothe's experiment with impure graphite caused the Germans to turn away from graphite and to heavy water. It was not until September 1941 that the next major experiment took place in the Virus House, but this time with 40 gallons of heavy water. This time Heisenberg found enough neutron activity to know that his research was headed in the right direction.

Suggested readings: David Irving, *The German Atomic Bomb: The History of Nuclear Research in Nazi Germany* (New York: Simon and Schuster, 1967); David Irving, *The Virus House* (London: William Kimber, 1967).

Vladisvostok Agreement. *See* Strategic Arms Limitation Treaty I

Von Laue, Max Felix

Max von Laue (1879–1960) was one of Germany's foremost physicists in the first half of the twentieth century, and his work on the German atomic bomb project made it a creditable military threat to the Allies. Von Laue was born in Pfaffendorf, Germany, on October 9, 1879. His father was a civil servant. He married Magdalene Deger in 1910, and they had a son and a daughter. His son is prominent American historian Theodore von Laue. After attending the Universities of Strasbourg, Göttingen, and Munich, he received his doctorate in physics from the University of Berlin in 1903. His thesis adviser for his Ph.D. was Max Planck, and for his Habitation, Arnold Sommerfeld. After finishing his education, von Laue taught at the University of Berlin from 1906 to 1909. In 1909, he moved to the University of Munich. After three years at Munich, von Laue obtained a position at the University of Zurich. Returning to Germany in 1914, von Laue received a professorship at the University of Frankfurt am Main, where he stayed during World War I. It was in the midst of these numerous moves that von Laue conducted research that led to his discovery of the diffraction of X-rays in crystals. For his research he was awarded the 1914 Nobel Prize in Physics. After the end of the war, von Laue moved to the University of Berlin, where he stayed from 1919 to 1943 and served both as a professor and as an administrator. One of his early contributions in Berlin beginning in the early 1920s was the Von Laue Colloquia in which speakers reported on the results of journal articles before a galaxy of great physicists, which often included Albert Einstein, Max Planck, Erwin Schrödinger, and younger physicists, such as Leo Szilard and Edward Teller, studying in Berlin.

It was during the 1920s that von Laue became involved in the controversies between proponents and opponents of Einstein's theory of relativity and Planck's quantum theory. Von Laue's pointed review of Johannes Stark's book *The Contemporary Crisis in German Physics* and defense of Einstein and Planck made him many enemies among conservative physicists. In 1933, von Laue almost single-handedly blocked Johannes Stark's candidacy for the prestigious Prussian Academy of Sciences. In his other dealings with the Nazis, however, von Laue behaved more circumspectively, but he never joined the Nazi Party, as so many of his scientific colleagues did. He continued to be anti-Nazi and helped those in political trouble. In 1937, he sent his son to the United States to remove him from Nazi influence.

Soon after World War II broke out, von Laue began to work with other German scientists on atomic energy. While he was part of the Heisenberg team investigating the possibility of a German atomic bomb, von Laue devoted his research to scientific topics with no military value. In 1943, He moved to Göttingen as director of the Max Planck Institute. His international reputation and his supposed participation on the bomb project resulted in his seizure by the Americans in 1945 and his lengthy interrogation at the Farm Hall with nine other German scientists.

After World War II, von Laue returned to research and teaching, but at the University of Göttingen. In 1951, he was appointed the director of the Fritz Haber Institute of Physical Chemistry in Berlin. He also became active in the German scientist movement opposed to work on nuclear weapons.

Von Laue retired in 1958 with the full honors of the German physics community and died in Berlin on April 24, 1960, as the result of an automobile accident.

Suggested readings: Alan D. Beyerchen, *Scientists Under Hitler: Politics and the Physics Community in the Third Reich* (New Haven, CT: Yale University Press, 1977); Henry A. Boorse, Lloyd Motz, and Jefferson Hane Weaver, *The Atomic Scientists: A Biographical History* (New York: Wiley Science Editions, 1989); John Daintith, Sarah Mitchell, Elizabeth Tootill, and Derek Gjertsen, *A Biographical Encyclopedia of Scientists*, 2nd ed., vol. 2 (Bristol, UK: Institute of Physics, 1994).

Von Neumann, John

John von Neumann (1903–1957) was a Hungarian-American mathematician who played an important role as a consultant on the building of the atomic and hydrogen bombs. He was born on December 28, 1903, in Budapest, Hungary. His father was a banker. Von Neumann's early education was at the Lutheran Gymnasium in Budapest with his friend Eugene Wigner. His advanced academic training in mathematics was at the University of Berlin, the Berlin Institute of Technology, and the University of Budapest. In 1925, he obtained a degree in chemical engineering from the Zurich Institute. Von Neumann received his doctorate from the University of Budapest in 1926. He studied at the University of Göttingen for a year, where he collaborated with J. Robert Oppenheimer. After teaching at the University of Berlin from 1927 to 1929, he moved to the University of Hamburg in 1929. Much of his early research was on set theory and the mathematical theory of games.

Von Neumann emigrated to the United States in 1930 to teach at Princeton University. Later in 1933, he moved to Princeton's Institute of Advanced Study as its youngest member. In 1943, he started serving as a consultant on mathematical problems for the Manhattan Project. He was frequently at Los Alamos working on theoretical problems and designing computers to do advanced calculations on implosion theory. He was also instrumental in recommending the use of the calculating capability of ENIAC (Electronic Numerical Integrator and Calculator) to work on advanced mathematics problems on the atomic bomb. Then he became active building and designing an improved computer, the Mathematical Analyzer, Numerical Integrator and Computer (MANIAC), which was used in computing the final figures for the hydrogen bomb. In the postwar period, von Neumann held a number of advisory posts with the U.S. government from 1945 to 1954. He became ill with cancer in 1955 and died on February 8, 1957, in Washington DC.

Suggested readings: John Daintith, Sarah Mitchell, Elizabeth Toothill, and Derek Gjerten, *A Biographical Encyclopedia of Scientists*, 2nd ed., vol. 2 (Bristol, UK: Institute of Physics, 1994); Norman Moss, *Men Who Play God: The Story of the H-Bomb and How the World Came to Live with It* (New York: Harper and Row, 1968); Robert W. Seidel, *Los Alamos and the Development of the Atomic Bomb* (Los Alamos, NM: Otowi Crossing Press, 1995).

W

Waltz Mill Reactor Accident

In 1957, the Atomic Energy Commission (AEC) requested that the Westinghouse Corporation build a nuclear test reactor. Westinghouse built a 60-megawatt reactor at Waltz Mill, Pennsylvania, which started operations on July 1, 1959. Its reactor had a low-pressure, low-temperature, water-cooled system. On April 3, 1960, there was a partial meltdown of some of the fuel rods. This meltdown caused radioactive krypton and xenon gases to be released into the atmosphere. After the reactor site had been evacuated, tests were run on radiation levels. Radiation levels were high at the reactor site but not elsewhere. The incident was reported to the AEC, but no public notice reached the national media. After repairs and improvements to the venting system, the reactor was restarted on October 1960. In 1962 Westinghouse shut down the Waltz Mill reactor permanently.

Suggested reading: John W. Simpson, *Nuclear Power from Underseas to Outer Space* (La Grange Park, IL: American Nuclear Society, 1995).

Washington Conference on Theoretical Physics

The Washington Conference on Theoretical Physics was an annual meeting of theoretical physicists held in Washington, DC., beginning in 1935 and lasting until World War II. It was the brainchild of Russian-American theoretical physicist George Gamow and had been one of the conditions of his employment at George Washington University. He envisaged the conference to serve the same function as a meeting place for ideas in the United States as the annual meetings at Niels Bohr's Institute for Theoretical Physics in Copenhagen. A cosponsor of the conference was Gamow's colleague and friend at George Washington University, Edward Teller. Each meeting had a specialized topic of discussion. The most famous of these gatherings was the fifth conference in 1939. Low-energy physics was the announced topic of the conference, but Niels Bohr had inside information on Otto Hahn's research in Germany and Lise Meitner and Otto Frisch's explanation of the splitting of the atom before publication of their letter in *Nature*. This conference's prearranged topic of discussion was ignored as the physicists eagerly discussed the news and pondered on the implications of the news. A series of quick experiments were performed within days, which authenticated Hahn's experiment.

Suggested reading: Stanley A. Blumberg and Gwinn Owens, *Energy and Conflict: The Life and Times of Edward Teller* (New York: Putnam's Sons, 1976).

Washington Declaration

The Washington Declaration was an agreement between the president of the United States, Harry S. Truman, the prime minister of the United Kingdom, Clement Attlee, and the prime minister of Canada, Mackenzie King, over future cooperation between their three countries on atomic energy. It was negotiated in November 1945 and issued on November 15, 1945. In the statement the leaders declared their willingness "to proceed with the exchange of fundamental scientific information and the interchange of scientists and scientific information for peaceful ends with any nation that will fully reciprocate." While reserving the right to control military uses of atomic energy, these leaders recommended that the United Nations establish a committee to study the peaceful uses of atomic energy. This statement coming only months after the end of World War II was a positive step toward the internationalization of atomic energy, but within months, this positive atmosphere changed, and the initial stages of the Cold War began.

Suggested reading: Margaret Gowing, *Independence and Deterrence: Britain and Atomic Energy, 1945–1952*, vol. 2 (New York: St. Martin's Press, 1974).

WASH-740 Report

The WASH-740 Report was a document prepared by the Atomic Energy Commission's (AEC) scientists at Brookhaven National Laboratory in 1956 and presented to the AEC in March 1957. In this report the scientists investigated the potential consequences of a major nuclear reactor accident. These scientists set up a hypothetical case in which a nuclear power reactor was located about 30 miles upwind of a large American city. If a reactor accident released a major part of the radioactive material into the atmosphere, scientists estimated that 3,000 people would die and another 43,000 would be injured. Property damage would be about $7 billion. These figures stunned the leadership of the AEC, who had hoped for more positive data.

Despite the assurances of the AEC's scientists that such an accident would never happen, private businesses made it plain that unless the government protected them from massive damage claims from such accidents, they would refuse to go into nuclear energy. This report led to the U.S. Congress passing the Price-Anderson Act of 1957 to limit liability for nuclear accidents. Many scientists found the WASH-740 Report overoptimistic and demanded another study to incorporate data about larger nuclear power stations.

Suggested reading: Daniel F. Ford, *The Cult of the Atom: The Secret Papers of the Atomic Energy Commission* (New York: Simon and Schuster, 1982).

WASH-740 Update

The WASH-740 Update was a secret report to the Atomic Energy Commission (AEC) on nuclear safety in 1964. It was to update the 1957 WASH-740 Report, and the AEC hoped the new report would be more positive in tone. A steering panel of senior officials under Clifford Beck of the AEC regulatory staff was entrusted with this study. Scientists at the Brookhaven National Laboratory in Long Island, New York, had the assignment to make detailed technical analysis on safety features. For almost six months the Brookhaven scientists made computer calculations, and they came to the conclusion that as many as 45,000 people might be killed in a major reactor accident. The difference between the two studies was that the new reactors were much larger and situated closer to large metropolitan areas. These results shocked the steering panel, and members questioned the conclusions of the report. Brookhaven scientists refused to back down from their analysis. Beck and the rest of the steering committee decided that the publication of this update would hurt the nuclear energy program. This conclusion was communicated to the commissioners of the AEC, and the update was never published. It was kept secret and never released except to certain key political figures.

Suggested reading: Daniel F. Ford, *The Cult of the Atom: The Secret Papers of the Atomic Energy Commission* (New York: Simon and Schuster, 1982).

WASH-1400. *See* Rasmussen Report

Weiss Report

The Weiss Report was a Public Health Department project studying the incidences of thyroid disease and leukemia deaths in Utah from 1958 to 1962. Edward Weiss, a scientist at the Public Health Department, found a disturbingly high rate of thyroid cancer and leukemia in southwestern Utah.

He contacted Gordon Dunning of the Atomic Energy Commission (AEC), who tried to discourage Weiss from continuing the Project. Weiss went ahead anyway, and the survey found an excessive number of cases of leukemia in southwestern Utah. AEC chairman Glenn Seaborg found a way to bury the report by sending it to President Lyndon Johnson, where the report vanished without a trace. The Weiss Report surfaced, however, in 1979 as a result of a combination of congressional investigations and Freedom of Information Act requests made to do so.

Suggested readings: Howard Ball, *Justice Downwind: America's Atomic Testing Program in the 1950s* (New York: Oxford University Press, 1986); Richard L. Miller, *Under the Cloud: The Decades of Nuclear Testing* (New York: Free Press, 1986).

Western Supplies Group

The Western Supplies Group was a secret international cartel of uranium suppliers organized and run by the United States in the late 1950s. It was founded in 1954, and the principal organizers were Secretary of State John Foster Dulles and the Special Assistant to the Secretary of State for Atomic Affairs Gerard Coad Smith. Membership included Australia, Belgium, Canada, France, Portugal, South Africa, the United Kingdom, and the United States. Formal meetings were called by the U.S. State Department, and its delegates to this meeting had ambassadorial rank. Members regularly exchanged information on uranium sales and planned for safeguards. All major exports of uranium were cleared and approved through this organization. Uranium supplies were controlled by this organization, but such a body could not prevent nuclear proliferation by other means.

Suggested reading: Peter Pringle and James Spigelman, *The Nuclear Barons* (London: Michael Joseph, 1981).

Westinghouse Electric Corporation

The Westinghouse Electric Corporation was a major U.S. dealer in uranium and operator of nuclear reactors. Westinghouse had been founded in 1886 and specialized in electrical products. This corporation obtained much of its early experience working with nuclear energy while designing nuclear propulsion systems for the U.S. Navy. Captain Hyman Rickover selected Westinghouse as his primary choice for building a nuclear propulsion reactor for submarines. Westinghouse built its naval ship reactors in conjunction with scientists at the Argonne National Laboratory and the National Reactor Testing Station under the auspices of the Atomic Energy Commission (AEC). As a result, they specialized in building light water nuclear reactors. Most of the nuclear research was done at the Bettis Atomic Power Laboratory in the Pittsburgh suburb of West Mifflin, Pennsylvania. By 1975, over half of the electricity producing nuclear plants were operated by Westing-house. It also had contracts to provide uranium to these plants at a fixed price, but in 1975 the price of uranium had skyrocketed. In September 1975, Westinghouse defaulted on their contract to deliver uranium at an average price of about $9.95 a pound when the market price at the time was between $25 and $40 a pound. The commitments held by Westing-house were estimated at more than $2 billion at the time of the default. Seventeen suits were brought by the 27 utility companies against Westing-house for failure to perform. These trials, in several separate trials, ended in 1981 with out-of-court settlements costing Westinghouse dearly. Legal costs alone ranged in the tens of millions of dollars. In the aftermath of these trials, the Antitrust Division of the U.S. Department of Justice become interested in the actions of the Uranium Cartel.

Suggested readings: Bertrand Goldschmidt, *The Atomic Complex: A Worldwide Political History of Nuclear Energy* (La Grange Park, IL: American Nuclear Society, 1982); Earle Gray, *The Great Uranium Cartel* (Toronto, Ontario: McClelland and Stewart, 1982); John W. Simpson, *Nuclear Power from Underseas to Outer Space* (La Grange Park, IL: American Nuclear Society, 1995).

Whiteshell Nuclear Research Establishment

In July 1960 the Canadian government established a second nuclear re-search center, the Whiteshell Nuclear Research Establishment (WNRE) at Whiteshell Provincial Park, about 65 miles northeast of Winnipeg, Mani-toba. A townsite of Pinawa was selected and houses built for personnel at the WNRE. An organic-cooled experimental reactor (OCDRE) was to be built for advance reactor research, but the technology for it was too un-proven for the Atomic Energy Control Board (AECB). A substitute was constructed instead, the OTR (Organic Test Reactor), which was similar to the OCDRE but cost half as much. Construction on the OTR started in 1963, and the reactor was in operation by November 1, 1965. This facility is still in operation, but much of its activities have been diverted away from atomic energy to other research efforts.

Suggested reading: Robert Bothwell, *Nucleus: The History of Atomic Energy of Canada Limited* (Toronto, Canada: University of Toronto Press, 1988).

Whyl Demonstration

In February 1975 the inhabitants of Whyl, a small village in the south-western Rhineland in West Germany, launched the first major protest against the construction of a nuclear power station. Whyl's 300 citizens started the protest by occupying the construction site, sitting down in front of bull-dozers and halting work. German police attempted to drive off the demonstrators without much success. Although the initial impetus for the protest was fear of losing their livelihoods because the vineyards would be affected by the humidity of the plant's cooling towers, it soon escalated into a protest over reactor safety. Once the news was exposed on the national media, supporters flocked to Whyl. It has been estimated that there were over 20,000 demonstrators taking part, and the site was occupied for a year. Moreover, the participants began a study program to educate and recruit antinuclear activists. A petition against the plant garnered 90,000 signatures. Against this type of organized opposition the state had little recourse but the law courts. A lawsuit ensued, but the nuclear reactor was never built. The success of this demonstration created a precedent for confrontations both in Germany and abroad.

Suggested readings: Peter Pringle and James Spigelman, *The Nuclear Barons* (London: Michael Joseph, 1981); Spencer R. Weart, *Nuclear Fear: A History of Images* (Cambridge: Harvard University Press, 1988).

Wigner, Eugene Paul

Eugene Wigner (1902–1995) was a Hungarian-American physicist whose theoretical work on quantum mechanics and other nuclear problems made him one of the most influential physicists of the twentieth century. He was born on November 17, 1902, in Budapest, Hungary. Wigner's father was a tannery manager. His three marriages— to Amelia Z. Frank in 1936, Mary Annette Wheeler in 1941, and Eileen C. P. Hamilton— resulted in a daughter and a son. After early schooling at the Lutheran Gymnasium in Budapest with his friend John von Neumann, Wigner received his education in chemical engineering in 1924 and his PH.D. in engineering in 1925 from the Berlin Institute of Technology. After subsequently working in a Budapest tannery for almost two years, he found a one-year assistantship in theoretical physics at the Berlin Institute of Technology in 1926. After several years as a professor at the University of Göttingen, Wigner returned to the Berlin Institute of Technology in 1930 and remained there until 1936. Wigner also had a half-time position at Princeton University from 1930 until 1936, and he commuted back and forth between Germany and the United States with his friend John von Neumann. After the Nazi government terminated his employment at Berlin in 1936, he emigrated to the United States and was appointed a professor at the University of Wisconsin. Wigner became an American citizen in 1937. He made his final academic move to Princeton University in 1938, where he was appointed to the chair of theoretical physics. Except for a leave of absence from 1942 to 1947, Wigner remained at Princeton until retirement in 1971.

Wigner made several major contributions to the theories on nuclear structure. In 1927, he proposed the idea of parity as a conserved property of

Eugene Wigner (left) and Leo Szilard (right) were Hungarian physicists active in warning the Americans about the dangers of the Germans acquiring atomic weapons.

nuclear reactions. This idea was only challenged in 1956 when Tsung Dao Lee and Chen Ning Yang argued that parity was not conserved in the weak interaction. In the 1930s, Wigner worked on neutron reaction theory. He collaborated with Gregory Breit in 1936 on the Breit-Wigner formula, which explains neutron absorption by a compound nucleus. Wigner also became involved in the planning and building of the first atomic reactor at the University of Chicago.

During World War II, Wigner played an active role in the Manhattan Project. His friendship with Leo Szilard and Albert Einstein led to the Einstein-Szilard Letter to President Roosevelt that warned about the possibility of an atomic bomb. Later, Wigner served at the Metallurgical Laboratory (Met Lab), Chicago, from 1942 to 1945 and at Oak Ridge, Tennessee, as director of the Clinton Laboratories from 1946 to 1947. It was while serving at the Met lab that he discovered a major change in the crystal structure of graphite that results from prolonged exposure to neutrons. This process was soon called the Wigner effect and had implications for operations of graphite moderated reactors. Wigner received the 1963 Nobel Prize for Physics for his contributions to the acceptance of quantum mechanics and the design of nuclear reactors.

Wigner was a beloved figure among fellow physicists. He was both brilliant and shy. His politeness was proverbial. He died on January 1, 1995, in Princeton, New Jersey.

Suggested reading: John Daintith, Sarah Mitchell, Elizabeth Tootill, and Derek Gjertsen, *A Biographical Encyclopedia of Scientists*, 2nd ed., vol. 2 (Bristol, UK: Institute of Physics, 1994).

Wilson, Charles Thomson Rees

Charles Wilson (1869–1959) was a British physicist whose development of the Wilson cloud chamber made it possible for early nuclear physicists to study subatomic particles. He was born near Glencorse, Scotland, on February 14, 1869. His father was a sheep farmer, but after his father's death in 1873, the family moved to Manchester. Wilson was educated at Owens College (later the University of Manchester), and his original interest was in biology. After receiving a bachelor's of Science in 1887, he became a student at Cambridge University and earned a Bachelor of Arts in 1892. The next four years Wilson spent teaching science at the Bradford Grammar School, before returning to Cambridge University to study physics under J. J. Thomson. In 1900, Wilson began work as a lecturer, but most of his attention was directed toward research. Then in 1925, he was appointed Jacksonian Professor of Natural Philosophy, and he held this position until 1934. Wilson retired from Cambridge in 1936, but continued to conduct physics experiments. He died in Carlops, Scotland, on November 15, 1959.

It was at Cambridge that Wilson started to investigate the possibility of duplicating cloud formation in the laboratory. In his experiments, Wilson showed that clouds need dust particles to seed the condensation, and that X-rays accelerated the process. He was able to prove that charged subatomic particles traveling through supersaturated air could form water droplets. By 1911, Wilson had perfected his cloud chamber. This Wilson cloud chamber became one of the primary research tools for studying subatomic particles in the first four decades of the twentieth century. In 1927, Wilson received the Nobel Prize for Physics for his work with the Wilson cloud chamber.

Suggested readings: David Abbott, ed., *The Biographical Dictionary of Scientists: Physicists* (London: Blond Educational, 1984); Henry A. Boorse, Lloyd Motz, and Jefferson Hane Weaver, *The Atomic Scientists: A Biographical History* (New York: Wiley Science Editions, 1989); John Daintith, Sarah Mitchell, Elizabeth Tootill, and Derek Gjertsen, *A Biographical Encyclopedia of Scientists*, 2nd ed., vol. 2 (Bristol, UK: Institute of Physics, 1994).

Windscale

Windscale is a British uranium-reprocessing plant near the Irish Sea and the scene of Great Britain's worst nuclear accident. The Windscale facility produced much of Great Britain's plutonium for military weapons. On October 8, 1957, a uranium fire started inside a reactor, unleashing radioactive fallout. A physicist was working on No. 1 pile to release excess energy that had built up in the matrix of graphite blocks surrounding the uranium nuclear fuel. Instead of cooling down, the temperature kept rising, and the physicist tried again to clear the excess energy. The temperature continued to increase, and when technicians went inside the reactor, they found the uranium fuel and the graphite burning. During the night of October 10, the fire reached a temperature of 1,300 degrees Centigrade. A fire brigade was called in and water was pumped into the reactor. Although this procedure risked producing a hydrogen explosion, the reactor began to cool down. Particularly harmful was the presence of iodine-131. More than a hundred farms in a radius of 500 square kilometers surrounding the plant had been contaminated by radioactive materials. Thousands of gallons of milk were dumped into the Irish sea, and dairy cattle had to be kept off pastures. It was the worst environmental disaster of its type in the history of the British Isles. News of this accident caused a renewed interest in reactor safety in the United States. The Windscale reactors were eventually closed down permanently as a result of this accident. Decommissioning of both reactors was finally started in 1987 and will take as long as several decades to complete. Because of much bad publicity about nuclear incidents, the *Windscale* name was changed in May 1981 to *Sellafield*.

Suggested readings: Crispin Aubrey, *Melt Down: The Collapse of the Nuclear Dream* (London: Collins and Brown, 1991); John May, *The Greenpeace Book of the Nuclear Age: The Hidden History, the Human Cost* (New York: Pantheon Books, 1989).

The World Set Free

British science fiction writer H. G. Wells published the futuristic novel *The World Set Free* in 1914. In it Wells took information gathered from a book, *Interpretations of Radium* (1909), by British physicist Frederick Soddy and foretold a scenario in which atomic energy had been mastered for both peaceful and military purposes. Despite these advances, a European war breaks out in 1958, and European cities are reduced to radioactive rubble. While literary critics dismissed the novel as nonsense, prominent European physicists soon realized that Wells was not far off in the potential of his projections. The picture of devastated cities and widespread radioactivity was an image that many scientists carried as possibilities even before many of them worked on the atomic bomb from 1943 to 1945. In particular, Leo Szilard, the Hungarian physicist, found the ideas in Wells's novel both suggestive and frightening.

Suggested reading: Richard Rhodes, *The Making of the Atomic Bomb* (New York: Simon and Schuster, 1986).

Y

Yellowcake

Yellowcake is the name for uranium ore in its raw state of uranium oxide. Miners named uranium ore yellowcake because of its yellow color and rough texture. Natural uranium is mined and then made into a concentrate containing about 85 percent uranium oxide. This ore is then sent to processing plants to be converted for use in reactors. Tailings are the waste products of this process and discharge radon-222 gas into the air. This gas, as well as the presence of thorium-230, makes these wastes dangerous for handlers. Uranium miners have long suffered health problems in the mining and processing of yellowcake. The most common and deadly health problem is lung cancer from inhaling radon. In October 1995 the Advisory Committee on Human Radiation Experiments recommended that the strict requirements for proving radon-induced lung cancer for uranium miners be loosened to include all yellowcake miners suffering from the disease.

Suggested reading: Jerome Price, *The Antinuclear Movement* (Boston: Twayne, 1982).

Yucca Mountain Waste Disposal Site

The Yucca Mountain in Nevada has been selected by the U.S. government as its preferred site for nuclear waste storage for highly radioactive materials. In 1957 the National Academy of Sciences recommended that high-level nuclear waste be buried in a safe area. In the 1982 Nuclear Waste Policy Act (NWPA), the Department of Energy (DOE) was given the task to select a number of potential nuclear waste sites for consideration. Yucca Mountain was picked as one of three sites. In the Nuclear Waste Policy Amendments Act of 1987, Congress mandated that only the Yucca Mountain site be seriously studied. While this study was under way, Nevada's state attorney general filed a court petition seeking to disallow the Yucca Mountain site. This petition made its way through the court system until it reached the U.S. Supreme Court. The Supreme Court has denied review until the Yucca Mountain site is formally approved as a waste site. After extensive review and at a cost of more than $2.5 billion, nothing still has been done by the U.S. government about initiating the final stages for approval. Indeed, the earliest date for a permanent geological facility for storage of high-level radioactive waste is 2010. In the meantime, the state of Nevada has renewed its legal challenges in using the site for waste disposal. The DOE had intended to start accepting radioactive waste in 1998, but legal and scientific delays prevented this from happening. Concurrently, nuclear power plant companies are suing the federal government and winning large sums that could ultimately be as high as $53 billion for the failure of the government to provide a place to store radioactive waste. These judgments are under appeal, but the government is caught between the problem of finding a safe place for storage and meeting its legal requirements to accept nuclear waste from private companies.

The DOE proposes to bury 77,000 tons of nuclear waste at Yucca Mountain, with nearly 1,500 workers busy on the project. These workers are employed by 17 contractors under the overall contractor, the TRW Environmental Safety System, five national laboratories, and the U.S. Geological Service. Besides making the area secure, scientists and engineers are concerned about earthquakes in the region and possible damage to the water table. Waste would be buried in a series of tunnels 990 feet below the top of the mountain and almost 900 feet above the water table. Waste would be put into specially tested containers and stored in the tunnels. Another area of concern is transportation of the nuclear waste to the Yucca Mountain facility. At the turn of the century, the Yucca Mountain site is still under consideration, but the federal government has yet to make it official.

Suggested reading: K. S. Shrader-Frechette, *Burying Uncertainty: Risk and the Case Against Geological Disposal of Nuclear Waste* (Berkeley: University of California Press, 1993).

Z

Zangger Committee

Inquiries about procedures for nuclear export controls from nuclear nations to nonnuclear states led to the formation of the Zangger Committee, or the Non-Proliferation Treaty Exporters Committee. Fears among the nuclear states about proliferation of nuclear weapons technology resulted in the formation of this committee in early 1974. It was in part a progression of the Nuclear Non-Proliferation Treaty (NPT) of 1970. The committee was named after its chair, Claude Zangger, a Swiss official. Representatives from advanced industrial countries held a series of secret meetings to determine policy. This committee established a "Trigger List" of special nuclear materials that could only be exported under International Atomic Energy Agency (IAEA) safeguards. News of a nuclear test in India in May 1974 showed that this committee was not going to be effective unless a more active approach was undertaken. A larger, less secret London group formed the London Suppliers Guidelines to fill this void. The Zangger Committee meets in Vienna twice a year, and the deliberations are kept secret. By 1994, the Zangger Committee consisted of representatives from 29 nations. In 1999 the Zangger Committee is still active and includes all nuclear weapons states (23 nations are presented on the Zangger Committee) except for France and China. Both France and China refuse to cooperate on nuclear export controls.

Suggested reading: Congressional Quarterly, *The Nuclear Age: Power, Proliferation and the Arms Race* (Washington, DC: Author, 1984); Richard Kokoski, *Technology and the Proliferation of Nuclear Weapons* (Oxford: Oxford University Press, 1995); Leonard S. Spector, Mark G. McDonough, and Evan S. Medeiros, *Tracking Nuclear Proliferation: A Guide to Maps and Charts, 1995* (Washington, DC: Carnegie Endowment for International Peace, 1995).

ZEEP

The Zero Energy Experimental Pile (ZEEP) was a zero energy experimental reactor designed and built by the British-Canadian team in 1944–1945 at the Chalk River research facility. It was a small heavy water reactor intended to produce fission products for further experiments. Lew Kowarski was the project chief of the ZEEP Project. Despite constant supply and building problems, the ZEEP began operations on September 5, 1945. It was the first successful reactor operating outside the United States. Experiments conducted with this reactor allowed Canadian scientists and engineers to design more efficient reactors—the NRX (National Research X-perimental) and NRU (National Research Universal). Because it was too small except for basic research, it was decommissioned in the late 1940s and is now kept as a museum piece.

Suggested reading: Robert Bothwell, *Nucleus: The History of Atomic Energy of Canada Limited* (Toronto, Canada: University of Toronto Press, 1988).

Zinn, Walter Henry

Walter Zinn (1906–   ) is a Canadian-American nuclear physicist whose work on the Manhattan Project and at the Argonne National Laboratory made major contributions to the development of nuclear energy. He was born on December 10, 1906, in Kitchener, Ontario, Canada. In 1938, he became an American citizen. Zinn obtained his doctorate from Columbia University in 1934. Much of his postdoctoral work was with Enrico Fermi and Leo Szilard, investigating atomic fission. In 1939, Zinn and Szilard demonstrated Otto Hahn's discovery of fission with experiments that proved uranium underwent fission when bombarded with neutrons. His work on fission led him to be recruited for the Manhattan Project and the construction of the atomic bomb. Zinn worked closely with Enrico Fermi at the Metallurgical Laboratory in Chicago to construct the first atomic reactor, Chicago Pile One. After the war, Zinn concentrated his activities as director of the Argonne National Laboratory on building atomic reactors. His position became more important when the Atomic Energy Commission centralized reactor research at the Argonne National Laboratory in early 1948. Zinn's specialty was designing breeder reactors, and his first such reactor appeared in 1951.

Suggested readings: John Daintith, Sarah Mitchell, Elizabeth Tootill, and Derek Gjertsen, *A Biographical Encyclopedia of Scientists*, 2nd ed., vol. 2 (Bristol, UK: Institute of Physics, 1994); Richard Rhodes, *The Making of the Atomic Bomb* (New York: Simon and Schuster, 1986).

Zirconium

Zirconium is the metallic element used as the cover base for highly enriched uranium fuel for nuclear reactors. The initial goal in using zirconium
was to provide a safe material for use in naval warship reactors. This was so
successful that zirconium was also used in the light water reactors (LWRs)
for civilian reactors. Zirconium was first identified by German chemist Martin Klaproth in 1789. Zirconium's ability to maintain strength at high operating temperatures with a corresponding high resistance to corrosion made
zirconium invaluable to scientists designing nuclear propulsion for warships,
especially submarines. To gain these characteristics, however, the 2 percent
hafnium had to be removed. Herbert Pomerance discovered a process to
separate the two elements at the Oak Ridge National Laboratory in 1947.
At first only a small amount of zirconium was available, but since the ore is
plentiful, this shortage was soon alleviated. A battle ensued between the
Atomic Energy Commission (AEC) and Admiral Hyman Rickover's nuclear
submarine group over the zirconium supply, until a contract was given to
the Carborundum Metals Company in the spring of 1951. Zircaloy, an alloy
of zirconium, is the standard tube used with uranium dioxide fuel in most
industrial reactors.

Suggested readings: Daniel F. Ford, *The Cult of the Atom: The Secret Papers of
the Atomic Energy Commission* (New York: Simon and Schuster, 1982); Theodore
Rockwell, *The Rickover Effect: How One Man Made a Difference* (Annapolis, MD:
Naval Institute Press, 1992).

Chronology of Atomic Energy

1780	Martin Heinrich Klaproth discovers the element uranium
1828	Jons Jakob Berzlius discovers the element thorium
1895	
November 8	Wilhelm Roentgen discovers the X-ray in Germany
December 28	Roentgen reports discovery of the X-ray to the Würzburg Physico-Medical Society
1896	
March 1	Antoine Becquerel discovers radioactivity in France
1897	Publication by J. J. Thomson of proof of the existence of the electron
1898	
December 26	French physical chemists Marie and Pierre Curie discover radium
1900	
January–February	Ernest Rutherford and Frederick Soddy trace spontaneous disintegration of radioactive elements
December 14	Max Planck introduces the quantum theory to the German Physical Society
1905	
February–June	Albert Einstein advances the theory of relativity
1908	Hans Geiger develops the Geiger Counter for detection of radioactivity
1909	
September	Robert Millikan's oil drop experiment confirms atomic theory of matter

1911

March 7   Ernest Rutherford announces the discovery of the nucleus of the atom

October 30   Meeting of First Solvay Conference on radioactivity and quantum theory

1913

March 6   Niels Bohr's paper on the structure of the atom published in *Philosophical Magazine*

1914   Publication of H. G. Wells's futuristic novel *The World Set Free*

1920

June 3   Ernest Rutherford proposes intrusion of neutron into nucleus

1921

March 3   Opening of the Institute for Theoretical Physics in Copenhagen, Denmark

November 23   Founding of the Leningrad Physico-Technical Institute (LFTI)

1923   Direct experimental confirmation that the photon is an elementary particle by Arthur H. Compton (Compton effect)

1924   Discovery of the wave nature of electrons by Louis de Broglie

1926   Erwin Schrödinger invents wave mechanics

May   Werner Heisenberg, Max Born, and Pascual Jordan work out the mathematics behind quantum mechanics

1927

February   Werner Heisenberg postulates the Uncertainty Principle

October   Fifth Solvay Conference where Einstein confronts Bohr-Heisenberg over quantum mechanics

1930   Discovery of the Great Bear Lake Eldorado Uranium Mine

1931   Ernest O. Lawrence and M. S. Livingston construct first cyclotron
                    Van de Graaff invents the first electrostatic accelerator

1932

January   Harold Urey announces the discovery of deuterium

February 17   Sir James Chadwick's letter to *Nature* identifies the existence of the neutron

April   John Cockcroft and E.T.S. Walton demonstrate the first artificial disintegration of nuclei

August 2	Carl Anderson discovers the positron

1933

September 12	Leo Szilard postulates the idea of chain reaction via neutron
October	Seventh Solvay Conference where participants discuss the structure of the proton

1934

February 10	Irene and Frederic Joliot-Curie announce the discovery of artificial radioactivity
May 10	Enrico Fermi assumes the existence of transuranic elements based on results of his experiments
October 22	Enrico Fermi and his Rome team discover slow neutrons

1936

January 27	Niels Bohr proposes liquid-drop model of atomic nucleus

1938

December 19	Results of Otto Hahn and Fritz Strassmann experiment on neutron irradiation of uranium

1939

January 6	Lise Meitner and Otto Frisch compose letter explaining uranium nuclear fission
January 15	Lise Meitner and Otto Frisch's letter announces the splitting of the atom by Hahn and Strassmann experiment
January 26	Washington Conference on Theoretical Physics where Niels Bohr discusses the Hahn-Strassmann experiment
February 7	Niels Bohr theorizes in a letter to *Physical Review* that it is U-235 that produces fission
March 3	Leo Szilard and Walter Zinn prove possibility of chain reaction in an experiment conducted in the United States
April 24	Paul Harteck and Wilhelm Groth inform German authorities about possibility of an atomic bomb
April 29	First meeting of the German Uranium Club is held
June 27	President Roosevelt approves formation of the National Defense Research Committee (NDRC)
July 3	Enrico Fermi and Leo Szilard propose the idea of a nuclear reactor based on uranium as fuel and graphite as a moderator
August 2	Einstein-Szilard Letter to President Roosevelt on atomic bomb project

September 26	German Army Ordnance Bureau under Kurt Diebner begins research on a uranium bomb
October 10	Alexander Sachs presents Einstein-Szilard Letter to President Roosevelt
October 21	First meeting of the Uranium Committee in the United States
November 1	American Uranium Committee recommends further support of atomic energy projects
November 15–20	Conference on Questions of the Physics of the Atomic Nucleus held at Kharkov, Soviet Union, which led to the Khariton-Zeldovich Letters

1940

February 1	Frisch-Peierls Memorandum on feasibility of an atomic weapon
March 9	Norsk-Hydro signs contract with France, delivering heavy water supplies to France
April 10	MAUD Committee holds first meeting
April 27	Edwin M. McMillan and Philip H. Abelson discover neptunium
June 9–14	French scientists Hans van Halban and Lew Kowarski escape from France to England with French supply of heavy water
June 27	National Defense Research Committee under Vannevar Bush established by order of President Roosevelt
October	Suzuki Report on the availability of uranium for the Japanese atomic bomb project

1941

January 28	Glenn T. Seaborg and associates at the University of California at Berkeley discover plutonium
June 6	MAUD Report on the possibility of building an atomic bomb accepted by British government
June 28	Office of Scientific Research and Development (OSRD) established by order of President Roosevelt
July 7	MAUD Report disclosed to officials in the United States
September 21–24	Bohr-Heisenberg conversation over German-Allied atomic bombs
December 6	President Roosevelt approves all-out push to build the atom bomb following a report from Vannevar Bush
December 18	Uranium project (S-1 Committee) placed under OSRD

1942

January 24	Metallurgical Laboratory transferred from Columbia University to the University of Chicago
April	Georgi Flerov sends letter to Joseph Stalin on need for Soviet work on an atomic weapon
June 17	President Roosevelt authorizes atomic bomb program
June 23	Leipzig atomic reactor explodes and burns, delaying German atomic program for many months
July	Berkeley Summer Colloquium of 1942 discusses the possibility of building a fission and fusion bomb
August 11	Colonel James C. Marshall creates the Manhattan Engineering District (MED)
September 19	War Production Board gives highest priority for supplies to the Manhattan Project
September 21	Military Policy Committee formed to oversee the building of the American atomic bomb
September 23	General Leslie R. Groves appointed military chief of the Manhattan Project
October	Igor Kurchatov named scientific director of Soviet atomic bomb project
November 16	Begin construction at the University of Chicago Pile One
December 2	Fermi group produces first self-sustaining chain reaction at the University of Chicago

1943

January 1	Founding of Los Alamos Scientific Laboratory in Los Alamos, New Mexico
February 1	Begin construction of uranium processing plants at Oak Ridge, Tennessee
February 27	Norwegian resistance attacks German heavy water factory in Norway
March 22	Begin construction of Hanford Plant for plutonium production in Richland, Washington
April	Seth Neddermeyer proposes the implosion method for firing an atomic bomb
July 2	Yashio Nishina reports to the Japanese army on the feasibility of an atomic bomb
August 19	Quebec Conference where Roosevelt and Churchill negotiate Quebec Agreement
August 27	Begin construction on the first Hanford reactor

September 29	Niels Bohr escapes from Denmark to Sweden before arrest by the Nazis
November 16	Allies bomb the German heavy water factory in Norway
November 29	Project to remodel the B-29 for delivery of an atomic bomb
December 14	Alsos Mission arrives in Italy to investigate the German atomic bomb program

1944

February 20	Norwegian resistance sinks the ship carrying German heavy water to Germany
March 1	First plutonium shipment from Oak Ridge arrives at Los Alamos
March 3	Air Force tests dropping atomic bombs by B-29s
June 13	Anglo-American Declaration of Trust signed to ensure cooperation on obtaining uranium and thorium ores
July 17	J. Robert Oppenheimer gives highest priority to the development of the implosion method
August	Air Force makes Colonel Paul W. Tibbets commander of the 509th Composite Group
August 1	American forces occupy Tinian Island and construction of six runways for B-29s under way
September 19	Hyde Park Meeting where Roosevelt and Churchill agree on atomic policy
September 27	First reactor at Hanford, Washington, begins operations
December	Allied scientists conclude from captive German documents that Germans are nowhere close to building an atomic bomb

1945

February 2	First plutonium shipment from Hanford to Los Alamos
April 4	Destruction of Riken Scientific Institute in Tokyo ends Japanese atomic bomb project
April 23	American troops of the Alsos Mission seize German reactor at Heigerloch, German
April 25	President Truman informed about Manhattan Project by Henry Stimson and General Leslie Groves
April 27	First meeting of the Target Committee to select Japanese cities for the two atomic bombs
May 2	Interim Committee on the Atomic Bomb formed by President Truman
May 9	First meeting of the Interim Committee on the Atomic Bomb

May 10	Target Committee meets in Los Alamos and recommends four cities as targets: Kyoto, Hiroshima, Yakohama and Kikura
May 30	Henry Stimson, Secretary of Defense, tells General Leslie Groves to drop Kyoto from the targets list
June 1	Interim Committee on the Atomic Bomb recommends to President Truman the use of the atomic bomb on Japanese cities
June 10	The 509th Composite Group arrives at Tinian Island
June 11	Franck Report delivered to secretary of war recommending notification before use of the atomic bomb
July 16	Successful test of Fat Man–type atomic bomb in the Trinity Test at Alamogordo Testing Ground, New Mexico Cruiser U.S.S. *Indianapolis* leaves for Tinian Island with Little Boy
July 17	Szilard and 70 other scientists petition President Truman against use of the atomic bomb
July 24	Truman informs Stalin about a new weapon at the Potsdam Conference
July 25	Truman orders delivery of atomic bomb on Japanese city
August 6	First atomic bomb (Thin Man) dropped on Hiroshima
August 9	Second atomic bomb (Fat Man) dropped on Nagasaki
August 10	Japan offers to surrender
August 12	Publication of the Smyth Report on the building of the atomic bomb
August 20	Lavrentii Beria named head of the Special Committee on the Atomic Bomb, with the mission to build an atomic bomb
August 21	Harry Dahlian dies in a nuclear experiment accident at the Los Alamos Scientific Laboratory
September 5	Igor Gouzenko defects and exposes Soviet atomic bomb spy ring
September 28	Scientific panel of the Interim Committee recommends against developing a hydrogen bomb
October 10	J. Robert Oppenheimer resigns from the Los Alamos Scientific Laboratory
October 18	Charles de Gaulle forms the French Commissariat à l'Énergie Atomique (CEA)
October 31	Formation of the Federation of American Scientists (FAS)
November 15	Washington Declaration issued on peaceful uses of atomic energy

| December 20 | Passage of the McMahon Bill establishing the Atomic Energy Act of 1946 |

1946

January	United Nations forms the Atomic Energy Commission (UNAEC)
March 12	Lilienthal Group presents report on internationalization of atomic energy to Acheson Committee
April 2	Soviet authorities select town of Sarov for its atomic weapons research center, code name Arzamas-16
May 21	Louis Slotin suffers fatal accident at Los Alamos Scientific Laboratory during demonstration of a chain reaction
June 14	Baruch Plan presented by Bernard Baruch before the United Nations
July 1	First atomic bomb test of Operations Crossroads at Bikini Atoll Argonne National Laboratory founded to conduct research on radiation and radioactive materials
July 25	Underwater atomic bomb test of Operations Crossroads reveals radioactive fallout problem
August 1	Atomic Energy Commission (AEC) established by act of U.S. Congress
December 25	Soviets start operations of first nuclear reactor at Laboratory No. 2 in Moscow

1947

January 2	First meeting of the Atomic Energy Commission (AEC)
September	Founding of the Long Range Detection Project to detect Soviet atomic tests by U.S. Air Force
November 16	Formation of the Atomic Bomb Casualty Commission by Presidential Directive

1948

| January 1 | AEC moves reactor work from Oak Ridge to Argonne National Laboratory |
| April 15 | First test of Operation Sandstone at Eniwetok Testing Site in the Marshall Islands in the Pacific |

1949

August 29	Union of Soviet Socialist Republics detonates an A-bomb
September 3	U.S. Air Force discovers evidence of a Soviet nuclear test
October 30	Recommendation by J. Robert Oppenheimer and other scientists against the development of the hydrogen bomb
November 18	Special Subcommittee on the Hydrogen Bomb formed by President Truman to study hydrogen bomb development

| November 29 | Lewis L. Strauss recommends to President Truman the building of the hydrogen bomb |

1950

January 21	President Truman orders construction of hydrogen bomb
February 3	In London Klaus Fuchs confesses to spying for the Soviet Union
April 1	British begin construction of nuclear weapons production center at Aldermaston
December 19	President Truman approves the Nevada Test Site for atomic bomb testing

1951

January 27	First atomic test at the Nevada Test Site (Operation Ranger)
February	Edward Teller and Stanislaw Ulam outline configuration for the hydrogen bomb
October 5	Panda Committee to design and build the hydrogen bomb meets for first time

1952

February 15	Consortium of European countries establish CERN
March	Apex Committee formed to oversee development of British atomic bomb
	Lawrence Livermore National Laboratory founded in Livermore, California, to work on hydrogen bomb development
April 1	Atomic Energy of Canada Limited formed by the Canadian government
May 7	Shot Easy at the Nevada Test Site shows for the first time the dangers of radioactive fallout in the United States
October 3	Great Britain explodes its first atomic bomb at Monte Bello Island, Australia
October 31	Detonation of a thermonuclear bomb (Mike nuclear test) by the United States at Eniwetok Atoll, Pacific Ocean
November 11	Test of last atomic bomb, King nuclear test, at Nevada Test Site
December 13	NRX reactor at Chalk River, Canada, has partial meltdown, killing an operator and injuring four others

1953

| June 19 | Ethel and Julius Rosenberg are executed in the United States as Soviet atomic spies |
| August 12 | Union of Soviet Socialist Republics detonates a prototype hydrogen bomb |

| December 8 | President Eisenhower delivers Atoms for Peace speech to the U.N. General Assembly |
| December 13 | J. Robert Oppenheimer declared a security risk |

1954

January 21	U.S. Navy launches first nuclear submarine, U.S.S. *Nautilus*, in Groton, Connecticut
February 17	Atomic Energy Act of 1954 receives President Eisenhower's endorsement
February 28	Bravo nuclear test at the Bikini Test Site spreads radioactive fallout and causes an international incident after Japanese tuna boat receives high-level radiation
April 12	J. Robert Oppenheimer loses his security clearance
June 27	Soviet Union opens first civilian nuclear power station at Obninsk, south of Moscow
September 6	Construction of Shippingport Nuclear Reactor started

1955

| August 8–20 | First U.N. International Conference on the Peaceful Uses of Atomic Energy held in Geneva, Switzerland |
| November 11 | Soviet Union tests an airborne hydrogen bomb |

1956

March	Joint Institute for Nuclear Research established in Dubna, Soviet Union
October 17	British claim first fully functional civilian nuclear power station at Calder Hall
October 26	International Atomic Energy Agency (IAEA) begins operations

1957

May 15	First British hydrogen bomb detonated at Christmas Island in the South Pacific
June 21	First meeting of SANE, or Committee for a Sane Nuclear Policy
June 24	Edward Teller and Lewis L. Strauss propose a hydrogen bomb with reduced radiation, or clean H-bomb
September	Advisory Committee on Reactor Safeguards (ACRS) formed
September 2	President Eisenhower signs Price-Anderson Amendment
September 29	Nuclear waste accident at Chelyabinsk-40 causes heavy casualties
October 1	International Atomic Energy Agency established
October 10	Windscale nuclear reactor accident in Great Britain
December 5	Soviet Union launches nuclear-powered icebreaker ship *Lenin*

| December 18 | Power generation from first American civilian power generation station at Shippingport Nuclear Power Plant |

1958

January 16	Campaign for Nuclear Disarmament (CND) founded in London
March 31	Soviet Union announces unilateral suspension of nuclear bomb tests
June	Alice Stewart, a British epidemiologist, publishes findings on carcinogenic effects of diagnostic radiation on children
August 22	United States and Great Britain announce one-year bomb test moratorium
October 14	Neptune Shot underground test creates an accidental crater at Nevada Test Site

1959

February 17	High levels of strontium-90 found in U.S. milk and in children's bones
July 21	First nuclear-powered merchant ship, the *Savannah*, launched at Camden, New Jersey
August	Federal Radiation Council (FRC) formed to advise the president of the United States on radiation matters
December 1	Twelve countries sign the Antarctic Treaty to prohibit nuclear devices or radioactive waste

1960

February 13	France explodes its first atomic bomb in the Sahara
April 3	Partial meltdown of reactor core at Waltz Mills test reactor for Westinghouse Corporation
May 1	A U-2 U.S. reconnaissance plane is shot down over the Soviet Union

1961

January 3	Test reactor (SL-1) accident kills three maintenance men
April	President Kennedy's reference to the need for fallout shelters starts national fad
July 4	Soviet Union launches a nuclear-powered submarine
September 1	Soviet Union resumes nuclear testing
September 15	United States resumes underground nuclear testing
October 30	Soviet Union detonates 50-megaton hydrogen bomb at Novaya Zemlya Test Site

1962

| March 4 | Nuclear power plant at McMurdo Sound in the Antarctic becomes operational under U.S. Navy administration |
| March 15 | Soviet Union proposes complete nuclear disarmament |

October 22	President Kennedy gives Soviet Union ultimatum concerning nuclear missiles in Cuba
1963	
June 20	Soviet Union and the United States sign hot line agreement
August 5	Soviet Union and the United States sign Nuclear Test Ban Treaty, or Moscow Treaty, prohibiting nuclear weapons testing in the atmosphere
October 7	President Kennedy oversees ratification of the Nuclear Test Ban Treaty
December	Israeli nuclear reactor at Dimona goes into operation
1964	
August 26	President Lyndon B. Johnson signs the Private Ownership of Special Nuclear Materials Act, allowing private industry to own nuclear materials
August 28	Successful test of a nuclear engine for space travel (KIWI B4-E)
October 15	China explodes first atomic bomb at Lop Nur Testing Site
1965	
April 3	First nuclear reactor engine operates in space
June 17	Phoebus-1A nuclear rocket tested for NERVA (Nuclear Engine for Rocket Vehicle Applications)
November	Atomic Energy Commission (AEC) gives liquid metal fast breeder reactor top priority and decides to begin experiments at a new facility
1966	
October 5	Fermi fast breeder reactor outside of Detroit, Michigan, suffers partial core meltdown
1967	
February 12	Treaty of Tlatelolco prohibiting nuclear weapons in Latin America
June 17	China detonates an airborne hydrogen bomb
1969	
June 12	Japan launches a nuclear-powered merchant ship, *Mutsu*
October 29	Gofman-Tamplin Manifesto challenges AEC's radiation protection standards
1970	
January 1	Passage of National Environmental Policy Act
March 5	Nuclear Non-Proliferation Treaty becomes effective

1971

| July 23 | *Calvert Cliffs* decision extends the provision of the National Environmental Policy Act to the AEC |

1972

February	Formation of the Uranium Cartel
May 18	Sea-Bed Treaty to prevent nuclear weapons on the ocean bottoms
May 26	Signatures on SALT I Treaty by the United States and Soviet Union
August 3	SALT I Treaty ratified by U.S. Senate

1974

May 18	India explodes first nuclear bomb
August	Rasmussen Report on nuclear safety published
October 11	President Gerald Ford abolishes the Atomic Energy Commission and the Joint Congressional Committee on Atomic Energy
November 13	Karen Silkwood, prominent antinuclear activist, dies in suspicious automobile accident

1975

| January 19 | Energy Research and Development Administration and Nuclear Regulatory Commission (NRC) replace the Atomic Energy Commission (AEC) |
| March 22 | Electrical fire at the Brown's Ferry Nuclear Power Station shows inadequacy of AEC's safety regulations |

1976

| January 27 | Nuclear Suppliers Group forms to establish new guidelines on nuclear exports |
| July 11 | Clamshell Alliance forms at Rye, New Hampshire |

1977

| July 12 | President Jimmy Carter decides to develop the neutron bomb |
| August 4 | President Carter combines the Energy Research and Development Administration with the Federal Energy Administration to create the Department of Energy |

1979

March	Formation of the National Association of Atomic Veterans
March 28	Three Mile Island Nuclear Plant accident
May 6	Between 75,000 and 125,000 antinuclear protesters march on Washington, DC
June 18	SALT II signed by representatives of the United States and the Soviet Union

September 22	Soviet Union reports evidence of a nuclear explosion off the coast of South Africa
1981	
June 7	Israeli Air Force attacks and destroys Iraqi Tammuz Nuclear Center
1983	
January 7	President Ronald Reagan signs the Nuclear Waste Policy Act of 1982
March 3	President Reagan proposes the Strategic Defense Initiative
1985	
July 10	French Secret Service operatives sink the antinuclear protest ship *Rainbow Warrior* in Auckland Harbor, New Zealand
August 6	Signing of the Treaty of Rarotonga establishes a South Pacific Nuclear Free Zone
August 10	Soviet nuclear submarine accident at Chazhma Bay, Soviet Union
1986	
April 25	Nuclear accident at the Chernobyl Nuclear Power Station
April 28	Swedish scientists note radioactive cloud from Chernobyl over Europe
October 3	Sinking of the Soviet nuclear submarine *K-219* near Bermuda
1988	
February 28	A Pakistani official admits that Pakistan has an atomic bomb
December 31	Pakistan and India sign agreement prohibiting any attacks on each other's nuclear installations
1989	
January	President Ronald Reagan signs paperwork shutting down unneeded nuclear weapons plants at six localities
April 7	A Komsomolets Mike–class Soviet nuclear-powered attack submarine sinks in a nuclear accident in the Barents and Norwegian seas
1991	
July	Boris Yeltsin, president of Russia, and George Bush, president of the United States, sign Strategic Arms Reduction Treaty (START)
September 27	President Bush initiates a unilateral cut of tactical nuclear weapons

1992

October 24	President Bush signs the Energy Policy Act reforming licensing procedures for nuclear power plants

1993

January	Boris Yeltsin, president of Russia, and George Bush, president of the United States, sign Strategic Arms Reduction Treaty (START-II)
March	South Africa acknowledges that it possesses nuclear bomb capability but claims that the arsenal of six had been dismantled by 1991
April 4	President William Clinton and President Boris Yeltsin agree at Vancouver summit to begin negotiations on multilateral nuclear test ban agreement
December 9	First demonstration of controlled fusion energy takes place with the Tokamak Fusion Test Reactor (TFTR) at Princeton University

1994

October	United States and North Korea reach a nuclear accord on nuclear weapons
December 15	U.N. General Assembly passes a resolution supporting multilateral negotiations on a Comprehensive Test Ban Treaty

1995

May 11	Non-Proliferation Treaty (NPT) made permanent by member states
May 15	China conducts underground nuclear test at Lop Nur Test Site
September 5	France resumes underground nuclear testing on Mururoa Atoll in the South Pacific
December 8	Accident at Japanese plutonium high-speed breeder reactor Monju involves leak of cooling liquid sodium
December 15	Signing of the Treaty of Bangkok by ASEAN countries, establishing a nuclear weapon free zone in southeast Asia

1996

January 27	France concludes its underground test series of six nuclear explosions
January 29	French President Jacques Chirac announces the end of French nuclear testing in the South Pacific
March 25	France, Great Britain, and the United States sign the South Pacific Nuclear Free Zone Treaty
April 11	African states (43 states) sign treaty making Africa a nuclear free zone

June 8	China conducts a nuclear test at its Lop Nur Test Site
July 29	China concludes its nuclear test series at its Lop Nur Test Site
September 24	United States is among 146 countries signing the Comprehensive Test Ban Treaty

1997

| July 2 | United States begins underground nuclear weapons tests at Nevada Test Site |
| September 18 | United States conducts a second underground nuclear weapons test at the Nevada Test Site |

1998

March 19	United States starts series of underground nuclear tests on radioactive plutonium
May 11–13	India conducts five underground nuclear tests at a test site in Rajasthan, near the Pakistan border, and at Pikhran
May 28–30	Pakistan responds to the Indian tests by staging six nuclear tests
December 24	Russia announces the end of a series of five underground nuclear tests at Novaya Zemlya from September 14 to December 13

1999

| February 26 | India refuses to sign the Comprehensive Test Ban Treaty |

Selected Bibliography

BIOGRAPHICAL DICTIONARIES

Abbott, David, ed. *The Biographical Dictionary of Scientists: Physicists.* London: Blond Educational, 1984.

Boorse, Henry A., Lloyd Motz, and Jefferson Hane Weaver. *The Atomic Scientists: A Biographical History.* New York: Wiley Science Editions, 1989.

Daintith, John, Sarah Mitchell, Elizabeth Tootill, and Derek Gjertsen. *A Biographical Encyclopedia of Scientists.* 2nd ed. 2 vols. Bristol, UK: Institute of Physics, 1994.

Gillispie, Charles Coulston. *Dictionary of Scientific Biography.* New York: Charles Scribner's Sons, 1975.

Schlessinger, Bernard S., and June H. Schlessinger, eds. *The Who's Who of Nobel Prize Winners.* Phoenix, AZ: Oryx Press, 1986.

Who's Who in Soviet Nuclear Science. Berkeley: University of California, Radiation Laboratory, 1960.

Williams, Trevor I., ed. *A Biographical Dictionary of Scientists.* 3rd ed. New York: Wiley, 1982.

BIOGRAPHIES

Andrade, Edward. *Rutherford and the Nature of the Atom.* Garden City, NY: Doubleday, 1964.

Bernstein, Jeremy. *Einstein.* New York: Viking Press, 1973.

———. *Hans Bethe, Prophet of Energy.* New York: Basic Books, 1980.

Blaedel, Niels. *Harmony and Unity: The Life of Niels Bohr.* Madison, WI: Science Tech, 1988.

Blumberg, Stanley A., and Gwinn Owens. *Energy and Conflict: The Life and Times of Edward Teller.* New York: Putnam's Sons, 1976.

Brandt, Keith. *Marie Curie, Brave Scientist.* Mahwah, NJ: Troll Associates, 1983.

Brian, Denis. *Einstein: A Life.* New York: Wiley, 1996.

Bunge, Mario Augusto, and William R. Shea, eds. *Rutherford and Physics at the Turn of the Century*. Kent, UK: Science History Publications, 1979.

Caroe, Gwendolyn M. *William Henry Bragg, 1862–1942: Man and Scientist*. Cambridge: Cambridge University Press, 1978.

Cassidy, David C. *Uncertainty: The Life and Science of Werner Heisenberg*. New York: W. H. Freeman, 1992.

Childs, Herbert. *An American Genius: The Life of Ernest Lawrence*. New York: Dutton, 1968.

Christman, Al. *Target Hiroshima: Deak Parsons and the Creation of the Atomic Bomb*. Annapolis, MD: Naval Institute Press, 1998.

Clark, Ronald William. *Einstein: The Life and Times*. New York: World Publishing, 1971.

De Haas-Lorentz, G. L. *H. A. Lorentz: Impressions of His Life and Work*. Amsterdam, Holland: North-Holland, 1957.

Dorozynski, Alexander. *The Man They Wouldn't Let Die*. New York: Macmillan, 1965.

Dyke, Richard Wayne. *Mr. Atomic Energy: Congressman Chet Holifield and Atomic Energy Affairs, 1945–1974*. Westport, CT: Greenwood Press, 1989.

Fischer, Klaus. *Changing Landscapes of Nuclear Physics: A Scientometric Study on the Social and Cognitive Position of German-speaking Emigrants within the Nuclear Physics Community, 1921–1947*. Berlin: Springer-Verlag, 1993.

Folsing, Albrecht. *Albert Einstein: A Biography*. New York: Viking, 1997.

Forsee, Aylesa. *Albert Einstein: Theoretical Physicist*. New York: Macmillan, 1993.

French, Anthony, and P. J. Kennedy, eds. *Niels Bohr: A Centenary Volume*. Cambridge: Harvard University Press, 1985.

Friedman, Alan J. *Einstein as Myth and Muse*. Cambridge: Cambridge University Press, 1985.

Giroud, Françoise. *Marie Curie, a Life*. New York: Holmes and Meier, 1986.

Gleick, James. *Genius: The Life and Science of Richard Feynman*. New York: Pantheon Books, 1992.

Goertzel, Ted George. *Linus Pauling: A Life in Science and Politics*. New York: Basic Books, 1995.

Goldsmith, Maurice. *Einstein, the First Hundred Years*. New York: Pergamon Press, 1980.

———. *Frederic Joliot-Curie: A Biography*. London: Lawrence and Wishart, 1976.

Goodchild, Peter. *J. Robert Oppenheimer: Shatterer of Worlds*. Boston: Houghton Mifflin, 1980.

Grady, Sean M. *Marie Curie*. San Diego, CA: Lucent Books, 1992.

Greenaway, Frank. *John Dalton and the Atom*. Ithaca, NY: Cornell University Press, 1966.

Hager, Thomas. *Force of Nature: The Life of Linus Pauling*. New York: Simon and Schuster, 1995.

Harrod, Roy. *The Prof: A Personal Memoir of Lord Cherwell*. London: Macmillan, 1959.

Hartcup, Guy, and T. E. Allibone. *Cockcroft and the Atom*. Bristol, UK: Adam Hilger, 1984.

Hershberg, James. *James B. Conant: Harvard to Hiroshima and the Making of the Nuclear Age*. New York: Knopf, 1993.

Kargon, Robert H. *The Rise of Robert Millikan: Portrait of a Life in American Science.* Ithaca, NY: Cornell University Press, 1982.

Kauffman, George B., ed. *Frederick Soddy (1877–1956): Early Pioneer in Radiochemistry.* Dordrecht, Holland: D. Reidel, 1986.

Knight, Amy. *Beria: Stalin's First Lieutenant.* Princeton, NJ: Princeton University Press, 1993.

Kragh, Helge. *Dirac: A Scientific Biography.* Cambridge: Cambridge University Press, 1990.

Kunetka, James W. *Oppenheimer: The Years of Risk.* Englewood Cliffs, NJ: Prentice-Hall, 1982.

Lanouette, William. *Genius in the Shadows: A Biography of Leo Szilard: The Man Behind the Bomb.* New York: Scribner's Sons, 1992.

Latil, Pierre de. *Enrico Fermi: The Man and His Theories.* New York: P. S. Eriksson, 1964.

Laurikainen, Kalervo Vihtori. *Beyond the Atom: The Philosophical Thought of Wolfgang Pauli.* Berlin: Springer-Verlag, 1988.

Lawrence, William. *The General and the Bomb: A Biography of General Leslie R. Groves, Director of the Manhattan Project.* New York: Dodd, Mead and Company, 1988.

Livanova, Anna. *Landau: A Great Physicist and Teacher.* London: Pergamon Press, 1980.

Massey, Harrie Stewart Wilson. *The New Age in Physics.* 2nd ed. New York: Basic Books, 1966.

Mehra, Jagdish. *Erwin Schrodinger and the Rise of Wave Mechanics.* New York: Springer-Verlag, 1987.

Mendelssohn, Kurt. *The World of Walther Nernst: The Rise and Fall of German Science, 1864–1941.* Pittsburgh: University of Pittsburgh Press, 1973.

Moore, Ruth E. *Niels Bohr: The Man, His Science and the World They Changed.* New York: Knopf, 1966.

Moore, Walter John. *A Life of Erwin Schrödinger.* Cambridge: Cambridge University Press, 1994.

Moss, Norman. *Klaus Fuchs: The Man Who Stole the Atom Bomb.* New York: St. Martin's Press, 1987.

Neuse, Steven M. *David E. Lilienthal: The Journey of an American Liberal.* Knoxville: University of Tennessee Press, 1996.

Nitske, Robert W. *The Life of Wilhelm Conrad Roentgen: Discoverer of the X-ray.* Tucson: University of Arizona Press, 1971.

O'Shea, William R., ed. *Otto Hahn and the Rise of Nuclear Physics.* Dordrecht, Holland: D. Reidel, 1983.

Pais, Abraham. *Niels Bohr's Times: In Physics, Philosophy, and Polity.* Oxford: Oxford University Press, 1991.

———. *Subtle Is the Lord: The Science and Life of Albert Einstein.* Oxford: Oxford University Press, 1982.

Pauli, Wolfgang. *Niels Bohr and the Development of Physics: Essays Dedicated to Niels Bohr on the Occasion of His Seventieth Birthday.* New York: McGraw-Hill, 1955.

Pfau, Richard. *No Sacrifice Too Great: The Life of Lewis L. Strauss.* Charlottesville: University Press of Virginia, 1984.

Pflaum, Rosalynd. *Grand Obsession: Madame Curie and Her World.* New York: Doubleday, 1989.

Pyenson, Lewis. *The Young Einstein: The Advent of Relativity.* Bristol, UK: Adam Hilger, 1985.

Quinn, Susan. *Marie Curie: A Life.* New York: Simon and Schuster, 1995.

Reid, Robert William. *Marie Curie.* New York: Saturday Review Press, 1974.

Reminiscences about a Great Physicist: Paul Adrien Maurice. Cambridge: Cambridge University Press, 1987.

Rigden, John S. *Rabi: Scientist and Citizen.* New York: Basic Books, 1987.

Robert, J. S. *The Life of J. J. Thomson.* Cambridge: Cambridge University Press, 1943.

Rowland, John. *Ernest Rutherford, Atom Pioneer.* New York: Philosophical Library, 1957.

Rozental, Stefan, ed. *Niels Bohr: His Life and Work as Seen by His Friends and Colleagues.* Amsterdam, Holland: North-Holland, 1967.

Rumel, Jack. *Robert Oppenheimer: Dark Prince.* New York: Facts on File, 1992.

Sakharov, Andrei, Sidney D. Drell, and Sergei P. Kaptiza. *Sakharov Remembered: A Tribute by Friends and Colleagues.* New York: American Institute of Physics, 1991.

Segre, Emilio. *Enrico Fermi: Physicist.* Chicago: University of Chicago Press, 1970.

Serafini, Anthony. *Linus Pauling: A Man and His Science.* New York: Paragon House, 1989.

Shea, R., ed. *Otto Hahn and the Rise of Nuclear Physics.* Dordrecht, Holland: D. Reidel, 1983.

Sime, Ruth Lewin. *Lise Meitner: A Life in Physics.* Berkeley: University of California Press, 1996.

Veglahn, Nancy. *Mysterious Rays: Marie Curie's World.* New York: Coward, McCann and Geoghegn, 1977.

White, Michael. *Einstein: A Life in Science.* New York: Dutton, 1994.

Whitman, Willson. *David Lilienthal: Public Servant in a Power Age.* New York: Holt, 1948.

Williams, Robert Chadwell. *Klaus Fuchs, Atom Spy.* Cambridge: Harvard University Press, 1987.

Wilson, David. *Rutherford, Simple Genius.* Cambridge, MA: MIT Press, 1983.

DICTIONARY

Bergaust, Erik, ed. *The Illustrated Nuclear Encyclopedia.* New York: Putnam's Sons, 1971.

DOCUMENTS

Cantelon, Philip L., Richard G. Hewlett, and Robert C. William, eds. *The American Atom: A Documentary History of Nuclear Policies from the Discovery of Fission to the Present.* 2nd ed. Philadelphia: University of Pennsylvania Press, 1991.

Stoff, Michael B., Jonathan F. Fanton, and R. Hal Williams, eds. *The Manhattan*

Project: A Documentary Introduction to the Atomic Age. New York: McGraw-Hill, 1991.

Truman, Harry S. *Harry S. Truman and the Bomb: A Documentary History.* Worland, WY: High Plains Publishing, 1996.

ELECTRONIC RESOURCES

Atomic Physics Links. http://www.phy.llnl.gov/N_Div/atomic.html
General Atomic Fusion Group Educational Home Page. http.//deno-www.gat.com/
Nuklinks. http://www.engin.umich.edu
Plasma Gate. http://plasma-gate.weizmann.ac.il/API.html
Worldatom: International Atomic Energy Agency. http://www.iaea.or.at

GLOSSARY

Alter, Harry, and Subcommittee on Nuclear Terminology and Units. *Glossary of Terms in Nuclear Science and Technology.* La Grange, IL: American Nuclear Society, 1986.

MEMOIRS

Alvarez, Luis W. *Alvarez: Adventures of a Physicist.* New York: Basic Books, 1987.
Born, Max. *My Life: Recollections of a Nobel Laureate.* New York: Scribner, 1978.
Budker, Gersh Itskovich. *G. I. Budker: Reflections and Remembrances.* Woodbury, NY: AIP Press, 1994.
Compton, Arthur Holly. *Atomic Quest: A Personal Narrative.* New York: Oxford University Press, 1956.
Curie, Eve. *Madame Curie.* Garden City, NY: Doubleday, Doran, 1938.
———. *Pierre Curie.* New York: Dover Publications, 1963.
Ermenc, Joseph J., ed. *Atomic Bomb Scientists: Memoirs, 1939–1945: Interviews with Werner Karl Heisenberg, Paul Harteck, Lew Kowarski, Leslie R. Groves, Aristid von Grosse, C. E. Larson.* Westport, CT: Meckler, 1989.
Fermi, Laura. *Atoms in the Family: My Life with Enrico Fermi.* Chicago: University of Chicago Press, 1954.
Feynman, Richard P. *Surely You're Joking, Mr. Feynman! Adventures of a Curious Character.* New York: Norton, 1984.
Frisch, Otto Robert. *What Little I Remember.* Cambridge: Cambridge University Press, 1979.
Gamow, George. *Biography of Physics.* New York: Harper, 1961.
———. *My World Line: An Informal Autobiography.* New York: Viking Press, 1970.
Goldschmidt, Bertrand. *Atom Rivals.* New Brunswick, NJ: Rutgers University Press, 1990.
Goudsmit, Samuel A. *Alsos.* New York: Henry Schuman, 1947.
Groves, Leslie R. *Now It Can Be Told: The Story of the Manhattan Project.* New York: Harper, 1962.
Heisenberg, Elisabeth. *Inner Exile: Recollections of a Life with Werner Heisenberg.* Boston: Birkhauser, 1984.

Johnston, Marjorie, ed. *The Cosmos of Arthur Holly Compton*. New York: Knopf, 1967.

Kapitsa, Pyotr L. *Peter Kapitza on Life and Science*. New York: Macmillan, 1968.

Lapp, Ralph E. *Atoms and People*. New York: Harper, 1956.

Lilienthal, David Eli. *Change, Hope and the Bomb*. Princeton, NJ: Princeton University Press, 1963.

Medvedev, Grigori. *No Breathing Room: The Aftermath of Chernobyl*. New York: Basic Books, 1993.

Millikan, Robert A. *The Autobiography of Robert A. Millikan*. New York: Prentice-Hall, 1950.

Nichols, Kenneth D. *The Road to Trinity*. New York: Morrow, 1987.

Oliphant, Mark. *Rutherford: Recollections of the Cambridge Days*. Amsterdam, Holland: Elsevier, 1972.

Peierls, Rudolf Ernst. *Bird of Passage: Recollections of a Physicist*. Princeton, NJ: Princeton University Press, 1985.

Riehl, Nikolaus, and Frederick Seitz. *Stalin's Captive: Nikolaus Riehl and the Soviet Race for the Bomb*. Washington, DC: American Chemical Society, 1996.

Roche, John, ed. *Physicists Look Back: Studies in the History of Physics*. Bristol, UK: Hilger, 1990.

Seaborg, Glenn Theodore. *The Plutonium Story: The Journals of Professor Glenn T. Seaborg, 1939–1946*. Columbus, OH: Battelle Press, 1994.

Simpson, John W. *Nuclear Power from Underseas to Outer Space*. La Grange Park, IL: American Nuclear Society, 1995.

Speer, Albert. *Inside the Third Reich: Memoirs*. New York: Macmillan, 1970.

Thomson, J. J. *Recollections and Reflections*. New York: Macmillan, 1937.

Trower, Peter W., ed. *Discovering Alvarez: Selected Works of Luis W. Alvarez*. Chicago: University of Chicago Press, 1988.

Weinberg, Alvin Martin. *The First Nuclear Era: The Life and Times of a Technological Fixer*. New York: AIP Press, 1994.

Wheeler, John Archibald. *At Home in the Universe*. New York: American Institute of Physics, 1994.

MONOGRAPHS

Agassi, Joseph. *Radiation Theory and the Quantum Revolution*. Basel, Switzerland: Birkhauser Verlag, 1993.

Albrecht, Ulrich. *The Soviet Armaments Industry*. Chur, Switzerland: Harwood Academic Publishers, 1993.

Alexander, Ronni. *Putting the Earth First: Alternatives to Nuclear Security in Pacific Island States*. Honolulu: Matsunaga Institute for Peace, University of Hawaii, 1994.

Allardice, Corbin. *Atomic Power: An Appraisal Including Atomic Energy in Economic Development*. New York: Pergamon Press, 1957.

Allardice, Corbin, and Edward R. Trapnell. *The Atomic Energy Commission*. New York: Praeger, 1974.

Allen, Thomas B. *Code-name Downfall: The Secret Plan to Invade Japan and Why Truman Dropped the Bomb*. New York: Simon and Schuster, 1995.

Alperovitz, Gar. *The Decision to Use the Atomic Bomb and the Architecture of an American Myth.* New York: Knopf, 1995.

Angelo, Joseph A., and David Buden. *Space Nuclear Power.* Malabar, FL: Orbit Book Company, 1985.

The Antarctic Treaty System. Canberra: Australian Government Publishing Services, 1983.

Arnett, Eric, ed. *Nuclear Weapons after the Comprehensive Test Ban: Implications for Modernization and Proliferation.* Oxford: Oxford University Press, 1996.

Arnikar, Hari. *Isotopes in the Atomic Age.* New York: Wiley, 1989.

Atomic Energy of Canada Limited. *Canada Enters the Nuclear Age: A Technical History of Atomic Energy of Canada Limited.* Montreal, Canada: McGill-Queen's University Press, 1997.

Aubrey, Crispin. *Melt Down: The Collapse of the Nuclear Dream.* London: Collins and Brown, 1991.

Aviv, Dan, and Yossi Melman. *Every Spy a Prince: The Complete History of Israel's Intelligence Community.* Boston: Houghton Mifflin, 1990.

Babin, Ronald. *The Nuclear Power Game.* Montreal, Canada: Black Rose Books, 1985.

Badash, Lawrence. *Kapitza, Rutherford, and the Kremlin.* New Haven, CT: Yale University Press, 1985.

Bailey, C. C. *The Aftermath of Chernobyl: History's Worst Nuclear Reactor Accident.* Dubuque, IA: Kendall-Hunt, 1989.

Ball, Howard. *Cancer Factories: America's Tragic Quest for Uranium Self-Sufficiency.* Westport, CT: Greenwood Press, 1993.

———. *Justice Downwind: America's Atomic Testing Program in the 1950s.* New York: Oxford University Press, 1986.

Ball, S. J. *The Cold War: An International History, 1947–1991.* London: Arnold, 1998.

Balogh, Brian. *Chain Reaction: Expert Debate and Public Participation in American Commercial Nuclear Power, 1945–1975.* Cambridge: Cambridge University Press, 1991.

Barnaby, Frank. *How Nuclear Weapons Spread: Nuclear-Weapon Proliferation in the 1990s.* London: Routledge, 1993.

Beck, Peter. *The International Politics of Antarctica.* New York: St. Martin's Press, 1986.

Bedford, Henry. *Seabrook Station: Citizen Politics and Nuclear Power.* Amherst: University of Massachusetts Press, 1990.

Bennett, Paul R. *Russian Negotiating Strategy: Analytic Case Studies from SALT to START.* Commack, NY: Nova Science Publishers, 1997.

Bernstein, Jeremy. *Hitler's Uranium Club: The Secret Recordings at Farm Hall.* Woodbury, NY: American Institute of Physics, 1996.

Beyer, Don E. *The Manhattan Project: America Makes the First Atomic Bomb.* New York: F. Watts, 1991.

Beyerchen, Alan D. *Scientists Under Hitler: Politics and the Physics Community in the Third Reich.* New Haven, CT: Yale University Press, 1977.

Bhatia, Shyam. *India's Nuclear Bomb.* Ghaziabad, India: Vikas House, 1979.

Bickel, Lennard. *The Deadly Element: The Story of Uranium.* New York: Stein and Day, 1979.

Birks, J. B. *Rutherford at Manchester*. New York: W. A. Benjamin, 1962.

Blackaby, Frank, Jozef Goldblat, and Sverre Lodgaard. *No-First-Use*. London: Taylor and Francis, 1984.

Blakeway, Denys, and Sue Lloyd-Roberts. *Fields of Thunder: Testing Britain's Bomb*. London: George Allen and Unwin, 1985.

Bodansky, David. *Nuclear Energy: Principles, Practices, and Prospects*. Woodbury, NY: American Institute of Physics, 1996.

Bothwell, Robert. *Nucleus: The History of Atomic Energy of Canada Limited*. Toronto, Canada: University of Toronto Press, 1988.

Botti, Timothy J. *Ace in the Hole: Why the United States Did Not Use Nuclear Weapons in the Cold War, 1945 to 1965*. Westport, CT: Greenwood Press, 1996.

Brookes, Leonard G., and Homa Motamen-Scobie, eds. *The Economics of Nuclear Energy*. London: Chapman and Hall, 1984.

Brooks, Geoffrey. *Hitler's Nuclear Weapons: The Development and Attempted Deployment of Radiological Armaments by Nazi Germany*. London: Leo Cooper, 1992.

Brown, Laurie M. *Twentieth Century Physics*. 3 vols. Philadelphia, PA: Institute of Physics, 1995.

Brown, Laurie M., and Lillian Hoddeson, eds. *Birth of Particle Physics*. Cambridge: Cambridge University Press, 1983.

Browne, Corinne, and Robert Munroe. *Time Bomb: Understanding the Threat of Nuclear Power*. New York: William Morrow, 1981.

Bupp, Irvin C., and Jean-Claude Derian. *Light Water: How the Nuclear Dream Dissolved*. New York: Basic Books, 1978.

Burke, Colin B. *Information and Secrecy: Vannevar Bush, Ultra and the Other Memex*. Metuchen, NJ: Scarecrow Press, 1994.

Caldwell, Dan. *The Dynamics of Domestic Politics and Arms Control: The SALT II Treaty Ratification Debate*. Columbia: University of South Carolina Press, 1991.

Carlisle, Rodney P. *Supplying the Nuclear Arsenal: American Production Reactors, 1942–1992*. Baltimore, MD: Johns Hopkins University Press, 1996.

Cathcart, Brian. *Test of Greatness: Britain's Struggle for the Atom Bomb*. London: Murray, 1994.

Cave Brown, Anthony, and Charles Brown MacDonald, eds. *Secret History of the Atomic Bomb*. New York: Dial Press, 1977.

Center for Strategic and International Studies. *The Nuclear Black Market*. Washington, DC: Author, 1996.

Clark, Ronald William. *The Birth of the Bomb*. New York: Horizon Press, 1961.

———. *The Greatest Power on Earth: The International Race for Nuclear Supremacy*. New York: Harper and Row, 1980.

Close, F. E. *Too Hot to Handle: The Race for Cold Fusion*. Princeton, NJ: Princeton University Press, 1991.

Cochran, Thomas B., Robert S. Norris, and Oleg A. Bukharin. *Making the Russian Bomb: From Stalin to Yeltsin*. Boulder, CO: Westview Press, 1995.

Cohen, Etahn M. *Ideology, Interest Group Formation and the New Left: The Case of the Clamshell Alliance*. New York: Garland Publishing, 1988.

Cohen, S. T. *The Neutron Bomb: Political, Technological and Military Issues*. Cambridge, MA: Institute for Foreign Policy Analysis, 1978.

Committee for the Compilation of Materials on Damage Caused by the Atomic Bombs in Hiroshima and Nagasaki. *Hiroshima and Nagasaki: The Physical, Medical, and Social Effects of the Atomic Bombings*. New York: Basic Books, 1981.

Congressional Quarterly. *The Nuclear Age: Power, Proliferation and the Arms Race*. Washington, DC: Author, 1984.

Cortright, David, and Amitabb Mattoo. *India and the Bomb: Public Opinion and Nuclear Options*. Notre Dame, IN: University of Notre Dame Press, 1996.

Crease, Robert P. *The Second Creation: Makers of the Revolution in Twentieth-Century Physics*. New York: Macmillan, 1986.

Crowther, James Gerald. *The Cavendish Laboratory, 1874–1974*. New York: Science History Publications, 1974.

Curtis, Charles P. *The Oppenheimer Case: The Trial of a Security System*. New York: Simon and Schuster, 1955.

Davenport, Elaine, Paul Eddy, and Peter Gillman. *The Plumbat Affair*. Philadelphia, PA: J. B. Lippincott, 1978.

Davis, Nuel Pharr. *Lawrence and Oppenheimer*. New York: Simon and Schuster, 1968.

Dawson, Frank G. *Nuclear Power: Development and Management of a Technology*. Seattle: University of Washington Press, 1976.

Deitch, Kenneth M., ed. *The Manhattan Project: A Secret Wartime Mission*. Lowell, MA: Discovery Enterprises, 1995.

Dibblin, Jane. *Day of Two Suns: U.S. Nuclear Testing and the Pacific Islanders*. London: Virago Press, 1988.

Dowling, John, and Evans M. Harrell, eds. *Civil Defense: A Choice of Disasters*. New York: American Institute of Physics, 1987.

Drell, Sidney D. *In the Shadow of the Bomb: Physics and Arms Control*. New York: American Institute of Physics, 1993.

Duffy, Gloria. *Soviet Nuclear Energy: Domestic and International Policies*. Santa Monica, CA: Rand, 1979.

Durch, William J. *The Future of the ABM Treaty*. London: International Institute for Strategic Studies, 1987.

Ebel, Robert E. *Chernobyl and Its Aftermath: A Chronology of Events*. Washington, DC: Center for Strategic and International Studies, 1994.

Egen, Greg. *The Origins of the United States' Non-Proliferation Policy: A Study Project*. Washington, DC: Atomic Industrial Forum, 1978.

Eggleston, Wilfrid. *Canada's Nuclear Story*. Toronto, Canada: Clarke, Irwin, 1965.

Elliot, David. *The Politics of Nuclear Power*. London: Pluto Press, 1978.

Endicott, John E. *Japan's Nuclear Option: Political, Technical, and Strategic Factors*. New York: Praeger, 1975.

Engdahl, Sylvia Louise. *The Subnuclear Zoo: New Discoveries in High Energy Physics*. New York: Atheneum, 1977.

Eubank, Keith. *The Bomb*. Malabar, FL: Krieger, 1991.

Fermi, Rachel. *Picturing the Bomb: Photographs from the Secret World of the Manhattan Project*. New York: Abrams, 1995.

Ferrara, Grace M. ed. *Atomic Energy and the Safety Controversy*. New York: Facts on File, 1978.

Fierz, Markus. *Theoretical Physics in the Twentieth Century: A Memorial Volume to Wolfgang Pauli.* New York: Interscience Publishers, 1960.

Finch, Ron. *Exporting Danger: A History of the Canadian Nuclear Energy Export Programme.* Montreal, Canada: Black Rose Books, 1986.

Fischer, David. *History of the International Atomic Energy Agency: The First Forty Years.* Vienna, Austria: International Atomic Energy Agency, 1997.

Flam, Helena, ed. *States and Anti-Nuclear Movements.* Edinburgh, Scotland: Edinburgh University Press, 1994.

Ford, Daniel F. *The Cult of the Atom: The Secret Papers of the Atomic Energy Commission.* New York: Simon and Schuster, 1982.

————. *Meltdown.* New York: Simon and Schuster, 1986.

————. *Three Mile Island: Thirty Minutes to Meltdown.* New York: Viking Press, 1982.

Fowler, T. Kenneth. *The Fusion Quest.* Baltimore, MD: Johns Hopkins University Press, 1997.

Fradkin, Philip L. *Fallout: An American Nuclear Tragedy.* Tucson: University of Arizona Press, 1989.

Freeman, Leslie J. *Nuclear Witnesses: Insiders Speak Out.* New York: Norton, 1981.

Fritzsch, Harald. *Quarks: The Stuff of Matter.* New York: Basic Books, 1983.

Fuller, John Grant. *We Almost Lost Detroit.* New York: Reader's Digest Press, 1975.

Furman, Necah Stewart. *Sandia National Laboratories: The Postwar Decade.* Albuquerque: University of New Mexico Press, 1990.

Gallagher, Thomas. *Assault in Norway: The True Story of the Telemark Raid.* London: Purnell Books Services, 1975.

Gantz, Kenneth. *Nuclear Flight: The United States Air Force Programs for Atomic Jets, Missiles, and Rockets.* New York: Duell, Sloan, and Pearce, 1960.

Garber, Marjorie B., and Rebecca L. Walkowitz, eds. *Secret Agents: The Rosenberg Case, McCarthyism and Fifties America.* New York: Routledge, 1995.

George, James L. *The New Nuclear Rules: Strategy and Arms Control after JNF and START.* New York: St. Martin's Press, 1990.

Gibson, James N. *Nuclear Weapons of the United States: An Illustrated History.* Atglen, PA: Schiffer Publishing, 1996.

Gofman, John W., and Arthur R. Tamplin. *Poisoned Power: The Case against Nuclear Power Plants before and after Three Mile Island.* Emmaus, PA: Rodale Press, 1979.

Goldschmidt, Bertrand. *The Atomic Adventure, Its Political and Technical Aspects.* Oxford; Pergamon Press, 1964.

————. *The Atomic Complex: A Worldwide Political History of Nuclear Energy.* La Grange Park, IL: American Nuclear Society, 1982.

Goldsmith, Maurice. *Europe's Giant Accelerator: The Story of the CERN 400 GeV Proton Synchrotron.* London: Taylor and Francis, 1977.

Gowing, Margaret. *The Atom Bomb.* London: Butterworths, 1979.

————. *Independence and Deterrence: Britain and Atomic Energy, 1945–1952.* 2 vols. New York: St. Martin's Press, 1974.

Gray, Earle. *The Great Uranium Cartel.* Toronto, Canada: McClelland and Stewart, 1982.

Gray, Mike. *The Warning: Accident at Three Mile Island.* New York: Norton, 1982.

Green, Harold P., and Alan Rosenthal. *The Joint Committee on Atomic Energy: A*

Study in Fusion of Governmental Power. Washington, DC: George Washington University, 1961.

Greene, Owen, Ian Percival, and Irene Ridge. *Nuclear Winter: The Evidence and the Risks*. Cambridge: Polity Press, 1985.

Greenhalgh, Geoffrey. *The Future of Nuclear Power*. London: Graham and Trotman, 1988.

Groueff, Stephane. *Manhattan Project: The Untold Story of the Making of the Atomic Bomb*. London: Collins, 1967.

Hacker, Barton C. *Elements of Controversy: The Atomic Energy Commission and Radiation Safety in Nuclear Weapons Testing, 1947–1974*. Berkeley: University of California Press, 1994.

Hart, Ivor Blashka. *The Great Physicists*. Freeport, NY: Books for Libraries Press, 1970.

Hawkins, David, Edith C. Truslow, and Ralph Carlisle Smith, eds. *Project Y, the Los Alamos Story*. Los Angeles: Tomash Publishers, 1983.

Haynes, Viktor, and Marko Bohcun. *The Chernobyl Disaster*. London: Hogarth Press, 1988.

Hendry, John. *The Creation of Quantum Mechanics and the Bohr-Pauli Dialogue*. Dordrecht, Holland: D. Reidel, 1984.

———, ed. *Cambridge Physics in the Thirties*. Bristol, UK: Adam Hilger, 1984.

Hertsgaard, Mark. *Nuclear Inc.: The Men and Money behind Nuclear Energy*. New York: Pantheon Books, 1983.

Hevly, Bruce, and John M. Findlay, eds. *The Atomic West*. Seattle, WA: Center for the Study of the Pacific Northwest, 1998.

Hewlett, Richard G., and Francis Duncan. *Nuclear Navy, 1946–1962*. Chicago: University of Chicago Press, 1974.

Hinnawi, Essam E., ed. *Nuclear Energy and the Environment*. Oxford: Pergamon Press, 1980.

A History of the Cavendish Laboratory, 1871–1910. London: Longmans, Green, and Co., 1910.

Hoddeson, Lillian, Paul W. Henriksen, Roger A. Meade, and Catherine Westfall. *Critical Assembly: A Technical History of Los Alamos during the Oppenheimer Years, 1943–1945*. Cambridge: Cambridge University Press, 1993.

Hogan, J. Michael. *The Nuclear Freeze Campaign: Rhetoric and Foreign Policy in the Telepolitical Age*. East Lansing: Michigan State University Press, 1994.

Holl, Jack M. *Argonne National Laboratory, 1946–96*. Urbana: University of Illinois Press, 1997.

Holloway, David. *Stalin and the Bomb: The Soviet Union and Atomic Energy, 1939–1956*. New Haven, CT: Yale University Press, 1994.

Holloway, Rachel L. *In the Matter of J. Robert Oppenheimer: Politics, Rhetoric, and Self-Defense*. Westport, CT: Praeger, 1993.

Houts, Peter S. *The Three Mile Island Crisis: Psychological, Social, and Economic Impacts on the Surrounding Population*. University Park: Pennsylvania State University Press, 1988.

Huchthausen, Peter, Igor Kurdin, and R. Alan White. *Hostile Waters*. New York: St. Martin's Press, 1997.

Huizenga, John R. *Cold Fusion: The Scientific Fiasco of the Century*. Rochester, NY: University of Rochester Press, 1992.

Hund, Friedrich. *The History of Quantum Theory.* New York: Barnes & Noble, 1974.

Hyde, H. Montgomery. *The Atom Bomb Spies.* New York: Atheneum, 1980.

Hyland, William. *The Cold War: Fifty Years of Conflict.* New York: Times Books, 1991.

International Atomic Energy Agency: Personal Reflections. Vienna, Austria: International Atomic Energy Agency, 1997.

International Conference on Nuclear Decommissioning. *Nuclear Decommissioning: The Strategic, Practical, and Environmental Considerations.* London: Mechanical Engineering Publications Limited, 1995.

International Physicians for the Prevention of Nuclear War. *Radioactive Heaven and Earth: The Health and Environmental Effects of Nuclear Weapons Testing in, on, and above the Earth.* New York: Apex Press, 1991.

Irving, David. *The German Atomic Bomb: The History of Nuclear Research in Nazi Germany.* New York: Simon and Schuster, 1967.

————. *The Virus House.* London: William Kimber, 1967.

Jammer, Max. *The Conceptual Development of Quantum Mechanics.* New York: McGraw-Hill, 1966.

Jasper, James M. *Nuclear Politics: Energy and the State in the United States, Sweden, and France.* Princeton, NJ: Princeton University Press, 1990.

Johnson, Leland, and Daniel Schaffer. *Oak Ridge National Laboratory: The First Fifty Years.* Knoxville: University of Tennessee Press, 1994.

Jones, Vincent C. *Manhattan, the Army and the Atomic Bomb.* Washington, DC: Center of Military History, U.S. Army, 1985.

Joppke, Christian. *Mobilizing Against Nuclear Energy: A Comparison of Germany and the United States.* Berkeley: University of California Press, 1993.

Josephson, Paul R. *Physics and Politics in Revolutionary Russia.* Berkeley: University of California Press, 1991.

Jungk, Robert. *Brighter Than a Thousand Suns: A Personal History of the Atomic Scientists.* New York: Harcourt Brace, 1958.

————. *The New Tyranny: How Nuclear Power Enslaves Us.* New York: F. Jordan Books, 1979.

Kapur, Ashok. *Pakistan's Nuclear Development.* London: Croom Helm, 1987.

Katz, Milton S. *Ban the Bomb: A History of SANE, the Committee for a Sane Nuclear Policy, 1957–1985.* Westport, CT: Greenwood Press, 1986.

Keisling, Bill. *Three Mile Island: Turning Point.* Seattle, WA: Veritas Books, 1980.

Keller, Alex. *The Infancy of Atomic Physics: Hercules in His Cradle.* Oxford: Clarendon Press, 1983.

Kevles, Daniel J. *The Physicists: The History of a Scientific Community in Modern America.* Cambridge, MA: Harvard University Press, 1987.

King, Michael. *Death of the Rainbow Warrior.* Auckland, New Zealand: Penguin Books, 1986.

Kohn, Howard. *Who Killed Karen Silkwood?* New York: Summit Books, 1981.

Kokoski, Richard. *Technology and the Proliferation of Nuclear Weapons.* Oxford: Oxford University Press, 1995.

Kramish, Arnold. *Atomic Energy in the Soviet Union.* Stanford, CA: Stanford University Press, 1959.

Kuhn, Thomas S. *Black-Body Theory and the Quantum Discontinuity, 1894–1912.* Oxford; Clarendon Press, 1978.

Kunetka, James W. *City of Fire: Los Alamos and the Birth of the Atomic Age, 1943–1945.* Englewood Cliffs, NJ: Prentice-Hall, 1978.

Kurzman, Dan. *Blood and Water: Sabotaging Hitler's Bomb.* New York: Henry Holt, 1997.

Lamont, Lansing. *Day of Trinity.* New York: Atheneum, 1965.

Laurence, William Leonard. *Men and Atoms: The Discovery, the Uses, and the Future of Atomic Energy.* New York: Simon and Schuster, 1959.

Leeuwen, Marianne van, ed. *The Future of the International Nuclear Non-Proliferation Regime.* Dordrecht, Holland: Martinus Nijhoff, 1995.

Leppzer, Robert, ed. *Voices from Three Mile Island: The People Speak Out.* Trumansberg, NY: Crossing Press, 1980.

Lewis, Richard S. *The Nuclear-Power Rebellion: Citizens vs. the Atomic Industrial Establishment.* New York: Viking Press, 1972.

Loeb, Paul. *Nuclear Culture: Living and Working in the World's Largest Atomic Complex.* New York: Coward, McCann and Geoghegan, 1982.

MacPherson, Malcolm C. *Time Bomb: Fermi, Heisenberg, and the Race for the Atomic Bomb.* New York: E. P. Dutton, 1986.

Major, John. *The Oppenheimer Hearing.* New York: Stein and Day, 1971.

Marples, David R. *Chernobyl and Nuclear Power in the USSR.* New York: St. Martin's Press, 1986.

Marples, David R., and Marilyn J. Young. *Nuclear Energy and Security in the Former Soviet Union.* Boulder, CO: Westview Press, 1997.

Martin, Daniel. *Three Mile Island: Prologue or Epilogue?* Cambridge, MA: Ballinger, 1980.

Mattausch, John. *A Commitment to Campaign: A Sociological Study of CND.* Manchester, UK: Manchester University Press, 1989.

May, John. *The Greenpeace Book of the Nuclear Age: The Hidden History, the Human Cost.* New York: Pantheon Books, 1989.

McCusker, Brian. *The Quest for Quarks.* Cambridge: Cambridge University Press, 1983.

McKay, Alwyn. *The Making of the Atomic Age.* Oxford: Oxford University Press, 1984.

Medvedev, Grigori. *The Truth about Chernobyl.* New York: Basic Books, 1991.

Medvedev, Zhores. *The Legacy of Chernobyl.* New York: Norton, 1990.

Megaw, W. J. *How Safe?: Three Mile Island, Chernobyl and Beyond.* Toronto, Ontario: Stoddart, 1987.

Mehra, Jadish. *The Solvay Conferences on Physics: Aspects of the Development of Physics since 1911.* Dordrecht, Holland: D. Reidel, 1975.

Meyer, David S. *A Winter of Discontent: The Nuclear Freeze and American Politics.* New York: Praeger, 1990.

Mihalka, Michael. *International Arrangements for Uranium Enrichment.* Santa Monica, CA: Rand, 1979.

Miller, Robert L. *Under the Cloud: The Decades of Nuclear Testing.* New York: Free Press, 1986.

Mladenovic, Milorad. *History of Early Nuclear Physics (1896–1931).* River Edge, NJ: World Scientific, 1992.

Mojtabai, A. G. *Blessed Assurance: At Home with the Bomb in Amarillo, Texas.* Syracuse, NY: Syracuse University Press, 1997.

Morgan, Michael, and Susan Leggett, eds. *Mainstream(s) and Margins: Cultural Politics in the 90s.* Westport, CT: Greewood Press, 1996.

Moss, Norman. *Men Who Play God: The Story of the H-Bomb and How the World Came to Live with It.* New York: Harper and Row, 1968.

———. *The Politics of Uranium.* London: Deutsch, 1981.

Nardo, Don. *Chernobyl.* San Diego, CA: Lucent Books, 1990.

Nizer, Louis. *The Implosion Conspiracy.* Garden City, NY: Doubleday, 1973.

Nordhaus, William D. *The Swedish Nuclear Dilemma: Energy and the Environment.* Washington, DC: Resources for the Future, 1997.

Nuclear Power Reactors in the World. Vienna, Austria: International Atomic Energy Agency, 1997.

O'Neill, Dan. *The Firecracker Boys.* New York: St. Martin's, 1994.

Orlans, Harold. *Contracting for Atoms: A Study of Public Policy Issues Posed by the Atomic Energy Commission's Contracting for Research, Development, and Managerial Services.* Washington, DC: Brookings Institution, 1967.

Owen, Anthony David. *The Economics of Uranium.* New York: Praeger, 1985.

Pacific War Research Society. *The Day Man Lost: Hiroshima, 6 August 1945.* Tokyo: Kodansha International, 1972.

Payne, Keith B., and Colin S. Gray, eds. *The Nuclear Freeze Controversy.* Lanham, MD: University Press of America, 1984.

Pendergrass, Connie Baack. *Public Power, Politics, and Technology in the Eisenhower and Kennedy Years: The Hanford Dual-Purpose Reactor Controversy, 1956–1962.* New York: Arno Press, 1975.

Pickering, Andrew. *Constructing Quarks: A Sociological History of Particle Physics.* Edinburgh, Scotland: Edinburgh University Press, 1984.

Pilat, Joseph F., Robert E. Pendley, and Charles K. Ebinger, eds. *Atoms for Peace: An Analysis after 30 Years.* Boulder, CO: Westview Press, 1985.

Pitt, David, and Gordon Thompson, eds. *Nuclear-Free Zones.* London: Croom Helm, 1987.

Pligt, J. van der. *Nuclear Energy and the Public.* Oxford: Blackwell, 1992.

Pochin, Edward. *Nuclear Radiation: Risks and Benefits.* Oxford: Clarendon Press, 1983.

Powers, Thomas. *Heisenberg's War: The Secret History of the German Bomb.* New York: Knopf, 1993.

Price, Jerome. *The Antinuclear Movement.* Boston, MA: Twayne, 1982.

Pringle, Peter, and James Spigelman. *The Nuclear Barons.* London: Michael Joseph, 1981.

Pry, Peter. *Israel's Nuclear Arsenal.* Boulder, CO: Westview Press, 1984.

Purcell, John Francis. *The Best-Kept Secret: The Story of the Atomic Bomb.* New York: Vanguard Press, 1963.

Radosh, Ronald. *The Rosenberg File: A Search for the Truth.* New York: Holt, Rinehart and Winston, 1983.

Raviv, Dan, and Yossi Melman. *Every Spy a Prince: The Complete History of Israel's Intelligence Community.* Boston: Houghton Mifflin, 1990.

Reader, Mark, Ronald A. Hardert, and Gerald L. Moulton, eds. *Atom's Eve: Ending of the Nuclear Age: An Anthology.* New York: McGraw-Hill, 1980.

Rees, Joseph V. *Hostages of Each Other: The Transformation of Nuclear Safety Since Three Mile Island.* Chicago: University of Chicago Press, 1994.

Renneburg, Monika, and Mark Walker, eds. *Science, Technology, and National Socialism*. Cambridge: Cambridge University Press, 1994.

Rhodes, Richard. *Dark Sun: The Making of the Hydrogen Bomb*. New York: Touchstone, 1995.

———. *The Making of the Atomic Bomb*. New York: Simon and Schuster, 1986.

Riordan, Michael. *The Hunting of the Quark: A True Story of Modern Physics*. New York: Simon and Schuster, 1987.

Roberts, Alan. *Hazards of Nuclear Power*. Nottingham, UK: Spokesman Books, 1977.

Rockwell, Theodore. *The Rickover Effect: How One Man Made a Difference*. Annapolis, MD: Naval Institute Press, 1992.

Roff, Sue Rabbit. *Hotspots: The Legacy of Hiroshima and Nagasaki*. London: Cassell, 1996.

Romer, Alfred. *Radiochemistry and the Discovery of Isotopes*. New York: Dover Publications, 1970.

Romero, Philip J. *Nuclear Winter: Implications for U.S. and Soviet Nuclear Strategy*. Santa Monica, CA: Rand, 1984.

Rosenberg, Howard L. *Atomic Soldiers: American Victims of Nuclear Experiments*. Boston, MA: Beacon Press, 1980.

Rosenthal, Debra. *At the Heart of the Bomb: The Dangerous Allure of Weapons Work*. Reading, MA: Addison-Wesley, 1990.

Rothblat, Joseph. *Nuclear Radiation in Warfare*. London: Taylor and Francis, 1981.

———. *Pugwash—The First Ten Years: History of the Conferences of Science and World Affairs*. New York: Humanities Press, 1967.

Rudig, Wolfgang. *Anti-Nuclear Movements: A World Survey of Opposition to Nuclear Energy*. Harlow, UK: Longman Current Affairs, 1990.

Russ, Harlow W. *Project Alberta: The Preparation of Atomic Bombs for Use in World War II*. Los Alamos, NM: Exceptional Books, 1990.

Ryutova-Kemoklidze, Margarita. *The Quantum Generation: Highlights and Tragedies of the Golden Age of Physics*. Berlin: Springer-Verlag, 1995.

Sachs, Mendel. *Einstein Versus Bohr: The Continuing Controversies in Physics*. La Salle, IL: Open Court, 1988.

Saffer, Thomas H., and Orville E. Kelly. *Countdown Zero*. New York: Putnam's Sons, 1982.

Schoenberger, Walter Smith. *Decision of Destiny*. Athens: Ohio University Press, 1969.

Schonland, Basil Ferdinand. *The Atomists (1805–1933)*. Oxford, UK: Clarendon Press, 1968.

Schull, William J. *Effects of Atomic Radiation: A Half-Century of Studies from Hiroshima and Nagasaki*. New York: Wiley-Liss, 1995.

Segre, Emilio. *From X-Rays to Quarks: Modern Physicists and Their Discoveries*. San Francisco: W. H. Freeman, 1980.

Seidel, Robert W. *Los Alamos and the Development of the Atomic Bomb*. Los Alamos, NM: Otowi Crossing Press, 1995.

Serafini, Anthony. *Legends in Their Own Time: A Century of American Physical Scientists*. New York: Plenum Press, 1993.

Shapiro, Fred C. *Radwaste*. New York: Random House, 1981.

Sherwin, Martin J. *A World Destroyed: The Atomic Bomb and the Grand Alliance*. New York: Knopf, 1995.

Shrader-Frechette, K. S. *Burying Uncertainty: Risk and the Case Against Geological Disposal of Nuclear Waste*. Berkeley: University of California Press, 1993.

Shroyer, Jo Ann. *Quarks, Critters, and Chaos: What Science Terms Really Mean*. New York: Prentice Hall, 1993.

Silver, L. Ray. *Fallout from Chernobyl*. Toronto, Ontario: Deneau, 1987.

Simpson, John W. *Nuclear Power from Underseas to Outer Space*. La Grange Park, IL: American Nuclear Society, 1995.

Smyth, Henry DeWolf. *Atomic Energy for Military Purposes: The Official Report on the Development of the Atomic Bomb Under the Auspices of the United States Government 1940–1945*. Princeton, NJ: Princeton University Press, 1945.

Snow, C. P. *The Physicists*. Boston, MA: Little, Brown, 1981.

———. *Science and Government*. Cambridge: Harvard University Press, 1961.

Spector, Leonard S., Mark G. McDonough, and Evan S. Medeiros. *Tracking Nuclear Proliferation: A Guide in Maps and Charts, 1995*. Washington, DC: Carnegie Endowment for International Peace, 1995.

Starr, Philip. *Three Mile Island Sourcebook: Annotations of a Disaster*. New York: Garland, 1983.

Stephens, Mark. *Three Mile Island*. New York: Random House, 1980.

Stern, Philip M. *The Oppenheimer Case: Security on Trial*. New York: Harper and Row, 1969.

Stever, Donald. *Seabrook and the Nuclear Regulatory Commission: The Licensing of a Nuclear Power Plant*. Hanover, NH: University Press of New England, 1980.

Stokley, James. *The New World of the Atom*. New York: Washburn, 1957.

Strickland, Donald A. *Scientists in Politics: The Atomic Scientists Movement, 1945–46*. West Lafayette, IN: Purdue University Studies, 1968.

Strout, Cushing. *Conscience, Science and Security: The Case of Dr. J. Robert Oppenheimer*. Chicago: Rand McNally, 1963.

Stuewer, Roger H. *The Compton Effect: Turning Point in Physics*. New York: Science History Publications, 1975.

———, ed. *Nuclear Physics in Retrospect: Proceedings of a Symposium on the 1930s*. Minneapolis: University of Minnesota Press, 1979.

Stutzle, Walther, Bhupendrea Jasani, and Regina Cowen, eds. *The ABM Treaty: To Defend or Not to Defend*. Oxford: Oxford University Press, 1987.

Sylves, Richard Terry. *The Nuclear Oracles: A Political History of the General Advisory Committee of the Atomic Energy Commission, 1947–1977*. Ames: Iowa State University Press, 1987.

Szasz, Ferenc Morton. *The Day the Sun Rose Twice: The Story of the Trinity Site Nuclear Explosion, July 16, 1945*. Albuquerque: University of New Mexico Press, 1984.

Takaki, Ronald T. *Hiroshima: Why America Dropped the Atomic Bomb*. Boston: Little, Brown, 1995.

Taubes, Gary. *Bad Science: The Short Life and Weird Times of Cold Fusion*. New York: Random House, 1993.

Taylor, June H., and Michael D. Yokell. *Yellowcake: The International Uranium Cartel*. New York: Pergamon Press, 1979.

Taylor, Raymond W., and Samuel W. Taylor. *Uranium Fever, or No Talk Under $1 Million*. New York: Macmillan, 1970.

Taylor, Richard. *Against the Bomb: The British Peace Movement, 1958–1965*. Oxford, UK: Clarendon Press, 1988.

Thakur, Ramesh, ed. *Nuclear Weapons-Free Zones*. New York: St. Martin's Press, 1998.

Thayer, Harry. *Management of the Hanford Engineer Works in World War II: How the Corps, DuPont and the Metallurgical Laboratory Fast Tracked the Original Plutonium Works*. New York: ASCE Press, 1996.

Thomas, Gordon. Enola Gay: *Mission to Hiroshima*. Loughborough, UK: White Owl Press, 1995.

Thomas, Gordon, and Max Morgan-Witts. *Ruin from the Air: The Atomic Mission to Hiroshima*. London: Hamish Hamilton, 1977.

Thomas, S. D. *The Realities of Nuclear Power: International Economic and Regulatory Experience*. Cambridge: Cambridge University Press, 1988.

U.S. Department of the Army. *Trinity Site, 1945–1995*. Washington, DC: Author, 1995.

Van Cleave, William R., and S. T. Cohen. *Tactical Nuclear Weapons: An Examination of the Issues*. New York: Crane, Russak, 1978.

Wainstock, Dennis. *The Decision to Drop the Atomic Bomb*. Westport, CT: Praeger, 1996.

Walker, J. Samuel. *Containing the Atom: Nuclear Regulation in a Changing Environment, 1963–1971*. Berkeley: University of California Press, 1992.

———. *A Short History of Nuclear Regulation, 1946–1990*. Washington, DC: U.S. Nuclear Regulatory Commission, 1993.

Walker, Mark. *Nazi Science: Myth, Truth, and the German Atomic Bomb*. New York: Plenum Press, 1995.

Walker, Martin. *The Cold War: A History*. New York: Holt, 1994.

Walters, Roland W. *South Africa and the Bomb: Responsibility and Deterrence*. Lexington, MA: Lexington Books, 1987.

Wapner, Paul. *Environmental Activism and World Civic Politics*. Albany: State University of New York Press, 1996.

Wasserman, Sherri L. *The Neutron Bomb Controversy: A Study in Alliance Politics*. New York: Praeger, 1983.

Weale, Adrian, ed. *Eye-Witness Hiroshima*. New York: Carroll and Graf, 1995.

Weart, Spencer R. *Nuclear Fear: A History of Images*. Cambridge, MA: Harvard University Press, 1988.

———. *Scientists in Power*. Cambridge, MA: Harvard University Press, 1979.

Weisgall, Jonathan M. *Operation Crossroads: The Atomic Tests at Bikini Atoll*. Annapolis, MD: Naval Institute Press, 1994.

Whitaker, Andrew. *Einstein, Bohr, and the Quantum Dilemma*. Cambridge: Cambridge University Press, 1996.

Williams, Phil, ed. *The Nuclear Debate: Issues and Politics*. London: Royal Institute of International Affairs, 1984.

Wittner, Lawrence S. *The Struggle Against the Bomb: One World or None: A History of the World Nuclear Disarmament Movement Through 1953*. Vol. 1. Stanford, CA: Stanford University Press, 1993.

Wolfson, Richard. *Nuclear Choices: A Citizen's Guide to Nuclear Technology.* Cambridge, MA: MIT Press, 1991.

Wood, Alexander. *The Cavendish Laboratory.* Cambridge: Cambridge University Press, 1946.

Woodbury, David. *Atoms for Peace.* New York: Dodd, Mead, 1995.

Wyden, Peter. *Day One: Before Hiroshima and After.* New York: Simon and Schuster, 1984.

Yaroshinska, Alla. *Chernobyl, the Forbidden Truth.* Lincoln: University of Nebraska Press, 1995.

York, Herbert F. *The Advisors: Oppenheimer, Teller, and the Superbomb.* San Francisco: W. H. Freeman, 1976.

Ziegler, Charles A., and David Jacobson. *Spying without Spies: Origins of America's Secret Nuclear Surveillance System.* Westport, CT: Praeger, 1995.

MOVIE

Nichols, Mike. *Silkwood.* Los Angeles: ABC Motion Pictures, 1984.

REPORTS

International Atomic Energy Agency. *The Radiological Accident at the Irradiation Facility in Nesvizh.* Vienna, Austria: Author, 1996.

———. *The Radiological Accident in Goiania.* Vienna, Austria: Author, 1988.

———. *The Radiological Accident in San Salvador: A Report.* Vienna, Austria: Author, 1990.

———. *The Radiological Accident in Soreq.* Vienna, Austria: Author, 1993.

———. *The Radiological Accident in Tammiku.* Vienna, Austria: Author, 1998.

SOURCEBOOKS

Kruschke, Earl R., and Byron M. Jackson. *Nuclear Energy Policy: A Reference Handbook.* Santa Barbara, CA: ABC-CLIO, 1990.

Schwartz, Richard Alan. *The Cold War Reference Guide: A General History and Annotated Chronology with Selected Biographies.* Jefferson, NC: McFarland, 1997.

Index

Entries in **bold** are main entries.

340th Bomb Squadron, 368
393rd Heavy Bombardment Squadron, 368
509th Composite Group, 121
9812th Special Engineer Detachment (SED), 221; Manhattan Project, 220–222

A-26 (Reconnaissance Aircraft), 8
Aachen Institute of Technology, 108
Abelson, Philip Hauge, 1–2, 230, 267; Life of, 1–2; Naval Research Laboratory, 244; Thermal diffusion process, 226
Academie Française, 104
Acheson, Dean, 212; Acheson Board, 2; Special Subcommittee on the Hydrogen Bomb, 344
Acheson Board, 2, 211
Acheson-Lilienthal Report. *See* Acheson Board; Lilienthal Group
Adamson, K. F., Uranium Committee, 386
Adenauer, Konrad, 149–150
Advanced gas-cooled reactor (AGR), 28
Advisory Committee on Atomic Energy, 2–3, 146
Advisory Committee on Human Radiation Experiments, Uranium Committee, 406
Advisory Committee on Reactor Safeguards (ACRS), 3, Atomic Energy Commission, 30; China Syndrome, 85; Fermi I, 131; Reactor Safeguard Committee, 308
Advisory Committee on Uranium. *See* Uranium Committee
AEG-Telefunken, 147
Aerojet General, 242
Afghanistan War, 319
AFOAT-1 (Air Force Office of Atomic Testing), 217
Africa, Treaty of Pelindaba, 371
African Nuclear Weapons Free Zone Treaty. *See* Treaty of Pelindaba
Agency for the Prohibition of Nuclear Weapons in Latin America and the Caribbean (OPANAL), 373–374
Aiken (South Carolina), 263, 322
Aioi Bridge, 121
Aircraft Nuclear Propulsion, 4
Akers, Wallace Allen, 4–5, 88; Role in British mission to Manhattan Project, 4–5
Akers Research laboratory, 5
Alamo (Nevada), 330

Alamogordo. *See* Trinity

Alamogordo Test Site, 13–14, 40, 247; Trinity, 374

Alaska, Project Chariot, 294

Albuquerque (New Mexico), 319

Aldermaston, 5–6, 21, 62

Aldermaston March, Ban the Bomb Movement, 41; Campaign for Nuclear Disarmament, 70

Aleksandrov, Anatoli P., 201

Alexander, Earl, 18

Algoma Basin (Canada), Uranium Mining, 388

All-Russian Scientific Research Institute of Experimental Physics, 23, 80

Allen v. United States of America, 6; Shot Harry, 331–332

Allied spies against Germany, Paul Rosbaud, 313

Allier, Jacques, 227; Allier Affair, 7

Allier Affair, 7

Allison, Samuel, 130, 229

Almendariz, Pedro, 100

Alpha particles, 169, 189, 302, 316; Contributions of Hans Geiger, 143–144; Contributions of Ernest Rutherford, 317; Work of Patrick Blackett, 49; Work of W. H. Bragg, 57

Alpha radiation, 124, 290, 393

Alpher, Ralph, 45

Alsos, 150

Alsos Mission, 8–9, 150; Farm Hall, 126; Paul Rosbaud, 313

Altshuler, Lev, 23

Alvarez, Luis Walter, 9–10, 148

Amaldi, Enrico, 291

Amarillo (Texas), 278

Ambio, 265

Amboy (California), 286

American Bar Association, 314

American Export Line, 95

American Nuclear Energy Council, 34

American Nuclear Program, 10–12; American Nuclear Testing program, 13–14; Atomic Bomb, 24–25; Atomic Energy Act of 1946, 27; Atomic Energy Act of 1954, 28; Atomic Energy Commission, 29–31;

Bikini Tests, 47–48; Bravo Nuclear Test, 58–59; Buster-Jangle Nuclear Tests, 67; *Calvert Cliffs* decision, 11; Crossroads Nuclear Test, 101–108; Eniwetok Atoll Test Sites, 120–121; Greenhouse Nuclear Tests, 151–152; Hardtack Nuclear Tests, 159–160; History of, 10–12; Hydrogen Bomb, 167; King Nuclear Test, 197; Knapp Report, 198–199; Los Alamos National Laboratory, 220–222; Mike Nuclear Test, 235; National Reactor Testing Station, 20, 242; Nevada Test Site, 247–248; Panda Committee, 277; Plumbbob Nuclear Tests, 287–288; Power Demonstration Reactor program, 292; Radon Daughters, 304–305; Ranger Nuclear Tests, 306; Rasmussen Report, 306–307; Reactor Safeguard Committee, 308; Sandi National Laboratories, 319–320; Sandstone Nuclear Tests, 320; Savannah River Project, 322; Seabrook Nuclear Power Plant, 326–327; Shippingport Nuclear Power Plant, 329–330; Shot Easy, 330; Shot Harry, 330–331; SL-1 Reactor Accident, 332; SNAP, 333; Teapot Nuclear Tests, 360–361; Tritium, 374–375; Tumbler-Snapper Nuclear Tests, 377–378; Upshot-Knothole Nuclear Tests, 382–383 Uranium Mining, 387–388

American Nuclear Society, 12–13

American nuclear stockpile, 11

American Nuclear Testing Program, 13–14, Atomic Veterans, 37; Bikini Tests, 47–48, Bravo Nuclear Test, 58–59; Buster-Jangle Nuclear Tests, 67; Crossroads Nuclear Tests, 101–102; Eniwetok Atoll Test Site, 12–121; Greenhouse Nuclear Tests, 151–152; Hardtack Nuclear Tests, 159–160; History of, 13–14; Hydrogen Bomb, 167; King Nuclear Test, 197; Knapp Report, 198–199; Los Alamos National Laboratory, 220–222; Mike Nuclear Test, 235; National Reactor

Testing, Station, 20, 242; Nevada Test Site, 247–248; Panda Committee, 277; Plumbbob Nuclear Tests, 287–288; Ranger Nuclear Tests, 306; Sandstone Nuclear Tests, 320; Science Based Stockpile Stewardship Program, 380; Shot Easy, 330; Shot Harry, 330–331; Teapot Nuclear Tests, 360–361; Trinity, 374; Tritium, 374–375; Tumbler-Snapper Nuclear Tests, 377–378; Upshot-Knothole Nuclear Tests, 382–383

American Physical Society, 46; Attack on Rasmussen Report, 307; Role of *The Physical Review*, 284

American Relief Administration (ARA), 352

American Telephone & Telegraph (AT&T), 319

Americium-241, 159, 263

Ames Research Center, 266

Amchitka Island (Alaska), 152

Analog computer, 65

Anders, William, Nuclear Regulatory Commission, 257

Anderson, Carl David, 14–15, 49

Anderson, Clinton, 293

Anderson, Hubert, Chicago Pile One, 84

Anderson, Sir John, 5, 88, 146; Advisory Committee on Atomic Energy, 2–3; Groves-Anderson Memorandum, 154

Anglo-American Declaration of Trust, 15

Anglo-Canadian Atomic Energy Commission, 91

Anglo-Canadian atomic program, 111, 291

Annalen der Physik, 114; Einstein writings, 114–116

Antarctic Treaty, 15–16; Nuclear Free Zones, 255

Anti-Americanism, 70

Antiballistic Missile Treaty, 16, 350

Antiballistic missiles, 16

Antinuclear Movement. *See* Antinuclear

Movements in the United States; Ban the Bomb Movement; SANE

Antinuclear Movements in the United States, 17, 254–255

Apex Committee, 17–18

Apollo (Pennsylvania), 266

Appel des 400, 18

Appropriation Committees (U.S. Congress), 223

Apsara Reactor (India), 170

Arab-Israeli War of 1973, 181, 350, 385

Arab states, 350

Arco (Idaho), 242

Ardenne, Manfred von, 18–19

Argentina, 72, 255, 262, Antarctic Treaty, 16; South American Nuclear Program, 337–338; Treaty of Tlatelolco, 373–374

Argentine-Brazilian Common System of Accounting and Control (ABACC), 374

Argentine Nuclear Program. *See* South American Nuclear Program

Argonne National Laboratory, 19–20, 60, 63, 130, 181, 234, 242, 269, 322, 409; Chicago Pile One, 84–85; Contributions of Zinn, 20; History of, 19–20; Role of Compton, 19–20; Relationship with National Reactor Testing Station, 20; Rivalry with Oak Ridge National Laboratory, 269; Work with Westinghouse Electric Corporation, 400

Arisue, Seizo, 164

Arisue Mission, Hiroshima, 164

Arizona (State), Uranium Mining, 388

Armament Research Department (ARD), 20–21, 62, 283

Armament Research Establishment (ARE), 21

Armed Forces Special Weapons Project, 247

Arms race, 93

Army Ordinance Bureau (Germany), 21, 111, 163; Uranium Club, 386

Artic (Ship), 95

Artificial radioactivity, 129, 304; Dis-

covery by Frederick and Irene Joliot-
Curie, 189, 191
Aryan Physics Movement, 21–22,
162, 208, 336; Anti-Einstein motiva-
tion, 22; Attacks on Arnold Sommer-
feld, 336; Role of Johannes Stark,
346–347
Arzamas-16, 23; Sakharov's work, 318
Asmara Chemical Company (West Ger-
many), 287
Associated University Inc. (AUI), 63–
64
Association of Manhattan Project Scien-
tists (AMPS), 127–128
Association of Oak Ridge Scientists at
Clinton Laboratory (AORSCL), 128
Association of Southeast Asian Nations
(ASEAN), 370
Aston, Francis William, 23–24
Atlanta (Georgia), 172
Atlantic Richfield Company, 264
Atmospheric nuclear testing, 217, 239
Atomic Bomb, 24–25, 87, 127, 132,
134, 139–141, 164–165, 173, 183–
184, 191, 198, 209, 217, 219, 221,
235, 240–241, 270–273, 280, 289,
312, 319, 326, 336–337, 341, 346,
355–356, 362–364, 368, 376–377,
386, 395, 409; Atomic Bomb Vic-
tims, 26–27; Berkeley Summer Col-
loquium of 1942, 44; British Atomic
bomb tests at Maralinga, 226–227;
British origins, 214, 227–228; Chain
Reaction, 77; Chinese Atomic Bomb,
217; Greenhouse Nuclear Tests, 151–
153; Interim Committee on the
Atomic Bomb, 173; Japanese Atomic
Bomb Program, 183; King Nuclear
Test, 197; Los Alamos National Laboratory, 220–222; Manhattan Project, 223–226; Military Policy
Committee, 235–236; National Defense Research Committee, 240–241;
Nevada Test Site, 247–248; Role of
Franklin Delano Roosevelt, 312; Role
of Leo Szilard, 355–356; Sandia National laboratories, 319–320; Sandstone Nuclear Tests, 320; Shot Easy,
330; Soviet Atomic Spying, 338–340;

Special Committee on the Atomic
Bomb, 342–344; Teapot Nuclear
Tests, 360–361; Top Policy Group,
370; Truman's role, 375–376; Ura-
nium Committee, 386–387; U-235,
390
Atomic Bomb Casualty Commission
(ABCC), 25–26
Atomic Bomb Victims, 26–27
Atomic Committee, 100, 270
Atomic Energy Act of 1946, 10, 27–
28, 212, 230, 283; Creation of Joint
Committee on Atomic Energy, 187;
Role of Chet Holifield, 166
Atomic Energy Act of 1954, 28, 292;
Expansion of Joint Committee on
Atomic Energy, 187–188; Power
Demonstration Reactor program, 292
Atomic Energy Authority (Great Brit-
ain), 18, 28–29, 62, 76, 160, 214
Atomic Energy Bill of 1946, 29, 187
Atomic Energy Commission (AEC),
6, 10–11, 13, 29–31, 60, 64, 100,
119, 125, 130–131, 144–145, 153–
154, 161, 172, 205, 210, 216, 260,
266, 268, 273–274, 305, 310, 215,
319, 326–327, 329, 332, 352, 366,
376, 409–410; Abolition of, 119–
120; Advisory Committee on Reactor
Safeguard, 3; Argonne National Laboratory, 20; Argonne National Laboratory, 20; Atomic Bomb Casualty
Commission, 26; Atomic Energy Act
of 1946, 27; Atomic Energy Act of
1954, 28; Atomic Veterans, 37; Bikini Tests, 47–48; Bravo Nuclear
Test, 59; Brockett Report, 63; *Calvert Cliffs* Decision, 69; Fallout, 124;
Emergency Core Cooling Systems,
119; Gofman-Tamplin Manifesto,
148–149; Greenhouse Nuclear Tests,
151–152; Hydrogen Bomb, 167;
Joint Committee on Atomic Energy,
187–188; Knapp Report, 198; National Reactor Testing Station, 20,
242; NERVA, 246; Nevada Test Site,
248; Nuclear Waste Disposal, 263–
264; Panda Committee, 277; Plowshare, 286; Plumbbob Nuclear

Tests, 287–289; Power Demonstration Reactor Program, 292; Radon Daughters, 304–305; Price-Anderson Bill, 293; Private Ownership of Nuclear Materials, 293–294; Rasmussen Report, 306–307; Reactor Safeguard Committee, 308–309; Replacement by Nuclear Regulatory Commission, 257–258; Role of Chet Holifield, 165–167; Role of David Lilienthal, 210–211; Role of Brien McMahon, 229–230; Role of Glenn Seaborg, 326; Role of Lewis Strauss, 352–353; Savannah River Project, 322; Seaborg Report, 326; Seabrook Nuclear Power Plant, 326–327; Shot Easy, 330; Shot Harry, 330–331; SNAP, 333; Strontium-90, 353; Support of Harry Truman, 376; Teapot Nuclear Tests, 360–361; Tumbler-Snapper Nuclear Tests, 377–378; Upshot-Knothole Nuclear Tests, 382–383; Uranium Mining, 388; Waltz Mill Reactor Accident, 397; WASH-740 Report, 398–399; WASH-740 Update, 399; Weiss Report, 399–400; Work with Westinghouse Electronic Corporation, 400

Atomic Energy Commission (India), 47, Indian Nuclear Program, 170–171

Atomic Energy Commission, United Nations (UNAEC), 31–33, 41; Bikini Tests, 48

Atomic Energy Control Act, 33

Atomic Energy Control Board, 33; Whiteshell Nuclear Research Establishment, 401

Atomic Energy Corporation (South Africa), 336; South African Nuclear Program, 336

Atomic Energy Council (Great Britain), 292

Atomic Energy of Canada Limited, 33–34, 72–73, 337

Atomic Energy Research Establishment (AERE), 62, 91; Harwell Laboratory, 160

Atomic Industrial Forum (AIF), 34–35; Loss of influence, 34–35

Atomic Reactors. See Chicago Pile One; Gas-Cooled Reactors; Light Water Reactors

Atomic Scientists' Association, 35, 295

Atomic Scientists' News, The, 35

Atomic Scientists of Chicago, 35–37, 65, 127; Role of Leo Szilard, 35–36

Atomic spies, 140–141, 291, 314–315

Atomic Structure and Spectral Lines, 336; Arnold Sommerfeld, 335–336

Atomic theory of matter, 107

Atomic Veterans, 37

Atomic Weapons Research Establishment, 5–6

Atomic weights, 107, 180

Atomindex, 176

Atoms for Peace, 37–38, 181, 353; Relationship of International Atomic Energy Agency, 174

Attlee, Clement, 146, 237

Attlee government, 61, 292; Washington Declaration, 398

Aubert, Axel, Allier Affair, 7

Auckland (New Zealand), 306

Auger, Pierre, 156

Australia, 86, 392; Antarctic Treaty, 16; Australian Atomic Weapons Test Safety Committee, 38; Maralinga Proving Grounds, 226–227; Monte Bello Atomic Test Site, 237–238; Treaty of Rarotonga, 372–373; Uranium supplies, 384; Uranium Cartel, 385; Western Supplies Group, 400

Australian Atomic Weapons Test Safety Committee (AWTSC), 38–39

Australian government, 385; Australian Atomic Weapons Test Safety Committee, 38

Australian Labor Party, 372

Australian Royal Commission, 38, 227

Axis spies, 259

B-2 Bomber, 168
B-17 Bomber, 368
B-29 Bomber, 49–50, 121, 163, 217, 240, 368

B-36 Bomber, 168, 382, 384

B-43 (Hydrogen bomb), 168

B-50 (Bomber), 383

B-58 Bomber Hustler, 168

B-83 (Hydrogen bomb), 168

Babcock and Wilcox, 366; Turnkey Nuclear Energy Plants, 379

Bacher, Robert, 30, 217, 225, 301

Bagge, Erich, 21

Baghdad (Iraq), 179

Bainbridge, Kenneth, Life of, 40–41

Baku Polytechnic Institute, 200

Balkan countries, 82, 255

Ban on nuclear testing, 97

Ban the Bomb Movement, 41

Bangalore Institute of Science, 47

Bangkok (Thailand), 371

Bard, Ralph A., 173

Baruch, Bernard, 213; Acheson Plan, 2

Baruch Plan, 2, 41–42, 213; Defeat of, 33

Battle of Jutland, 49

Battle of the Falklands, 49

BDM Corporation, 262

Beaverlodge Lake (Canada), 388

Bechtel Corporation, 337

Beck, Clifford, WASH-740 Update, 399

Becquerel, Antoine Henri, 42, 103, 382, 384; Life of, 42

Becquerel (Unit), 382

BEIR V (Biological Effects of Ionizing Radiation), 96–97

Bela Kun Regime (Hungary), 355, 362

Belarus, 161, 348

Belgian Syndicate, 117

Belgium, 75, 123, 216, 245; Antarctic Treaty, 16; Western Supplies Group, 400

Bell Telephone Research Laboratories, 241

Bellman, Robert, 96

Bellona Foundation, 249

Beloyarsk Fast Breeder Reactor, 342

Beloyarsk Nuclear Reactor Fire, 43

Belyushy Guba (Russia), 252

Ben-Gurion, David, 181

Ben-Gurion government, 181

Bennett, Bill, 33

Bergman, Ernest David, Israeli Nuclear Program, 181

Beria, Lavrentii, 43–44, 195, 200, 318, 346; Dispute with Kaptisa, 44; Life of, 43–44; Red Specialists, 309; Semipaltinsk-21 Test Site, 327; Soviet Atomic Spying, 338–340; Soviet Nuclear Program, 340–341; Special Committee on the Atomic Bomb, 343–344

Berkeley Summer Colliquium of 1942, 44, 167, 270

Berlin Crisis, 89, 125, 346

Berlin Institute of technology, 402

Bermuda, 199

Bernard, Chester, 212

Beta decay, 129

Beta radiation, 124, 302, 316, 393

Bethe, Hans Albrecht, 45–46, 118, 128, 132, 140, 169, 225, 335, 362; Life of, 45–46; Relationship with Edward Teller, 46; Target Committee, 359

Bethe's bible, 45

Bettes Laboratory (Pittsburgh, Pennsylvania), 244, 400

Bevin, Ernest, 146

Bhabha, Homi Jehangir, 46–47, 170; Life of, 46–47; Relationship of Canadian scientists, 47

Bhutan, 97

Bhutto, Ali, 276

Big Bang Theory, Role of Hans Bethe, 46

Bihar (India), 170

Bikini Atoll, 13, 12, 160, 222, 280, 283; Bikini Tests, 47–48; Bravo Nuclear Test, 58–59; Crossroads Nuclear Tests, 101–102

Bikini Tests, 47–48, 222, 368; Role of William Penney, 283

Billingham Research Laboratory (ICI), 5

Biomedical Research Division (Lawrence Radiation Laboratory), 148–149

Birks, John, Nuclear Winter, 265

The Black Corps (Das Schwarze Korps), 347

Blackbody problem, 285, 297

Blackett, Patrick Maynard Stuart, 15, 49, 70, 75, 139; Life of, 49

Blandy, William H. (Spike), 101, 280

Blind River (Canada), Canadian Nuclear Energy program, 73; Uranium Mining, 387–388

Bloch, Felix, 9, 44

Blokintsev, D. I., 269

Board of Atomic Energy Research, 47

Board of Governors (IAEA), 174–175

Board of Military Advisors, 27

Bock, Frederick, 49

Bock's Car, 49–50, 240; Bombing of Nagasaki, 49–50

Bohr, Niels Henrik David, 50–52, 74, 88, 139, 162, 172, 183, 202, 258–259, 281, 308, 389, 397–398; Animosity of Churchill towards, 312; Friendship of Ernest Rutherford, 50; Institute for Theoretical Physics, 51; Life of, 50–52; Solvay Conference of 1927, 335; Provision in Roosevelt-Churchill Hyde Park Aide—Memoir, 312; U-235 problem, 390

Bohr Festival, Influence of, 52

Bohr-Heisenberg Conversation, September 1941, 53

Bohr's atom, 50, 238, 336

Boiling-water reactors, 210

Bolivia, Treaty of Tlatelolco, 373–374

Boltzmann, Ludwig, 231

Borden, William Liscum, 274

Born, Max, 53–54, 55, 113, 128, 140, 245, 296; Life of, 53–54; Role in development of quantum mechanics, 54, 296

Born-Haber cycle, 54

Bothe, Walther Wilhelm Georg Franz, 54–55, 393; Life of, 54–55; Mistake on graphite, 55

Bowen, Harold G., 244

Bradbury, Norris, 55–57, 277, 330, 377; Feud with Teller, 363; Life of, 55–57; Los Alamos National Laboratory, 221–222

Bragg, William Henry, 57–58, 280; Life of, 57–58

Bragg, William Lawrence, Life of, 58

Bragg's law, 58

Bravo Nuclear Test, 58–59, 166, 222, 305, 315; Fallout, 125; History of, 58–59; International incident, 58–59

Brazil, 255; Antarctic Treaty, 16; South American Nuclear Program, 337–338; Treaty of Tlatelolco, 373–374

Brazil Nuclear Program. *See* South American Nuclear Program

Breeder Reactors, 3, 18, 29, 59–60, 110, 145, 175, 185, 409; Beloyarsk Nuclear Reactor Fire, 43; Fermi I, 131; Japanese Nuclear Program, 185; National Reactor Testing Station, 20, 242, 270; Soviet Nuclear Program, 342

Breit, Gregory, 259, 403

Breit-Rabi theory, 300

Breit-Wigner formula, 403

Brezhnev, Leonid, 350–351

Briggs, Lyman J., 241, 270; Uranium Committee, 386–387

Britanov, Igor, 199

British Admiralty, 57, 355

British Association of Scientific Workers, 35, 49

British atomic bomb, 146, 214, 291–292; Aldermaston, 5–6, Apex Committee, 17–18; Maralinga Proving Grounds, 226–227; Monte Bello Atomic Test Site, 237–238; Role of William Penney, 283

British commandos, Attack on Norsk-Hydro Plant, 251

British government, Advisory Committee on Atomic Energy, 3; Attitude towards Cavendish Laboratory, 74; Gen 75, 146

British hydrogen bomb, British Nuclear Program, 60–62; Christmas Island Tests, 86–87

British Intelligence (MI6), 126, 313

British mission to Manhattan Project, 5, 91, 315, 376; Contributions of William Penney, 283; Importance of James Chadwick, 76

British Nuclear Program, 4–5, 60–62, 160, 214, 226–227, 237–238; Aldermaston, 5–6; Apex Committee, 17–18; Armament Research Department, 20; Atomic Energy Authroity, 28–29; Atomic Energy Bill of 1946, 29; Christmas Island Tests, 86–87; Collapse of Anglo-American cooperation, 61; Contributions of William Penney, 282–283; Harwell Laboratory, 160; Lord Portal, 291–292; Monte Bello Atomic Test Site, 237–238; Windscale, 404–405

British scientists, 61, 260, 369; Effect of compartmentalization, 62

Brockett, George, Brockett Report, 63

Brockett Report, 63, 242; Criticism of emergency cooling procedures, 63; National Reactor Testing Station, 242

Brodie, Robert B., Target Committee, 359

Brookhaven National Laboratory, 10, 63–64, 293, 301; History of, 63–64; Strontium-90, 353; WASH-740 Report, 398–399; WASH-740 Update, 399

Brookhaven Science Association (BSA), 64

Brower, David, 137

Brown, Harold, 205

Brown's Ferry Nuclear Power Station Incident, 64

Bruce, 72

Brues, Austin, 26, 36

Brues-Henshaw Report, 26

Brunei, Treaty of Bangkok, 371

Brunner Mond, 5

Brussels (Belgium), 334

Bryukhanov, Viktor, Chernobyl Nuclear Power Station Accident, 81

Bubble Chamber, 148; Luis Alverez's version, 10

Budapest Institute of Technology, 362

Bulgaria, 189

Bulletin of the Atomic Scientists, 37, 65, 128, 315

Bureau of Ordnance, 352

Bureau of Ships, 92, 244

Bureau of Standards, 169, 270, 386

Burger Supreme Court, 70

Burns and Roe, 366

Bush, George, 13, 301, 354

Bush, Vannevar, 40, 65–67, 94, 100, 173, 216–217, 223, 278, 312, 387; Acheson Board, 2; Life of, 65–67; Lilienthal Group, 213; Military Policy Committee, 235–236; National Defense Research Committee, 240–241; Office of Scientific Research and Development, 270; Top Policy Group, 370

Bush administration, 302

Buster-Jangle Nuclear Tests, 67–68

Byrnes, Jimmy, 2, 173, 213

Cairncross, John, 338

Cairo (Egypt), 371

Calcium, Resemblance to strontium-90, 353

Calder Hall Reactor, 62

California, 331

California Institute of Technology (Caltech), 14, 132, 148, 169, 230, 236–237, 271–272, 281

Calutron, 205

Calvert Cliffs Coordinating Committee, 69

Calvert Cliffs Decision, 11, 69–70

Cambodia, Treaty of Bangkok, 371

Cambridge (Massachusetts), 381

Cambridge University, 24, 45, 47, 49–50, 53–54, 57–58, 71, 73, 76, 91, 98, 113, 194, 271, 282, 316, 364, 404

Camp Upton, 64

Campaign for Nuclear Disarmament (CND), 41, 70–71

Canada, 31, 71–73, 94, 101, 117, 145, 162, 171, 175, 216, 253–254, 261, 273, 276, 297–298, 398; Combined Policy Committee, 94–95; Groves-Anderson Memorandum, 154; Soviet

Atomic Spying, 339; Uranium supplies, 384; Uranium Cartel, 385; Western Supplies Group, 400

Canadian General Electric, 34, 72

Canadian government, 33–34

Canadian National Research Council, 71

Canadian Nuclear Energy Program, 72–73, 162, 253–254; Canadian National Research Council, 71; CANDU Reactor, 73; Chalk River Facility, 77–78; Eldorado Mine, 117; History of, 72–73; NRU Reactor, 253; NRX Reactor, 253–254; Whiteshell Nuclear Research Establishment, 401; ZEEP, 409

CANDU Reactor (Canadian Deuterium-Uranium Reactor), 47, 73, 78, 253

Cape Kerauden (Australia), 286

Cape Thompson (Alaska), 294

Carbon-14, 281

Carborundum Metals Company, Zirconium, 410

Carlsbad, New Mexio, 14

Carnegie Institution, 1–2, 65, 241, 244

Carter, Jimmy, 196, 258, 323, 351; Three Mile Island, 367

Carter administration, 174, 176, 246–247, 256, 323, 351, 357

Case Institute of Technology, 148

Cathode rays, 208, 236, 365, 393

Cavendish, Henry, 74

Cavendish, William, 73–74

Cavendish Chair of Physics, 73–74

Cavendish Laboratory, 24, 40, 47, 49–50, 53, 57–58, 73–74, 76, 90–91, 98, 156, 172, 194, 228, 236, 258, 271, 316; History of, 73–75; Impact of Ernest Rutherford, 75; Influence of J. J. Thomson, 364–365

Celle (Germany), 310

Center Party (Sweden), 354

Centers for Disease Control, 37

Central Intelligence Agency (CIA), 181, 333

Central Intelligence Group (CIG), 216

Central Power Supply Management (Sweden), 354

CERN (Conseil Europeen pour la Recherche Nucleaire), 75, 301, 354

Cesium-137, 303

Chadwick, James, 5, 75, 76, 227, 315; Life of, 76

Chagai Test Site (Pakistan), Pakistani Nuclear Program, 277

Chain Reaction, 77, 130, 180, 190, 197, 207, 219, 228, 271, 308, 340, 355, 384, 390, 393; Chicago Pile One, 84–85; Fission, 133–134

Chalk River Facility, 71, 77–78, 91, 291, 322; NRU Reactor, 253; NRX Reactor, 253–254; ZEEP, 409

Challenger (Space shuttle), 132

Chamberlain government, 227

Chashma Nuclear Power Station (Pakistan), 276

Chazhma Bay Submarine Accident, 78–79; K-431 submarine, 78; Soviet Nuclear Submarine Accidents, 343

Chelyabinsk-40, 79–80; Nuclear accident, 80, 161; Soviet Nuclear Program, 340–342

Chelyabinsk-65, 80; Soviet Nuclear Program, 340–342

Chelyabinsk-70, 80; Soviet Nuclear Program, 340–342

Chemical bonding, 281

Chemical Division (Los Alamos), 225

Chemical-Separation Section (Metallurgical Laboratory), 325

Chemical Warfare Service, 99

Cherenkov, Pavel, 358

Cherenkov radiation effect, 358

Chernobyl decommissioning activities, 110

Chernobyl Nuclear Power Station Accident, 81–84, 161, 231, 294–295, 303, 342; Casualties, 82; Fallout, 124–125; History of, 81–84; Psychological Impact of Nuclear Accidents, 294–295; Soviet Nuclear Program, 340–342

Chernobyl RBMK Reactors, 342

Cherwell, Lord. *See* Lindemann, Frederick Alexander

Chicago Pile One, 84–85, 117, 403, 409; Chain Reaction, 77; Contributions of Enrico Fermi, 130; History of, 84–85; Metallurgical laboratory, 234

Chichester, Guy, 89

Chile, 255, 338; Antarctic Treaty, 16; Treaty of Tlateloloco, 373–374

China, 109, 251, 256, 263, 276, 408; Chinese atomic bomb, 217, Chinese hydrogen bomb, 217; Chinese-Indian War, 171; Chinese Nuclear Program, 85; Lop Nur Nuclear Weapons Test Site, 217–218; Treaty of Pelindaba, 372

China Syndrome, 3, 85

China Syndrome (Movie), 85

Chinese Army, 171

Chinese atomic bomb, 217; Chinese Nuclear Program, 85–86; Lop Nur Nuclear Weapons Test Site, 217–218

Chinese hydrogen bomb, 217; Chinese Nuclear Program, 85–86; Lop Nur Nuclear Weapons Test Site, 217–218

Chinese-Indian War, Influence on Indian Nuclear Program, 171

Chinese Nuclear Program, 85–86, 217–218, 333; Break with Soviet Union, 85–86; History of, 85–86; Lop Nur Nuclear Weapons Test Site, 217–218

Chinese scientists, 85

Christiansen, C., 50

Christmas Island Tests, 14, 70, 86–87

Church, A. E., 322

Churchill, Winston, 18, 52, 71, 88, 94, 127, 259, 297, 312, 376; British Nuclear Program, 61–62; Friendship with Franklin Delano Roosevelt, 312; Relationship with Frederick Lindemann, 213–214; Roosevelt-Churchill Hyde Park Aide—Memoir, 313–314

Churchill government, 18

Cirus Reactor (India), 170

Cisler, Walker, Atomic Energy Forum, 34

City College of New York, 299, 314

Civil Defense, 88–89, 125, 288, 361, 382; Support from Chet Holifield, 166; Teapot Nuclear Tests, 361; Upshot-Knothale Nuclear Tests, 382

Civilian Nuclear Power: A Report to the President 1962, Seaborg Report, 326

Clamshell Alliance, 89–90

Clarendon Laboratory, 90, 213, 355, 376; Role of Frederick Lindemann, 90, History of, 90

Clark, Ramsey, 96

Clayton, William L., 173

Clifton, R. B., 90

Clinton, William, 97

Clinton Laboratories. *See* Oak Ridge National Laboratory

Club of Five, Uranium Cartel, 385

Cobalt-60, 303

Cockcroft, John Douglas, 62, 71, 75, 78, 91, 160, 227, 292, 339

Cockcroft-Walton accelerator, 220

Code 390, 92; Naval Research Laboratory, 244

Code 390–590, 92; Naval Research Laboratory, 244

Code 490, 92; Naval Research Laboratory, 244

Code 590, 92; Naval Research Laboratory, 244

Cohen, Samuel T., 246

Cold Fusion, 93

Cold War, 13, 17, 93–94, 118, 265, 373; First Strike Doctrine, 94

Cole, Sterling, 28

College de France, 291

College of Aberystwyth, 334

College of Wooster, 98

College Sevigne, 191

Collins, Canon, 70

Colorado (State), uranium Mining, 388

Columbia University, 63, 130, 132, 140, 151, 209, 223, 234, 236, 270, 299–301, 310, 389, 409

Combined Development Trust (CDT), 298

Combined Policy Committee, 94–95, 154, 298

Combustion Engineering, 242; Turnkey Nuclear Energy Plants, 379

Commercial Nuclear Ships, 95, 112

Commercial spent fuel, 263

Commissariat à l'Énergie Atomique (ICEA), 95–96, 108, 111, 118, 136, 191; PEON Commission, 283–284

Committee for a Sane Nuclear Policy. *See* SANE

Committee for Interplanetary Flight, 195

Committee for Nuclear Responsibility, 96; Role of Lenore Marshall, 96

Committee on Biological Effects of Ionizing Radiation Report, 96–97

Committee on Social and Political Implications, 36

Committee on the Scientific Survey of Air Defence (Great Britain), 7, 227

Communist Party, 140, 191, 273, 314

Compartmentalization, 259–260; Effect on British scientists, 62; Groves implementation, 259–260; Nuclear Secrecy, 258–260

Comprehensive Test Ban Treaty, 14, 86, 97–98, 239, 255, 349

Compton, Karl T., 173, 241

Compton, Arthur Holly, 14, 55, 98–99, 115–116, 135–136, 173, 185–186, 224, 235, 237, 270, 272, 312; Argonne National Laboratory, 19–20; Chicago Pile One, 84–85; Life of, 98–99; Metallurgical Laboratory, 234–235; X-ray research, 98–99

Compton effect, 98, 144, 183

Conant, James Bryant, 85, 94, 99–100, 173, 198, 223, 236, 241; Acheson Board, 2; Atomic Committee, 100; General Advisory Committee, 100; Lilienthal Group, 213; National Defense Research Committee, 99–100; Office of Scientific Research and Development, 270; Relationship with Vannevar Bush, 100; Top Policy Group, 370

Conference of States Parties, 98

Conference on Questions of the Physics of the Atomic Nucleus (Kharkov, 1939), 196

The Conqueror (Movie), 100; Nevada Test Site, 248

Conservative Party (Great Britain), 62

Containment strategy, 94

The Contemporary Crisis in German Physics: Critique of book by Max von Laue, 394; Johannes Stark, 347

Controller of Productions, Atomic Energy (Great Britain), 292

Cook Islands, Treaty of Rarotong, 372

Coordinating Council of the Friends of the Earth International, 138

Copenhagen, 172, 183

Cornell University, 45, 109, 132, 266, 284, 299, 348–349

Cosmos 954, 100–101; Loss of, 101

Cosmotron, 64

Council of Foreign Ministers, Atomic Energy Commission United Nations, 31

Council of 100, 70

Council of Scientific and Industrial Research (CSIR) (India), 170

Cowpuncher Committee, Implosion crisis, 169–170

Cray computer, 6

Creys-Malville, 18

Crimean State University, 199, 358

Cripps, Stafford, 146

Critical Mass, 101; Role of Ralph Nader, 101

Critical Mass, 101

Crossroads Nuclear Tests, 13, 48, 101–102, 280, 283, 368, 377

Crud (Chalk River unidentified deposit), 78

Cruise Missiles, 174, 351

Crutzen, Paul, Nuclear Winter, 265

Cuba, 97, 255, 333; Treay of Tlatelolco, 373–374

Cuban Missile Crisis, 70, 373

Curie, Irene, 183, 191–192, 302

Curie, Marie Sklodowska, 102–103, 186, 189, 191, 302, 317, 381, 391; Radium Institute, 304

Curie, Pierre, 103–104, 186, 191, 302, 304, 381

Curie (Unit), 381

Curie point, 103

Cyclic accelerators, 160

Cyclotron, 19, 76, 105–106, 183, 200, 230, 245, 289, 315, 352; French cyclotron, 106; German cyclotron, 55; History of, 105–106; Japanese cyclotron, 183; Lawrence Radiation Laboratory, 206; Role of Ernest O. Lawrence, 204–205; Role of Edwin McMillan, 230; Soviet cyclotron, 340

Cyrus, 287

Czechoslovakia, 189, 216, 255; Uranium supplies, 384

Daghlian, Harry, 303; Los Alamos Atomic Accidents, 218–219

Dahlgren Proving Ground, 56

Dallet, Joe, 273

Dalton, Hugh, 146

Dalton, John, 107

Daniel, Farrington, 268

Darien (Connecticut), 229

Dartmouth University, 196

Darwin, Charles, 238

Dautry, Raoul, 96; Allier Affair, 7

De Broglie, Louis-Victor Pierre Raymond, 108, 113, 323–324

De Klerk, F. W., Ending of South African Nuclear Program, 337

De Kronig, R. L., 282

Deaf Smith (Texas), 264

Dean, Gordon, 30, 352

Debye, Peter, 108–109, 335, 346; Life of, 108–109; Warning to Allies, 109

Decatur, 64

Declared Nuclear Weapon States, 109

Decommissioning of Nuclear Plants, 109–110, 159, 278, 330

Decommissioning of nuclear submarines, 79, 343

Dee, P. I., 75

De Gaulle, Charles, 111, 118, 136, 239

Delta Mine (Utah), Uranium Mining, 388

Demonstrations against nuclear reactors: Campaign for Nuclear Disarmament, 70–71; Clamshell Alliance, 89–90; Seabrook Nuclear Power Plant, 326–327; Whyl Demonstration, 401–402

Dempster, Arthur Jeffrey, U-235, 290

Denmark, 75, 259, 308; Chernobyl Nuclear Power Station Accident, 82

Dennison, D. M., Target Committee, 359

Denuclearization of Latin America (COPREDAL), 373

Department of Defense (DOD), 349–377

Department of Energy (DOE), 31, 64, 110, 119, 206, 264, 319, 349, 380, 406–407; Pantex, 278

Department of Radiation Physics (Kaiser-Wilhelm-Society), 232

DePauw College, 211

Derry, J. A., Target Committee, 359

Detroit Edison Company, 3, 34, 60, 131

Deuterium. *See* Heavy Water

Deuterium-tritium mix, 167–168, 203, 369, 375

Deutsche Physik, 208

Diebner, Kurt, 111–112, 147, 310; Life of, 111–112; Role in Nazi Atomic Bomb Program, 110–111

Dill, Sir John, 94

Dimona, 112, 181; Based on French model, 112; Plumbat Affair, 287; Venunu Affair, 392

Dirac, Paul Adrien Maurice, 15, 47, 112–113; Life of, 112–113; Role in Quantum Mechanics, 113

Directorate for Nuclear Energy (Russia), 231

Disarmament Movement, 17

Dixon-Yates Deal, 352

Doan, Richard, 268

Dobson-Lindemann theory of the upper atomosphere, 213

Doepel, Robert, 207

Don't Make a Wave Committee
(DMWC), 152
Douglas Point, 72–73
Dr. Lee's Professor of Experimental
Philosophy (Oxford U.), 213, 334
Dublin Institute for Advanced Studies,
324
Dubna (Russia), 188, 291
Dugway (Utah), 247
Dulles, John Foster, Western Supplies
Group, 400
Dunning, Gordon, Weiss Report, 400
Du Pont Corporation, 145, 153, 158,
234, 264, 308, 322
Dupuis, Rene, 33
Duquesne Light Company of Pitts-
burgh (Pennsylvania), 329–330
Dyson, Freeman, 132

East German Central Institute for Nu-
clear Physics, 141
East Germany, 141, 189, 216, 255
Eastman Kodak Company, 306
Eaton, Cyrus, S., 295
Ecole des Pontes-et-Chaussees, 42
Ecole Normale Superieure, 392
Ecole Polytechnique, 42
Eden, Sir Anthony, 18
Edsall, John, 96
Edward Teller–Stanisilaw Ulam thesis,
58
Ecuador, Treaty of Tlatelolco, 373–374
Ehrenfest, Paul, 128, 178, 272
Ehrlich, Paul, 96
Einstein, Albert, 22, 50, 54, 103,
108, 114–116, 162, 218, 237, 284–
285, 295–296, 314, 355, 403; At-
tacks by Philip Lenard, 208; Jewish
Physics, 186; Solvay Conferences of
1927, 335; Support by Max von
Laue, 394
Einstein-Szilard Letter, 116–117,
259, 312, 356, 386, 403
Eisenhower, Dwight David, 127, 174,
247, 352–353, 376; Atomic Energy
Act of 1954, 28; Atoms for Peace,
37–38

Eisenhower administration, 88, 166,
198, 274, 329, 352, 357
Eldorado Mine, 117; Canadian Nu-
clear Energy Program, 72–73; Ura-
nium Mining, 387–388
Electric Boat Company (Groton, Con-
necticut), 92, 243
Electricité de France (EDF), 96, 118,
233; PEON Commission, 283–284
Electromagnetic pulse, 160
Electromagnetic radiation, 285, 297
Electromagnetic separation of uranium,
207, 224; Oak Ridge National labo-
ratory, 267; Y-12 Plant, 267
Electron, 208, 236–237, 336, 364;
Contributions of Niels Bohr, 50;
Bothe's research, 55; Lorentz's con-
tributions, 218
Electron spin, 150
Elektrostal (Russia), 341
Elugelab (Eniwetok Atoll), 121, 235
Ely (Nevada), 330
Emergency Committee of Atomic
Scientists (ECAS), 118, 242, 381;
Three Mile Island, 366–367
Emergency Core Cooling System
(ECCS), 119, 242; Brockett Report,
63; Chernobyl Nuclear Power Station
Accident, 81–84; National Reactor
Testing Station, 242; Three Mile Is-
land, 365
Emu Test Site, 38, 227
Energy Reorganization Act, 119–120
Energy Research Development Admin-
istration, 31, 119
Enhanced radiation weapon (ERW),
246–247
ENIAC (Electronic Numerical Integra-
tor and Calculator), 120; Role of von
Neumann, 395
Eniwetok Atoll Test Site, 13–14, 120–
121, 151, 160, 235, 277, 320
Enola Gay, 121, 368; Hiroshima, 163–
164
Enterprise (Ship), 203
Environmental laws, 11
Environmental Protection Agency, 127
Enyu Island, 58–59

Equatorial Guinea, Treaty of Pelindaba, 372

Esau, Abraham, 147, 245; Reich's Research Council, 309; Uranium Club, 385–386

ESECOM (Environmental, Safety, and Economic Committee on Magnetic Confinement Fusion Reactors), 122

European Atomic Energy Community, 123, 136, 147, 287; Hostility of France, 123

EUROTOM. See European Atomic Energy Community

Evans, Ward V., 274

Executive Council of the Comprehensive Test Ban Agreement, 98

Experimental Breeder Reactor (EBR-1), 20

Experimental Breeder Reactor (EBR-2), 20

Experimental Physics Division (Los Alamos), 225

The Expert Opinion, 231

Explosives Division (Los Alamos), 56, 198, 279

Exxon, 222

F-15 (Fighter), 359
F-16 (Fighter), 359
Fairchild Corporation, 4
Fallon, Nevada, 14
Fallout, 58–59, 68, 124–125, 127, 149, 160, 164, 210, 222, 237–238, 286–287, 305, 327, 361, 363; Allen v. United States of America, 6; Australian Atomic Weapons Test Safety Committee, 38; Bravo Nuclear Test, 58–59; Chernobyl Nuclear Power Station Accident, 83; Hardtack Nuclear Tests, 159–160; Knapp Report, 198; Nevada Test Site, 247–248: Ranger Nuclear Tests, 306; Schaffer Report, 323; Shot Easy, 330; Shot Harry, 330–331; Research of Sternglass, 349; Strontium-90; Teapot Nuclear Tests, 360–361; Trinity, 374; Upshot-Knothole Nuclear Tests, 382–383

Fallout Shelters, 125; Civil Defense, 89; Kennedy administration, 125

Fallout Studies Branch (AEC), 198

Fangatauf Test Site, 137, 238–239

Faraday, Michael, 316

Farm Hall, 112, 125–126, 147, 150, 163, 394

Farmington, New Mexico, 14

Farnborough (Great Britain), 213

Farrell, Thomas F., 25; Target Committee, 359–360

Farwerke Hoechst A. G., 147

Fat Man, 13, 25, 127, 237, 240, 283, 290; Expertise of William Penney, 283; Implosion crisis, 169–170; Trinity, 374; Use on Nagasaki, 49–50, 240

Fault-free analysis, Rasmussen Report, 306–307

F-Division (Los Alamos), 130

Federal Bureau of Investigation (FBI), 266, 274

Federal Civil Defense Agency, 88–89

Federal Communications Act of 1934, 28

Federal Emergency Management Agency (FEMA), 89

Federal Emergency Response Agency, 196

Federal Institute of Technology (Switzerland), 281

Federal Radiation Council, 127

Federation of American Scientists, 35, 65, 127–128, 229; Formation to oppose secrecy, 259

Fellowship of Reconciliation (United States), 255

Ferebee, Thomas, 121

Fergana Valley (Russia), 340

Fermi, Enrico, 45, 77, 116, 128–131, 139, 151, 158, 167, 173, 185, 223, 229, 232, 291, 324–325, 355, 362, 386, 409; Chicago Pile One, 84–85; General Advisory Committee, 130; Life of, 128–131

Fermi I, 3, 60, 110, 131, 166; Power Demonstration Reactor program, 292

Fermilab Tevatron, 354

Feynman, Richard Phillips, 132; Life
  of, 132; Role at Los Alamos, 132
Fiji, Treaty of Rarotonga, 372
First Armored Division, 378
First Idea, 318
First Strike Doctrine, 133; Cold War,
  94
Fischer, Emil, 155
Fisher Institute (Germany), 231
Fission, 19, 59, 130, 133–134, 139,
  141, 156, 189–190, 233, 244, 353,
  355, 390, 409; Chain Reaction, 77;
  Discovery by Otto Hahn, 133; His-
  tory of, 133–134; Interpretation by
  Lise Meitner and Otto Frisch, 133–
  134; Uranium, 384
Fitch, Val, 221
Fleishmann, Martin, 93
Flerov, Georgi Nikolaevich, 134, 340;
  Importance, 134; Letter to Stalin,
  234; Soviet Nuclear Program, 340
Flerov's letter, 154
Florida State University, 113
Fomin, Nikolai, Chernobyl Nuclear
  Power Station, 81
Fondation Francqui (Belgium), 324
Ford, Gerald, 351
Ford administration, 337
Ford Motor Company, 222
Forrestal, James, 216–217, 352
Forsberg, Randall, Nuclear Freeze
  Movement, 254–255
Fort de Chatillon Reactor, 136
Fort Halstead, 20, 62
Fort St. Vrain Nuclear Generating Sta-
  tion, 110
France, 60, 75, 109, 123, 145, 161,
  179, 181, 216, 250, 256, 259, 261–
  263, 266, 276, 291, 359, 408; Acci-
  dent, 81; Antarctic Treaty, 16; Cher-
  nobyl Nuclear Power Station, 83;
  Messmer Plan, 233–234; European
  Atomic Energy Community, 123;
  Radium Institute, 304; Treaty of Pe-
  lindaba, 372; Uranium Cartel, 385;
  Western Supplies Group, 400
Franck, James, 35–36, 134–135, 136,

185; Life of, 134–135; Franck Re-
  port, 135–136
Franck Report, 36, 99, 135–136, 326,
  356
Frank, Ilya, 358
Frankford Arsenal, 132
Frankland, P. R., 23
Free University of Poland, 315
Freedom of Information Act, 400
French atomic scientists, 271; Allier Af-
  fair, 7; Halban Affair, 156–157; Ra-
  dium Institute, 304
French cyclotron, 106
French hydrogen bomb, 239
French nuclear industry, 18, 95, 283–
  284
French Nuclear Program, 18, 136–
  137, 373; Appel des 400, 18; Com-
  missariat à l'Énergie Atomique, 95–
  96; Electricité de France, 118;
  History of, 136–137; Messmer Plan,
  233–234; Mururoa Atoll, 238–239;
  PEON Commission, 283; Superphe-
  nix, 60
French Resistance, 190
French Secret Service (Direction Gene-
  rale de la Securite Exterieure)
  (DGSE), Allier Affair, 7; Rainbow
  Warrior, 305–306
Frenchman's Flats, 67, 306, 383
Frendel, Iakov, 178
Friends of the Earth (FOE), 137–138
Friends of the Earth International
  (FOEI), 138
Frick, Wilelm, 347
Frisch, Otto Robert, 133, 138–139,
  156, 219, 233, 282, 355, 398; Frisch-
  Peierls Memorandum, 139–140; In-
  terpretation of fission, 139; Life of,
  138–139
Frisch-Peierls Memorandum, 139–
  140, 282; Response of MAUD Com-
  mittee, 227–228
Fritz Haber Institute of Physical Chem-
  istry (Germany), 394
Fromm, Friedrich, 345
Fuchs, Emil Julius Klaus, 140–141,
  170, 291, 314, 338–239, 341; Help

on British atomic bomb, 61; Implosion crisis, 169–170; Life of, 140–141; Rosenberg Case, 314; Soviet Atomic Spies, 338–340

Full-Length Emergency Cooling Heat Transfer Tests (FLECHTs), 63; Response to Brockett Report, 63

Fusion, 19, 44, 141–142, 349, 363; Cold fusion; ESCOM, 122; International Thermonuclear Experimental Reactor, 177; Tokamak Fusion Reactor, 369; Project Sherwood, 377

Fusion reactors, International Thermonuclear Experimental Reactor, 177; Project Sherwood, 377; Tokamak Fusion Reactor, 369

Gabon (Africa), 271

Gaherty, Geoffrey, 33

Gallium, 290

Gamma radiation, 124, 219, 302, 381–381; Discovery by Villard, 392–393

Gamow, George, 45, 362; Sponsor of the Washington Conference of Theoretical Physics, 397–398

Gandhi, Mahatma, 170

Garching (Germany), 177

Garcia Robles, Alfonso, 373

Gas-Cooled Reactors, 143

Gaseous diffusion uranium process, 85, 110, 140, 267, 390; Decommissioning of, 110; K-25 Plant, 267; Kellex Corporation, 195–196; Oak Ridge National Laboratory, 267; Role of Urey, 389

Gas-graphite reactor, 118, 136

Geiger, Hans, 55, 76, 143–144, 316; Life of, 143–144

Geiger counter, 9, 100, 298, 330; Development of, 144

Gen 75, 146; Advisory Committee on Atomic Energy, 2; Atomic Energy Commission, 30

Gen 163, 146

Gen 173, 61

A General Account of the Development of Atomic Energy for Military Purposes, 339; Smyth Report, 333; Soviet Atomic Spying, 339–340

General Accounting Office (U.S. Congress), 322

General Advisory Committee (AEC), 100, 130, 144, 167, 187, 273–274, 301; Reactor Safeguard Committee, 308–309; Relationship with Special Subcommittee of the Hydrogen Bomb, 344

General Atomics, 242

General Electric, 63, 92, 145, 158–159, 185, 212, 242, 244, 293; Air Nuclear Propulsion, 4; Breeder Reactors, 59–60; Hanford Plant, 145; Impact of Price-Anderson Bill, 293; Three Mile Island, 366; Turnkey Nuclear Energy Plants, 378–379

Geneva Conference of 1955, 38, 47, 52, 145–146, 174, 260, 301

Gentilly I, 72

Gentner, Wolfgang, 191

George Washington University, 362; Washington Conference on Theoretical Physics, 397–398

Gerlach, Walther, 45, 146–147, 313; Farm Hall, 147; Life of, 146–147; Reich's Research Council, 309–310

German Atomic Bomb Program. See Nazi Atomic Bomb Program

German Bureau of Standards, 111, 385

German cyclotron, 106

German National Physical Laboratory (Berlin), 138

German Nuclear Program, 147–148, 156; Geneva Conference of 1955, 147; History of, 147–148; Role of Karl Winnacker, 147–148

German physicists, 54, 125–126, 162, 346–347, 402

German Physics Society, 347

German Research Foundation, 347

German scientists, Alsos Mission, 8–9

German scientists in Russia, 269. See also Ardenne, Manfred von

Germany, 60, 75, 123, 183 216, 250, 258, 276, 361; Antarctic Treaty, 16

Gestapo, 191, 313

Ghauri missile, 277
Gilbert Associates, 366
Ginsburg, Vitaly, 358
Glaser, Donald A., 148; Bubble
  Chamber, 148; Life of, 148
Gnome (New Mexico), 286
Godmanchester, 125
Goering, Herman, 345
Gofman, John, 96, 148
Gofman-Tamplin Manifesto, 148–149;
  Response to Sternglass, 149
Goiania (Brazil), 303
Gold, Harry, 140, 314–315, 338–339
Goldschimdt, Bertrand, 156
Goldstine, Herman, 120
Gorbachev, Mikhail, 174, 177, 319
Gordon, Andrew, 33
Gore, Albert, 166
Gore-Holifield Bill, 166
Gorki (Russia), 319
Göttingen Manifesto, 149–150
Goudsmit, Samuel, 150; Alsos Mis-
  sion, 8; Farm Hall, 125–126; Life of,
  150
Gouzanko, Igor, 314; Soviet Atomic
  Spying, 339
Grand Valley, Colorado, 14
Graphite, 55, 84, 130, 151, 158, 223,
  225, 234, 245, 251, 393; Uranium
  Committee, 386
Gray, Gordon, 274
Gray, Mike, 85
Gray (Unit), 381
Gray Committee, 46, 274, 352
Great Artiste, 10, 163
Great Bear Lake, 72, 117, 328; Ura-
  nium Mining, 387–388; Uranium,
  384
Great Bear Lake Mine. See Eldorado
  Mine
Greece, 75, 82
Greenglass, David, 170, 314–315, 338–
  339
Greenglass, Ruth 315
Greenhouse Nuclear Tests, 151–152,
  167
Greenpeace, 152, 239; Rainbow War-
  rior, 305–306

Greenpeace International, 152; Rain-
  bow Warrior, 305–306
Greenpeace New Zealand, 152
Greenwood, Arthur, 146
Gregory, E. E., 153
Grossman, Marcel, 115
Groth, Wilhelm, 21
Ground Experimental Engine (XE),
  246
Groves, Leslie R., 1, 76, 152–154,
  156–157, 173, 195, 229, 235, 251,
  267–268, 270, 272, 278, 301, 355,
  359–360; Acheson Board, 2; Alsos
  Mission, 8; Atomic Bomb Casualty
  Commission, 25; Atomic Energy
  Commission, 153–154; Groves-
  Anderson memorandum, 154; In-
  terim Committee on the Atomic
  Bomb, 173; Life of, 152–154; Lilien-
  thal Group, 213; Los Alamos Na-
  tional Laboratory, 220–222;
  Manhattan Project, 223–226; Mili-
  tary Policy Committee, 236; Rela-
  tionship with David Lilienthal, 212;
  Selection for Manhattan Project, 66–
  67; Thermal diffusion process, 1; Un-
  happiness with Leo Szilard, 356
Groves-Anderson Memorandum, 154
Gueron, Jules, 111, 156
Gulf Mineral Resource Corporation,
  Uranium Mining, 388
Gundremmingen (Germany), 147
Gunn, Ross, 244

Hahn, Otto, 1, 19, 50, 139, 155–156,
  190, 193, 313, 355, 384, 390, 398,
  409; Fission, 133; Life of, 155–156;
  Work with Lise Meitner, 231–233
Haigerloch (Germany), 163, 245, 310
Halban, Hans von, 71, 156–157, 190,
  227–228
Halban Affair, 156–157; Allier Affair,
  7; French atomic scientists, 177;
  Hostility of Leslie Groves, 157
Halban team, 61, 71, 156–157, 227–
  228
Hanauer, Stephen, 63
Handbook on Implosion Technique, 140

Handford (Washington), 145, 157–
159, 225–226, 264
Hanford Plant, 157–159 225–226,
267, 289, 322, 375; General Electric,
145; Nuclear waste disposal, 263–
264; Reactor Poisoning, 307–308
Hankey, Lord, 338
Hardtack Nuclear Tests, 159–160
Hardtack I, 159–160
Hardtack II, 160
Harmon Report, 133
Harteck, Paul, 21
Harvard University, 40, 63, 65, 99,
198, 211, 220, 241, 270–271
Harwell Laboratory, 61–62, 75, 91,
140, 160, 291
Hasenhorl, Friedrich, 323
Hatch, Orin, 301
Hattiesburg, Mississippi, 14
Hawkes, Jacquetta, 70
Hayward, Susan, 100
Healey, Denis, 70
Health Impact of Nuclear Accidents,
161
Hearn, Richard, 34
Heavy Water, 7, 44, 71, 78, 111–112,
151, 156, 161–162, 180–181, 190,
197, 207, 223, 227, 245, 322, 345
393, 409; Discovery by Harold Urey,
389; Norsk-Hydro Plant, 250–251,
253–254; Telmark Raid, 361–362
Heavy water reactors, 72, 171, 322
Hechinger (Germany), 163, 310
Heisenberg, Werner Karl, 8, 19, 52–
54, 109, 111, 113, 147, 150, 162–
163, 165, 193, 208, 223, 245, 259,
282, 296, 299, 310, 313, 324, 335–
336, 345, 362, 394; Aryan Physics
Movement, 22, 162, 347; Bohr-
Heisenberg Conversation, September
1941, 53; Friendship with Niels
Bohr, 162; Leipzig Reactor, 207;
Life of, 162–163; Quantum Mechan-
ics, 296; Relationship with Albert
Speer, 345; Role in Nazi Atomic
Bomb Program, 245; Solvay Confer-
ence of 1927, 335; Virus House, 303
Heisenberg Uncertainty Principle, 162
HELEN, 6

Helmholtz, Hermann von, 236, 285
Henshaw, Paul, 26
HERALD, 6
Hernu, Henri, *Rainbow Warrior*, 306
Hertz, Gustav, 135
Hertz, Heinrich, 208
Hickenlooper, Bourke, 28, 30, 298
High energy physics, 284, 354
High Energy Physics Advisory Panel
(HEPAP), 354
High-level nuclear waste, 263
High Voltage Engineering Corporation,
391
High Energy Research (HER), 6, 21
High-particle linear accelerators, 75
High-powered lasers, 203
Hilberry, Norman, 151
Hill, David, 36
Himmler, Heinrich, 22, 162
Hinton, Christopher, 292; Role in Brit-
ish Atomic Bomb Program, 61
Hiroshima, 13, 24, 36, 163–164, 240,
280, 333, 344, 360, 368, 375, 390;
Atomic Bomb Casualty Commission,
25–26; *Enola Gay*, 121; Little Boy,
214–215; Radiation Sickness, 303;
Target Committee of the Manhattan
Project, 359–360
Hiroshima Maidens, 26
Hitler, Adolf, 165, 208, 313, 345,
347–348; Disagreement with Max
Planck, 285–286
Hoffman, Gerhard, 111
Hogness, Thorfin R., 185
Holdren, John, 122–123
Holifield, Chester Earl, 30, 165–167,
229, 333
Holloway, Marshall, 277
Hoover, G. C., Uranium Committee,
386
Hoover, Herbert, 352
Hoover, J. Edgar, 272
Hopkins, Harry, 65
The Hot Chamber, 231
House Committee on Un-American
Activities, 274
House of Commons, 138; Atomic En-
ergy bill of 1946, 29
Houtermans, Fritz, 19

Howe, C. D., 94

Hughes, Donald, 135

Human Resources and Research Office, 37

Hungary, 189

Hussein, Saddam, 179

Hutchins, Robert Maynard, 84

Hyde Park Aide—Memoir. *See* Roosevelt-Churchill Hyde Park Aide—Memoir

Hydro (Ship), 251, 361

Hydrogen Bomb, 11, 25, 46, 56–57, 100, 141, 144, 167–168, 187, 205, 210, 212, 221, 273–274, 288, 301, 305, 318, 320, 346, 352, 358, 374–375, 377, 388, 395; Berkeley Summer Colloquium of 1942, 44–45 Bravo Nuclear Test, 58–59; British Hydrogen bomb, 86–87; Chinese Hydrogen Bomb, 85, 217; Christmas Island Tests, 86–87; French Hydrogen Bomb, 137, 239; Greenhouse Nuclear Tests, 151–152; Hardtack Nuclear Tests, 159–160; History of, 167–168; Influence of Edward Teller, 167, 363–363; Joint Committee on Atomic Energy, 187; Lawrence Livermore National Laboratory, 206; Mike Nuclear Test, 235; Neutron bomb, 246–247; Nevada Test Site, 247–248; Novaya Zemlya, 252; Opposition of General Advisory Committee, 144; Opposition of J. Robert Oppenheimer, 373; Panda Committee, 277; Role of Andrei Sakharov, 318–319; Role of Lewis Strauss, 352; Sandia National Laboratories, 320; Savannah River Project, 322; Shot Harry, 330–331; Soviet Atomic Spying, 339; Soviet Nuclear Program, 342; Special Subcommittee on the Hydrogen Bomb, 344; Tritium, 374–375

I. G. Farben, 250

Idaho (Ship), 278

Idaho Falls (Idaho), 263

Idaho National Engineering Labora-

tory. *See* National Reactor Testing Station

Imperial Chemical Industries, 5, 88, 90

Imperial College of Science and Technology, 49, 282–283, 364

Imperial Japanese Army Air Force, Japanese Atomic Bomb Program, 183–184

Imperial Japanese Navy, Japanese Atomic Bomb Program, 183–184

Imperial Physical-Technical Institute (Germany), 347

Implosion, 9, 127, 140, 169–170, 225, 278, 283, 290, 395; Contributions of George Kistiakowsky, 198; History of, 169–170; Role of Seth Neddermeyer, 169; Support of James Tuck, 376–377

India, 72–73, 97, 109, 216, 257, 380, 408; Antarctic Treaty, 16; Indian Nuclear Program, 170–171; Pakistani Nuclear Program, 276–277; Indian Nuclear Testing, 177, 277

Indian Nuclear Program, 170–171; Indian Nuclear Testing, 171, 277; Role of Homi Bhabha, 46–47

Indonesia, Treaty of Bangkok, 370

Inertial Confinement Fusion. *See* Laser Fusion

Inertial Confinement Fusion (ICF) program, 203

Institute for Advanced Study (Princeton), 115, 273, 395

Institute for Experimental Physics (Germany), 311

Institute for Nuclear Physics (University of Chicago), 130, 209, 377

Institute for Theoretical Physics (Copenhagen), 51, 172, 202, 258, 389, 397; History of, 172; Influence of Niels Bohr, 172

Institute of Atomic Energy (Russia), 134

Institute of Chemical Physics (Russia), 200

Institute of Geophysics (University of California), 210

Institute of Nuclear Power Opera-

tions (INPO) (United States), 172–173

Institute of Organic Chemistry (Germany), 155

Institute of Physical Chemistry (Russia), 196

Institute of Physical Problems (Russia), 178, 194, 202

Institute of Theoretical Physics (Germany), 336

Integral fast reactor, 242

Intercontinental Ballistic Missile Defense System, 46

Intercontinental ballistic missiles, 16, 174, 351

Interim Committee on the Atomic Bomb, 173; Role of Harry Truman, 375

Intermediate Nuclear Forces Treaty, 174

Intermediate Power Breeder Reactor (IPBR), 59

International Atomic Development Authority, 212

International Atomic Energy Agency (IAEA), 174–175, 176, 177, 180, 216, 256, 303, 359, 371–373; Declared Nuclear Weapon States, 109; International Nuclear Information System, 176; Iraqi Nuclear Program 180; North Korean Nuclear Program, 251–252; Nuclear Nonproliferation Act, 256; Radiological Accidents, 303; Treaty of Bangkok, 371; Treaty of Pelindaba, 272; Treaty of Tlatelolco, 373; Zangger Committee, 408

International Commission on Radiological Protection (ICRP), 175

International Congress of Anthropological and Ethnological Sciences, 51

International Nuclear Fuel Cycle Evaluation Program, 176; Breeder Reactors, 59–60

International Nuclear Information System (INIS), 176

International Test Ban Conference, 46

International Thermonuclear Experimental Reactor (ITER), 177, 369

International X-Ray and Radium Commission, 175

Internationalize nuclear energy, 31, 46, 88, 118; Baruch Plan, 41–42; Lilienthal Group, 212–213; Washington Declaration, 398

The Interpretation of Radium, Frederick Soddy, 334; Influence on H. G. Wells, 405

Ioffe, Abraham Feodorovich, 177–178, 194, 200; Life of, 177–178; Role in Leningrad Physico-Technical Institute, 209; Soviet Nuclear Program, 341

Iodine-131, 31, 124, 198, 405

Iodine-135, 308

Iowa State University, 266

Iran-Iraq War, 359

Iranian government, 179

Iranian Nuclear Program, 179

Iranian revolution, 179

Iraq, 175, 261, 359

Iraqi government, 359

Iraqi Nuclear Program, 179–180; Tammuz Nuclear Center Attack, 359

Iraqi nuclear reactors, 359

Isfahan Nuclear Reactor, Iranian Nuclear Program, 179

Islamabad (Pakistan), 276

Isotope, 180, 334, 364, 374, 384, 390; Identification of isotopes, 24; Research of George Thomson, 364; Strontium-90, 353

Israel, 175, 179–180, 266, 277, 336, 350; Tammuz Nuclear Center Attack, 359; Vanunu Affair, 392

Israel Atomic Energy Commission (IAEC), 181; Israeli Nuclear Program, 180–182

Israeli Air Force, 175, 179; Tammuz Nuclear Center Attack, 359

Israeli Defense Force (IDF), 181, 392

Israeli Nuclear Arsenal, 181–182

Israeli Nuclear Program, 112, 180–182; Israel Atomic Energy Commission, 181; Plumbat Affair, 287; Six Day War, 181; Vanunu Affair, 392

Israeli Secret Service (Mossad), Plumbat Affair, 287; Vanunu Affair, 392
Italy, 8, 75, 123, 216, 259, 287
Iwo Jima, 121

Jackson, Henry M., 350
Japan, 60, 93, 216, 261, 312, 360; Antarctic Treaty, 16; Japanese Atomic Bomb Program, 183–184; Japanese Atomic Energy Commission, 184; Japanese Nuclear Program, 184–185
Japan Tokamak 60 (JT-60), 369
Japanese Atomic Bomb Program, 183–184
Japanese Atomic Energy Commission (JAEC), 184
Japanese Nuclear Program, 184–185; Japanese Atomic Energy Commission, 184; Breeder Reactors, 185
Jefferies, Jay, 185
Jefferies Report, 185–186
Jenkins, Bruce, 6
Jewish German physicists, 54, 114–116, 208, 213, 231–232; Hostility of Aryan Physics Movement, 21–22; Jewish Physics, 186; Opportunities at Clarendon Laboratory, 90
Jewish Physics, 186, 208, 347; Accusation directed against Albert Einstein, 115
Joachimsthal Mines, 186–187, 304; Radon Daughters, 304–305; Uranium, 384; Uranium Mining, 387–388
Joe One, 217
Johns Hopkins University, 389
Johnson, E. C., 228
Johnson, Ellis, 217
Johnson, Leonard E., 242
Johnson, Louis, 217; Special Subcommittee on the Hydrogen Bomb, 344
Johnson, Lyndon, 294, 350; Plowshare, 286; Weiss Report, 400
Johnston Atoll, 14
Joint Committee on Atomic Energy (JCAE), 27–28, 144, 150, 165–167, 187–188, 243, 274, 294, 329, 333, 344; Atomic Energy Commission, 30; Contributions of Chet Holifield, 166–167; Gofman-Tamplin Manifesto, 149; Knapp Report, 198; Private Ownership of Special Nuclear Materials Act, 293–294
Joint European Torus (JET), 369
Joint Institute for Nuclear Research (Russia), 188–189
Joint Task Force 7, 320
Joliet-Curie, Frederic, 96, 106, 156, 178, 189–191, 227, 291, 304, 340; Allier Affair, 7; Dismissal from Commissariat a l'Energie Atomique, 96, 191; Life of, 189–191
Joliet-Curie, Irene, 191–192
Jordan, Pascual, 296
Jornada del Muerto (Journey of Death), 374
Journal of Atomic Scientists, 315

K-8 (Submarine), Soviet Nuclear Submarine Accidents, 343
K-25 (Oak Ridge), 196, 267
K-219 (Submarine), 199; Soviet Nuclear Submarine Accidents, 343
K-431 (Submarine), Chazhma Bay Submarine Accident, 78–79; Soviet Nuclear Submarine Accidents, 343
Kaiser Wilhelm Institute for Chemistry, 193
Kaiser Wilhelm Institute for Physical Chemistry and Electrochemistry, 155, 193, 231
Kaiser Wilhelm Institute for Physics, 21, 109, 111, 163, 193, 245, 309, 341; Virus House, 393
Kaiser-Wilhelm-Society, 193–194; Nazi Atomic Bomb Program, 245; Nazi racial laws impact, 193
Kalahari Desert Nuclear Test Site, 336; South African Nuclear Program, 336–337
Kalkar (Germany), 148
Kamchatka Peninsula, 217
Kapitsa, Pyotr Leonidovich, 44, 75, 178, 194–195, 202, 258; Dispute with Lavrentii Beria, 44, 195; Life of,

194–195; Relationship with Ernest
Rutherford, 194; Special Committee
on the Atomic Bomb, 343–344
Karlsruhe Nuclear Research Center,
148
Kasli, 79–80
Kasli Disaster. See Chelyabinsk-40
Kazakhstan, 327, 348; Closing of
Semipalatinsk-21 Test Site, 327
Kazan (Russia), 340
Kehler, Randy, Nuclear Freeze Move-
ment, 254–255
Keirn, Donald, Air Nuclear Propul-
sion, 4
Keith, Dobie, 195, 225
Kellex Corporation, 140, 195–196;
Gaseous-diffusion reactor, 195–196;
Manhattan Project, 225; Role of Do-
bie Keith, 195
Kellogg (M. S.) Company, 195
Kelly, Orville, 37
Kemeny Commission, 196, 367;
Three Mile Island, 365–368
Kennan, George, 70
Kennedy, John, 89, 125, 166, 263,
326, 333
Kennedy, Joseph W., 289
Kennedy, Ted, 301
Kennedy administration, 166, 357
Kenney, Joe, 225
Kerr-McGee Nuclear Fuel Plant: Settle-
ment of Silkwood Case, 332; Silk-
wood Case, 331–332
Khan, Abdul Qadeer, 276
Khariton, Yuki, 196; Arzamas-16, 23
Khariton-Zeldovich Letters, 196–197;
Soviet Nuclear Program, 340–343
Kharkov Institute of Mechanical Engi-
neering (Soviet Union), 202
Khrushchev, Nikita, 319; Red Special-
ists, 309
King, Mackenzie, Washington Declara-
tion, 398
King Joseph Institute of Technology,
355
King Nuclear Test, 197
Kirchhoff, G. R., 285
Kiribati, Treaty of Rarotonga, 372

Kirkland Air Force Base, 319
Kistiakowsky, George Bogdan, 40,
56, 128, 197–198, 279; Implosion
crisis, 198; Life of 197–198
KIWI. See NERVA
KIWI-A, 246
KIWI-B4-E, 246
KIWI experimental reactor, 246
Klaproth, Martin, 186; Discovery of
uranium, 384; Discovery of zirco-
nium, 410
Klein, Felix, 335
Knapp, Harold, 198–199
Knapp Report, 198–199; Fallout, 124–
125
Knolls Atomic Power Laboratory
(KAPL), 145; General Electric, 145
Kokura (Japan), 49, 240, 360
Konopinski, Emil, 44
Korean War, 13, 88, 158–159, 346,
376
Korff, Serge, 209
Kowarski, Lew, 71, 78, 156, 190, 228,
409
Kremer, Simon Davidovich, 140
Kremlev, 23
Krishnan, K. S., 170
Krypton-90, 353, 397
Kuboyama, Aikichi, 222
Kundt, August, 311
Kurchatov, Igor, 79, 178, 199–200,
269, 327; Laboratory No. 2, 201;
Life of, 199–200; Red Specialists,
309; Soviet Nuclear Program, 341–
343; Obninsk Atomic Energy
Station, 269; Semipalatinsk-21 Test
Site, 327; Special Committee on the
Atomic Bomb, 343–344; Tokamak
Fusion Reactors, 369; Working with
Lavrentii Beria, 200
Kurchatov Institute of Atomic Energy
(Soviet Union). See Laboratory No. 2
Kursk RBMK Reactors, 342
Kvasnikov, Leonid, Soviet Atomic Spy-
ing, 338
Kyoto (Japan), 360
Kyshtym (Russia), 79

LaBine, Gilbert A. (Lucky), 117
Labor Party (Great Britain), 71
Laboratory No. 2 (Soviet Union), 201
Lacoste, Pierre, *Rainbow Warrior*, 306
Lagoona Beach, Michigan, 131
Lake Tinnsjo, 361
Lamb, Willis, 113
Lanchow Gaseous Diffusion Uranium
    Enrichment Plant, 85
Landau, Lev Davidovich, 202; Life of,
    202; Work on Soviet Atomic Bomb,
    202
Langevin, Paul, 189, 191
Lansdale, John, Alsos Mission, 8
Laos, Treaty of Bangkok, 371
Lapp, Ralph, 222
Large Ship Reactor (LSR), 202–203
Laser Fusion, 203–204; Fusion, 141;
    Stockpile Stewardship Program, 349
Las Vegas Bombing and Gunnery
    Range, 247
Latin America, Treaty of Tlatelolco,
    373–374
Lauritsen, Charles C., Target Commit-
    tee, 359
Lawrence, Ernest Orlando, 10–11,
    57, 173, 204–205, 206, 221, 224,
    229, 270, 352, 387; Cyclotron, 105–
    106; Life of, 204–205
**Lawrence Livermore National Labo-
    ratory**, 10, 14, 57, 59, 80, 167, 203,
    205–206, 221, 247; Influence of Ed-
    ward Teller, 363; Laser Fusion, 203–
    204; Neutron bomb, 247; Plumbbob
    Nuclear Tests, 287–288; Under-
    ground Nuclear Testing, 380;
    Upshot-Knothole Nuclear Tests, 382
Lawrence Radiation Laboratory, 148,
    206–207, 230, 360
Lebanon, 97
Lebedev Physical Institute (Moscow),
    178, 319
Lee, Tsung Dao, 403
Leeds University, 57
Lehman Corporation, 116
Leipzig Laboratory, 207
Leipzig Reactor, 207, 245
LeMay, Curtis, 121, 163

Lenard, Philip Edward Anton, 186,
    207–208; Alliance with Johannes
    Stark, 347; Life of, 207–208
Lenin (Ship), 95
Leningrad Institute of Chemical Physics
    (Russia), 200, 209
Leningrad Institute of Electrical Physics
    (Russia), 209
**Leningrad Physico-Technical Insti-
    tute** (LFTI), 134, 194, 200, 209;
    Contributions of Ioffe, 178; Purge of
    1936, 209
Leningrad RBMK Reactor, 342
Leukemia, 303; Atomic Veterans, 37;
    Weiss Report, 399–400
Leventhal, Paul, Nuclear Control Insti-
    tute, 254
Levine, Saul, 307
Lewis, G. N., 324
Lewis Committee, 220
Libby, Willard Frank, 209–210; Life
    of, 209–210; Radiocarbon dating,
    209
Libya, 97
Light Water Reactors (LWR), 118–
    119, 136, 141, 147, 181, 184–185,
    210, 262; Contributions of Hyman
    Rickover, 310; Role of Zirconium,
    410; Three Mile Island, 366; West-
    inghouse Electric Corporation, 400
Lilienthal, David Eli, 153–154, 166,
    187, 210–212; Atomic Energy Com-
    mission, 29–30; Hostility of Groves,
    153–154; Life of, 210–212;
    Lilienthal Group, 211–212; Relation-
    ship with Joint Committee on
    Atomic Energy, 187; Special Sub-
    committee on the Hydrogen Bomb,
    344
Lilienthal Group, 212–213; Acheson
    Board, 2; Baruch Plan, 41
Lillienthal Report, 2
Limited Test Ban Treaty. *See* Nuclear
    Test Ban Treaty
Lindemann, Frederick Alexander
    (Lord Cherwell), 18, 88, 213–214,
    376; Clarendon Laboratory, 90–91;
    Feud with Henry Tizard, 214;

Friendship with Winston Churchill, 88; Life of, 213–214

Lindemann electrometer, 213

Lindemann glass, 213

Lindemann melting-point, 213

Linear accelerator, 10, 160

Liquid-drop theory of the atomic nucleus, Niels Bohr's contribution, 50

Liquid Metal Fast Breeder Reactor. *See* Breeder Reactors

Liquid metal reactors, 60

Liquid oxygen, 194

Lithium-6-deutride, 58, 168

Little Boy (Mark-1), 25, 49, 214–215, 390; History of, 214–215; Nagasaki, 290; Role of Deak Parsons, 280

Livermore (California), 205

Livingston, M. S., 105

Llewelin, J. J., 94

Lloyds of London, 293

Lockheed Martin, 319

London, 138, 160, 216

London Suppliers Guidelines, 216; Nuclear Nonproliferation Act, 256; Relationship with Zangger Committee, 408

Long Beach (Ship), 203

Long Island, New York, 63

Long Range Detection Project, 216–217

Long-range intermediate nuclear forces, 174

Lop Nur Nuclear Weapons Test Site, 171, 217–218; Chinese Nuclear Program, 85–86

Lorentz, Henrik Anton, 218; Life of, 218

Lorentz Transformations, 218

Los Alamos (New Mexico), 225

Los Alamos Atomic Accidents, 218–219; Radiological Accidents, 303

Los Alamos National Laboratory, 9, 14, 40, 46, 55–59, 61, 119, 130, 132, 139–141, 167–169, 196–198, 220–222, 271–273, 278–283, 301, 303, 319, 362–363, 376–377, 390, 395; Accidents at Los Alamos, 218–219; Fusion Research, 377; History of, 220–222; Mike Nuclear Test, 235; Nevada Test Site, 247–248; Oppenheimer as director, 272–273; Panda Committee, 277; Rivalry with Lawrence Livermore National Laboratory, 306; Role of George Kistiakowsky, 198; Role of Deak Parsons, 278–279; Sandstone Nuclear Tests, 320; Soviet Atomic Spying, 338–340; Teapot Nuclear Tests, 360–361; Tumbler-Snapper Nuclear Tests, 377–378; Underground Nuclear Testing, 380

Los Alamos Primer, 225

Los Alamos Scientific Laboratory. *See* Los Alamos National Laboratory

Los Angeles, California, 165

Low-level nuclear waste, 263

Loyola University, 274

Lucky Dragon (Ship), 222; Bravo Nuclear Test, 59; Fallout, 125

Lukinskii, P. I., 178

Lum, James, 268

Luxembourg, 123

M-39 (Hydrogen bomb), 168

MacHarg, Ian, 96

Machon 2, Vanunu Affair, 392

Mackenzie, Chalmers Jack, Atomic Energy of Canada Limited, 33

Madagascar, Treaty of Pelindaba, 372

Magnetic fusion reactors, 122, 177, 206

Magnetic separation process, 390

Maiman, Theodore, Laser Fusion, 203

Makhnev, V. A., Special Committee on the Atomic Bomb, 343–344

Malaysia, Treaty of Bangkok, 370

Malden Island, 86–87

Malenkov, Georgi, Special Committee on the Atomic Bomb, 343–344

Malyshev, Vyacheslav: Red Specialists, 309; Soviet Nuclear Program, 342

Manchester New College, 107

Manhattan Project, 1, 13, 24, 35–37, 46, 52, 56, 72, 76, 85, 100, 106, 117, 135, 139, 150–151, 157, 162, 167, 173, 197–198, 214–215, 223–

226, 244, 278, 282–283, 297, 328,
346, 356, 378, 389, 395, 403, 409;
Argonne National Laboratory, 20;
End of Manhattan Project, 212; Fat
Man, 127; Groves in command, 153;
Handford Plant, 157–159; Hiroshima, 163–164; History of, 223–
226; Influence of MAUD Report,
227–228; Implosion, 169–170;
Interim Committee on the Atomic
Bomb, 173; Jefferies Report, 185–
186; Kellex Corporation, 195–196;
Little Boy, 214–215; Los Alamos
National Laboratory, 220–222; Met-
allurgical Laboratory of Chicago, 234–
235; Military Policy Committee, 235–
236; Nagasaki, 240; Nuclear Secrecy,
29; Origin of, 66–67; Plutonium,
289–290; Role of Klaus Fuchs, 140;
Role of George Kistiakowsky, 198;
Role of J. Robert Oppenheimer, 272–
273; Smyth Report, 332–333; Soviet
Atomic Spying, 338–340; Target
Committee, 359–360; Top Policy
Committee, 370

MANIAC (Mathematical Analyzer, Numerical Integrator and Computer),
120; Role of von Neumann, 395

Maralinga Proving Grounds, 62, 86,
226–227

March 4th Movement. *See* Union of
Concerned Scientists

Mark-1 (Atomic bomb) (Little Boy),
214–215

Mark-1 reactor, 92

Mark-2 reactor, 92

Mark-3 (Atomic bomb), 102, 127

Mark-4 (Atomic bomb), 25, 127, 320

Mark-5 (Atomic bomb), 25

Mark-6 (Atomic bomb), 25

Mark-7 (Atomic bomb), 25

Mark-12 (Atomic bomb), 25

Mark-14 (Hydrogen bomb), 168

Mark-15 (Hydrogen bomb), 168

Mark-17 (Hydrogen bomb), 168

Mark-28 (Hydrogen bomb), 168

Mark-53 (Hydrogen bomb), 168

Mark A reactor, 92

Mark B reactor, 92

Mars exploration, 246

Marsden, Ernest, 316

Marshall, George C., 370

Marshall, Lenore, Committee for Nuclear Responsibility, 96

Marshall Plan, 298

Martin, Kingsley, 70

Martin Marietta Corporation, 269

Mason & Hanger Corporation, 278

Mason College. *See* University of Birmingham

Mass spectrometry, 40, 364

Massachusetts Institute of Technology
(MIT), 9, 40, 56, 63, 65, 132, 150,
153, 198, 240, 300, 307, 381, 391

Massue, Huet, 33

Matsu (Ship), 95

Matsutaro, Shoriki, Japanese Nuclear
Program, 184

Matthias, Franklin T., 158

MAUD Committee, 40, 139, 156,
214, 227–228, 241, 370; History of,
227–228; Role of George Thomson,
364; Soviet knowledge about MAUD
Committee, 341

MAUD Report, 66, 88, 228, 338, 341

Mauritius, 97

Max Planck Institute, 55, 194, 394

Maximum credible accident theory, Reactor Safeguard Committee, 308

Maxwell, James Clerk, 73, 218, 285,
297, 300

May, Alan Nunn, 339; Soviet Atomic
Spying, 338–340

May, Andrew J., 228

May-Johnson Bill, 27, 67, 127, 153,
166, 228–229, 259; History of, 228–
229

Mayak Chemical Combine, 79

McCarthy, Joe, 274

McCarthy, Richard Max, 96

McCloy, John J., Acheson Board, 2

McCluskey, Harold, 159

McCone, John A., 30

McCullough, Rogers, 3

McGill University, 155, 316, 325

McLean, Donald, 95, 338

McMahon, Brien, 27, 167, 187, 229–230, 243, 344; Life of, 229–230
McMahon Act. *See* Atomic Energy Act of 1946
McMahon Bill, 153; Atomic Energy Act of 1946, 27
McMillan, Edwin Mattison, 1, 230; Life of, 230; Work with Glenn Seaborg, 325
McMurdo Air Base, 16
McNaughton, A.G.L., 33
McTaggart, David, 152
Medical Research Center, 64
Medvedev, A. Zhores, 80, 161
Medvedev, Grigori, 230–231
Medvedev Case, 230–231
Meitner, Lise, 133, 138–139, 155–156, 193, 231–233, 355, 398; Life of, 231–233
Melba, 337; South African Nuclear Program, 336–337
Memphis (Tennessee), 352
Menzies, Robert Gordon, 237
Meshappen Nuclear Power Plant, 233
Meson, 15
Messmer, Pierre, 233–234
Messmer Plan, 18, 233–234
Metallurgical Economy, 313
Metallurgical Laboratory (Met Lab), 20, 35, 66, 99, 135, 225, 234–235, 289, 325, 403, 409; Chicago Pile One, 84–85; Jefferies Report, 185–186; History of, 234–235; Opposition to compartmentalization, 259; Role of Arthur Compton, 99; Soviet Atomic Spying, 239
Metropolitan Edison (United States), Three Mile Island, 365–366
Metropolitan Vickers, 91
Mexico, Treaty of Tlatelolco, 373–374
Mi Vida Mine, Uranium Mining, 388
Michelson, A. A., 98, 236
Mike Nuclear Test, 121, 235; Panda Committee, 277
Mikhailov, Viktor, 23
Military Advisory Committee (Manhattan Project), 223–224

Military Liaison Committee (Atomic Energy Commission), 20, 154, 280
Military Policy Committee, 235–236, 270
Millikan, Robert Andrews, 14, 236–237; Life of, 236–237
Ministerial Atomic Energy Committee, 146
Ministry for Atomic Questions (Germany), 147
Mirzayanov, Vil S., 249
MIT Radiation Laboratory, 40, 300
Molotov, Vyacheslav, 43, 200, 341
Mond Laboratory, 194
Monsanto Chemical Company, 185, 212, 268
Monte Bello Atomic Test Site (Australia), 6, 18, 38, 62, 86, 227, 237–238; History of, 237–238; Role of Penney, 283
Moon, P. B., 227
Moorehead, Agnes, 100
Morgan, A. E., 211
Morgan, Sir Frederick, 62, 292
Morgan, Thomas Alfred, 274
Morrison, Herbert, 146
Moscow, 16, 31, 263, 319
Moscow State University, 202, 318
Moscow Treaty. *See* Nuclear Test Ban Treaty
Moseley, Henry G. J., 238; Life of, 238
Mott, Nevill, 35, 140
Muller, Walther, 144
Muller, Wilhelm, 347
Mulliken, Robert S., 185
Multiple independently targetable reentry vehicles (MIRVs), 351
Murdoch, Rupert, Vanunu Affair, 392
Murmansk (Russia), 249
Murphree, Eger V., 270
Mururoa Atoll, 238–239; Test Site, 137, 238–239; French Nuclear Program, 136–137
Mutual assured destruction (MAD), 16; Cold War, 94
MV Greenpeace (Ship), 239

MX Missile System, 321
Myanmar, Treaty of Bangkok, 371

Nabarlek (Australia), Uranium Mining, 388
Nader, Ralph, *Critical Mass*, 101
Nagaski, 13, 24, 173, 226, 240, 280, 283, 290, 303, 360, 375; Atomic Bomb Casualty Commission, 25–26; Attack on, 240; *Bock's Car*, 49–50; Fat Man, 127; Radiation Sickness, 303; Target Committee of the Manhattan Project, 359–360
Nahal Soreq Reactor, 112; Israeli Nuclear Program, 181
Naka (Japan), 177
Namu Island, 58
National Academy of Science, 26, 241; Recommendation on nuclear waste, 406
National Advisory Committee for Aeronautics (NACA), 65
National Aeronautics and Space Administration (NASA), 246
National Association of Atomic Veterans, 37
National Carbon Company, 151; Production of graphite, 151
National Council for the Abolition of Nuclear Weapons Tests (NCANWT), 70
National Defense Research Committee (NDRC), 66, 99, 198, 236, 240–241, 244; History of, 240–241; Franklin Delano Roosevelt, 241; Leadership of Vannevar Bush, 65–67; Office of Scientic Research and Development, 66; Role of James Conant, 99–100; Uranium Committee, 386–387
National Environmental Policy Act, 69; *Calvert Cliffs* Decision, 69–70
National Ignition Facility. *See* Laser Fusion
National Ignition Facility (NIF), 203
National Organization of Women (NOW), 332

National Physics Laboratory (India), 179
National Reactor Testing Station (NRTS), 20, 60, 63, 92, 119, 169, 242, 243; Brockett Report, 63; History of, 242; Price-Anderson Bill, 293; Reactor Safeguard Committee, 308; SL-1 Reactor Accident, 332; Relationship with Argonne National Laboratory, 20; Testing of Nautilus reactor, 20; Work with Westinghouse Electric Corporation, 400
National Recovery Administration (NRA), 116
National Research Council, 26, 259
National Security Council, 205; Special Subcommittee on the Hydrogen Bomb, 344
Nature, 76, 129, 139, 199, 233, 284, 398
Nauru, Treaty of Rarotonga, 372
Nautilus (Submarine), 20, 92, 243–244; History of, 243–244; Role of Hyman Rickover, 310
Navajo Indians, 305; Radon Daughters, 304–305
Naval Ordnance Laboratory, 349
Naval Postgraduate School, 310
Naval Research Laboratory, 1, 226, 244; History of, 244
Nazarbayev, Nursultan A., 327
Nazi Atomic Bomb Program, 55, 108, 148, 162–163, 165, 186–187, 194, 207, 245, 313, 393–394; Army Ordinance Bureau, 21; Bohr-Heisenber Conversation, September 1941, 53; Contributions of Werner Heisenberg, 163; Heavy Water, 161–162; History of, 245; Joachimsthal Mines, 186–187; Leipzig Reactor, 207; Norsk-Hydro Plant, 250–251; Reich's Research Council, 309–310; Role of Manfred von Ardenne, 18–19; Role of Walther Bothe, 55; Role of Walther Gerlach, 147; Role of Albert Speer, 345; Telemark Raid, 361–362; Uranium Club, 385; Virus

House, 393; Warning about German research, 109

Nazi Party, 111, 347

Nazi racial laws, 54, 68, 135, 138, 193, 232, 285, 313, 336, 355; Impact on Kaiser-Wilhelm-Society, 193; Jewish Physics, 186; Response of Clarendon Laboratory, 90

NDP (Nuclear Power Demonstration), 34

Neddermeyer, Seth, 16, 198, 279, 376; Implosion crisis, 169–179

Ne'eman, Yuval, 181

Negev Desert, 181

Nehru, Jawaharlal, 47, 170

Nelson, Curtis, 322

Neptunium, 1, 263, 289, 325; Discovery of, 230

Nerst, Walther, 213, 236

Nerst-Lindemann theory of specific head, 213

NERVA (Nuclear Engine for Rocket Vehicle Applications), 222, 246

Nesvizh (Belarus), 303

Netherlands, 75, 123, 216

Neutron Bomb, 246–247, 250, 281, 321; Plumbbob Nuclear Tests, 289; Tactical Nuclear Weapons, 357

Neutrons, 9, 45, 76–77, 124, 133–134, 161, 180, 190, 219, 225, 308, 316, 355, 375, 384, 393, 403; Breeder Reactors, 59–60; Contributions of Enrico Fermi, 130; Discovery by James Chadwick, 76; Experiments at Los Alamos National Laboratory, 289–290; Metallurgical Laboratory, 289

Nevada (State), 264; Legislation against Yucca Mountain nuclear waste site, 407

Nevada Test Site, 6, 10, 13–14, 100, 124, 197, 206, 221, 247–249, 320–321, 323, 380; Buster-Jangle Nuclear Tests, 67–68; Hardtack Nuclear Tests, 159–160; History of, 247–249; Plumbob Nuclear Tests, 287–289; Ranger Nuclear Tests, 306; Shot Easy, 330; Shot Harry, 330–331;

Teapot Nuclear Tests, 360–361; Tumbler-Snapper Nuclear Tests, 377; Upshot-Knothole Nuclear Tests, 382–382

New Hampshire Site Evaluation Committee, Seabrook Nuclear Power Plant, 326–327

New Jersey Bell Telephone Company, 212

New London, 65

New Mexico (State), Uranium Mining, 388

New Scientist, 161

New York City, 34, 97, 174, 299

New York Times, 321

New Zealand, 305; Antarctic Treaty, 16; Sinking of Rainbow Warrior, 306; Treaty of Rarotonga, 372

Nichols, Edward L., 284

Nichols, Kenneth, 328

Nickson, James J., 36, 135

Nieh, Jung-chen, 86

Nier, Alfred Otto Carl, U-235, 390

Nigata (Japan), 360

Nikitin, Aleksandr K., 249

Nikitin Affair, 249

Nishina, Yoshio, 183

Niue, Treaty of Rarotonga, 372

Nixon, Richard, 70, 127

Nixon administration, 246, 286

No. 91, 163

Nobel Institute (Sweden), 232

Noble Prize, 10, 15, 42, 46, 49–50, 57–58, 74, 76, 91, 99, 103, 108–109, 113, 130, 134, 148, 156, 162, 185, 189, 191, 195, 202, 204, 208, 218, 230, 237, 281, 284–285, 295, 300, 311, 315–317, 319, 324–325, 334, 346, 358, 365, 389, 403, 405

No-First-Use, 250

Non-Proliferation Treaty. See Nuclear Non-Proliferation Treaty

Non-Proliferation Treaty Exporters Committee. See Zangger Committee

Nonviolent civil disobedience, 89

Normal Superior School, 102

Norman Bridge Laboratory (Caltech), 237

Norsk-Hydro Plant, 250–251, 361; Allier Affair, 7; Heavy Water, 162; Nazi Atomic Bomb Program, 245; Telemark Raid, 361–362

North Atlantic Treaty Organization (NATO), 41, 150, 247, 255; First Strike Doctrine, 133; No-First-Use, 250

North Korean Nuclear Program, 251–252; International Atomic Energy Agency, 251

Norway, 162, 181, 190, 361; Antarctic Treaty, 16; Chernobyl Nuclear Power Station Accident, 82

Norwegian Special Forces, Attack on Norsk-Hydro Plant, 251

Novaya Zemlya (Russia), 252; Soviet Union Program, 342

NRU Reactor (National Research Universal), 72, 78, 253, 409

NRX Reactor (National Research X-perimental), 72, 78, 253–254, 409

Nuclear accidents. See Chernobyl Nuclear Power Station Accident; Three Mile Island

Nuclear arms race, 93

Nuclear Black Market Task Force, 261

Nuclear Control Institute (United States), 254; Role of Paul Leventhal, 254

Nuclear delivery systems, 174

Nuclear Energy Propulsion, Aircraft. See Aircraft Nuclear Propulsion

Nuclear Free and Independent Pacific Movement (Hawaii), 152

Nuclear Freeze Movement, 254–255, 321

Nuclear Free Zones (NFZ), 255; Antarctic Treaty, 15–16; Treaty of Bangkok, 370; Treaty of Pelindaba, 371; Treaty of Rarotonga, 372–373

Nuclear incidents, 161, 218–219, 254; Brown's Ferry Nuclear Power Station Incident, 64; Chalk River Reactor accident, 78; Chelybinsk-40 nuclear waste disaster, 70–80; Chernobyl Nuclear Power Station Accident, 81–84; Fermi I accident, 131; Hanford accident, 159; Leipzig Reactor, 207; Los Alamos Atomic Accidents, 218–219; NRX nuclear incident, 254; Psychological Impact of Nuclear Accidents, 294–295; Radiological Accidents, 303; SL-1 Reactor Accident, 332; Three Mile Island, 365–367; Waltz Mills Reactor Accident, 397; Windscale reactor accident, 404–405

Nuclear neurosis, 294

Nuclear Nonproliferation Act, 256; International Atomic Energy Act, 176; London Supplies Groups, 216

Nuclear Non-Proliferation Treaty (NPT), 179, 181, 216, 251–252, 256–257, 276, 337, 372; Refusal of India to sign, 171; Refusal of Pakistan to sign, 276; Zangger Committee, 407

Nuclear Power Division. See Code 390–590

Nuclear-Powered Ships. See Commercial Nuclear Ships

Nuclear-propelled bomber, 4

Nuclear propulsion, 92, 310

Nuclear Reactions Laboratory of the Joint Institute for Nuclear Research (Dubna), 134

Nuclear Reactor. See Breeder Reactors; Chicago Pile One; Light Water Reactors

Nuclear reactor meltdown, 83; Chernobyl Nuclear Power Station Accident, 82–83; Three Mile Island, 367

Nuclear Regulatory Commission (NRC), 31, 64, 110, 119, 196, 257–258, 307, 327; History of 257–258; Reaction to Brown's Ferry Nuclear Power Station Incident, 64; Recommendations of Kemeny Commission, 196, 367; Rejection of Rasmussen Report, 307; Seabrook Nuclear Power Plant, 327; Three Mile Island, 366–367

Nuclear Research Group (Reich's Research Council), 147; Role of Walther Gerlach, 147

Nuclear Research Laboratory (Dubna), 85

Nuclear rocket program, 246

Nuclear Secrecy, 258–260, 356; Turner Paper, 378

Nuclear ship propulsion, 1, 183, 202, 210, 242, 244; Code 390–590, 92; Contributions of Hyman Rickover, 310; Large Ship Reactor, 202–203; Naval Research Laboratory, 244

Nuclear Smuggling, 260–261; Importance of Russia, 260–261

Nuclear submarines, 210, 242, 310; Code 390–590, 92; Contributions of Hyman Rickover, 310; National Reactor Testing Station, 242; Naval Research Laboratory, 244

Nuclear Suppliers Group (NSG), 216, 261

Nuclear Terrorism, 254, 262; Plutonium-239, 289–290; Uranium-235, 390

Nuclear Test Ban Treaty, 13, 262–263, 321; Support from SANE, 321; Underground Nuclear Testing, 380

Nuclear waste, 79, 290, 343, 370; ESECOM, 122–123; Plutonium waste problem, 290; Savannah River Project, 322; Yucca Mountain Waste Disposal Site, 406–407

Nuclear waste accident, Chelyabinsk-40, 79–80

Nuclear Waste Disposal, 79, 263–264; Yucca Mountain Waste Disposal Site, 406–407

Nuclear Waste Policy Act (NWPA), 264, 406

Nuclear weapon design, 80; Arzamas-16, Chelyabinsk-70, 80; Lawrence Livermore National Laboratory, 205–206; Sandia National Laboratories, 319–320

Nuclear weapon states, 109

Nuclear Weapons Accidents, 265

Nuclear weapons industry, 11

Nuclear Winter, 265–266

Nukey Poo, 16

NUMEC (Nuclear Materials and Equipment Corporation), 266; Israeli Nuclear Program, 180–181

Oak Ridge (Tennessee), 267

Oak Ridge Gaseous Diffusion Plant, 110

Oak Ridge National Laboratory, 1, 157, 181, 195–196, 205, 207, 225–226, 229, 244, 267–269, 310, 375, 389–390, 410; Air Nuclear Propulsion, 4; History of, 267–269; Opposition to May-Johnson Bill, 229; Role of Leslie Groves, 267; Role of Eugene Wigner, 403

Oak Ridge School of Reactor Technology, 92

Oberlin College, 236

Obninsk (Russia), 342

Obninsk Atomic Energy Station, 269–270

Obninsk Physico-Power Institute (Russia), 269–270

Occhialini, Guiseppe, 15

Odessa Polytechnical Institute, 358

Office of Atomic Testing. *See* AFOAT-1 (Air Force Office of Atomic Testing)

Office of Chief of Naval Operations, 236

Office of Environmental Management (EM), 110

Office of Materials Stabilization and Facility Management (EM-60), 110

Office of Naval Research, 216

Office of Scientific Research and Development (OSRD), 15, 66, 99–100, 198, 236, 241, 270, 272, 278; History of, 270; Role of Vannevar Bush, 66

O'Gorman, Mervyn, 213

Ohnesorge, Wilhelm, 19, 165

Oil embargo, 118

Okinawa, 50, 310

Oklo Mine (Africa), 271

Oliphant, Marcus, 75, 139

Ontario Hydro, 34

Oolan (Belgium), 287, 328

Operation Alert, Civil Defense, 88–89

Operation Crossroads. *See* Crossroads Nuclear Tests
Operation Dominic, 87
Operation Gasbuggy, Plowshare, 286
Operation Grapple, 87
Operation Hurricane. *See* Monte Bello Atomic Test Site
Operation Plowshare. *See* Plowshare
Operation Ranger, 247
Operation Rio Blanco, Plowshare, 286
Operation Sandstone, 121
Operation Upshot-Knothole, 330
The Operators, 231
Oppenheimer, Frank, 273
Oppenheimer, J. Robert, 1, 8, 10, 46, 55–57, 130, 144, 153 169, 173, 205, 210, 217, 229, 259, 271–273, 279–280, 395; Thermal diffusion process, 1; Berkeley Summer Colloquium of 1942, 44; Hydrogen Bomb, 167; Friendship of Deak Parsons, 280; Lilienthal Group, 212–213; Los Alamos National Laboratory, 220–222; Manhattan Project, 224–226; Life of, 271–273; Friendship of Rabi, 301; Feud with Strauss, 352; Target Committee, 359–360; Relationship with Teller, 362–363
Oppenheimer Affair, 274–275
Oppenheimer's security clearance, 10, 46, 130, 210, 260, 273, 301; Feud with Lewis Strauss, 353; History of, 274–275; Lawrence's testimony against Oppenheimer, 205
Oranienburg, 9
Ordinance Division (Los Alamos), 198, 278–279
Oregon State University, 281
Organic-Cooled Experimental Reactor (OCDRE): Whiteshell Nuclear Research Establishment, 401
Organic Test Reactor (OTR), Whiteshell Nuclear Research Establishment, 401
Organization of African Unity, 371
Organization of Petroleum Exporting Counries (OPEC), 354

Otto Hahn (Ship), 95
Overseas Press Club, 321
Owens, Wayne, 301
Oxford University, 5, 90, 160, 213–214, 238, 282, 324, 334, 376, 391
Oyster Creek Nuclear Plant, 379
Ozdzhonikidze, Sergei, 178
Ozersk, 79
Ozone layer, 265

Paducah Gasesous Diffusion Plant, 110
Pakistan, 109, 171, 276–277, 380
Pakistani Atomic Energy Commission (PAEC), 276–277
Pakistani Nuclear Program, 276–277, 380
Panda Committee, 235, 277
Pantex, 278
Papua New Guinea, Treaty of Rarotonga, 372
Paris (France), 385
Parsons, William S. (Deak), 56, 121, 198, 278–280; Target Committee, 359
Particle accelerator, 64, 376; Cyclotron, 105–106; Role of Van de Graaff, 391
Pash, Boris T., 150; Alsos Mission, 8
Pasko, Grigory, 249
Pasteur Institute, 304
Patent Compensation Board, Atomic Energy Commission, 30
Pauli, Wolfgang, 129, 335
Pauling, Linus Carl, 96, 118, 280–281; Influence on Andre Sakharov, 281
Pavlov, Nikolai I., 201
Peace Action, 321
Pearl Harbor, 133, 241
Peenemunde (Germany), 313
Peierls, Rudolf Ernst, 45, 140, 281–282, 335; Frisch-Peierls Memorandum, 139; Life of, 281–282
Pelindaba (South Africa), 336, 372
Pelindaba Nuclear Complex, 336
Penney, William, 6, 61, 217, 282–283; Apex Committee, 18; Armament Research Department, 20; Life of, 282–283; Target Committee, 359–360

Pennsylvania (State), 367
Pentagon, 153
Pentomic organization, 357; Tactical Nuclear Weapons, 357
PEON Commission (Production d'Electricité d'Origine Nucléaire), 283–284
People's Commissar of Internal Affairs (NKVD), 43, 79, 195, 273, 308, 344, 346; Soviet Atomic Spying, 338–339; Soviet Nuclear Program, 341
Pereira, Fernando, 306
Perrin, Jean Baptiste, 191
Perrin, Michael, 292
Persian Gulf War, 180
Pervukhin, Mikhail G., 309; Soviet Nuclear Program, 341–342; Special Committee on the Atomic Bomb, 343–344
Petrograd Polytechical Institute, 209
Petrzhak, Konstantin S., 134
Philadelphia Navy Yard, 1
Philippines, 370
Phillips Petroleum, 242
Phoebus-1A, 246
Phoebus-2A, 246
Phoebus reactor series, 246
Phoebus Reactors. *See* NERVA
Phyllis Cormack (Ship), 152
Physical and Chemical Research Institute (Japan), 183
The Physical Review, 284; Bohr's analysis of U-235, 390; Role in promoting secrecy, 284; Turner Paper, 378, 390
Physico-Technical Institute (Kharkov, Soviet Union), 134, 178
Physics Institute of the Academy of Science (FIAN), 318, 342, 358
Physics of Explosives Committee (Great Britain), 282
Physikalische Technische Reichsanstalt, 144
Pick, Vernon, Uranium Mining, 388
Pickering, 72
Piezoelectricity, 103
Pike, Summer, 29–30

Pinawa (Canada), Whiteshell Nuclear Research Establishment, 401
Pitchblende, 186, 304
Pittsburgh (Pennsylvania), 329
Placzek, George, 139
Planck, Max Karl Ernst Ludwig, 55, 162, 208, 231, 236, 285–286, 323, 355; Life of, 285–286; Development of Quantum Theory, 297; Role in development of Quantum Mechanics, 296; Support by Max von Laue, 394
Planck's constant, 285, 297
Plowshare, 2, 286–287, 294
Plumbat Affair, 181, 287; History of, 287; Israeli Nuclear Program, 181
Plumbbob Nuclear Tests, 287–289
Plutonium, 289–291. *See also* Plutonium-239; Plutonium-240
Plutonium-239, 19, 24, 44, 47, 59, 79–80, 112, 127, 133, 145, 207, 210, 223, 225–227, 230, 234, 242, 251, 254–255, 260, 263, 267, 277, 280, 320, 331, 374, 378, 385; Discovery by Glenn Seaborg, 324–325; Hanford Plant production, 157–159; History of, 289–290; Hydrogen Bomb, 167–168; Implosion crisis, 169–170; Savannah River Project, 322; Uranium Committee, 387; Windscale accident, 404–405
Plutonium-240, Implosion, 169
Plym (Ship), 237
Poincare, Henri, 42, 236
Poison gas, 155
Poisoned Power: The Case Against Nuclear Power Plants, 149; Gofman-Tamplin Manifesto, 149
Pokhran (India), 171
Pokrovsko-Streshnevo (Soviet Union), 201
Poland, 189, 216, 255; Antarctic Treaty, 16
Policy Committee (Great Britain), 228
Politburo, 309
Polonium, 305, Discovery of, 103; Radon Daughters, 304–305
Polytechnic Institute (Switzerland), 311

Pomerance, Herbert, Contributions on zirconium, 410
Pons, B. Stanley, Cold Fusion, 93
Pontecorvo, Bruno, 291, 338; Soviet Atomic Spying, 338–340
Port Hope Refinery (Canada), 328; Eldorado Mine, 117
Port Radium, 117
Portal, Charles, Lord, 62, 283, 291–292
Portsmouth Gaseous Diffusion Plant, 110
Portugal, Western Supplies Group, 400
Positron, 49, 113, 316; Discovery by Carl Anderson, 14
Potsdam Meeting, 341, 374
Powell, Dick, 100
Power Demonstraion Reactor Program, 292
Power Reactor Development Compnay, 292
Powers, Francis Gary, 80
Poynting, J. H., 23
Preminin, Sergei, 199
Pressured water reactors, 92, 143, 203, 210, 366
Pressurized Water Reactors. *See* Light Water Reactors
Price, Melvin, 229, 293
Price-Anderson Bill, 138, 166, 293, 307; Influence of WASH-740 Report, 398–399; Opposition of Chet Holifield, 166; Rasmussen Report, 306–307
Priestly, J. B., 70
Princeton University, 98, 115, 120, 132, 197–198, 230, 333, 369, 378, 391, 395, 402
Private Ownership of Special Nuclear Materials Act, 293–294
Professional Community of University Physicists, 347
Project Alberta, 280
Project Carryall, Plowshare, 286
Project Chariot, 286, 294; Idea of Edward Teller, 286
Project Nutmeg, 247
Project Rulison, 286

Project Sherwood, 206; Role of James Tuck, 377
Project TR, 40
The Prospectus on Nucleonics, 185; Jeffries Report, 185
Protons, 10, 45, 316; Experiments of John Cockcroft, 91
Psychological Impact of Nuclear Accidents, 294–295
Public Health Department, Weiss Report, 399–400
Public Service Company of New Hampshire, 326–327
Puerto Rico, 110
Pugwash (Canada), 295
Pugwash Movement, 195, 295; Role of Joseph Rotblat, 315
Pupin, Michael, 236
Purnell, William, 223, 236

Quantum electrodynamics (QED), 132
Quantum Mechanics, 98, 113, 202, 236, 271, 281–282, 284, 296, 323, 335; Alternative theory by Erwin Schrödinger, 324; Contributions of Werner Heisenberg, 162; Opposition of Johannes Stark, 347; Solvay Conference of 1927, 335
Quantum Theory, 22, 108, 208, 282, 296–297, 324; Development by Max Planck, 285; Opposition of Johannes Stark, 346–347
Quebec Conference, 71, 94, 154, 297–298; Combined Policy Committee, 94–95; Uranium supplies, 384–385
Quincke, Georg, 207

Rabbit Lake (Canada), Uranium Mining, 388
Rabi, Isidor Isaac, 40, 225, 299–301, 335; Founding of CERN, 75; Life of, 299–301; Role in Brookhaven National Laboratory, 63
Rabi field, 299
Rabinowitch, Eugene, 36, 65, 135
Rad (Radiation Absorbed Dose), 381
Radar project, 40, 223, 300, 376

Radiation Effects Research Foundation (RERF), 26

Radiation dose standards, 149, 281, 302–303, 348–349; Committee on Biological Effects of Ionizing Radiation, 96–97; Gofman-Tamplin Manifesto, 148–149

Radiation exposure, 97, 246, 349; Chernobyl Nuclear Power Station Accident, 82–83; Committee on Biological Effects of Ionizing Radiation, 96–97; International Commission on Radiological Protection, 175

Radiation Exposure Compensation Act, 301–302; Uranium Mining, 387–388

Radiation gamma wave, 25

Radiation Laboratory (MIT), 105, 150

Radiation Sickness, 124, 164, 219, 221, 231, 302–303, 305, 393; Chazhma Bay Submarine Accident, 78; Chelyabinsk-40, 79; Chernobyl Nuclear Power Station Accident, 82–84; Fallout, 124; Los Alamos Atomic Accidents, 218–219; Medvedev Case, 230–231

Radiation Therapy Facility, 64

Radioactive cleanup, 159; Eniwetok Atoll Test Site, 121

Radioactive fallout. *See* Fallout

Radioactive waste, 242, 249; Hanford Plant, 154; Nuclear Waste Disposal, 263–264; Savannah River Project, 322

Radioactivity, 155, 189, 222, 231, 289, 316, 334, 381; Discovery by Antoine Becquerel, 42; Fallout, 124

Radioactivity Laboratory, 103

Radiocarbon dating technique, 209; Contribution of Willard Libby, 209

Radiological Accidents, 303; Los Alamos Atomic Accidents, 218

Radiothorium, Discovery by Otto Hahn, 155

Radium, 117, 186, 302, 328; Discovery of, 103

Radium Institute (France), 103, 189, 191–192, 292, 304; History of, 304

Radium Institute (Soviet Union), 106, 178, 200

Radon Daughters, 304–305; Uranium Mining, 387–388

Radon-222 gas, 304, 406

RAF (Royal Air Force) Physical Laboratory, 213

Rainbow Warrior (Ship), 239, 305–306; Greenpeace International, 152

Rajasthan (India), 171

Ramsey, Norman, 359

Ramsey, Sir William, 155, 334

Rancho Seco Nuclear Power Plant, 85

Rand Corporation, 127

Rand Institute, 246

Ranger Nuclear Tests, 306

Rapacki, Adam, Nuclear Free Zones, 255

Rappoport, Iakov, 79

Rasmussen, Norman C., Rasmussen Report, 307

Rasmussen Report, 306–307; Nuclear Regulatory Commission, 257–258; Role of Norman Rasmussen, 307

Ray, Dixy Lee, 31

Ray, Maud, 228

Rayleigh, Lord, 74

RBMK (Reactors High-Boiling Channel Type) Reactors, 43, 81, 84, 270; Soviet Nuclear Program, 342

RBMK-1000, 81–84

Reactor Poisoning, 307–308

Reagan, Ronald, 174, 177, 247, 258, 321, 351, 354

Reagan administration, 46, 174, 177, 255–256, 323, 348, 357, 375

The Reactor Union, 231

Reactor Safeguard Committee, 308; Advisory Committee on Reactor Safeguards, 3

Rebound, 380

Red Specialists, 309; Soviet Nuclear Program, 341–342

Redstone rockets, 160

Reggan Test Site (Algeria), 137, 238

Reich's Research Council, 21, 147, 245, 309–310, 345; History of, 309–

310; Nazi Atomic Bomb Program, 245

Rem (Roentgen equivalent man), 382

Remington-Rand, 154

Republican Party (United States), 212

Review of Modern Physics, Articles by Hans Bethe, 45

Reykjavik Summit, 174

Rholphton, 72

Richards, T. W., 99

Richardson, Owen W., 98

Richland (Washington), 263

Rickover, Hyman George, 78, 166, 203, 210, 242, 278, 280, 310–311, 410; Code 390–590, 92; Life of, 310; *Nautilus*, 243–244; National Reactor Testing Station, 242; Role in Shippingport Nuclear Power Plant, 329; Relationship with Westinghouse Electric Corporation, 400; Support of Chet Holifield, 166

Riehl, Nikolaus, 341

Rifle (Colorado), 14, 286

Rolsdag (Sweden), 354–355

Risley, 61

RKO Studios, 100

Robb, Roger, 274

Roberts, Richard B., 386

Rockefeller Foundation, 106, 138, 204, 282

Rockwell International, 264

Roentgen, Wilhelm Conrad, 177–178, 258, 311, 381

Rolling Stone, 85

Rongelap Island; 59, *Rainbow Warrior*, 305–306

Roosevelt, Franklin Delano, 52, 66–67, 88, 94, 214, 223, 259, 270, 297, 312, 352, 403; Einstein-Szilard Letter, 115–117, 355; Military Policy Committee, 235–236; National Defense Research Committee, 241; Role in Atomic Energy, 312; Roosevelt-Churchill Hyde Park Aide—Memoir, 312–313; Top Policy Group, 370; Uranium Committee, 386–387

Roosevelt-Churchill Hyde Park Aide—Memoir, 312–313

Rosbaud, Paul, 313; Relationship with Uranium Club, 385–386

Rosenberg, Ethel, 170, 314–315, 338

Rosenberg, Julius, 170, 314–315, 338–339

Rosenberg Case, 314–315; Soviet Atomic Spying, 238–340

Rotblat, Joseph, 35, 315–316; Life of, 315–316; Pugwash Movement, 295

Rover Program, 246

Royal Air Force, 5, 291, 364

Royal Engineers, 238

Royal Institution (London), 57–58, 107

Royal Irish Academy (Ireland), 324

Royal Naval College, 49

Royal Navy, Career of Patrick Blackett, 49

Royal Swedish Academy of Sciences, 265

Rubidium-90, 353

Rucker, Nelson, Oak Ridge National Laboratory, 269

Ruhleben, 76

Rumania, 189

Rusinow, Lev, 134; Soviet Nuclear Program, 340

Russell, Bertrand, 70, 295

Russell-Einstein Manifesto, 295; Origin of Pugwash Movement, 295

Russia, 60, 109, 177, 260–261; Chernobyl Nuclear Power Station Accident, 82–84; Comprehensive Test Ban Treaty, 97–98; International Theronuclear Experimental Reactor, 177; Nakitin Affair, 249; START Treaty, 348; Treaty of Pelindaba, 372

Russian Air Force, 82

Russian Association of Physicists, 178

Russian Federal Security Service (FSB), 249

Russian government, 249

The Russian Northern Fleet, Sources of Radioactive Contamination, 249

Russian Revolution, 178, 197

Ruthenium, 82

Rutherford, Ernest, 50, 58, 76, 90–91, 143–144, 155, 194, 238, 316–

317, 355, 365; Cavendish Laboratory, 74–75; Life of, 316–317; Work
with Soddy, 334
Rye, New Hampshire, 89

S-1. *See* Atomic Committee
S-50, 268
Saaesum-137, 79
Sabath, Adolph, 310
Sachs, Alexander, 312, 386; Role in
Einstein-Szilard Letter, 116–117, 312
Safeguard Missile System, 16
Safeguards Division of the European
Atomic Energy Community, 123
Sagan, Karl, Nuclear Winter, 266
Saint Olaf College, 204
Sakharov, Andrei, 318–319; Life of,
318–319, 369; Influence of Linus
Pauling, 281; Soviet Nuclear Pro-
gram, 342; Relationship with Tamm,
358
Salt Lake City (Utah), 330
San Diego (California), 177
San Salvador (El Salvador), 303
Sandia National Laboratories, 14,
319–320; Underground Nuclear
Testing, 380
Sandstone Nuclear Tests, 320
Sandys, Duncan, 18
SANE, 321
Sarov, 23, 80
Sarov. *See* Arzamas-16
Savannah (Ship), 95
Savannah River Project, 264, 278,
322–323; History of, 322; Nuclear
Waste Disposal, 263–264; Tritium
production, 375
Schaffer, William G., 323
Schaffer Report, 323
Schawlow, Arthur, Laser Fusion, 203
Scheersberg (Ship), 287
Schlesinger, James R., 30
School of Industrial Physics and Chem-
istry, 103, 189
Schrödinger, Erwin, 113, 281, 296,
323–324, 394; Life of, 323–324
Schrödinger equation, 45, 113
Schumann, Erich, 21, 245

Schuster, Arthur, 143
Science, 2
Science Based Stockpile Stewardship
Program, 380; American Nuclear
Testing Program, 13–14
Scientific-Technical Council (Special
Committee on the Atomic Bomb),
344
Scully, V. W., 33
Sea of Japan, 249
Seaborg, Glenn Theodore, 30, 36,
207, 230, 289, 324–326; Advisory
Committee on Reactor Safeguards, 3;
Life of, 324–326; Weiss Report, 400
Seaborg Report, 326
Seabrook Nuclear Power Plant, 89,
326–327; Clamshell Alliance, 89–90
Seawolf (Submarine), 92
Second Physical Institute (University of
Gottingen), 135
Seeman, Colonel, Target Committee,
359–360
Segre, Emilio, 169
Seismological Institute (Soviet Union),
201
Semenov, Nikolai, 178
Semipalatinsk-21 Test Site, 327; So-
viet Nuclear Program, 342
Sengier, Edgar, Shinkolobwe Uranium
Mine, 328
Serber, Robert, 44, 225
Severnyj (Russia), 252
Seversk (Russia), 370
Shah of Iran, 179
Shapiro, Zalman, 266
Sharp, Richard, 328
Shchelkin, Kirill I., 80
Shevchnko Fast Breeder Reactor, 342
Shima Hospital, 164
Shinkolobwe Uranium Mine, 117,
287, 328; History of, 328; Uranium,
384; Uranium Mining, 387–388
Shippingport Nuclear Power Plant,
11, 329–330; Decommissioning of,
110; History of, 329–330
Shivwits Indians, 100
Shock wave, 25
Short-range nuclear forces, 174

Shot Able, 67, 102, 306, 377
Shot Annie, 382
Shot Apple-1, 361
Shot Apple-2, 361
Shot Badger, 382
Shot Baker, 67, 102, 306
Shot Baker-2, 306
Shot Bell, 360
Shot Boltzmann, 287
Shot Charleston, 289
Shot Charlie, 67, 102, 377–378
Shot Climax, 383
Shot Coulomb B., 288–289
Shot Diablo, 288
Shot Dixie, 382
Shot Dog, 67, 151, 378
Shot Doppler, 288
Shot Easy, 67, 151, 306, 330, 378
Shot Encore, 383
Shot ESS, 361
Shot Fir, 160
Shot Fizeau, 289
Shot Fox, 306, 378
Shot Franklin, 287
Shot Franklin Prime, 288
Shot Galileo, 288
Shot George, 151, 378
Shot Grable, 383
Shot HA, 361
Shot Harry, 247, 330–331, 383
Shot Hood, 288
Shot Hornet, 360
Shot How, 378
Shot Humboldt, 160
Shot Item, 151
Shot John, 288
Shot Kepler, 288
Shot Laplace, 289
Shot Lassen, 287
Shot MET, 361
Shot Moth, 360
Shot Nancy, 382
Shot Newton, 289
Shot Nutmeg, 160
Shot Oak, 160
Shot Orange, 160
Shot Owens, 288
Shot Post, 361

Shot Priscilla, 287–288
Shot Rainier, 289
Shot Ray, 382
Shot Ruth, 382
Shot Santa Fe, 160
Shot Shasta, 288
Shot Simon, 382
Shot Smokey, 288
Shot Socorro, 160
Shot Stokes, 288
Shot Sugar, 67
Shot Sycamore, 160
Shot Tesla, 360
Shot Uncle, 67
Shot Wash, 360
Shot Wasp, 361
Shot Wheeler, 288
Shot Whitney, 289
Shot Wilson, 287
Shot X-Ray, 320
Shot Yoke, 320
Shot Yucca, 160
Shot Zebra, 320
Shot Zucchini, 361
Siberia (Ship), 95
Siberian Chemical Combine, 370;
    Tomsk-7, 370
Siegbaln, Manne, 232
Sieman, 147
Sierra Club, 137, 331
Sievert (Unit), 382
Silkwood, Karen, 331–332
Silkwood Case, 331–332
Silverplate, 368
Simpson, John, 36
Sinclair, Upton, 166
Sinelnikov, Kirill, 200
Singapoore, Treaty of Bangkok, 371
Six Day War, 181
SL-1 Reactor Accident (Stationary
    Low-Power), 332
Slotin, Louis, 303; Los Alamos Atomic
    Accidents, 218–219
Slow neutrons, 130
Smith, Gerard Coad, Western Supplies
    Group, 400
Smyth, Henry DeWolf, 339; Smyth Re-
    port, 332–333

Smyth Report, 332–333; History of, 332–333; Reactor Poisoning, 308

SNAP (Special Nuclear Auxiliary Power), 333

Snow, C. P., 316

Snow's Canyon, 100

Societe Generale Minerais (SGM) (Belgium), 287

Soddy, Frederick, 316, 334, 405; Life of, 334

Sodium-beryllium reactor, 92

Solomon Islands, Treaty of Rarotonga, 372

Solvay, Ernest, 334; Solvay Congresses, 334–335

Solvay Conference, 1927, 334–335

Solvay Congresses, 218, 258, 334–335

Somalia, Treaty of Pelindaba, 372

Sommerfeld, Arnold, 45, 162, 281–282, 296, 335–336, 362, 394; Life of, 335–336; Feud with Johannes Stark, 346

Soreq (Israel), 303

South Africa, 179, 371–372; Antarctic Treaty, 16; South African Nuclear Program, 336–337; Uranium supplies, 384; Uranium Cartel, 385; Uranium Mining, 388; Western Supplies Group, 400

South African government, 337

South African Nuclear Program, 336–337

South America, 337–338

South American Nuclear Program, 337–338; Argentina, 337–338; Brazil, 337–338; Treaty of Tlatelolco, 373–374

Southeast Asia Nuclear Weapon Free Zone. *See* Treaty of Bangkok

South Pacific Forum, 372

South Pacific Nuclear Free Zone Treaty. *See* Treaty of Rarotonga

Soviet Air Force, 134

Soviet Atomic Bomb Program. *See* Soviet Nuclear Program

Soviet Atomic Spying, 338–340; Beria's role, 43–44; Bruno Ponte-

corvo, 291; Klaus Fuchs, 140–141; Rosenberg Case, 314–315

Soviet cyclotron, 106

Soviet hydrogen bomb, 318

Soviet navy, 249, 343

Soviet Nuclear Program 43, 195, 200, 202, 308–309, 340–343, 358; Arzamas-16, 23; Beloyarsk Nuclear Reactor Fire, 43; Beria as head, 43–44; Chelyabinsk-40, 79–80; Chazhma Bay Submarine Accident, 78; Chernobyl Nuclear Power Station Accident, 81–84; History of, 340–343; International Thermonuclear Experimental Reactor, 177; Khariton-Zeldovich Letter, 196–197; Laboratory No. 2, 201; Medvedev Case, 230–231; Novaya Zemlya, 252; Reactor Poisoning, 308; Role of Igor Kurchatov, 200; Role of Andre Sakharov, 318–319; Role of Joseph Stalin, 345–346; Role of Igor Tamm, 358; Semipalalinsk-21 Test Site, 327; Soviet Atomic Spying, 338–340; Tokamak Fusion Reactor, 369; Tomsk-7, 370

Soviet Nuclear Submarine Accidents, 343; *K-219* sinking, 199; *K-431* disaster, 78–79

Soviet Nuclear Tests, 216–217; Novaya Zemlya, 252; Semipalalinsk-21 Test Site, 327

Soviet Radar Ocean Reconnaissance Satellite (RORSAT), 100–101

Soviet scientists, 340–341

Soviet Union, 4, 19, 33, 52, 85, 88, 101, 133, 145, 157, 161, 167, 170, 174–176, 187, 191, 194, 202, 213, 217, 250–251, 255, 258–261, 274, 291, 305, 312, 318, 340, 345–346, 348, 353, 359, 371, 380, 388; Antarctic Treaty, 16; Antiballistic Missile Treaty, 16; Atomic Spies, 140–141; Baruch Plan, 41–42; Cold War, 93–94; Intermediate Nuclear Forces Treaty, 174; Novaya Zemlya, 252; Nuclear Non-Proliferation Treaty, 256–257; Nuclear Test Ban Treaty,

262–263; Soviet Atomic Spying, 238–240; Soviet Nuclear Program, 340–341; Soviet Nuclear Submarine Accidents, 343; Special Committee on the Atomic Bomb, 343–344; Strategic Arms Limitation Treaty, 350; Strategic Arms Limitation Treaty II, 351; Tactical Nuclear Weapons, 357–358; Tokamak Fusion Reactor, 369; Uranium supplies, 384

Spain, 305

Spanish Civil War, 190

Special Committee for the Problem of Uranium (Russia), 340; Soviet Nuclear Program, 340

Special Committee on the Atomic Bomb (Russia), 43–44, 341, 343–344; History of, 343–344; Role of Beria, 43–44; Soviet Nuclear Program, 341

Special Subcommittee on the Hydrogen Bomb (United States), 344; Hydrogen Bomb, 167–168

Special Weapons Office (U.S. Navy), 280

Spectrograph, Work of Alston, 24

Speer, Albert, 165, 345; Role in Nazi Atomic Bomb Program, 345

Sperry Corporation, 274

Spies. See Rosenberg Case; Soviet Atomic Spying

Spock, Benjamin, SANE, 321

Springdale, Arkansas, 165

Sproul, Robert, 206

SS (Schutzstaffel), 22, 162, 347

St. George (Utah), 100, 198, 330

St. Louis, 198

St. Petersburg (Russia), 249

St. Petersburg Polytechnical Institute, 178

St. Petersburg Technological Institute, 177

Stagg Football Stadium, 84, 234

Stalin, Joseph, 2, 23, 43, 93, 195, 200, 345–346, 374; Flerov's letter, 134; Relationship wih Beria, 43–44; Role in Soviet Nuclear Program, 345–346; Soviet Atomic Spying, 339; Soviet Nuclear Program, 340–341; Special Committee on the Atomic Bomb, 343–344; Role of Stalin, 345–346

Standard Oil Development Company of New Jersey, 270

Stanford University, 56

Stark, Johannes, 162, 208, 336, 346–347; Life of, 346–347; Feud with von Laue, 394

Stark effect, 346

START, 348

START II, 348

Staten Island Warehouse, 328

Stearns, Joyce, 135; Target Committee, 359

Steen, Charles, Uranium Mining, 388

Stern, Otto, 138–139, 146, 299

Stern-Gerlach experiment, 300

Sternglass, Ernest J., 149, 233, 348–349; Life of, 348–349

Stevenson, Adlai, 353

Stewart, Alice, 349

Stewart, Balfour, 365

Stimson, Henry, 8, 94, 136, 173, 236, 312, 375; Decision on Kyoto, 360; Top Policy Group, 370

Stockholm (Sweden), 232

Stockholm International Peace Research Institute (SIPRI), 171, 181, 255, 277

Stockpile Stewardship Program, 349

Stone, Robert S., 185

Strassmann, Fritz, 139, 155–155, 384

Strategic Arms Limitation Treaty I, 350–351

Strategic Arms Limitation Treaty II, 348, 351

Strategic Arms Reduction Treaty. See START

Strauss, Lewis L., 30, 116, 167, 216, 273, 275, 351–353; Relationship with David Lilienthal, 212; Opposition to Oppenheimer, 274–275; Power Demonstration Reactor Program, 292; Life of, 351–353

Strontium-90, 31, 79, 198, 248, 330, 353; History of, 353

Styer, Wilhelm, 223, 236
Subcritical nuclear tests, 14
Suez Canal Crisis, 136
Sunday Times, 112; Vanunu Affair, 392
Sung, Kim Il, 252
Superconducting Super Collider
(SSC), 354
Superphenix, Breeder Reactor, 60
Supply Division of the European
Atomic Energy Community, 123
Supporters of Silkwood, 332
Supreme Council of the National Econ-
omy (Soviet Union), 178
Surplus Facilities Management Program
(SFMP), 110
Survival Under Atomic Attack, Civil
Defense, 89
Sverdlov Communist University, 358
Swann, W. F., 204
Sweden, 75, 82, 139, 155, 212, 233,
354–355
Swedish Nuclear Power Program,
354–355
Sweeney, Charles, 49
Swiss Patent Office, 114
Switzerland, 75, 216
Synchrocyclotron, 230
Synthetic ammonia production, 250
Syria, 97
Szilard, Leo, 35–36, 115, 118, 151,
258–259, 284, 355–356, 378, 386,
394, 409; Einstein-Szilard Letter,
116–117; Franck Report, 135–139;
Life of, 355–356; Friendship with
Strauss, 352; Nuclear Secrecy, 258–
259; Wells' interpretation, 405

Tactical Nuclear Weapons (TNWs),
13, 67–68, 250, 277, 357–358;
Atomic Veterans, 37; Buster-Jangle
Nuclear Tests, 67; History of, 357–
358; Neutron Bomb, 246–247, 360,
377–378, 382–393; Plumbbob
Nuclear Tests, 288; Soviet Nuclear
Program, 342; Upshot-Knothole Nu-
clear Tests, 382
Taiwan, 72
Takeuchi, Tadeshi, 183

Tamm, Igor Evgenevich, 318–319,
358–359, 369; Life of, 358; Soviet
Nuclear Program, 342
Tammiku (Estonia), 303
Tammuz Nuclear Center Attack, 179,
359; Crisis in International Atomic
Energy Agency over attack, 175;
Iraqi Nuclear Program, 179–180
Tamplin, Arthur, 148
Tanzania, 97
Tarapur (India), 170
**Target Committee of the Manhattan
Project**, 283, 359–360
Task Force on Compensation for
Radiation-Related Illness, 323; Schaf-
fer Report, 323
Tata Institute of Fundamental Re-
search, 47, 170
Taylor, Ted, 378; King Nuclear Test,
197
T-Division (Los Alamos), 377
Teapot Nuclear Tests, 360–361
Technical Secretariat of the Compre-
hensive Test Ban Agreement, 98
Technical University of Charlottenberg
(Berlin), 144, 313, 355
Tel Aviv (Israel), 181
Telemark Raid, 361–362
Teller, Edward, 10–11, 44, 46, 57,
116, 141, 144, 202, 205, 206, 221,
277, 281, 288, 301, 308, 362–363,
377, 386, 394, 397; Bravo Nuclear
Test, 58; Greenhouse Nuclear Tests,
151–152; Hydrogen Bomb, 167–168;
Opposition toward Oppenheimer,
274–275; Operation Chariot, 294;
Life of, 362–363; Friendship with
Strauss, 352; Reactor Safeguard
Committee, 308; Upshot-Knothole
Nuclear Tests, 382
Tempiute (Nevada), 330
Tennessee Valley Authority (TVA), 29,
64, 211–212, 267; Dixon-Yates Deal,
352
Terrorism. *See* Nuclear Terrorism
Test Ban Treaty. *See* Comprehensive
Test Ban Treaty
Texas (Ship), 278

Teyler, Laboratory (Netherlands), 218
Thailand, Treaty of Bangkok, 371
Thatcher government, Renewal of British nuclear program, 62
Theoretical Megaton Group. *See* Panda Committee
Theoretical Physics Division (Los Alamos), 46, 132, 225, 362
Theory of Relativity, 22, 108, 113, 115, 162, 186, 208, 284, 335, 394; Opposition of Johannes Stark, 346
Theory of Special Relativity, 218
Thermal diffusion process, 244, 268, 390; Development by Philip Abelson, 1; Manhattan Project, 226; Naval Research Laboratory, 244; Oak Ridge National Laboratory, 268; S-50 Plant, 268
Thermal Oxide Reprocessing Plant (THORP), 138
Thermal radiation, 25
Thin Man. *See* Little Boy
Third World, 38
Thomas, Charles A., 185, 212
Thomas, Trevor, 321
Thomson, George Paget, 227, 364; Life of, 364
Thomson, Joseph John, 24, 50, 57, 74, 208, 316, 364–365, 404; Life of, 364–365
Thomson, Meldrim, 327
Thomson atom, 365
Thomson Committee. *See* MAUD Committee
Thorium-230, 15, 154–155, 298, 406
Thornburgh, Richard, 366
Three Mile Island, 11, 17, 34, 85, 89, 119, 145, 161, 172, 254, 258, 294–295, 307, 365–368; History of, 365–368; Kemeny Commission, 196; Psychological Impact of Nuclear Accidents, 294–295
Thresher (Submarine), 268
Thyroid cancer, Weiss Report, 399–400
Tibbets, Paul Warfield, 121, 368; Life of, 368
Tinian, 280, 368
Titterton, Sir Ernest, 38

Tizard, Sir Henry, 7, 213, 227, 292, 328
Tizard Committee, 139, 214
Tokai-Mura (Japan), 184
Tokamak Fusion Reactor, 177, 369; History of, 369
Tokamak Fusion Test Reactor, (TFTR), 369
Tokamak reactors, 177
Tolman, Richard, Target Committee, 359
Tomsk (Russia), 370
Tomsk-7, 370
Tonga, Treaty of Rarotonga, 372
Top Policy Group, 370
Townes, Charles, Laser Fusion, 203
Townsend, John, 391
Tracerlab, Long Range Detection Project, 217
Transuranic wastes, 263
Transuranium elements, 324–325; Research on by Glenn Seaborg, 324–325; Turner Paper, 378
Treason for Spying. *See* Rosenberg Case; Soviet Atomic Spying
Treaty of Bangkok, 370–371; Nuclear Free Zones, 255
Treaty of Pelindaba, 371–372; Nuclear Free Zones, 255
Treaty of Rarotonga, 372–373; Nuclear Free Zones, 255
Treaty of Tlatelolco, 255, 338, 373–374; Nuclear Free Zones, 255
Trigger List, Zangger Committee, 408
Trinity, 13, 24–25, 170, 226, 247, 374; History of, 374
Trinity College, 74–75
Tritium, 58, 167, 322, 374–375
Trump, John, 391
The Truth About Chernobyl, Medvedev Case, 231
Trivelpiece, Alvin, 177
Trombay (India), 47, 170
Truman, Harry, 10–11, 36, 136, 154, 173, 187, 212, 229, 247, 319, 322, 352, 375–376; Atomic Energy Act of 1946, 27; Contribution to passage of Atomic Energy Act of 1946, 27; Hy-

drogen Bomb, 167; Potsdam Meeting, 341; Role in Atomic Bomb, 375–376; Role in atomic energy, 375–376; Special Subcommittee on the Hydrogen Bomb, 344; Washington Declaration, 398

Truman administration, 2

Trunk, Anne, 196

TRW Environmental Safety System, Yucca Mountain nuclear waste site, 407

Tube Alloys. *See* Uranium

Tuck, James L., Life of, 376–377

Tufts University, 65

Tumbler-Snapper Nuclear Tests, 377–378

Turkey, 287

Turkmeniya (Russia), 318

Turner, Louis, 378

Turner Paper, 378

Turnkey Nuclear Energy Plants, 11, 145, 378–379; General Electric, 145; History of, 378–379; Westinghouse Electric Corporation, 400–401

Tuvalu, Treaty of Rarotonga, 372

Tuve, Merle, 244

Tuwaitha (Iraq), 179, 359

U-2 plane, 80

U-235, 1, 19, 21, 24, 44, 50, 77, 86, 101, 106, 133, 139, 190, 195, 197, 205, 215, 225–227, 244–245, 262, 267, 282, 339, 390; Little Boy, 214–215

Uhlenbeck George, 150

Ujelang, 121

Ukraine, 161, 348

Ukrainian government, 295

Ukrainian Health Ministry, 161

Ulam, Stanislaw, 46, 120, 277, 377; Bravo Nuclear Test, 58; Work with Teller, 363

Ultra, 226

Ultraviolet rays, 265

Underground Nuclear Testing, 217, 263, 380–381; History of, 380–381; Nuclear Test Ban Treaty, 262–263;

Science Based Stockpile Stewardship, 380

UNESCO, 75

Unilateral nuclear disarmament, 70

Union Carbide, 151, 196, 268–269; Oak Ridge National Laboratory, 268–269

Union Miniere du Haute Katanga, Shinkolobwe Uranium Mine, 328

Union of Concerned Scientists, 381; Complaints about emergency core cooling systems, 119

United Auto Workers, 3

United Kingdom, 27, 28–29, 75, 82, 94, 109, 138 145, 154 161, 163 175, 190, 216, 238, 256–257, 261, 298, 328, 380, 398; Anglo-American Declaration of Trust, 15; Antarctic Treaty, 16; Christmas Island Tests, 86–87; Combined Policy Committee, 94–95; Groves-Anderson Memorandum, 154; Monte Bello Atomic Test Site, 237–238; Nuclear Test Ban Treaty, 262–263; Roosevelt-Churchill Hyde Park Aide—Memoir, 312–313; Soviet Atomic Spying, 339; Treaty of Pelindaba, 372; Treaty of Rarotonga, 373; Uranium Cartel, 385; Western Supplies Group, 400

United Nations, 2, 37, 97, 164, 174, 240, 298; Atomic Energy Commission United Nations, 31; Comprehensive Test Ban Treaty, 97–98

United Nations General Assembly, 97, 371

United States, 94, 109, 145–146, 154, 161, 163, 171, 174–177, 181, 185, 197, 205, 216, 238, 246, 261–262, 276, 291, 298, 328, 333, 336, 340, 346, 354–355, 359, 380–381, 398; American Nuclear Program, 10–12; American Nuclear Testing, 13–14; Anglo-American Declaration of Trusts, 15; Antarctic Treaty, 16; Antiballistic Missile Treaty, 16; Cold War, 93–94; Comprehensive Test Ban Treaty, 97–98; Eldorado Mine, 117; Groves-Anderson Memorandum, 154; Hardtack Nuclear Tests, 159–160; Intermediate Nuclear Forces Treaty, 174; International Thermonuclear Experimental Reactor, 177; Kellex Corporation, 195–196; Knapp Report, 198–199; MAUD Report, 227–228; Nuclear Non-Proliferation Treaty, 256–257; Nuclear Test Ban Treaty, 262–263; Plumbbob Nuclear Tests, 287–288; Roosevelt-Churchill Hyde Park Aide—Memoir, 312–313; Soviet Atomic Spying, 339; START Treaty, 348; Strategic Arms Limitation Treaty I, 350; Strategic Arms Limitation Treaty II, 351; Strontium-90 scare, 353; Tactical Nuclear Weapon, 357; Treaty of Pelindaba, 372; Treaty of Rarotonga, 373; Uranium supplies, 384; Western Supplies Group, 400

Units of Measurement of Radiation Standards, 381–382

University of Aachen, 325, 346
University of Aberdeen, 334, 364
University of Adelaide, 57–58
University of Alabama, 391
University of Baku, 202
University of Berlin, 53, 55, 115, 134, 197, 213, 231, 281, 285, 323, 355, 394–395
University of Bern, 115
University of Birmingham, 23, 282
University of Bonn, 208
University of Breslau, 53, 208, 323
University of Bristol, 45, 113, 140
University of Budapest, 395
University of California at Berkeley, 1, 44, 56, 105, 122, 148–149, 200, 204–205, 209–210, 230, 270, 272, 289, 324–325, 363, 389; Lawrence Radiation laboratory, 206–207; Los Alamos National Laboratory, 221
University of California at La Jolla, 389
University of California at Los Angeles, 324
University of Chicago, 20, 84, 98, 130, 135, 204, 220, 223, 236, 356, 363,

377, 389–390; Metallurgical Laboratory, 234–235
University of Claustal, 335
University of Copenhagen, 50, 162, 172
University of Edinburgh, 54, 140, 358
University of Erlangen, 143
University of Florence, 128
University of Florida, 368
University of Frankfurt, 45, 53–54, 146, 394
University of Giessen, 55, 311
University of Glasgow, 334
University of Gottingen, 52–54, 128, 135, 162, 271, 335, 346, 362, 394–395, 402
University of Graz, 324
University of Greifswald, 346
University of Grenoble, 18
University of Halle, 111
University of Hamburg, 138, 395
University of Hanover, 346
University of Heidelberg, 53, 55, 134, 207–208
University of Illinois at Urbana-Champaign, 220
University of Insbruck, 111
University of Jena, 323
University of Karlsruhe, 362
University of Kharkov, 202
University of Kiel, 140, 144, 208, 285
University of Kiev, 197
University of Konigsberg, 335
University of Leipzig, 140, 162, 281, 309, 362
University of Leningrad, 202
University of Leyden, 128, 218, 272
University of Liverpool, 76, 291, 315
University of London, 49, 57, 219, 315, 334
University of Manchester, 45, 49–50, 58, 74, 76, 91, 143, 238, 316, 365, 404
University of Manitoba, 219
University of Maryland, 177
University of Michigan, 129, 148, 150
University of Minnesota, 98, 204
University of Montana, 389

Univeristy of Montpellier, 392
University of Moscow, 358
University of Munich, 45, 108, 143, 146, 162, 177, 189, 192, 281, 285, 296, 311, 335, 346–347, 362, 394
University of New Zealand, 316
University of North Carolina, 274
University of Paris–Sorbonne, 102–103, 108, 304, 391
University of Pennsylvania, 120
University of Pisa, 128, 291
University of Pittsburgh, 233
University of Pittsburgh Medical School, 349
University of Prague, 115
University of Rome, 128–129, 291
University of South Dakota, 204
University of Southhampton, 93
University of Strasbourg, 8, 311, 394
University of Stuttgart, 45, 323
University of Tubingen, 45, 144, 146
University of Utah, 93
University of Vienna, 138, 231, 323–324
University of Warsaw, 315
University of Washington (Seattle), 153, 282
University of Washington (St. Louis), 98, 235
University of Wisconsin, 221, 282, 402
University of Wurzburg, 311, 346
University of Zurich, 53, 109, 115, 272, 311, 323, 394
Upshot-Knothole Nuclear Tests, 124, 382–384; History of, 382–384
Uranium, 5, 10, 15, 59–60, 71–73, 78, 84, 86, 103, 111–112, 116, 130, 136, 154, 160, 163, 179 181, 183–184, 186–187, 197, 205, 210, 215, 222–223, 225, 230, 233–235, 242, 245, 251, 263, 266, 271, 287, 297–298, 324, 336–337, 354, 364, 384–385; Accidents, 218–219; Chain Reaction, 77; Eldorado Mine, 117; History of, 384–385; Leipzig Reactor, 207; Oklo Mine, 271; Shinkolobwe Uranium Mine, 328; Soviet Nuclear program, 341; Uranium

Club, 384–386; Uranium Committee, 386–387; Uranium Mining, 387–388; Western Supplies Group, 400; Westinghouse as supplier, 401; Yellowcake, 406
Uranium-233, 133
Uranium-238, 76, 168, 195, 267; Making of Plutonium, 289; History of, 384–385
Uranium Cartel, 385, 388; Westinghouse default, 401
Uranium Club, 313, 385–386
Uranium Committee, 117, 223, 241; History of, 386–387
Uranium enrichment plants, 171; Windscale accident, 404–405
Uranium mill tailings, 263
Uranium miners, 302, 304–305; Dangers of radon and thorium, 406
Uranium Mining, 302, 304–305, 387–388; Eldorado Mine, 117; Joachimsthal Mines, 186; Radon Daughters, 304–305; Shinkolobwe Uranium Mine, 328
Urey, Harold Clayton, 96, 118, 128, 140, 270, 314, 386, 389; Discovery of deuterium, 161; Isotope, 180; Life of, 389; Opposition against Rosenberg Case, 314
U.S. Air Force, 101, 112, 306; Aircraft Nuclear Propulsion, 4; Long Range Detection Project, 217
U.S. Army, Selection for Manhattan Project, 66–67; Buster-Jangle Nuclear Tests, 67
U.S. Army Corp of Engineers, 66, 153, 267
U.S. Army Signal Corps, 314
U.S. Congress, 98, 166–167, 187, 233, 255, 286, 298, 301–302, 310, 322, 352, 370, 399; Atomic Energy Act of 1946, 27; Atomic Energy Act of 1954, 28; Atomic Energy Commission, 29–31; *Calvert Cliffs* Decision, 69; Hanford Reactor, 159; Joint Committee on Atomic Energy, 187–188; May-Johnson Bill, 228–229; Nuclear Nonproliferation Act, 256;

Nuclear Waste Policy Act, 264; Plow-share, 286; Price-Anderson bill, 293; Private Ownership of Nuclear Materials, 293–294; Quebec Conference, 298; Radiation Exposure Compensation Act, 301–302; Stockpile Stewardship Program, 349; Uranium Mining, 388; Yucca Mountain Site, 406–407

U.S. Department of Defense, 265

U.S. Department of State, 333; Bikini Tests, 48; Delay of SNAP, 333; Western Supplies Group, 400

U.S. Geological Service, 407

U.S. House of Representatives, 166–167, 187

U.S. Joint Chiefs of Staff, 133

U.S. Justice Department, Uranium Cartel, 385; Westinghouse default, 401

U.S. Military Academy (West Point), 153

U.S. National Conference of Catholic Bishops, 250

U.S. Naval Academy, 278, 310

U.S. Navy, 1, 56, 92, 105, 116, 120, 199, 216, 268, 310; Code 390–590, 92; Contributions of Hyman Rickover, 310–311; Crossroads Nuclear Tests, 101–102; Importance of Deak Parsons, 279–280; *Nautilus*, 243–244; Naval Research Laboratory, 244

U.S. Senate, 98, 174, 187, 233, 255, 348; Strategic Arms Limitation Treaty II, 351

U.S. Supreme Court, 332; *Allen v. United States of America*, 6

Utah, *Allen v. United States of America*, 6; Fallout, 248; Knapp Test Site, 248–249; Nevada Test Site, 248–249; Shot Easy, 330; Shot Harry, 330–331; Uranium Mining, 388; Weiss Report, 399–400

VM-4 nuclear submarine reactor, 199

Van de Graaff, Robert Jemison, 391–392; Life of, 391

Van de Graaff accelerator, 19, 160, 220; Development of, 391

Van Vleck, John H., 44

Vancouver, British Columbia, 97, 152

Vandenberg, Hoyt, 217

Vandenburg, Arthur, 27, 298

Vannikov, Boris, 79, 341; Red Specialists, 309; Soviet Nuclear Program, 341–342; Special Committee on the Atomic Bomb, 343–344

Vanuatu, Treaty of Rarotonga, 372

Vanunu, Mordecai, 392

Vanunu Affair, 112, 392; History of, 392

Velikhov, Evgenii, 177

Vemork (Norway), 250

Veterans Administration, Atomic Veterans, 37

Victoria University, 50, 376

Vienna (Austria), 97, 174, 176, 408

Vietnam, Treaty of Bangkok, 371

Villard, Paul, 392–393; Life of, 392–393

VIPER, 6

Virus House, 194, 393; History of, 393

Vladivostok (Russia), 249

Vladivostok Agreement, 351

Von Laue, Max Felix, 54, 259, 335, 355, 393–395, 402–404; Life of, 393–395

Von Neumann, John, 120, 130, 169, 279, 376, 395–396, 402; Life of, 395; Feud with Johannes Stark, 391; Target Committee, 359–360

Voznesenskii, Nikolai, 344

VVER (Water-Water Power Reactors), 270; Soviet Nuclear Program, 342

Wahl, Arthur C., 289

Wakefield, 141

Wald, George, 96

Wallace, Henry, 66; Top Policy Group, 370

Walton, E.T.S., 91

Waltz Mill Reactor Accident, 397

Wang, Kan-chang, 86

Warren, Stafford, 25

Warsaw Pact, 246, 250

WASH-740 Report, 293, 398–399;

Price-Anderson Bill, 293; WASH-740
Update, 399
WASH-740 Update, 64, 399; WASH-
740 Report, 398–399
WASH-1400. *See* Rasmussen Report
Washington, D.C., 34, 127–128, 154,
173–174, 254, 280, 397–398
**Washington Conference on Theoreti-
cal Physics**, 397–398
Washington Declaration, 31, 398
Washington State University, 1
Watson, Edwin M., 386
Watson, James D., 96
Wave theory of the atom, 108, 296,
323–324
Waxahachie (Texas), 354
Waymack, William, 30
Wayne, John, 100
Weapon grade plutonium, 112, 171,
251–252, 255, 260–262, 269, 287;
Chelyabinsk-40, 79–80
Weapons Systems Evaluation Group
(WSEG), 280
Weinberg, Alvin, Oak Ridge National
Laboratory, 269
Weiss, Edward, Weiss Report, 399–400
Weiss Report, 399–400
Weisskopf, Victor, 118, 128
Weizsacker, Ernst von, 19; Virus
House, 393
Wells, H. G., *The World Set Free*, 405
Wells Survey Company, 291
Wendover Air Base, 368
West Germany, 255–256, 261; German
Nuclear Program, 147–148
Western Samoa, Treaty of Rarotonga,
372
Western Supplies Group, 400
Westinghouse Electric Corporation,
63, 92, 145, 147, 202, 242, 293,
310, 322, 329, 349, 400–401; Im-
pact of Price-Anderson Bill, 293;
Light Water Reactors, 210; Role in
Atomic Energy, 400–401; Savannah
River Project, 322; Three Mile Is-
land, 366; Turnkey Nuclear Energy
Plants, 378–379; Uranium default,

385; Waltz Mill Reactor Accident,
397
Westinghouse Laboratories, 98
Wheeler, John Archibald, 132, 158;
Reactor Poisoning, 307–308
Whitaker, Martin, 268
White Jew, 347; Aryan Physics Move-
ment, 22
**Whiteshell Nuclear Research Estab-
lishment**, 72, 401
Whiteshell Provincial Park (Canada),
401
Whyl (Germany), 401
Whyl Demonstration, 401–402
Wigner, Eugene Paul, 115, 268, 386,
395, 402–404; Life of, 402–404;
Oak Ridge National Laboratory, 268;
Wigner's disease, 403
Wigner's disease, 403; Hanford Plant,
158
Wilson, Charles Thomson Rees, 14,
404; Life of, 404
Wilson, R. B., Target Committee, 359
Wilson, Robert R., 377
Wilson Cloud Chamber, 14, 49, 98,
148; Development of, 404
Windscale, 61, 138, 404–405
Winnacker, Karl, German Nuclear Pro-
gram, 147–148
Winne, Harry, 212
Wirtz, Karl, Virus House, 393; German
Nuclear Program, 148
Wisconsin Public Service Commission,
211
Woolwich Arsenal, 21
Workingmen's College of Birkbeck
(London), 139
The World Set Free, 405
Wormwood Scrubs, 141
Wurzburg Physico-Medical Society
(Germany), 311
Wyoming County (Pennsylvania), 233

X-10 (Oak Ridge), 153; Oak Ridge
National Laboratory, 267
XE-prime prototype nuclear rocket en-
gine, 246

Xenon-135, 308, 347; Hanford Plant, 158; Reactor Poisoning, 307–308
X-ray scatter, 98
X-ray spectrometer, 57
X-rays, 42, 302, 311, 316, 381, 391, 393–394; Contributions of W. H. Bragg, 57–58

Y-12, 267
Yale University, 204, 244
Yang, Chen Ning, 403
Yankee Atomic Electric Company, 292
Yankee-class ballistic missile nuclear submarine, 199
Yankee Nuclear Power Plant, 292; Power Demonstration Reactor program, 292
Yellowcake, 406
Yeltsin, Boris, 97, 252
Yeltsin government, 260
Yongbyon Nuclear Complex (North Korea), 251–252; North Korean Nuclear Program, 251–252
York, Herb, 205
Yucca Mountain Waste Disposal Site, 67, 249, 264, 406–407; Nuclear

Waste Disposal, 263–264; Nuclear Waste Policy Act, 264
Yugoslavia, 75
Yuzhnyi (Russia), 252

Zangger, Cluade, Zangger Committee, 408
Zangger Committee, 408
Zaveniagin, Avraami, 79; Red Specialists, 309; Special Committee on the Atomic Bomb, 342–344
Z-Division (Los Alamos), 319
ZEEP, 71, 78, 409
Zeitschrift fur Physik, 144
Zeldovich, Yakov B., 196, 318
Zerlina Reactor, 170
Zernov, Pavel M., 23
Zincke, Theodor, 155
Zinn, Walter Henry, 20, 322, 409; Chicago Pile One, 84; Life of, 409
Zircaloy, 73, 410
Zirconium, 329, 410
Zoe (Zero Oxide eau lourde), 136
Zukerman, Veniamin, 23
Zurich Institute of Technology, 395
Zurich Polytechnical School, 114–115

About the Author

STEPHEN E. ATKINS is Associate University Librarian for Collection Management at the Sterling C. Evans Library at Texas A&M University. He is the author of *Terrorism: A Handbook* (1992), *Arms Control and Disarmament, Defense and Military, International Security and Peace: An Annotated Guide to Sources, 1980–1987* (1989), and *The Academic Library in the American University* (1991), as well as numerous journal articles on arms control.